SRA
Connecting
Math Concepts

Columbus, Ohio

The **McGraw·Hill** Companies

www.sra4kids.com

 **SRA**
McGraw-Hill

Send all inquiries to:
SRA/McGraw-Hill
8787 Orion Place
Columbus, OH 43240-4027

Printed in the United States of America.

ISBN 0-02-684473-7

1 2 3 4 5 6 7 8 9 0 PBM 08 07 06 05 04 03 02

The McGraw·Hill Companies

Lesson 1

Objectives

- **Read the program overview.** (Exercise 1)
- **Classify a fraction that is more than 1 or less than 1.** (Exercise 2)
  Note: Students apply the rule that, if the top number is larger than the bottom number, the fraction is more than 1 unit. If the top number is smaller than the bottom number, the fraction is less than 1 unit.
- **Identify a picture that illustrates a fraction.** (Exercise 3)
- **Identify and correct an incorrect answer to a problem that adds or subtracts fractions.** (Exercise 4)
  Note: Students evaluate a set of problems, some of which have mistakes, such as:

$$\frac{5}{3} - \frac{2}{3} = \frac{3}{0}$$

$$\frac{5}{7} + \frac{3}{7} = \frac{8}{14}$$

- **Work a problem that adds or subtracts fractions.** (Exercise 5)
  Note: All problems have fractions with like denominators. Students follow the procedure of copying the denominator and either adding or subtracting the numerators.
- **Write a column problem to find the missing number in a number family.** (Exercise 6)

  Note: Each family has either the big number (at the end of the arrow) missing:

  16 18 ⟶ ■

  or one of the small numbers missing:

  7 ■ ⟶ 47

  ■ 256 ⟶ 421

  Students add to find the missing big number or subtract to find a missing small number.
- **Identify and correct an incorrect answer to a number family problem.** (Exercise 7)
- **Write the missing factor in a multiplication fact.** (Exercise 8)
  Note: Problems are of the form:

  ■ x 4 = 20 or 6 x ■ = 18

EXERCISE 1 PROGRAM OVERVIEW

a. Find page 1 in your textbook.

b. I'll read what it says. Follow along: This program will teach you a great deal about doing mathematics. When something new is introduced, I'll show you how to work the problems. Later, you'll work the problems on your own, without help.

- Remember, everything that is introduced is important. It is knowledge you will need to work difficult problems that will be introduced later.
- You must follow my directions. I'll sometimes direct you to work **part** of a problem, sometimes a **whole** problem, and sometimes a group of problems.
- Listen very carefully to the directions. Work quickly and accurately. Most important of all, work hard. You'll be rewarded with math skills that will surprise you.

EXERCISE 2 FRACTION ANALYSIS
More than/Less than 1

a. Find lesson 1, part 1, on page 2 of your textbook.

- (Teacher reference:)

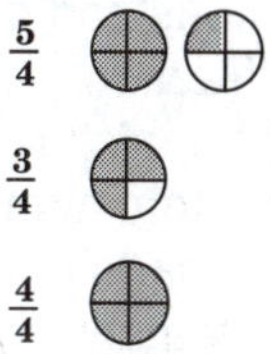

- I'll read what it says. Follow along: Some fractions are more than 1 whole unit. Some fractions are less than 1 whole unit.
- You can figure out whether a fraction is more than 1 or less than 1 by comparing the top number with the bottom number. The bottom number tells the number of parts in each unit.
- If the top number is **larger** than the bottom number, the fraction is **more than** 1 unit.
- You can see a fraction that is more than 1. The bottom number is 4, and the top number is more than 4. All the parts of 1 unit are shaded and 1 part of the next unit is shaded.
- If the top number is **smaller** than the bottom number, the fraction is **less than** 1 unit.
- You can see a fraction that is less than 1. The bottom number is 4, and the top number is less than 4. So less than 1 unit is shaded.
- If the top number is the **same** as the bottom number, the fraction **equals** 1 whole unit. The fraction 4/4 equals 1 whole unit.
- Remember, the bottom number tells how many parts each **unit** is divided into. The top number tells the number of shaded parts.

b. Look at the fractions in part 2.
- Some of them are more than 1 and some are less than 1.
- Use lined paper. Write **part 2.**
- Copy all the fractions that are more than 1. Raise your hand when you're finished.
  (Observe students and give feedback. Make sure students wrote the letter for each fraction.)
c. (Write on the board:)

b. $\dfrac{3}{2}$	c. $\dfrac{56}{13}$	f. $\dfrac{7}{6}$

- Here's what you should have. The fractions that are more than 1 are 3/2, 56/13 and 7/6. The other fractions are either equal to 1 unit or are less than 1 unit.
- Check your answers. If your answer is wrong, cross it out and write the correct answer above it or next to it. Do not erase.

EXERCISE 3 FRACTION ANALYSIS
Pictures

a. Find part 3.
- You can see fractions and pictures of fractions.
- You'll copy each fraction. After each fraction, write the letter of the picture for that fraction. Remember, the bottom number of the fraction tells the number of parts in each unit. The top number shows how many parts are colored.
b. Read the first fraction. (Signal.) *7-fourths*.
- Copy 7/4 and write the letter of the picture for that fraction. Raise your hand when you're finished. (Observe students and give feedback.)
- Everybody, which picture shows 7/4? (Signal.) *C.*
- Yes, picture C. There are four parts in each unit. Seven parts are shaded.
c. Copy the next fraction and write the letter of the picture. Raise your hand when you're finished. (Observe students and give feedback.)
- Everybody, which picture shows 3/5? (Signal.) *D.*
d. Your turn: Write the rest of the fractions and the letters of the corresponding pictures. Raise your hand when you're finished. (Observe students and give feedback.)
e. Check your work.
- The next fraction is 1/6. Which picture is that? (Signal.) *B.*
- The next fraction is 6/6. Which picture is that? (Signal.) *F.*
- 6/6 is 1 whole unit.
f. Raise your hand if you got everything right.

EXERCISE 4 FRACTION OPERATIONS
Addition/Subtraction Workcheck

a. Find part 4.
- These problems add fractions or subtract fractions. Here's how to work these problems. The bottom number tells about the number of parts in each unit. You do not add or subtract that number. You just copy it in the answer. You add or subtract on top.
b. Problem A. What's the bottom number in the answer? (Signal.) *7.*
- Say the problem for the top numbers. (Signal.) *9 plus 3.*
- Problem B. What's the bottom number in the answer? (Signal.) *10.*
- Say the problem for the top numbers. (Signal.) *5 minus 1.*
- Problem C. What's the bottom number in the answer? (Signal.) *3.*
- Say the problem for the top numbers. (Signal.) *20 plus 56.*
c. Find part 5.
- A student worked the problems in part 5. He got three of the problems wrong because he did not follow the correct procedure. He added or subtracted the bottom numbers.
d. Find the three problems he worked incorrectly. Copy them and work them correctly. Box your answers. Raise your hand when you're finished. (Observe students and give feedback.)
  Key:

$$b. \quad \frac{3}{5} + \frac{24}{5} = \boxed{\frac{27}{5}}$$

$$d. \quad \frac{14}{9} - \frac{8}{9} = \boxed{\frac{6}{9}}$$

$$e. \quad \frac{12}{13} + \frac{6}{13} = \boxed{\frac{18}{13}}$$

e. Turn to page 5 and find part J. √
- Problem A: 13/7 minus 2/7. Did the student work that problem correctly? (Signal.) *Yes.*
- Problem B: 3/5 plus 24/5. Did the student work that problem correctly? (Signal.) *No.*
- So you should have copied the problem and worked it.
- Problem C: 1/8 plus 7/8. Did the student work that problem correctly? (Signal.) *Yes.*
- Problem D: 14/9 minus 8/9. Was it worked correctly? (Signal.) *No.*
- So you copied it and worked it.

- Problem E: 12/13 plus 6/13. Was it worked correctly? (Signal.) *No.*
- So you copied it and worked it.
- Problem F: 5/20 minus 4/20. Was it worked correctly? (Signal.) *Yes.*

f. You should have copied and worked problems B, D and E.

g. Problem B: 3/5 plus 24/5. The student wrote the answer as 27/10. What's the correct answer? (Signal.) *27-fifths.*
- You should have 27/5 as your answer. 3/5 plus 24/5 equals 27-**fifths.**

h. Problem D: 14/9 minus 8/9. The student wrote the answer as 6 over zero. That's wrong. What's the correct answer? (Signal.) *6-ninths.*
- You should have 6/9 as your answer. 14/9 minus 8/9 equals 6-**ninths.**

i. Problem E: 12/13 plus 6/13. The student wrote the answer as 18/26. What's the correct answer? (Signal.) *18-thirteenths.*
- You should have written 18/13 as your answer. 12/13 plus 6/13 equals 18-**thirteenths.**

EXERCISE 5 FRACTION OPERATIONS
Addition/Subtraction

a. Find part 6.
- You're going to work these problems. Remember, don't add or subtract the bottom numbers.

b. Copy problem A and work it. Raise your hand when you've finished problem A.
(Observe students and give feedback.)
- (Write on the board:)

$$\text{a.} \quad \frac{9}{7} + \frac{21}{7} = \boxed{\frac{30}{7}}$$

- Here's what you should have.

c. Copy problem B and work it. Remember, subtract the top numbers and write the answer. Raise your hand when you're finished.
(Observe students and give feedback.)
- (Write on the board:)

$$\text{b.} \quad \frac{14}{3} - \frac{14}{3} = \boxed{\frac{0}{3}}$$

- Here's what you should have. The answer is zero-thirds. That's zero.

d. Work the rest of the problems in part 6. Raise your hand when you're finished.
(Observe students and give feedback.)

e. (Write on the board:)

$$\text{c.} \quad \frac{23}{12} - \frac{4}{12} = \boxed{\frac{19}{12}}$$

$$\text{d.} \quad \frac{7}{5} + \frac{5}{5} = \boxed{\frac{12}{5}}$$

$$\text{e.} \quad \frac{1}{32} - \frac{1}{32} = \boxed{\frac{0}{32}}$$

f. Check your work.
- Here's what you should have for problems C through E.
- Problem C: 23/12 minus 4/12. Everybody, what's the answer? (Signal.) *19-twelfths.*
- Problem D: 7/5 plus 5/5. Everybody, what's the answer? (Signal.) *12-fifths.*
- Problem E: 1/32 minus 1/32. Everybody, what's the answer? (Signal.) *Zero-thirty-seconds.*

g. Raise your hand if you got all of them right.

EXERCISE 6 NUMBER FAMILIES

a. Find part 7.
- I'll read what it says. Follow along: A lot of problems you'll work are based on number families. A number family is made up of three numbers that always go together in addition and subtraction problems.
- The family is made up of two small numbers and a big number. The big number is at the end of the arrow.
- You can see a family with the big number missing. The small numbers are 40 and 160. To find the **big** number, you **add the small numbers.** 40 plus 160 equals 200. The big number is 200.
- Below, you can see a family with a small number missing. To find a missing **small** number, you **subtract.** You start with the big number and subtract the small number that is shown. The subtraction for this family is 200 minus 160. So the missing small number is 40.
- Below, you can see a family with the other small number missing. You subtract to find that number. The subtraction for this family is 200 minus 40. So the missing small number is 160.

b. Later, you will use number families to figure out answers to very difficult word problems.

c. Find part 8.

- Look at the number families A through D. Each of these number families has a missing number. In some problems, the big number is missing. That's the number at the end of the arrow.
- In some problems, one of the small numbers is missing. That's one of the numbers on top of the arrow. Remember the rules: To find the big number, you add the small numbers. To find the small number, you **start** with the big number and subtract.
d. Touch family A.
- Is a small number or the big number missing in that family? (Signal.) *The big number.*
- What do you do to find the missing big number? (Signal.) *Add.*
- Say the addition problem. (Signal.) *116 plus 318.*
e. Touch family B.
- Is a small number or the big number missing? (Signal.) *A small number.*
- What do you do to find the missing small number? (Signal.) *Subtract.*
- Say the subtraction problem. (Signal.) *470 minus 207.*
f. Touch family C.
- Is a small number or the big number missing? (Signal.) *A small number.*
- What do you do to find the missing small number? (Signal.) *Subtract.*
- Say the subtraction problem. (Signal.) *321 minus 256.*
g. Touch problem D.
- Is a small number or the big number missing? (Signal.) *The big number.*
- What do you do to find the missing big number? (Signal.) *Add.*
- Say the addition problem. (Signal.) *867 plus 135.*
h. Your turn: Write the column problem and the answer for family A. Do not copy the number family. Just write the problem in a column. Box your answer. Raise your hand when you're finished. (Observe students and give feedback.)
- (Write on the board:)

$$\begin{array}{r} \textbf{a.} \quad 116 \\ + 318 \\ \hline \boxed{434} \end{array}$$

- You added: 116 plus 318. The answer is 434. That's the big number.

i. Your turn again: Write the number problem and the answer for family B. Box the answer. Raise your hand when you're finished. (Observe students and give feedback.)
- (Write on the board:)

$$\begin{array}{r} \textbf{b.} \quad 470 \\ - 207 \\ \hline \boxed{263} \end{array}$$

- The missing small number is 263.
j. Your turn: Write the problems and answers for the rest of the families. Remember, if the big number in a family is missing, you add to find it. If a small number is missing, you start with the big number and subtract. Raise your hand when you're finished. (Observe students and give feedback.)
k. Check your work.
- Family C. A small number is missing. Read the subtraction problem and the answer. (Signal.) *321 minus 256 equals 65.*
- Family D. The big number is missing. Read the addition problem and the answer. (Signal.) *867 plus 135 equals 1002.*

EXERCISE 7 NUMBER FAMILIES
Workcheck

a. Find part 9.
- This is the work a student did for number families. The student made some silly mistakes.
b. Problem A: Look at the number family. It has a missing small number.
- Look at the problem the student wrote. It adds. That's wrong. That's what you do when the big number is missing, not a small number.
c. Your turn: Look at each number family and the problem the student worked. Figure out which of the problems the student worked incorrectly and work them correctly. Don't do anything for the problems the student worked correctly. Raise your hand when you're finished. (Observe students and give feedback.)
Key:

$$\begin{array}{r} a. \quad 777 \\ - 213 \\ \hline \boxed{564} \end{array} \qquad \begin{array}{r} c. \quad 176 \\ + 408 \\ \hline \boxed{584} \end{array} \qquad \begin{array}{r} d. \quad 504 \\ - 263 \\ \hline \boxed{241} \end{array}$$

d. You should have worked problems for number families A, C and D.
- For problem A, a small number is missing. You worked the problem 777 minus 213. The missing number is 564.

- For problem C, the big number is missing. You worked the problem 176 plus 408. The missing number is 584.
- For problem D, a small number is missing. You worked the problem 504 minus 263. The missing number is 241. The student tried to work the problem 263 minus 504. Boo.

e. Raise your hand if you got everything right.

EXERCISE 8 MULTIPLICATION
Missing Factor

a. Find part 10.
- This is a review of multiplication facts.
- Each problem has the missing value as either the first number or the second number. You're going to copy each equation and write the missing value.
b. Problem A says: 2 times some value equals 14. Everybody, what's the missing value? (Signal.) *7*.
- Write the complete equation for problem A, then write equations for the rest of the items in part 10. Raise your hand when you're finished. (Observe students and give feedback.)

Key:

a. $2 \times \boxed{7} = 14$ d. $6 \times \boxed{5} = 30$

b. $\boxed{4} \times 5 = 20$ e. $6 \times \boxed{1} = 6$

c. $\boxed{9} \times 10 = 90$

c. Find part K on page 5 of your textbook. √
- That shows what you should have for the items in part 10.
d. Raise your hand if you got everything right.

EXERCISE 9 INDEPENDENT WORK

a. Find part 11.
- You'll work the problems in part 11 independently.
b. (Before beginning the next lesson, check the students' independent work.)

Lesson 2

Objectives

- Identify a picture that illustrates a fraction. (Exercise 1)
- Work a problem that adds or subtracts fractions. (Exercise 2)
- Write a column problem to find the missing number in a number family. (Exercise 3)
- Write the missing factor in a multiplication fact. (Exercise 4)
- Identify and correct an incorrect answer to a problem that adds or subtracts fractions. (Exercise 5)
- **Work a word problem that requires multiplication or division.** (Exercise 6) *Note:* Problems tell about 1 and ask about more than 1 or tell about more than 1 and ask about 1. For example: If each brick weighs 45 ounces, how much do 11 bricks weigh?
- **Classify a fraction that is more than 1, less than 1 or equal to 1.** (Exercise 7)

EXERCISE 1 FRACTION ANALYSIS
Pictures

a. Open your textbook to lesson 2 and find part 1.
- You're going to match fractions with pictures that show the fractions.
b. Look at the sample fraction: 5 over 1.
- There's **1 part** in each whole unit. The unit is the same size as the part.
- The top number is 5, so 5 units are colored. 5 over 1 is the same as 5 whole units.
c. Your turn: Use lined paper. Copy each fraction in part 1. Write the letter of the picture for that fraction. Raise your hand when you're finished. (Observe students and give feedback.)
d. Check your work.
- The first fraction is 3/3. Picture C shows 3/3.
- The next fraction is 6/4. Picture A shows 6/4.
- The next fraction is 3 over 1. Picture F shows 3 over 1.
- The next fraction is 3/8. Picture D shows 3/8.
- The last fraction is 1/6. Picture B shows 1/6.
e. Raise your hand if you got everything right.

EXERCISE 2 FRACTION OPERATIONS
Addition/Subtraction

a. Find part 2.
* These are problems that add or subtract fractions.
b. Copy problem A and work it. Raise your hand when you're finished.
  (Observe students and give feedback.)
* (Write on the board:)

$$a. \quad \frac{15}{8} - \frac{3}{8} = \boxed{\frac{12}{8}}$$

* Here's what you should have. 15/8 minus 3/8 equals 12/8.
c. Your turn: Copy the rest of the problems and write the correct answers. Remember, do not add or subtract the bottom numbers. Raise your hand when you're finished.
  (Observe students and give feedback.)
  Key:

$$a. \quad \frac{15}{8} - \frac{3}{8} = \boxed{\frac{12}{8}}$$

$$b. \quad \frac{2}{5} + \frac{9}{5} = \boxed{\frac{11}{5}}$$

$$c. \quad \frac{3}{8} - \frac{3}{8} = \boxed{\frac{0}{8}}$$

$$d. \quad \frac{27}{7} - \frac{6}{7} = \boxed{\frac{21}{7}}$$

$$e. \quad \frac{6}{12} - \frac{1}{12} = \boxed{\frac{5}{12}}$$

$$f. \quad \frac{5}{9} + \frac{9}{9} = \boxed{\frac{14}{9}}$$

d. Find part J on page 8 of your textbook. √
* Check your answers. If your answer is wrong, cross it out and write the correct answer above it or next to it. Do not erase.

EXERCISE 3 NUMBER FAMILIES

a. Find part 3.
b. Touch family A.
* Is the missing number in that family the big number or a small number? (Signal.) *A small number.*

* Do you add or subtract to find that number? (Signal.) *Subtract.*
* Say the subtraction problem. (Signal.) *371 minus 248.*
* Touch family B.
* Is the missing number in that family the big number or a small number? (Signal.) *The big number.*
* Do you add or subtract to find that number? (Signal.) *Add.*
* Say the addition problem. (Signal.) *567 plus 98.*
* Touch family C.
* Is the missing number in that family the big number or a small number? (Signal.) *A small number.*
* Do you add or subtract to find that number? (Signal.) *Subtract.*
* Say the subtraction problem. (Signal.) *286 minus 104.*
c. Your turn: Write the column problem and the answer for each family. Do not copy the number families. Just write each problem in a column. Box your answers. Raise your hand when you're finished.
  (Observe students and give feedback.)
d. Check your work. Read each number problem and the answer.
* Problem A. (Signal.) *371 minus 248 equals 123.*
* The missing number in family A is 123.
* Problem B. (Signal.) *567 plus 98 equals 665.*
* The missing number in family B is 665.
* Problem C. (Signal.) *286 minus 104 equals 182.*
* The missing number is 182.
* Problem D. (Signal.) *179 minus 83 equals 96.*
* The missing number is 96.
e. Raise your hand if you found all the missing numbers.

EXERCISE 4 MULTIPLICATION
Missing Factor

a. Find part 4.
b. The rules in the box are reminders for the kinds of multiplication problems that students miss most often:
* Any number times 1 equals the number.
* Any number times zero equals zero.
c. Write the complete equations for the problems in part 4. Don't get fooled. Raise your hand when you're finished.
  (Observe students and give feedback.)

Key:

a. $15 \times \boxed{1} = 15$

b. $7 \times \boxed{0} = 0$

c. $9 \times \boxed{10} = 90$

d. $\boxed{0} \times 9 = 0$

e. $\boxed{7} \times 5 = 35$

f. $\boxed{1} \times 5 = 5$

g. $\boxed{6} \times 8 = 48$

d. Find part K on page 8 of your textbook. √
• That shows what you should have for problems A through G.
• Equation A: Read it. (Signal.) *15 times 1 equals 15.*
• Equation B: Read it. (Signal.) *7 times zero equals zero.*
• Equation C: Read it. (Signal.) *9 times 10 equals 90.*
• Equation D: Read it. (Signal.) *Zero times 9 equals zero.*
• Equation E: Read it. (Signal.) *7 times 5 equals 35.*
• Equation F: Read it. (Signal.) *1 times 5 equals 5.*
• Equation G: Read it. (Signal.) *6 times 8 equals 48.*
e. Raise your hand if you got everything right.

EXERCISE 5 FRACTION OPERATIONS
Addition/Subtraction Workcheck

a. Find part 5.
• These are problems a student worked. The student may have made some mistakes.
• Go over the work for each problem. If the student wrote the wrong answer, copy the problem and show the correct answer. Raise your hand when you're finished.
(Observe students and give feedback.)
b. Look at part 5.
• Problem A: 12/12 minus 5/12. Is the answer correct? (Signal.) *No.*
• So you should have copied and worked the problem.
• Problem B: 26 over 1 plus 5 over 1. Is the answer correct? (Signal.) *Yes.*
• Problem C: 17/5 minus 17/5. Is the answer correct? (Signal.) *Yes.*
• Problem D: 4/9 plus 4/9. Is the answer correct? (Signal.) *Yes.*
• Problem E: 19/6 minus 6/6. Is the answer correct? (Signal.) *Yes.*
• Problem F: 4/14 plus 18/14. Is the answer correct? (Signal.) *No.*
• 22/28 is not the correct answer, so you should have copied and worked the problem.

c. You should have copied and worked problems A and F.
d. Problem A: 12/12 minus 5/12. The student wrote the answer as 7 over zero. What's the correct answer? (Signal.) *7-twelfths.*
• You should have written 7/12 as your answer. 12/12 minus 5/12 equals 7-**twelfths.**
e. Problem F: 4/14 plus 18/14. The student wrote 22/28 as the answer. Everybody, what's the correct answer? (Signal.) *22-fourteenths.*
• You should have written 22/14 as your answer. 4/14 plus 18/14 equals 22-**fourteenths.**
f. Raise your hand if you got everything right.

EXERCISE 6 PROBLEM SOLVING
Multiplication/Division

a. Find part 6.
• I'll read what it says. Follow along: Here's a word problem that you can solve by multiplication: If each of the bricks weighs 64 ounces, how much do 12 bricks weigh?
• The first part of the problem tells you that 1 brick weighs 64 ounces.
• The problem asks about the weight of more than 1 brick.
• To find the weight of more than 1 brick, you multiply.
• You find the weight of 12 bricks by working the problem: 64 times 12.
• The next problem you can solve using division: If 15 identical tiles weigh 390 grams, how much does each tile weigh?
• The problem tells what 15 tiles weigh and asks about the weight of 1 tile.
• To find the weight of 1 tile, you divide. The problem you work is 390 divided by 15.
• Remember, if the problem tells about **1** and asks about **more** than 1, you work a **multiplication** problem. If the problem tells about **more** than 1 and asks about **1,** you work a **division** problem.
b. Find part 7.
• For each problem, you'll write the name of the operation you'll use to solve the problem. Don't solve it. Just write the name of the operation.
c. Problem A: Each stamp costs 19 cents. How much do 9 stamps cost?
• The problem tells about each stamp and asks about **more** than 1 stamp. What operation do you use to work this problem? (Signal.) *Multiplication.*
• Write **multiplication** for item A. Then name the operations for the rest of the items in part 7. Raise your hand when you're finished.
(Observe students and give feedback.)

Key:

a. multiplication

b. division

c. division

d. multiplication

d. Problem B: There were 678 toothpicks in groups that were the same size. There were 6 groups. How many toothpicks were in each group?
• What operation do you use to work the problem? (Signal.) *Division.*
• The problem **asks** about 1 group. You should have written: division.
e. Problem C: 16 worms were arranged in equal-sized groups. There were 8 groups of worms. How many worms were there in each group?
• What operation do you use to work the problem? (Signal.) *Division.*
• The problem asks about **1** group, so you divide.
f. Problem D: A machine made 7 buttonholes each minute. How many buttonholes would the machine make in 45 minutes?
• What operation do you use to work the problem? (Signal.) *Multiplication.*
• The problem tells about **1** minute and asks about **more** than 1 minute, so you multiply.
g. Your turn: Work all the problems in part 7. Remember to show the answer as a number and a unit name. Raise your hand when you're finished. (Observe students and give feedback.)
h. (Write on the board:)

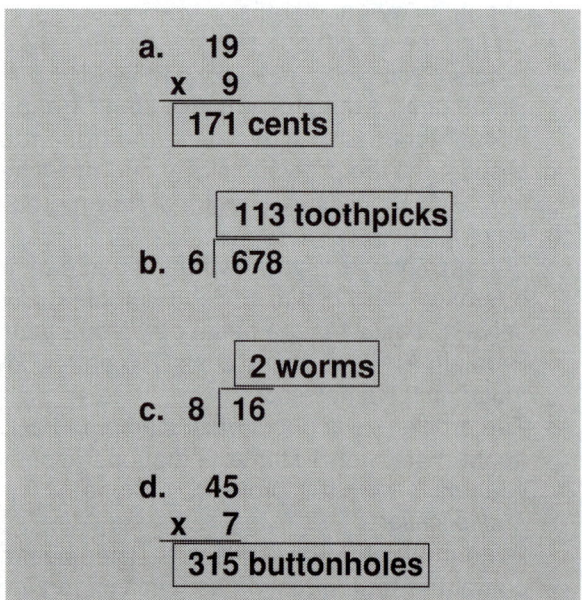

a. 19
x 9
171 cents

113 toothpicks
b. 6) 678

2 worms
c. 8) 16

d. 45
x 7
315 buttonholes

• Here's what you should have for each problem. Fix up any mistakes.

EXERCISE 7 FRACTION ANALYSIS
More than/Less than 1

a. Find part 8.
• You're going to figure out whether each fraction is more than 1 unit, less than 1 unit, or equal to 1 unit.
• If the top number is larger than the bottom number, the fraction is more than 1 unit. If the top number is less than the bottom number, what do you know about the fraction? (Signal.) *It's less than 1 unit.*
• And if the top and bottom numbers are the same size, the fraction equals 1. It doesn't matter how big the numbers are. 250 over 250 equals 1 because the fraction has 250 parts in each unit and 250 parts are shaded.
• Some of the fractions in part 8 are more than 1. Some are less than 1. And some equal 1.
b. Copy each fraction. After it write the words **more than 1, less than 1** or **equals 1.** Raise your hand when you're finished. (Observe students and give feedback.)
c. Check your work:
• Fraction A: 12/11. It's more than 1.
• Fraction B: 2/2. It equals 1.
• Fraction C: 3/2. It's more than 1.
• Fraction D: 30/30. It equals 1.
• Fraction E: 8 over 1. It's more than 1. It equals 8.
• Fraction F: 1/8. It's less than 1.

EXERCISE 8 INDEPENDENT WORK

a. Find part 9.
• You'll work the problems in part 9 independently.
b. (Before beginning the next lesson, check the students' independent work.)

Lesson 3

Objectives

- Write a column problem to find the missing number in a number family. (Exercise 1)

- Identify and correct an incorrect answer to a number family problem. (Exercise 2)

- Classify a fraction that is more than 1, less than 1 or equal to 1. (Exercise 3)

- Work a word problem that requires multiplication or division. (Exercise 4)

- **Identify and work an addition or subtraction fraction problem that can be worked as it is written.** (Exercise 5)
  Note: The set of problems includes problems that can't be worked without finding a common denominator. Students do not work these problems.

- **Write the missing factor or addend in a multiplication or addition fact.** (Exercise 6)
  Note: For all problems, either the first or second number is missing.

EXERCISE 1 NUMBER FAMILIES

a. Open your textbook to lesson 3 and find part 1.
b. Your turn: Use lined paper. Write the column problem for each family. Figure out the missing number. Box your answer. Raise your hand when you're finished.
(Observe students and give feedback.)
c. Check your work. Read each number problem and the answer.
- Problem A. (Signal.) *110 plus 89 equals 199.*
- The missing number in family A is 199.
- Problem B. (Signal.) *680 minus 529 equals 151.*
- The missing number in family B is 151.
- Problem C. (Signal.) *458 minus 350 equals 108.*
- The missing number in family C is 108.
- Problem D. (Signal.) *65 plus 128 equals 193.*
- The missing number in family D is 193.
d. Raise your hand if you got all the missing numbers correct.

EXERCISE 2 NUMBER FAMILIES
Workcheck

a. Find part 2.
- A student worked these problems and some of the work is silly. The student added to find a missing small number. The student subtracted to find a missing big number. You'll show the correct work for the problems the student worked incorrectly.
b. For each item the student worked incorrectly, copy the family. Then write the column problem to figure out the missing number. Box your answer. Don't do anything for items the student worked correctly. Raise your hand when you're finished. (Observe students and give feedback.)
c. (Write on the board:)

a.	238 − 115 ☐ 123 ☐	**b.**	601 + 298 ☐ 899 ☐
e.	891 − 523 ☐ 368 ☐	**f.**	1075 − 350 ☐ 725 ☐

- Here are the column problems you should have written.
d. You should have worked problems A, B, E and F.
- For problem A, a small number is missing in the family. The student worked an addition problem. You should have worked the problem: 238 minus 115. The answer is 123.
- For problem B, the big number is missing. The student worked a subtraction problem. You should have worked the problem: 601 plus 298. Or 298 plus 601. The answer is 899.
- For problem E, a small number is missing. The student wrote the wrong subtraction problem. You should have worked the problem: 891 minus 523. The answer is 368.
- For problem F, a small number is missing. The student worked an addition problem. You should have worked the problem: 1075 minus 350. The answer is 725.
e. Raise your hand if you corrected all of the mistakes.

EXERCISE 3 FRACTION ANALYSIS
More than/Less than 1

a. Find part 3.
- You're going to figure out whether each fraction is more than 1 unit, less than 1 unit, or equal to 1 unit.
- Remember, if the top number is larger than the bottom number, the fraction is more than 1 unit. If the top number is less than the bottom number, what do you know about the fraction? (Signal.) *It's less than 1 unit.*

- And if the top and bottom numbers are the same size, the fraction equals 1.
b. Copy each fraction. After it write the words **more than 1, less than 1** or **equals 1.** Raise your hand when you're finished.
   (Observe students and give feedback.)
c. Check your work.
- Fraction A: 13/14. It's less than 1.
- Fraction B: 2/3. It's less than 1.
- Fraction C: 24/24. It equals 1.
- Fraction D: 20/4. It's more than 1.
- Fraction E: 4/4. It equals 1.
- Fraction F: 2/9. It's less than 1.

EXERCISE 4 PROBLEM SOLVING
Multiplication/Division

a. Find part 4.
- To work some of the problems in part 4, you multiply. To work other problems, you divide. If the problem tells about 1 and asks about **more** than 1, you multiply. If the problem tells about more than 1 and asks about **1,** you divide.
b. Write the name of the operation you will use for each problem. Don't work the problems, just write **multiplication** or **division** to indicate how you'd solve the problem. Raise your hand when you've done that much.
   (Observe students and give feedback.)
c. Check your work.
- Problem A: Roy wants to make stacks of magazines that have the same number in each stack. He wants to make 5 stacks. He has 265 magazines. How many magazines will be in each stack? What operation do you use to solve that problem? (Signal.) *Division.*
- The problem tells about 5 stacks and asks about 1 stack, so you divide.
- Problem B: Edna has 14 identical coins. Each coin is worth 50 dollars. How much are all the coins worth? What operation? (Signal.) *Multiplication.*
- The problem tells about each coin and asks about 14 coins, so you multiply.
- Problem C: Baseball cards weigh 3 grams each. Tim has 89 cards. What is the weight of all of his cards? What operation? (Signal.) *Multiplication.*
- Problem D: A cake is divided into 8 equal pieces. If the entire cake weighs 96 ounces, how much does each piece weigh? What operation? (Signal.) *Division.*

d. Your turn: Work the problems. Box the answers. Remember to show the answer as a number and a **unit name** or with a dollar sign. Raise your hand when you're finished.
   (Observe students and give feedback.)
e. (Write on the board:)

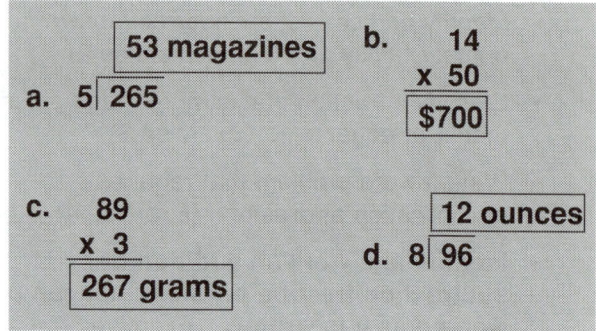

- Here's what you should have for each item. Raise your hand if you got everything right.

EXERCISE 5 FRACTION OPERATIONS
Addition/Subtraction

a. Find part 5.
- Some of these problems can be worked the way they are written. Others can't.
- Remember, if the bottom numbers are not the same, you can't copy the bottom number in your answer. So you can't work the problem the way it is written.
b. Problem A. The bottom number of the first fraction is 3. The bottom number of the other fraction is 5.
- You can't work that problem the way it is written, because you can't copy the same bottom number in the answer.
c. Problem B. Look at the bottom numbers of the fractions. Can you work that problem the way it is written? (Signal.) *Yes.*
- Yes, the bottom numbers are the same, so you can work the problem.
d. Your turn: Copy all the problems you can work the way they are written and work them. Don't copy any problems that you can't work the way they are written. Raise your hand when you're finished. (Observe students and give feedback.)

e. (Write on the board:)

$$b. \frac{17}{4} - \frac{5}{4} = \boxed{\frac{12}{4}}$$

$$c. \frac{4}{8} + \frac{8}{8} = \boxed{\frac{12}{8}}$$

$$e. \frac{9}{3} + \frac{14}{3} = \boxed{\frac{23}{3}}$$

$$h. \frac{27}{2} - \frac{27}{2} = \boxed{\frac{0}{2}}$$

- Here's what you should have. You can work problems B, C, E and H the way they are written. You can't work any of the other problems the way they are written.
- Check your answers. Fix any mistakes.

EXERCISE 6 MULTIPLICATION/ADDITION
Missing Small Value

a. Find part 6.
- All these problems have a missing first number or a missing middle number. Some of the problems add. Some multiply.
b. The rule in the box is a reminder for the kind of addition problem that students miss most often: Any number plus zero equals the number.
c. Your turn: Copy the problems in part 6 and write the missing numbers.
- Make sure you copy the correct sign in each problem. Raise your hand when you're finished. (Observe students and give feedback.)
d. (Write on the board:)

a. $9 + \boxed{6} = 15$	e. $\boxed{0} + 19 = 19$
b. $6 + \boxed{10} = 16$	f. $\boxed{0} \times 19 = 0$
c. $6 \times \boxed{0} = 0$	g. $17 + \boxed{0} = 17$
d. $\boxed{1} \times 19 = 19$	h. $5 \times \boxed{1} = 5$

- Check your work. Here's what you should have for each problem.
- Raise your hand if you got everything right.

EXERCISE 7 INDEPENDENT WORK

a. Do the independent work for lesson 3.
b. (Before beginning the next lesson, check the students' independent work.)

Lesson 4

Objectives

- Identify and work an addition or subtraction fraction problem that can be worked as it is written. (Exercise 1)

- **Work a problem that requires multiplying fractions.** (Exercise 2)
  Note: Students follow the procedure of multiplying the top numbers and writing that number as the numerator of the answer; then multiplying the bottom numbers and writing that number in the denominator of the answer.

- **Identify number families that are written correctly and figure out the missing number.** (Exercise 3)
  Note: Some number families are impossible to work because the small number shown is bigger than the big number.

- **Work a mixed set of multiplication problems, some of which have a missing factor.** (Exercise 4)
  Note: Problems have either the middle number or the last number missing. Parentheses are used instead of times signs. For example:
  $$3 (8) = \blacksquare$$
  $$3 (\blacksquare) = 21$$

- Classify a fraction that is more than 1, less than 1 or equal to 1. (Exercise 5)

- Work a word problem that requires multiplication or division. (Exercise 6)

EXERCISE 1 FRACTION OPERATIONS
Addition/Subtraction

a. Open your textbook to lesson 4 and find part 1.
b. These problems add or subtract fractions. Remember, you add or subtract on the top. You copy the bottom number in the answer. If the bottom numbers are not the same, you can't work the problem the way it is written. Copy all the problems that can be worked the way they are written and work them. Raise your hand when you're finished.
(Observe students and give feedback.)

c. (Write on the board:)

$$a. \ \frac{35}{4} - \frac{3}{4} = \boxed{\frac{32}{4}}$$

$$c. \ \frac{12}{5} - \frac{0}{5} = \boxed{\frac{12}{5}}$$

$$d. \ \frac{2}{7} - \frac{1}{7} = \boxed{\frac{1}{7}}$$

$$e. \ \frac{3}{8} + \frac{5}{8} = \boxed{\frac{8}{8}}$$

- Here's what you should have. You can't work problems B and F the way they are written.
d. Raise your hand if you got everything right.

EXERCISE 2 FRACTION OPERATIONS
Multiplication

a. Find part 2.
- I'll read what it says. Follow along: When you add or subtract fractions, you don't add or subtract the bottom numbers. You just copy the bottom number in the answer.
- When you **multiply** fractions, you do not follow the same procedure.
- You multiply the top numbers together and write the answer on top.
- Then you multiply the bottom numbers together and write the answer on the bottom.
- You can see the problem 2/5 times 3/4.
- On top, you work the problem 2 times 3. You write the answer on top.
- On the bottom, you work the problem 5 times 4 and write the answer on the bottom.
- 2/5 times 3/4 equals 6/20.
b. Find part 3.
c. Problem A is 2/3 times 7/8. Say the problem for the top numbers. (Signal.) *2 times 7.*
- Say the problem for the bottom numbers. (Signal.) *3 times 8.*
- Problem B is 1/5 times 9/3. Say the problem for the top numbers. (Signal.) *1 times 9.*
- Say the problem for the bottom numbers. (Signal.) *5 times 3.*
- Problem C: 5/4 times 2/6. Say the problem for the top numbers. (Signal.) *5 times 2.*

- Say the problem for the bottom numbers. (Signal.) *4 times 6.*
  (Repeat step c until firm.)
d. Find part 4.
- A student worked the problems in part 4. She got some of the problems wrong because she did not follow the correct procedure. She didn't multiply the top numbers correctly, or she did something strange with the bottom numbers.
e. Find the problems she worked incorrectly. Copy them and work them correctly. Don't do anything with the problems that are worked correctly. Raise your hand when you're finished. (Observe students and give feedback.)
Key:

$$a. \ \frac{3}{4} \times \frac{1}{5} = \boxed{\frac{3}{20}}$$

$$c. \ \frac{4}{5} \times \frac{8}{5} = \boxed{\frac{32}{25}}$$

$$d. \ \frac{1}{7} \times \frac{8}{1} = \boxed{\frac{8}{7}}$$

f. Check your work.
- You should have copied and worked problems A, C and D.
g. Problem A: 3/4 times 1/5. The student's answer is 3/9. Wrong. What's the correct answer? (Signal.) *3-twentieths.*
- You should have 3/20 as your answer.
h. Problem C: 4/5 times 8/5. The student wrote the answer as 32/5. What's the correct answer? (Signal.) *32-twenty-fifths.*
- You should have 32/25 as your answer.
i. Problem D: 1/7 times 8 over 1. The student wrote the answer as 9/8. What's the correct answer? (Signal.) *8-sevenths.*
- You should have 8/7 as your answer.
j. Find part 5.
- You're going to work these problems.
- Remember, multiply on top and write the answer on top. Multiply on the bottom and write the answer on the bottom.
- Raise your hand when you've copied all the problems and worked them.
  (Observe students and give feedback.)

a. $\dfrac{7}{3} \times \dfrac{4}{3} = \boxed{\dfrac{28}{9}}$

b. $\dfrac{8}{11} \times \dfrac{2}{2} = \boxed{\dfrac{16}{22}}$

c. $\dfrac{3}{6} \times \dfrac{4}{2} = \boxed{\dfrac{12}{12}}$

d. $\dfrac{5}{8} \times \dfrac{9}{2} = \boxed{\dfrac{45}{16}}$

k. Find part J on page 15 of your textbook. √
• That shows what you should have for each problem. Fix up any mistakes.
l. Raise your hand if you got everything right.

EXERCISE 3 NUMBER FAMILIES

a. (Write on the board:)

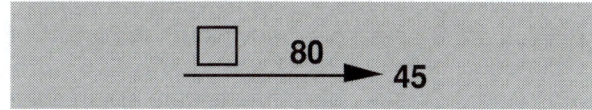

• Number families must have a big number that is larger than either of the small numbers.
• This number family is wrong. The big number shown is 45. The small number shown is 80. The small number is bigger than the big number so you can't find the missing number for this family.
b. Find part 6.
• Some of the number families in part 6 can be worked. Some are impossible to work because the small number that is shown is bigger than the big number.
c. Copy the families that are correct and figure out the missing numbers for those families. Don't copy and work problems for the families that are impossible. Remember to box your answer. Raise your hand when you're finished.
(Observe students and give feedback.)

a. $\underrightarrow{\quad 58 \quad \square \quad} 85$
$\begin{array}{r} 85 \\ -\ 58 \\ \hline \boxed{27} \end{array}$

b. $\underrightarrow{\quad 76 \quad 23 \quad} \square$
$\begin{array}{r} 76 \\ +\ 23 \\ \hline \boxed{99} \end{array}$

d. $\underrightarrow{\quad 64 \quad 128 \quad} \square$
$\begin{array}{r} 64 \\ +\ 128 \\ \hline \boxed{192} \end{array}$

f. $\underrightarrow{\quad 48 \quad \square \quad} 56$
$\begin{array}{r} 56 \\ -\ 48 \\ \hline \boxed{8} \end{array}$

d. Find part K on page 15 of your textbook. √
• That shows what you should have for families A, B, D and F.
• You should not have worked problems for families C and E. Those problems are impossible.

EXERCISE 4 MULTIPLICATION
Missing Factor

a. Find part 7.
• I'll read what it says. Follow along: Some problems show multiplication with a times sign. You can see the problem 3 times 6 written with a times sign.
• Sometimes, parentheses are used to show multiplication. You can see the same problem written with parentheses. You read that problem the same way you read the other problem: 3 times 6.
• You can also use parentheses to show fractions that are multiplied. You can see 3/4 times 5/8.
b. Find part 8.
• These problems have parentheses. Some problems show the value you're multiplying by. Other problems don't.
c. Sample problem 1 says: 4 times some value equals 20. What's the missing value? (Signal.) *5.*
• Yes, 4 times 5 equals 20.

- Sample problem 2 says: 9 times 8 equals some value. What value? (Signal.) *72.*
d. Copy problems A through E in part 8 and write the missing values to complete the equations. Raise your hand when you're finished. (Observe students and give feedback.)
e. I'll say each problem. You tell me the missing value.
- Problem A: 6 times some value equals 48. What's the missing value? (Signal.) *8.*
- Problem B: 4 times 8 equals some value. What's the missing value? (Signal.) *32.*
- Problem C: 20 times some value equals 20. What's the missing value? (Signal.) *1.*
- Problem D: 15 times 3 equals some value. What's the missing value? (Signal.) *45.*
- Problem E: 5 times some value equals 50. What's the missing value? (Signal.) *10.*
f. Raise your hand if you got everything right.

EXERCISE 5 FRACTION ANALYSIS
More than/Less than 1

a. Find part 9.
- Look at the fractions in part 9. Some are more than 1; some are equal to 1; some are less than 1.
b. Copy all the fractions that are more than 1 or are equal to 1. Raise your hand when you're finished. (Observe students and give feedback.)
c. (Write on the board:)

$$\frac{7}{7} \qquad \frac{3}{1} \qquad \frac{3}{2} \qquad \frac{3}{3} \qquad \frac{7}{3}$$

- Here's what you should have. The fractions that are more than 1 or equal to 1 are: 7/7, 3 over 1, 3/2, 3/3 and 7/3.
- The other fractions are less than 1 unit.
d. Some of the fractions you copied equal 1. Circle those fractions. Raise your hand when you're finished. (Observe students and give feedback.)
e. (Circle **7/7** and **3/3** on the board.)
- You should have circled 7/7 and 3/3. Both those fractions have the same number for the parts in each unit and the parts shaded. So they equal 1 whole unit.

EXERCISE 6 PROBLEM SOLVING
Multiplication/Division

a. Find part 10.
- Some of these problems can be worked by multiplication and some can be worked by division.

b. You're not going to write the word **multiplication** or **division;** you'll just work the problems. Remember to show the answer as a number and a unit name. Raise your hand when you're finished. (Observe students and give feedback.)
Key:

a. $6\overline{)384}$ $\boxed{64 \text{ words}}$

b. $\begin{array}{r} 75 \\ \times\ 11 \\ \hline 75 \\ +\ 750 \\ \hline \boxed{825 \text{ words}} \end{array}$

c. $7\overline{)56}$ $\boxed{8 \text{ bottles}}$

d. $\begin{array}{r} 28 \\ \times\ 4 \\ \hline \boxed{112 \text{ miles}} \end{array}$

c. Find part L on page 15 of your textbook. √
- That shows what you should have for each item.
- Problem A asks how many words the typist typed each minute. The answer is 64 words.
- Problem B asks how many words the typist typed in 11 minutes. The answer is 825 words.
- Problem C asks how many bottles are in each carton. The answer is 8 bottles.
- Problem D asks how far Mrs. Smith traveled in 4 bike rides. The answer is 112 miles.
d. Raise your hand if you got everything right.

EXERCISE 7 INDEPENDENT WORK

a. Do the independent work for lesson 4.
b. (Before beginning the next lesson, check the students' independent work.)

Lesson 5

Materials

- Each student will need a calculator for exercise 5.

Objectives

- Work a problem that requires multiplying fractions. (Exercise 1)

- Identify number families that are written correctly and figure out the missing number. (Exercise 2)

- Classify a fraction that is more than 1, less than 1 or equal to 1. (Exercise 3)

- **Make a number family for a diagram that shows a whole divided into two parts.** (Exercise 4)
  Note: Diagrams are of the form:

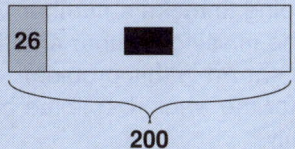

Students use the rules that: If the number for the whole bar is missing, they show that as the missing big number in the family. If a number for one of the parts is missing, they show that as a missing small number in the family.

- **Use division to figure out the missing factor in a multiplication equation.** (Exercise 5)
  Note: Problems are of the form:
  $$56 (\blacksquare) = 1904$$
  Students write the problem $56 \overline{|\ 1904}$ to solve it.

- Work a word problem that requires multiplication or division. (Exercise 6)

EXERCISE 1 FRACTION OPERATIONS
Multiplication

a. Open your textbook to lesson 5 and find part 1.
- You've worked problems that have parentheses. The parentheses work like a times sign.
b. The problems in part 1 show parentheses for multiplying fractions.
- Problem A is 7/3 times 7/3.

- Remember how to work these problems. You multiply on top and write the answer on top. You multiply on the bottom and write the answer on the bottom.
c. A student worked the problems in part 1 and got some of them wrong.
- Copy the problems the student worked incorrectly and work them correctly. Raise your hand when you're finished.
  (Observe students and give feedback.)
d. (Write on the board:)

$$\text{b. } \frac{5}{3} \left(\frac{0}{3} \right) = \boxed{\frac{0}{9}}$$

$$\text{c. } \frac{4}{5} \left(\frac{10}{5} \right) = \boxed{\frac{40}{25}}$$

- You should have worked problems B and C.
e. Problem B: 5/3 times 0/3. The student wrote 5/9 as the answer. The correct answer is 0/9.
- Raise your hand if you got it right.
f. Problem C: 4/5 times 10/5. The student wrote the answer as 40/5. The correct answer is 40/25.
- Raise your hand if you found both wrong answers.
g. Find part 2.
- Copy each problem and work it. Raise your hand when you're finished.
  (Observe students and give feedback.)
  Key:

$$a. \ \frac{7}{10} \left(\frac{3}{8} \right) = \boxed{\frac{21}{80}} \qquad b. \ \frac{1}{2} \left(\frac{1}{5} \right) = \boxed{\frac{1}{10}}$$

$$c. \ \frac{10}{7} \left(\frac{2}{7} \right) = \boxed{\frac{20}{49}} \qquad d. \ \frac{8}{13} \left(\frac{8}{1} \right) = \boxed{\frac{64}{13}}$$

h. Find part J on page 19 of your textbook. √
- That shows what you should have for each problem. Fix any mistakes.
i. Raise your hand if you got everything right.

EXERCISE 2 NUMBER FAMILIES

a. Find part 3.
- Remember, number families must have a big number that is larger than either of the small numbers.
- Some of the number families in part 3 can be worked. Some are impossible to work because the big number that is shown can't really be the big number.

b. Copy all the families that are correct and figure out the missing number. Don't work problems for the families that are impossible. Raise your hand when you're finished.
(Observe students and give feedback.)
Key:

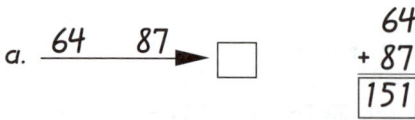

a. 64 87 → ☐   64 + 87 = 151

d. ☐ 45 → 386   386 − 45 = 341

e. 684 27 → ☐   684 + 27 = 711

c. Find part K. √
• That shows what you should have for families A, D and E.
d. You should not have worked problems for families B, C and F. Those problems are impossible.

EXERCISE 3 FRACTION ANALYSIS
More than/Less than 1

a. Find part 4.
• Look at the fractions in part 4. Some are more than 1; some are equal to 1; some are less than 1.
b. Copy all the fractions that are more than 1 or are equal to 1. Raise your hand when you're finished. (Observe students and give feedback.)
c. (Write on the board:)

$\frac{6}{5}$	$\frac{5}{5}$	$\frac{80}{72}$	$\frac{2}{1}$	$\frac{20}{20}$

• Here's what you should have. The fractions that are more than 1 or equal to 1 are: 6/5, 5/5, 80/72, 2 over 1 and 20/20.
• The other fractions are less than 1 unit.
d. Some of the fractions you copied equal 1. Circle those fractions. Raise your hand when you're finished.
e. (Circle **5/5** and **20/20** on the board.)
• You should have circled 5/5 and 20/20. Both those fractions have the same number for the parts in each unit and the parts shaded. So they equal 1 whole unit.

EXERCISE 4 NUMBER FAMILIES
Part/Whole

a. Find part 5.
• These are diagrams.
b. The sample diagram shows all three numbers. The shaded part of the bar is 38 units. The unshaded part is 67 units. The whole bar is 105 units.
• You can make a number family from the information given about the bar.
• The small numbers are for the shaded part and the unshaded part. The big number is for the whole bar.
• You can see the number family for the sample diagram. The small numbers are 38 and 67. The big number is 105.
c. Problem A. This bar has one of the numbers missing. Is the big number or a small number missing? (Signal.) *A small number.*
• Yes, the number for the shaded part is missing.
• You're going to make a number family with a box for the missing number and the two numbers shown in this problem. Remember, the number for the whole bar is the big number in your family.
• Make the family with a box and **two** numbers. Raise your hand when you've done that much. (Observe students and give feedback.)
• (Write on the board:)

a. ☐ 320 → 480

• Here's what you should have. The big number is 480. One of the small numbers is 320. It doesn't matter which small number you show as 320. The other small number is a box.
d. Your turn: Make the number family for problem B. Make the box and the two numbers the problem shows. Raise your hand when you're finished. (Observe students and give feedback.)
• (Write on the board:)

b. 111 222 → ☐

• Here's what you should have. The big number is missing. The small numbers are 111 and 222.
e. Your turn: Make number families for the rest of the problems in part 5. Raise your hand when you're finished.
(Observe students and give feedback.)

f. (Write on the board:)

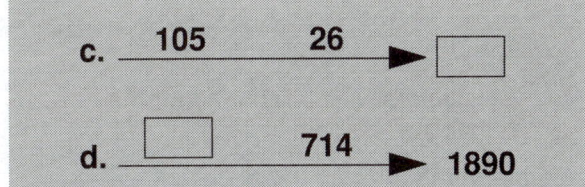

c. 105   26 → ☐

d. ☐   714 → 1890

- Here's what you should have for problems C and D.
- Raise your hand if you got all the families right.

g. Make sure you have the right families. Then figure out the missing number in each family. Remember to box your answers. Raise your hand when you're finished.
  (Observe students and give feedback.)

h. (Write on the board:)

a.	480	b.	111
	− 320		+ 222
	160		**333**
c.	105	d.	1890
	+ 26		− 714
	131		**1176**

- Check your work. Here's what you should have for each problem.
- Raise your hand if you got everything right.

EXERCISE 5 MULTIPLICATION
Missing Factor

Note: Each student will need a calculator for this exercise.

a. Find part 6.
- There's a little picture of a calculator at the beginning of part 6. That picture means that you'll use your calculator to work the problems. You'll see the calculator symbol in different lessons. Whenever you see it, you'll use your calculator. If you don't see it, you **won't** use your calculator.

b. I'll read what it says. Follow along: You can work multiplication problems that have the middle number missing.
- You can see the problem: 4 times some value equals 28.
- You use the multiplication fact that has the numbers 4 and 28: 4 times 7 equals 28.

- You can also work these problems as division problems.
- You can see the same problem written as a division problem: 28 divided by 4.
- When problems like these have large numbers, rewrite them as division problems.
- You can see the problem: 45 times some value equals 3690.
- You can see that problem written as a division problem: 3690 divided by 45.
- When you work that problem, you'll know what you'd multiply 45 by to get 3690.

c. Use your calculator. Work the problem 3690 divided by 45. Raise your hand when you know the answer.
  (Observe students and give feedback.)

- Everybody, what's 3690 divided by 45? (Signal.) *82.*
- That means that if you multiply 45 by 82, you'll end up with 3690.

d. Do it. Multiply 45 and 82 and see if you get 3690. Raise your hand when you're finished.
  (Observe students and give feedback.)

- It works. Remember, you can rewrite multiplication problems that have a missing middle number as division problems. The value that comes after the equal sign when you multiply is the value under the division sign.

e. Find part 7.
- Problem A: 49 times some number equals 343. Start with the number after the equal sign and say the division problem. (Signal.) *343 divided by 49.*
- Problem B: 46 times some number equals 2392. Say the division problem. (Signal.) *2392 divided by 46.*
- Problem C: 56 times some number equals 1904. Say the division problem. (Signal.) *1904 divided by 56.*
- Problem D: 28 times some number equals 532. Say the division problem. (Signal.) *532 divided by 28.*

f. Your turn: Copy each problem. Use your calculator. Work the division problems and write the answers. Raise your hand when you're finished. (Observe students and give feedback.)

g. (Write on the board:)

a. 49 (7) = 343

b. 46 (52) = 2392

c. 56 (34) = 1904

d. 28 (19) = 532

- Here's what you should have.
- Remember, to work multiplication problems with a middle missing value, you divide.

EXERCISE 6 PROBLEM SOLVING
Multiplication/Division

a. Find part 8.
- Work the problems in part 8. Remember, the answer has a number and a unit name. Raise your hand when you're finished.
  (Observe students and give feedback.)
  Key:

  a. 58
     $\times 11$
      58
    + 580
    ────
    638 eggs

  b. 9$\overline{)45}$ **5 ounces**

  c. 16
     $\times$ 4
    ────
    64 stops

  d. 4$\overline{)64}$ **16 books**

b. Find part L. √
- That shows what you should have for each problem.
- Problem A: How many eggs did the chickens on Foster Farm lay in all? (Signal.) *638.*
- Problem B: How many ounces did each piece of pie weigh? (Signal.) *5.*
- Problem C: How many stops did the truck make in all? (Signal.) *64.*
- Problem D: How many books were on each shelf? (Signal.) *16.*
c. Raise your hand if you got everything right.

EXERCISE 7 INDEPENDENT WORK

a. Do the independent work for lesson 5.
b. (Before beginning the next lesson, check the students' independent work.)

Lesson 6

Materials

Calculators will be required for the remaining lessons.

Objectives

- **Work a set of problems involving addition, subtraction and multiplication of fractions.** (Exercise 1)

- **Complete a table to show a multiplication equation and the corresponding division problem.** (Exercise 2)
  Note: The first column shows multiplication problems with a missing factor. Students write the corresponding division problems in the second column.

- **Complete an equation to show the whole number a fraction equals.** (Exercise 3)
  Note: Students apply the rule that if the top number is so many times the bottom number, the fraction equals the number of times.

- Make a number family for a diagram that shows a whole divided into two parts. (Exercise 4)

- **Figure out the missing factor in a fraction multiplication problem.** (Exercise 5)
  Note: Problems are of the form:
  $\frac{5}{3}\left(\frac{\blacksquare}{\blacksquare}\right) = \frac{30}{12}$. Students work the problem on top and write the number on top, then work the problem on the bottom and write that number on the bottom.

- Identify number families that are written correctly and figure out the missing number. (Exercise 6)

EXERCISE 1 FRACTION OPERATIONS
Addition/Subtraction/Multiplication

a. Open your textbook to lesson 6 and find part 1.
- Some of these problems add or subtract fractions. Others multiply fractions.
- The student who worked these problems got confused. She made four mistakes.
- Remember, look at the sign in the problem. If you multiply, you work on both the top and the bottom. If you add or subtract, you just copy the bottom number and work on top.

b. Your turn: Find the four mistakes. Copy those problems and write them with the correct answers. Raise your hand when you're finished. (Observe students and give feedback.)

c. (Write on the board:)

a. $\dfrac{1}{2}$ x $\dfrac{1}{2}$ = $\boxed{\dfrac{1}{4}}$

c. $\dfrac{2}{3}$ x $\dfrac{3}{2}$ = $\boxed{\dfrac{6}{6}}$

d. $\dfrac{15}{3}$ − $\dfrac{8}{3}$ = $\boxed{\dfrac{7}{3}}$

f. $\dfrac{3}{5}$ + $\dfrac{7}{5}$ = $\boxed{\dfrac{10}{5}}$

d. Check your work.
• Here's what you should have. The student worked problems A, C, D and F incorrectly. You should have copied those problems and worked them correctly.
• Problem A: 1/2 times 1/2. The student wrote the answer of 2/2. The correct answer is 1/4. On top, you work the problem 1 times 1. That's 1. On the bottom, you work the problem 2 times 2. That's 4.
• Problem C: 2/3 times 3/2. The student wrote the answer of 6/1. The correct answer is 6/6.
• Problem D: 15/3 minus 8/3. The student wrote the answer of 7 over zero. The correct answer is 7/3.
• Problem F: The problem is 3/5 plus 7/5. The student wrote the answer as 21/25. The correct answer is 10/5.

e. Find part 2.
• Some of these problems add or subtract. Some multiply.

f. Copy the problems and work them. Raise your hand when you're finished. (Observe students and give feedback.)
Key:

a. $\dfrac{5}{8}$ + $\dfrac{5}{8}$ = $\boxed{\dfrac{10}{8}}$ d. $\dfrac{17}{3}$ x $\dfrac{1}{5}$ = $\boxed{\dfrac{17}{15}}$

b. $\dfrac{5}{8}$ x $\dfrac{5}{8}$ = $\boxed{\dfrac{25}{64}}$ e. $\dfrac{3}{7}$ x $\dfrac{6}{7}$ = $\boxed{\dfrac{18}{49}}$

c. $\dfrac{5}{8}$ − $\dfrac{5}{8}$ = $\boxed{\dfrac{0}{8}}$ f. $\dfrac{6}{4}$ − $\dfrac{2}{4}$ = $\boxed{\dfrac{4}{4}}$

g. Check your work.
• Find part J on page 23.
• That shows what you should have for each problem.
• Problem A: 5/8 plus 5/8. What's the answer? (Signal.) *10-eighths.*
• Problem B: 5/8 times 5/8. What's the answer? (Signal.) *25-sixty-fourths.*
• Problem C: 5/8 minus 5/8. What's the answer? (Signal.) *zero-eighths.*
• Problem D: 17/3 times 1/5. What's the answer? (Signal.) *17-fifteenths.*
• Problem E: 3/7 times 6/7. What's the answer? (Signal.) *18-forty-ninths.*
• Problem F: 6/4 minus 2/4. What's the answer? (Signal.) *4-fourths.*

h. Raise your hand if you got everything right.

EXERCISE 2 MULTIPLICATION
Missing Factor

a. Find part 3.
• You've worked multiplication problems that have the middle number missing. You can work these problems as division problems.
• The first column of the table shows multiplication problems. The next column will show division problems.
• The top row is completed. The multiplication problem is 23 times some value equals 943. The division problem is 943 divided by 23. The answer to both problems is 41. You can see it written in red.
• The rest of the rows show only the multiplication problem.

b. Copy row A. Write the multiplication problem and the corresponding division problem. Then use your calculator to find the missing numbers. You do that by working the division problem, then writing the answer as the missing number in both problems.
• Raise your hand when you've written the problems and filled in the answers for row A. (Observe students and give feedback.)
• (Write on the board:)

Multiplication	Division
a. 36 (19) = 684	19 36⟌684

• Here's what you should have for row A.
• The multiplication problem is 36 times some value equals 684. The corresponding division problem is 684 divided by 36. The answer to both problems is 19.

c. Your turn: Work the rest of the rows in the table. Raise your hand when you're finished.
(Observe students and give feedback.)
d. (Write to show:)

	Multiplication	Division
a.	36 (19) = 684	$\dfrac{19}{36\overline{)684}}$
b.	14 (29) = 406	$\dfrac{29}{14\overline{)406}}$
c.	37 (24) = 888	$\dfrac{24}{37\overline{)888}}$
d.	45 (35) = 1575	$\dfrac{35}{45\overline{)1575}}$

• Here's what you should have for rows B, C and D.
• Raise your hand if you got everything right.

EXERCISE 3 FRACTIONS
Equal to Whole Numbers

a. Find part 4.
• Some fractions equal 1. Other fractions equal other whole numbers. Some fractions equal 2, some equal 3, some equal 30.
• Here's the rule about fractions that equal whole numbers: If the top number is 3 times the bottom number, the fraction equals 3 units. If the top number is 8 times the bottom number, the fraction equals 8 units. If the top number is 1 times the bottom number, the fraction equals 1.
• To figure out how many whole units a fraction equals, start with the bottom number and see how many times larger the top number is.
b. Look at the sample equation.
• What's the bottom number? (Signal.) 5.
• How many times would you multiply by 5 to get the top number? (Signal.) 3.
• 5 times 3 equals 15. So 15/5 equals 3 whole units.
c. Each fraction in part 4 equals a whole number.
• You'll copy each fraction, write an equal sign and the whole number it equals.
d. Problem A: The fraction is 12/6. Figure out how many times greater the top number is than the bottom number and write what 12/6 equals. Raise your hand when you're finished.
(Observe students and give feedback.)

• (Write on the board:)

$$a. \quad \frac{12}{6} = \boxed{2}$$

• Here's what you should have. 12/6 equals 2.
e. Copy fraction B and write the whole number it equals. Remember the equal sign. Raise your hand when you're finished.
(Observe students and give feedback.)
• (Write on the board:)

$$b. \quad \frac{18}{3} = \boxed{6}$$

• Here's what you should have. 18/3 equals 6 whole units. Raise your hand if you got it right.
f. Write equations for the rest of the fractions in part 4. Raise your hand when you're finished.
(Observe students and give feedback.)
• (Write on the board:)

$$c. \quad \frac{43}{1} = \boxed{43} \qquad d. \quad \frac{24}{8} = \boxed{3}$$

• Here are the equations you should have for fractions C and D. Raise your hand if you got everything right.
• Remember how to figure out the number of units a fraction equals. Start with the bottom number and see how many times greater the top number is.

EXERCISE 4 NUMBER FAMILIES
Part/Whole

a. Find part 5.
• These are bars that have a shaded part and an unshaded part. The values for those parts are the small numbers that go in a number family. The big number is the value for the total bar.
b. Your turn: Make the family for bar A. Don't figure out the missing number yet. Raise your hand when you've done that much.
(Observe students and give feedback.)
• (Write on the board:)

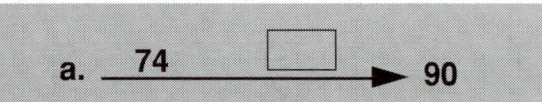

• Here's what you should have. The small numbers are 74 and a box. The big number is 90.

c. Your turn: Make number families for the rest of the items in part 5. Raise your hand when you're finished.
(Observe students and give feedback.)
d. (Write to show:)

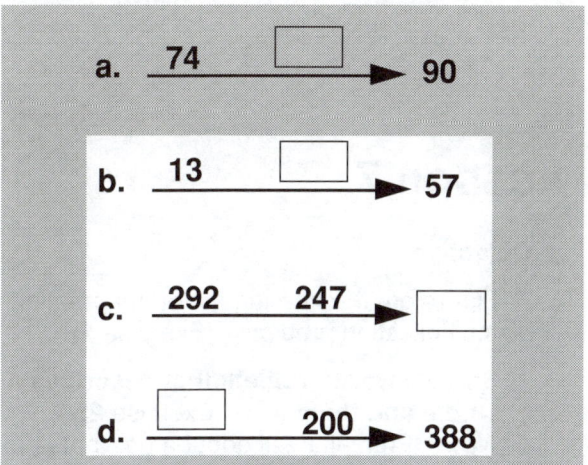

- Here's what you should have for items B, C and D.
- Raise your hand if you got everything right.
e. Your turn: Figure out the missing number in each family. Box the answer. Raise your hand when you're finished.
(Observe students and give feedback.)
f. (Write to show:)

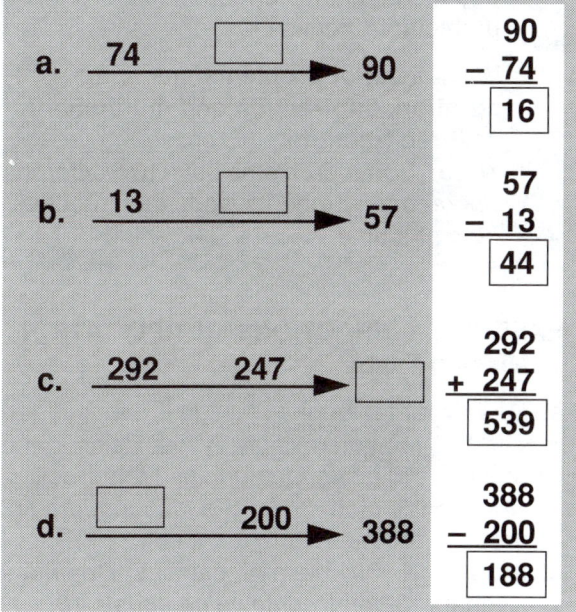

- Here's what you should have for each problem.
- Raise your hand if you got everything right.

EXERCISE 5 FRACTION MULTIPLICATION
Missing Factor

a. Find part 6.
- I'll read what it says. Follow along: You've worked problems like this one: 5 times some value equals 30.
- You multiply 5 by 6.
- You can work similar problems that involve fractions.
- You can see the problem: 5/3 times some value equals 30/12.
- You can write the missing value as a fraction. You do that by working the problem on the top and the problem on the bottom.
- Everybody, say the problem for the top numbers. (Signal.) *5 times some value equals 30.*
- Say the problem for the bottom numbers. (Signal.) *3 times some value equals 12.*
- The bottom equation shows the fraction you multiply 5/3 by to get 30/12. The fraction is 6/4.
b. Find part 7.
- These are problems with missing fractions. Remember, you find the missing fraction by working the problem for the top numbers and writing that value on top. You work the problem for the bottom numbers and write that number on the bottom.
c. Problem A. Say the problem for the top numbers. (Signal.) *2 times some value equals 6.*
- Say the problem for the bottom numbers. (Signal.) *7 times some value equals 14.*
- Your turn: Copy problem A and write the missing fraction. Raise your hand when you're finished. (Observe students and give feedback.)
- (Write on the board:)

$$\text{a. } \frac{2}{7}\left(\frac{3}{2}\right) = \frac{6}{14}$$

- Here's what you should have. The missing fraction is 3/2.
d. Your turn: Copy problem B and write the missing fraction. Raise your hand when you're finished. (Observe students and give feedback.)
- (Write on the board:)

$$\text{b. } \frac{7}{4}\left(\frac{5}{8}\right) = \frac{35}{32}$$

- Here's what you should have. 7/4 times some fraction equals 35/32. The missing fraction is 5/8.
e. Your turn: Work the rest of the problems in part 7. Raise your hand when you're finished. (Observe students and give feedback.)

f. (Write on the board:)

$$c. \quad \frac{1}{2} \left(\frac{2}{1} \right) = \frac{2}{2}$$

$$d. \quad \frac{5}{9} \left(\frac{3}{3} \right) = \frac{15}{27}$$

$$e. \quad \frac{10}{5} \left(\frac{3}{9} \right) = \frac{30}{45}$$

- Here's what you should have for problems C, D and E.
- Problem C: 1/2 times some fraction equals 2/2. What's the missing fraction? (Signal.) *2 over 1.*
- Problem D: 5/9 times some fraction equals 15/27. What's the missing fraction? (Signal.) *3-thirds.*
- Problem E: 10/5 times some fraction equals 30/45. What's the missing fraction? (Signal.) *3-ninths.*

g. Raise your hand if you got everything right.

EXERCISE 6 NUMBER FAMILIES
Part/Whole

a. Find part 8.
- Some of the number families are impossible. You can work the others. Copy the **families** that you can work and figure out the missing number. Raise your hand when you're finished.
(Observe students and give feedback.)
Key:

$$a. \quad \underrightarrow{264 \quad 19} \quad \square \qquad \begin{array}{r} 264 \\ + \ 19 \\ \hline \boxed{283} \end{array}$$

$$d. \quad \underrightarrow{25 \quad 98} \quad \square \qquad \begin{array}{r} 25 \\ + \ 98 \\ \hline \boxed{123} \end{array}$$

$$f. \quad \underrightarrow{213 \quad \boxed{}} \quad 286 \qquad \begin{array}{r} 286 \\ - \ 213 \\ \hline \boxed{73} \end{array}$$

b. Find part K. √
- You should have worked families A, D, and F.
- Raise your hand if you got everything right.

EXERCISE 7 INDEPENDENT WORK

a. Do the independent work for lesson 6.
b. (Before beginning the next lesson, check students' independent work.)

Lesson 7

Objectives

- Figure out the missing factor in a fraction multiplication problem. (Exercise 1)

- **Solve a word problem that describes a whole and the parts.** (Exercise 2)
Note: Problems tell about a bar that is divided into a shaded part and an unshaded part. Two numbers are given. Students make a number family, then figure out the missing number.

- Complete an equation to show the whole number a fraction equals. (Exercise 3)

- Complete a table to show a multiplication equation and the corresponding division problem. (Exercise 4)

- **Work a set of problems involving addition, subtraction and multiplication of three fractions.** (Exercise 5)
Note: Some problems have two different operations. Students work each problem a step at a time.

EXERCISE 1 FRACTION MULTIPLICATION
Missing Factor

a. Open your textbook to lesson 7 and find part 1.
- The fraction you multiply by is missing in these problems. Remember, you find that fraction by working the problem for the top numbers and the problem for the bottom numbers.
b. Problem A. Say the problem for the top numbers. (Signal.) *3 times some value equals 18.*
- Say the problem for the bottom numbers. (Signal.) *5 times some value equals 30.*
c. Copy all the problems in part 1 and work them. Raise your hand when you're finished.
(Observe students and give feedback.)

d. (Write on the board:)

$$a. \quad \frac{3}{5}\left(\frac{6}{6}\right) = \frac{18}{30}$$

$$b. \quad \frac{5}{1}\left(\frac{2}{1}\right) = \frac{10}{1}$$

$$c. \quad \frac{6}{5}\left(\frac{4}{4}\right) = \frac{24}{20}$$

$$d. \quad \frac{4}{7}\left(\frac{1}{2}\right) = \frac{4}{14}$$

$$e. \quad \frac{2}{5}\left(\frac{3}{2}\right) = \frac{6}{10}$$

- Check your work. Here's what you should have for each problem.
- Raise your hand if you got everything right.

EXERCISE 2 NUMBER FAMILIES
Part/Whole

a. Find part 2.
- (Teacher reference:)

Sample problem 1

A bar has a shaded part and an unshaded part. The whole bar is 381 inches long. The shaded part is 110 inches long. How long is the unshaded part?

shaded	unshaded	bar	381

110 ■ → 381   − 110 / 271 inches

- These are like bar problems you've worked, but they don't show the bar or the numbers. Each problem gives you information about two of the numbers.
b. Look at sample problem 1.
- It says: A bar has a shaded part and an unshaded part. The whole bar is 381 inches long. The shaded part is 110 inches long. How long is the unshaded part?
- You can see the number family with names and numbers. The three names are **shaded, unshaded** and **bar.** The numbers are shown for **shaded** and for **bar.**
- The calculation shows that you figure out the unshaded part by working the problem 381 minus 110. The answer is 271. The unshaded part is 271 inches long. Notice, the answer has a number and a unit name. The whole answer is boxed.

c. Sample problem 2 says: A bar has an unshaded part and a shaded part. The unshaded part is 56 feet long. The shaded part is 33 feet long. How long is the whole bar?
- The number family has the names **unshaded, shaded** and **bar.** Numbers are shown for **unshaded** and **shaded.** The calculation for the big number is 56 plus 33. The whole bar is 89 feet long.
d. Problem A: A bar has a shaded part and an unshaded part. The unshaded part is 49 meters long. The whole bar is 400 meters long. How long is the shaded part?
- Make the number family with three names and two numbers. Raise your hand when you've done that much.
(Observe students and give feedback.)
- (Write on the board:)

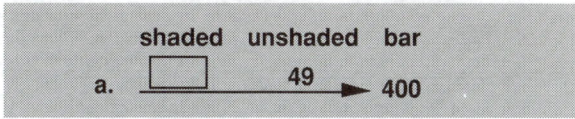

- Here's what you should have so far. The numbers are 49 for unshaded; 400 for the whole bar.
- Figure out the missing number. Write the unit name in your answer and box the answer. Raise your hand when you're finished.
(Observe students and give feedback.)
- (Write to show:)

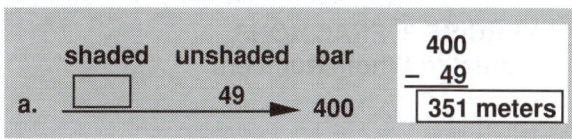

- Here's what you should have. The shaded part is 351 meters long.
e. Your turn. Work problem B. Remember to show your number family with three names and two numbers. Also remember to write the unit name in your answer and box the answer. Raise your hand when you've finished.
(Observe students and give feedback.)
- (Write on the board:)

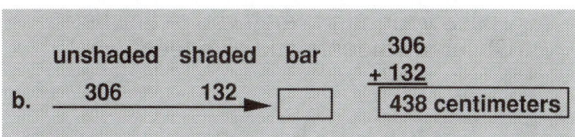

- Here's what you should have. The numbers are 306 for unshaded; 132 for shaded. The whole bar is 438 centimeters long. Raise your hand if you got everything right.
f. Your turn. Work problem C. Raise your hand when you're finished.
(Observe students and give feedback.)

• (Write on the board:)

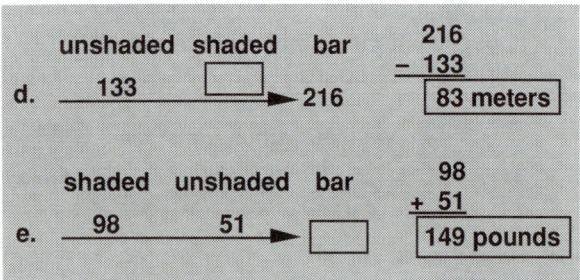

• Here's what you should have. The whole bar is 398 centimeters long. The shaded part is 311 centimeters. You figured out the unshaded part by working the problem: 398 minus 311. The answer is 87 centimeters.

g. Your turn. Work the rest of the problems in part 2. Raise your hand when you're finished.
(Observe students and give feedback.)

• (Write on the board:)

• Here's what you should have for problems D and E. The shaded part of the bar in problem D is 83 meters long. The whole bar in problem E weighs 149 pounds.

h. Raise your hand if you got everything right.

EXERCISE 3 FRACTIONS
Equal to Whole Numbers

a. Find part 3.
• The fractions in part 3 equal whole numbers because the top is so many times the bottom number.
• Fraction A: What's the bottom number? (Signal.) 1.
• How many times greater than 1 is the top? (Signal.) 7.
• So the fraction equals 7.

b. Write the equation for problem A, then write equations for the rest of the problems in part 3. Raise your hand when you're finished.
(Observe students and give feedback.)

c. (Write on the board:)

a. $\dfrac{7}{1}$ = $\boxed{7}$ d. $\dfrac{9}{9}$ = $\boxed{1}$

b. $\dfrac{14}{2}$ = $\boxed{7}$ e. $\dfrac{36}{9}$ = $\boxed{4}$

c. $\dfrac{63}{9}$ = $\boxed{7}$ f. $\dfrac{70}{10}$ = $\boxed{7}$

• Here are the equations you should have.
• Raise your hand if you got everything right.

EXERCISE 4 MULTIPLICATION
Missing Factor

a. Find part 4.
• You're going to complete this table to show multiplication problems in the first column and division problems in the second column.
• Remember, the number after the equal sign is the number that goes under the division sign.

b. Complete row A. Write the division problem. Use your calculator to work the division problem, then write the missing number for the multiplication problem. Raise your hand when you've completed row A.
(Observe students and give feedback.)

• (Write on the board:)

Multiplication	Division
a. 6 (94) = 564	94 6⟌564

• Here's what you should have for row A. 6 times 94 equals 564. 564 divided by 6 equals 94. Fix up any mistakes.

c. Your turn: Work the rest of the rows. Raise your hand when you're finished.
(Observe students and give feedback.)

d. (Write to show:)

	Multiplication	Division
a.	6 (94) = 564	$6\overline{)564}$ 94
b.	72 (11) = 792	$72\overline{)792}$ 11
c.	3 (306) = 918	$3\overline{)918}$ 306
d.	16 (44) = 704	$16\overline{)704}$ 44

- Check your work. Here's what you should have for the rest of the rows.
- Raise your hand if you got everything right.

EXERCISE 5 FRACTION OPERATIONS
Addition/Subtraction/Multiplication

a. Find part 5.
   Some problems have more than two values that are added, subtracted or multiplied. You work these problems a step at a time.
b. Look at the sample problem: 2 times 4 times 3. You first figure out 2 times 4. What's the answer? (Signal.) *8.*
- Now you work the problem 8 times 3. What's the answer? (Signal.) *24.*
- That's the answer to the whole problem. 2 times 4 times 3 equals 24.
c. Problem A adds and subtracts fractions with the same bottom number. So that's the number you'll write in the answer. You'll work the top numbers. 4 plus 9. What's the answer? (Signal.) *13.*
- 13 minus 5. What's the answer? (Signal.) *8.*
- That's the top number in the answer.
- Your turn: Copy problem A and work it. Raise your hand when you're finished.
   (Observe students and give feedback.)
- Check your work. 4/7 plus 9/7 minus 5/7 equals 8/7.
d. Copy problem B and work it. Raise your hand when you're finished.
   (Observe students and give feedback.)
- (Write on the board:)

$$\text{b. } \frac{2}{3} \times \frac{2}{2} \times \frac{9}{1} = \boxed{\frac{36}{6}}$$

- Here's what you should have. The problem on top is 2 times 2 times 9. That's 36. The problem on the bottom is 3 times 2 times 1. That's 6. The answer to problem B is 36/6.
e. Copy problem C and work it. Raise your hand when you're finished.
   (Observe students and give feedback.)
- (Write on the board:)

$$\text{c. } \frac{12}{8} - \frac{4}{8} + \frac{10}{8} = \boxed{\frac{18}{8}}$$

- Here's what you should have. The bottom number of the fraction is 8. On top, you worked the problem: 12 minus 4 plus 10. The answer is 18. The answer to problem C is 18/8.
- Raise your hand if you got it right.
f. Your turn: Work the rest of the problems in part 5. If you can't do everything in your head, you can work part of the problem on your paper. Raise your hand when you're finished.
   (Observe students and give feedback.)
- (Write on the board:)

$$\text{d. } \frac{5}{4} \left(\frac{2}{4} \right) \left(\frac{1}{2} \right) = \boxed{\frac{10}{32}}$$

$$\text{e. } \frac{14}{6} + \frac{7}{6} + \frac{7}{6} = \boxed{\frac{28}{6}}$$

$$\text{f. } \frac{1}{2} \times \frac{1}{3} \times \frac{1}{4} = \boxed{\frac{1}{24}}$$

- Here's what you should have. The answer to problem D is 10/32. The answer to problem E is 28/6. The answer to problem F is 1/24.

EXERCISE 6 INDEPENDENT WORK

a. Do the independent work for lesson 7.
b. (Before beginning the next lesson, check students' independent work.)

Lesson 8

EXERCISE 1 FRACTION MULTIPLICATION
Missing Factor/Product

a. Open your textbook to lesson 8 and find part 1.
- All these problems multiply the first fraction by another fraction. In some of the problems, the middle fraction is missing. In other problems, the last fraction is missing.
b. Work problem A. Raise your hand when you're finished.
- (Write on the board:)

$$\text{a. } \frac{4}{2}\left(\frac{7}{6}\right) = \boxed{\frac{28}{12}}$$

- Here's what you should have. On top, you multiplied 4 by 7. That's 28. On the bottom, you multiplied 2 by 6. That's 12. 4/2 times 7/6 equals 28/12.

c. Work problem B. Raise your hand when you're finished.
- (Write on the board:)

$$\text{b. } \frac{3}{1}\left(\frac{5}{1}\right) = \frac{15}{1}$$

- Here's what you should have. On top, you worked the problem: 3 times some value equals 15. The missing value is 5. On the bottom, you worked the problem 1 times some value equals 1. The missing value is 1. 3/1 times 5/1 equals 15/1.
d. Your turn: Work the rest of the problems in part 1. Raise your hand when you're finished. (Observe students and give feedback.)
e. (Write on the board:)

$$\text{c. } \frac{9}{4}\left(\frac{9}{3}\right) = \frac{81}{12}$$

$$\text{d. } \frac{3}{8}\left(\frac{6}{8}\right) = \boxed{\frac{18}{64}}$$

- Here's what you should have for problems C and D. Problem C: 9/4 times some fraction equals 81/12. The missing fraction is 9/3. Problem D: 3/8 times 6/8. The answer is 18/64.
- Raise your hand if you got everything right.

EXERCISE 2 NUMBER FAMILIES
Part/Whole

a. Find part 2.
- These are problems that tell about bars. For each problem, you make a number family with three names. The names are for the shaded part, the unshaded part and the whole bar. Then you'll put in the two numbers the problem gives and figure out the missing number.
b. Work problem A. Remember to write the unit name in your answer and box the answer. Raise your hand when you're finished. (Observe students and give feedback.)
- (Write on the board:)

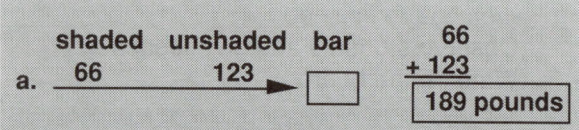

- Here's what you should have. The shaded part weighs 66 pounds. The unshaded part weighs 123 pounds. You figured out the whole bar by working the problem: 66 plus 123. The answer is 189 pounds.
c. Work problem B. Raise your hand when you're finished. (Observe students and give feedback.)
- (Write on the board:)

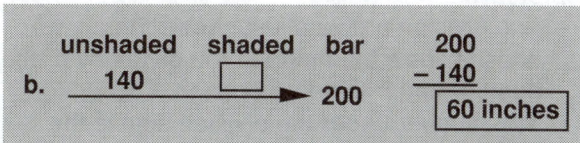

- Here's what you should have. The unshaded part is 140 inches long. The whole bar is 200 inches long. You worked the problem: 200 minus 140. The answer is 60 inches. That's the length of the shaded part.
d. Your turn. Work the rest of the problems in part 2. Raise your hand when you're finished.
  (Observe students and give feedback.)
  Key:

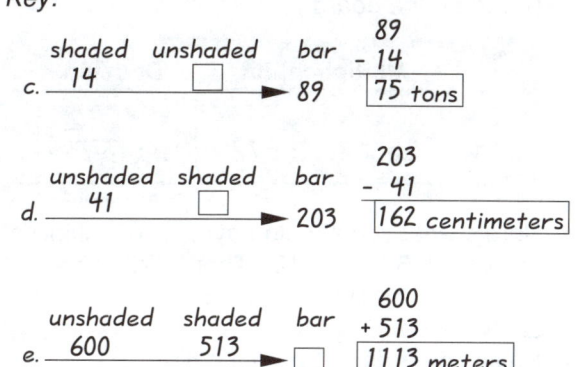

e. Find part J on page 30.
- That shows what you should have for problems C, D and E.
- Problem C: The unshaded part of the bar weighs 75 tons.
- Problem D: The shaded part of the bar is 162 centimeters long.
- Problem E: The bar is 1113 meters long.
- Raise your hand if you got everything right.

EXERCISE 3 FRACTIONS
Equivalence

a. Find part 3.
- I'll read what it says. Follow along: Some fractions are equivalent.
- If fractions are equivalent, you can write an equal sign between them.
- 1/2 and 4/8 are equivalent. So 1/2 equals 4/8.

- If fractions are equivalent, pictures of the fractions show exactly the same area that is shaded.
- You can see a picture of 1/2 and a picture of 4/8.
- The shaded areas are exactly the same size.
- You can figure out whether any two fractions are equivalent by finding the fraction you multiply the first fraction by to get the other fraction.
- If you multiply by a fraction that equals 1, the fractions you started with are equivalent.
- You can see the problem: 1/2 times some fraction equals 4/8. Below, you can see the fraction you multiply 1/2 by to get 4/8. That fraction is 4/4. 4/4 equals 1. 1/2 times 1 equals 4/8 so 1/2 and 4/8 are equivalent.
b. Find part 4.
- Each problem shows a pair of fractions. You'll write a problem that starts with the first fraction, multiplies by some value and ends with the second fraction.
c. (Write on the board:)

$$\text{a. } \frac{2}{5} \left(\quad\right) = \frac{8}{10}$$

- Here's what you write for problem A.
- Copy the problem and work it. Raise your hand when you're finished.
  (Observe students and give feedback.)
- (Write to show:)

$$\text{a. } \frac{2}{5} \left(\frac{4}{2}\right) = \frac{8}{10}$$

- Here's what you should have. You multiply 2/5 by 4/2 to get 8/10. You didn't multiply by 1. So 2/5 does **not** equal 8/10. Pictures of those fractions would not have the same-sized area shaded.
d. Work problem B and figure out the missing fraction. Raise your hand when you're finished.
  (Observe students and give feedback.)
- (Write on the board:)

$$\text{b. } \frac{4}{5} \left(\frac{2}{2}\right) = \frac{8}{10}$$

- Here's what you should have. You multiplied 4/5 times 2/2. 2/2 equals 1. So the pair of fractions you started with are equivalent.
- Write those two fractions below with an equal sign between them. √

- (Write to show:)

$$b. \quad \frac{4}{5} \left(\frac{2}{2} \right) = \frac{8}{10}$$

$$\frac{4}{5} = \frac{8}{10}$$

- Here's the simple equation for the equivalent fractions. 4/5 equals 8/10.
e. Work the multiplication problem for C. If you multiply by 1, write the simple equation for the equivalent fractions below. If you don't multiply by 1, don't write anything below. Raise your hand when you're finished.
  (Observe students and give feedback.)
- (Write on the board:)

$$c. \quad \frac{3}{4} \left(\frac{5}{5} \right) = \frac{15}{20}$$

$$\frac{3}{4} = \frac{15}{20}$$

- Here's what you should have. You multiply by 5/5. That's 1. So you wrote the simple equation below: 3/4 equals 15/20.
f. Your turn: Work the rest of the problems in part 4. If the fractions shown are equivalent, write the simple equation below. Raise your hand when you're finished.
  (Observe students and give feedback.)
g. (Write on the board:)

$$d. \quad \frac{4}{1} \left(\frac{3}{3} \right) = \frac{12}{3} \qquad e. \quad \frac{6}{5} \left(\frac{6}{7} \right) = \frac{36}{35}$$

$$\frac{4}{1} = \frac{12}{3}$$

- Here's what you should have for problems D and E.
- Problem D. You multiply by 3/3. So the fractions shown are equivalent. You should have written 4 over 1 equals 12/3. Both fractions equal 4.
- Problem E. You multiplied 6/5 by 6/7. You didn't multiply by 1, so the fractions are not equivalent. You should not have written 6/5 equals 36/35. Those fractions are not equal.

a. Find part 5.
- This is a table that shows multiplication in the first column and division in the second column.
- Look at row A. It shows the two numbers that are multiplied together, but it doesn't show the answer.
- The division problem just shows the answer. It doesn't show the number you divide by or the number under the division sign.
- The number under the division sign is the number that comes after the equal sign in the multiplication problem.
b. Work row A of the table. Use your calculator to figure out the missing number in the multiplication problem. Then complete the division problem. Raise your hand when you've written the multiplication equation and the division problem for row A.
  (Observe students and give feedback.)
- (Write on the board:)

	Multiplication	Division
a.	56 x 12 = 672	$\frac{12}{56 \overline{)672}}$

- Here's what you should have. The multiplication problem is 56 times 12. That's 672. The division problem is 672 divided by 56. The answer is 12.
- Raise your hand if you got it right.
c. Your turn: Work the rest of the rows. Remember, for row B, first work the multiplication problem. Then do the division problem. Raise your hand when you've finished the table.
  (Observe students and give feedback.)
d. (Write to show:)

	Multiplication	Division
a.	56 x 12 = 672	$\frac{12}{56 \overline{)672}}$
b.	13 x 61 = 793	$\frac{61}{13 \overline{)793}}$
c.	48 x 21 = 1008	$\frac{21}{48 \overline{)1008}}$
d.	74 x 13 = 962	$\frac{13}{74 \overline{)962}}$

- Here's what you should have for rows B, C and D.
- Raise your hand if you got everything correct.

EXERCISE 5 FRACTION OPERATIONS
Addition for Whole Numbers

a. Find part 6.
- These are equations that add the number 1.
- You're going to rewrite each equation. You'll rewrite the number 1 as a fraction that equals 1. Then you'll add the fractions and see if your answer equals the whole number shown for the original equation.

b. The sample problem says: 1 plus 1 plus 1 equals 3.
- Below, you can see each 1 written as a fraction.
- The fraction that equals 1 is 5/5. 5/5 plus 5/5 plus 5/5 equals 15/5. What whole number does 15/5 equal? (Signal.) *3.*
- 15/5 equals 3 because the top number of that fraction is 3 times the bottom number. So you get the same answer when you replace the whole numbers with fractions that equal those numbers.

c. Problem A. Below the equation with whole numbers, you can see the bottom number of each fraction you'll write.
- Listen: Write the bottom equation. Show the two fractions that equal 1. Remember the plus sign and the equal sign. Raise your hand when you've finished problem A.
  (Observe students and give feedback.)
- (Write on the board:)

$$\text{a. } \frac{7}{7} + \frac{7}{7} = \boxed{\frac{14}{7}}$$

- Here's what you should have. 7/7 plus 7/7. The answer is 14/7. 14/7 equals 2 because the top number is 2 times the bottom number.

d. Write the complete fraction equation for problem B. Show the fractions that equal 1. Then add and write the answer. Check to make sure that your answer equals 4. Raise your hand when you're finished.
  (Observe students and give feedback.)
- (Write on the board:)

$$\text{b. } \frac{3}{3} + \frac{3}{3} + \frac{3}{3} + \frac{3}{3} = \boxed{\frac{12}{3}}$$

- Here's what you should have. 3/3 plus 3/3 plus 3/3 plus 3/3 equals 12/3. The answer equals 4 because the top number is 4 times the bottom number.

e. Write the complete fraction equation for problem C. Raise your hand when you're finished.
  (Observe students and give feedback.)
- (Write on the board:)

$$\text{c. } \frac{10}{10} + \frac{10}{10} + \frac{10}{10} = \boxed{\frac{30}{10}}$$

- Here's what you should have. 10/10 plus 10/10 plus 10/10 equals 30/10. The answer equals 3 because the top number is 3 times the bottom number.

f. Work problem D. Raise your hand when you're finished. (Observe students and give feedback.)
- (Write on the board:)

$$\text{d. } \frac{8}{8} + \frac{8}{8} + \frac{8}{8} = \boxed{\frac{24}{8}}$$

- Here's what you should have. 8/8 plus 8/8 plus 8/8 equals 24/8. The answer equals 3 because the top number is 3 times the bottom number.

EXERCISE 6 INDEPENDENT WORK

a. Do the independent work for lesson 8.
b. (Before beginning the next lesson, check students' independent work.)

Lesson 9

Objectives

- **Work an addition or subtraction problem that has a whole number and a fraction.** (Exercise 1)
  Note: The term **denominator** is introduced.

- Determine whether a pair of fractions is equivalent. (Exercise 2)

- **Compute the missing number in each row of a number-family table.** (Exercise 3)

- Obtain a fraction that equals a whole number by substituting fractions in equations of the form: **1 + 1 + 1 = 3.** (Exercise 4)

- **Make a number family for a sentence that names binary categories.** (Exercise 5)
  Note: Each sentence tells about a whole that is divided into two parts. For example: **Some of the bricks were hot.** Students make a family with names:

  hot   not hot   bricks

  The big number

  tells about all the objects. The names for the small numbers tell about the two parts that constitute the whole.

- Complete a table to show a multiplication equation and the corresponding division problem. (Exercise 6)

EXERCISE 1 FRACTION OPERATIONS
Whole Numbers and Fractions

a. Open your textbook to lesson 9 and find part 1.
- I'll read what it says. Follow along: The bottom number of a fraction is called the **denominator.** You can see the denominator of 5/3. It's 3. From now on, I'll refer to the denominator.

b. If I say **denominator,** which part is that? (Signal.) *The bottom.*
(Repeat until firm.)

c. Find part 2.
- All these problems have a fraction and the number 1. You'll rewrite 1 as a fraction that has the same denominator as the fraction that is shown.

d. Look at the sample problem: 1 plus 2/5.
- What's the denominator of the fraction? (Signal.) *5.*

- So we have to rewrite 1 as a fraction with a denominator of 5. What fraction is that? (Signal.) *5-fifths.*

- So you'll rewrite 1 as 5/5. The equation below shows the problem 5/5 plus 2/5. The answer is 7/5. 1 plus 2/5 equals 7/5.

e. Problem A: 12/7 minus 1. What's the denominator of the fraction? (Signal.) *7.*

- You'll rewrite 1 as a fraction with that denominator. What fraction is that? (Signal.) *7-sevenths.*

- Rewrite problem A with a fraction for the whole number and write the answer. Raise your hand when you're finished.

- (Write on the board:)

$$\textbf{a.}\quad \frac{12}{7} - \frac{7}{7} = \boxed{\frac{5}{7}}$$

- Here's what you should have. 12/7 minus 7/7 equals 5/7. Everybody, what does 12/7 minus 1 equal? (Signal.) *5-sevenths.*

f. Problem B. What's the denominator of the fraction? (Signal.) *4.*

- Rewrite the problem so that 1 is a fraction with that denominator. Then write the answer. Raise your hand when you're finished.
(Observe students and give feedback.)

- (Write on the board:)

$$\textbf{b.}\quad \frac{4}{4} - \frac{3}{4} = \boxed{\frac{1}{4}}$$

- Here's what you should have for problem B. 4/4 minus 3/4 equals 1/4. Everybody, what does 1 minus 3/4 equal? (Signal.) *1-fourth.*

g. Work problem C. Raise your hand when you're finished. (Observe students and give feedback.)

- (Write on the board:)

$$\textbf{c.}\quad \frac{10}{10} + \frac{7}{10} = \boxed{\frac{17}{10}}$$

- Here's what you should have. 10/10 plus 7/10 equals 17/10. Everybody, what does 1 plus 7/10 equal? (Signal.) *17-tenths.*

h. Work problem D. Raise your hand when you're finished. (Observe students and give feedback.)

- (Write on the board:)

$$\textbf{d.}\quad \frac{6}{5} + \frac{5}{5} = \boxed{\frac{11}{5}}$$

- Here's what you should have. 6/5 plus 5/5 equals 11/5. Everybody, what does 6/5 plus 1 equal? (Signal.) *11-fifths.*
i. Raise your hand if you got everything right.

EXERCISE 2 FRACTIONS
Equivalence

a. Find part 3.
- These are pairs of fractions. Some pairs are equivalent. Others aren't. Remember, if a pair of fractions is equivalent, you multiply the first fraction by 1. And if you multiply by 1, you can write a simple equation that shows the equivalent fractions.
b. Work problem A. Raise your hand when you're finished. (Observe students and give feedback.) (Write on the board:)

$$\text{a.} \quad \frac{2}{7} \left(\frac{7}{2} \right) = \frac{14}{14}$$

- Here's what you should have. You multiply by 7/2. You didn't multiply by 1. So 2/7 does not equal 14/14. You should not have written 2/7 equals 14/14.
c. Work the rest of the problems in part 3. Remember, if the fractions are equivalent, write the simple equation below. Raise your hand when you're finished.
  (Observe students and give feedback.)
d. (Write on the board:)

$$\text{b.} \quad \frac{4}{3} \left(\frac{6}{6} \right) = \frac{24}{18} \qquad \text{c.} \quad \frac{7}{4} \left(\frac{3}{4} \right) = \frac{21}{16}$$

$$\frac{4}{3} = \frac{24}{18}$$

$$\text{d.} \quad \frac{2}{10} \left(\frac{4}{4} \right) = \frac{8}{40}$$

$$\frac{2}{10} = \frac{8}{40}$$

- Here's what you should have for each problem.
- Problems B and D have equivalent fractions. So you should have written a simple equation for those items. You shouldn't have a simple equation for problem C.
- Problem B: Read the simple equation. (Signal.) *4-thirds equals 24-eighteenths.*
- Problem D: Read the simple equation. (Signal.) *2-tenths equals 8-fortieths.*
e. Raise your hand if you got everything right.

EXERCISE 3 NUMBER FAMILIES
Tables

a. Find part 4.
  These are tables with missing numbers.
- Here's the rule about these tables: Each row works just like a number family. It has two small numbers and a big number. The big number is the total at the end of the row.
b. Table A. Touch the arrow for the top row.
- There's a number missing in the top row. Is that the big number or a small number? (Signal.) *A small number.*
- A small number is missing. Do you add or subtract to figure out the missing number? (Signal.) *Subtract.*
- Say the problem to find the missing number for the top row. (Signal.) *400 minus 60.*
c. Everybody, touch the next row.
- Is the missing number in that row the big number or a small number? (Signal.) *The big number.*
- Say the problem to find the missing number. (Signal.) *104 plus 68.*
d. Everybody, touch the bottom row.
e. Is the missing number in that row the big number or a small number? (Signal.) *A small number.*
- Say the problem to find the missing number. (Signal.) *572 minus 444.*
  (Repeat step e until firm.)
f. Your turn: Copy the table. Figure out the missing number in each row. Then write the missing numbers in the table. Raise your hand when you're finished.
  (Observe students and give feedback.)
- (Write on the board:)

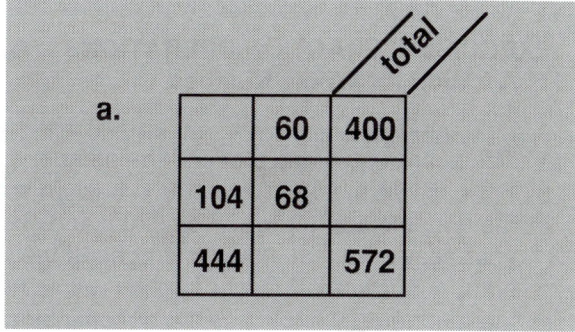

g. Here's the table you started with.
- Check your work.
- Everybody, touch the top row of your table. The problem is 400 minus 60. What's the missing number? (Signal.) *340.*
- (Write **340.**)

- Next row. The problem is 104 plus 68. What's the missing number? (Signal.) *172.*
- (Write **172.**)
- Bottom row. The problem is 572 minus 444. What's the missing number? (Signal.) *128.*
- (Write **128.**)

h. Your turn: Copy table B and write the missing number for each row. Raise your hand when you're finished.
   (Observe students and give feedback.)
- (Write on the board:)

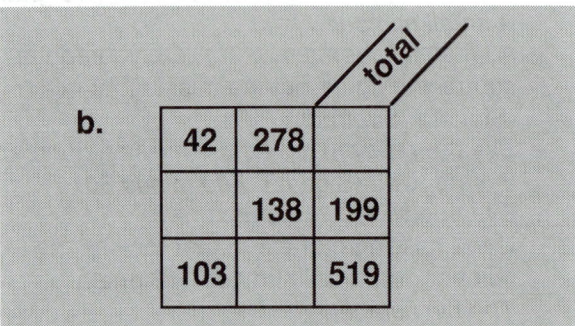

i. Here's the table you started with.
- Check your work.
- Touch the top row. The problem is 42 plus 278. What's the missing number? (Signal.) *320.*
- (Write **320.**)
- Touch the middle row. The problem is 199 minus 138. What's the missing number? (Signal.) *61.*
- (Write **61.**)
- Touch the bottom row. The problem is 519 minus 103. What's the missing number? (Signal.) *416.*
- (Write **416.**)

j. Raise your hand if you found out all three missing numbers.

EXERCISE 4 FRACTION OPERATIONS
Addition for Whole Numbers

a. Find part 5.
- These are equations that add the number 1.
b. For each item, you'll write 1 as a fraction with the denominator shown. You'll add and check your answer to make sure it is right. If the whole number answer is 5, your fraction will have a top number that is 5 times the denominator.
- If the whole number answer is 7, the top number of your fraction will be how many times the denominator? (Signal.) *7 times.*
c. Work problem A. Raise your hand when you're finished. (Observe students and give feedback.)

- (Write on the board:)

$$a. \quad \frac{6}{6} + \frac{6}{6} + \frac{6}{6} = \boxed{\frac{18}{6}}$$

- Here's what you should have. 6/6 plus 6/6 plus 6/6 equals 18/6. 18/6 equals 3 because the top number is 3 times the denominator.
d. Work the rest of the problems in part 5. Raise your hand when you're finished.
   (Observe students and give feedback.)
e. (Write on the board:)

$$b. \quad \frac{4}{4} + \frac{4}{4} + \frac{4}{4} + \frac{4}{4} = \boxed{\frac{16}{4}}$$

$$c. \quad \frac{9}{9} + \frac{9}{9} + \frac{9}{9} + \frac{9}{9} = \boxed{\frac{36}{9}}$$

- Here's what you should have.
- For problem B you added 4/4 plus 4/4 plus 4/4 plus 4/4. The answer is 16/4. What whole number equals 16/4? (Signal.) *4.*
- For problem C you added 9/9 plus 9/9 plus 9/9 plus 9/9. The answer is 36/9. What whole number equals 36/9? (Signal.) *4.*

EXERCISE 5 NUMBER FAMILIES
Part/Whole

a. Find part 6.
- All these sentences tell about parts of a whole. For each sentence, you'll make a number family with the names for the two parts and the name for the whole thing. The sentences are just like the bar problems you've worked. The name for the whole is always the big number. Two smaller parts make up the whole.
b. Sample sentence 1: Some of the children were girls.
- That sentence tells about the whole group and about one part of that group. What's the name for the whole group? (Signal.) *Children.*
- The name for one part of the children is given. What's the name for that part? (Signal.) *Girls.*
- What is the name for children that are not girls? (Signal.) *Boys.*
- So, what's the name for the other part of the family? (Signal.) *Boys.*
- You can see the family with the three names. The **small numbers** are **girls** and **boys.** The **big number** is **children.**

- You could have either **boys** or **girls** as the first small number. The problem names **girls.** That's why it's shown as the first small number.
- You can see the family for sample sentence 1.

c. Sample sentence 2: Some of the cars were dirty.
- That sentence tells about the whole group and about one part. What's the name for the whole group? (Signal.) *Cars.*
- What's the name for the part that the sentence tells about? (Signal.) *Dirty.*
- The name for the other part is **not dirty** or **clean.**
- You can see a number family for sample sentence 2.

d. Sentence A: Some of the glasses are full. What's the name for the whole group? (Signal.) *Glasses.*
- What's the name for the part the sentence tells about? (Signal.) *Full.*
- Yes, full glasses.
- Don't get fooled with the name of the other part. The glasses that are not full may **not** be empty. They may just be partly filled. So the correct name for the other part is **not full.**
- Make the family with the three names. Raise your hand when you're finished.
(Observe students and give feedback.)
- (Write on the board:)

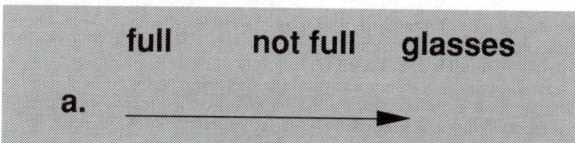

- Here's what you should have. The names are **full, not full** and **glasses. Glasses** is the name for the big number.

e. Sentence B: Some of the children wore shoes.
- What's the name for the whole group? (Signal.) *Children.*
- (Call on an individual student.) What's the name for the part the sentence tells about? (Idea: *Wore shoes.*)
- Yes, **wore shoes** or **shoes.**
- (Call on another student.) What's the name of the other part? (Idea: *Not shoes.*)
- Yes, **did not wear shoes** or **no shoes.**
- Make the family with the three names. Raise your hand when you're finished.
(Observe students and give feedback.)
- (Write on the board:)

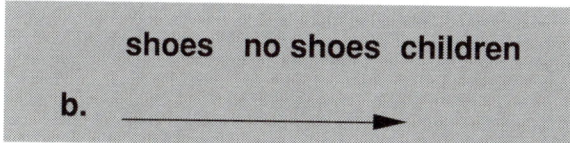

- Here's what you should have. The names are **shoes, no shoes** and **children.**

f. Sentence C: Some of the bricks were hot. Make the family. Don't get fooled. The name for the whole group is the name for the big number. The names for the parts are the names for the small numbers. Raise your hand when you're finished.
(Observe students and give feedback.)
- (Write on the board:)

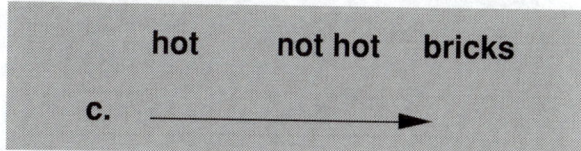

- Here's what you should have. The names are **hot, not hot** and **bricks.** If you wrote **cold** for the second small number, you got fooled. A brick that is not hot can be warm.

g. Make number families for the rest of the sentences in part 6. Raise your hand when you're finished.
(Observe students and give feedback.)
- (Write on the board:)

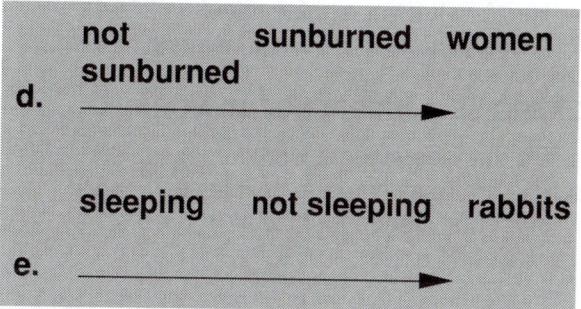

- Here's what you should have for sentences D and E.
- Sentence D: Some of the women were not sunburned. The names are **not sunburned, sunburned** and **women.** You could have **sunburned** as the first small number.
- Sentence E: Some of the rabbits are sleeping. The names are **sleeping, not sleeping** or **awake** and **rabbits.**

h. Raise your hand if you wrote all the correct names.

EXERCISE 6 MULTIPLICATION
Missing Factor/Product

a. Find part 7.
- The first column of the table shows multiplication; the second column shows division.

- Not all the rows show the same information. Remember, the number under the division sign is the number you end up with when you work a multiplication problem.
- b. Copy the table and complete all the rows. You can use your calculator. Raise your hand when you're finished.
  (Observe students and give feedback.)
- c. (Write on the board:)

	Multiplication	Division
a.	11 (39) = 429	$\dfrac{39}{11\overline{)429}}$
b.	26 (30) = 780	$\dfrac{30}{26\overline{)780}}$
c.	24 (29) = 696	$\dfrac{29}{24\overline{)696}}$
d.	39 (17) = 663	$\dfrac{17}{39\overline{)663}}$

- Here's what you should have.
- Raise your hand if you got everything right.

EXERCISE 7 INDEPENDENT WORK

- a. Do the independent work for lesson 9.
- b. (Before beginning the next lesson, check students' independent work.)

Lesson 10 – Test 1

Objectives

- **Write a pair of equivalent fractions from pictures.** (Exercise 1)

- Make a number family for a sentence that names binary categories. (Exercise 2)

- Work an addition or subtraction problem that has a whole number and a fraction. (Exercise 3)

- **Perform on a mastery test of skills presented in lessons 1 through 9.** (Exercise 4)

Note: Exercise 5 provides instructions for marking the test.

EXERCISE 1 FRACTIONS
Equivalence

- a. Open your textbook to lesson 10 and find part 1.
- Each item shows pictures of fractions.
- Touch the first picture for the sample item. √
- The fraction for that picture is 1/3.
- One of the other fractions in the sample item equals 1/3. It has the same size shaded area as 1/3. It's the last picture in the group.
- (Write on the board:)

$$\frac{1}{3} \quad = \text{—}$$

- Write 1/3, an equal sign and the fraction for the last picture. Leave space before the equal sign just like I've shown on the board. Raise your hand when you've done that much.
- (Write to show:)

$$\frac{1}{3} \quad = \frac{3}{9}$$

- Here's what you should have. The fraction shown in the last picture is 3/9.
- b. Now you'll prove that the fractions are equivalent.
- If they're equivalent, you'll multiply 1/3 by a fraction that equals 1 to get the fraction 3/9.
- Figure out the missing fraction. Raise your hand when you're finished.

- (Write to show:)

$$\frac{1}{3} \left(\boxed{\frac{3}{3}} \right) = \frac{3}{9}$$

- Here's what you should have. The missing fraction is 3/3. That fraction equals 1, so you've shown that 1/3 and 3/9 are equivalent.
c. Item A. Write the fraction for the first fraction and the other fraction that has the same shaded area. Show the equal sign between them. Raise your hand when you've done that much.
- (Write on the board:)

$$\text{a. } \frac{3}{4} = \frac{6}{8}$$

- Here's what you should have. The first fraction is 3/4. The other fraction is 6/8.
- Now prove that they're equivalent by showing that you multiply the first fraction by 1 to get the other fraction. Raise your hand when you're finished.
- (Write to show:)

$$\text{a. } \frac{3}{4} \left(\boxed{\frac{2}{2}} \right) = \frac{6}{8}$$

- Here's what you should have. You multiply 3/4 by 2/2. So the fractions 3/4 and 6/8 are equivalent.
d. Your turn: Write the complete equation for item B. Raise your hand when you're finished. (Observe students and give feedback.)
- (Write on the board:)

$$\text{b. } \frac{1}{2} \left(\frac{5}{5} \right) = \frac{5}{10}$$

- Here's what you should have. The first fraction is 1/2. The equivalent fraction is 5/10. You multiply by 5/5. That proves that 1/2 and 5/10 are equivalent.
e. Write the complete equation for item C. Raise your hand when you're finished. (Observe students and give feedback.)
- (Write on the board:)

$$\text{c. } \frac{3}{2} \left(\frac{4}{4} \right) = \frac{12}{8}$$

- Here's what you should have. The fractions are 3/2 and 12/8. You multiply by 4/4. So the fractions 3/2 and 12/8 are equivalent.

EXERCISE 2 NUMBER FAMILIES
Part/Whole

a. Find part 2.
- Each sentence tells about a number family. The whole group the sentence tells about is the name for the big number in the family. The parts are the small numbers. You have to figure out the part that is not named in the sentence.
b. Item A: Some of the roads are not paved.
- Make the family. Raise your hand when you're finished. (Observe students and give feedback.)
- (Write on the board:)

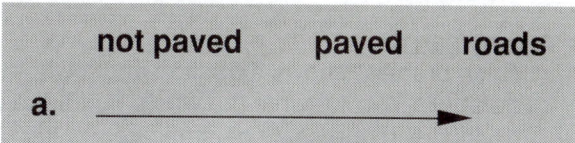

- Here's what you should have. The names for the small numbers are **not paved** and **paved.** The big number is **roads.**
c. Make the family for item B. Raise your hand when you're finished. (Observe students and give feedback.)
- (Write on the board:)

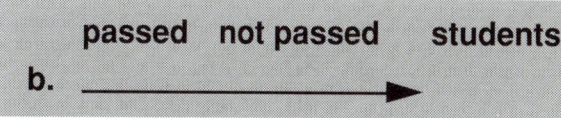

- Here's what you should have for item B. The sentence says: Many of the students passed the test. The names are **passed, not passed** or **failed** and **students.**
- Raise your hand if you got it right.
d. Your turn. Make families for the rest of the items in part 2. Raise your hand when you're finished. (Observe students and give feedback.)
- (Write on the board:)

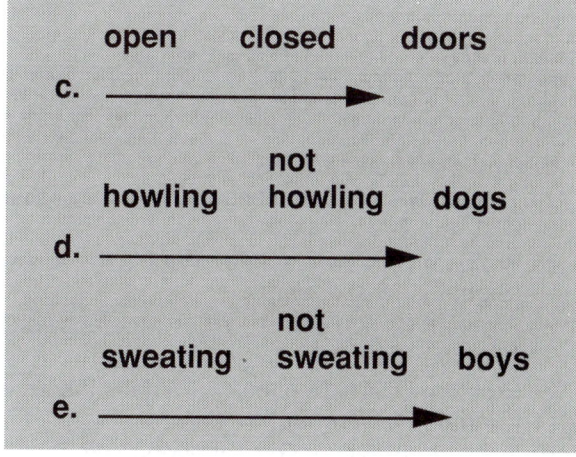

- Here are families for items C, D and E.

- Item C: A few of the doors are open. The names are **open, closed** or **not open** and **doors.**
- Item D: Lots of dogs were howling. The names are **howling, not howling** and **dogs.**
- Item E: A few boys are sweating. The names are **sweating, not sweating** or **dry** and **boys.**
- e. Raise your hand if you got everything right. Later, you'll use these kinds of number families to work difficult word problems.

EXERCISE 3 FRACTION OPERATIONS
Whole Numbers and Fractions

- a. Find part 3.
- Each problem shows a fraction and a whole number. You'll rewrite the whole number as a fraction with the same denominator as the other fraction.
- b. Problem A: 1 plus 7/20. What's the denominator of the fraction? (Signal.) *20.*
- Rewrite the equation so it has a fraction for the whole number, then write the answer. Raise your hand when you're finished. (Observe students and give feedback.)
- (Write on the board:)

$$\text{a.} \quad \frac{20}{20} + \frac{7}{20} = \boxed{\frac{27}{20}}$$

- Here's what you should have: 20/20 plus 7/20 equals 27/20.
- c. Your turn: Work the rest of the problems in part 3. Raise your hand when you're finished. (Observe students and give feedback.)
  Key:

$$b. \quad \frac{16}{16} - \frac{3}{16} = \boxed{\frac{13}{16}} \qquad c. \quad \frac{7}{3} + \frac{3}{3} = \boxed{\frac{10}{3}}$$

$$d. \quad \frac{16}{15} - \frac{15}{15} = \boxed{\frac{1}{15}}$$

- d. Find part J on page 35.
- That shows what you should have for problems B, C and D.
- Problem B: 1 minus 3/16. What's the answer? (Signal.) *13-sixteenths.*
- Problem C: 7/3 plus 1. What's the answer? (Signal.) *10-thirds.*
- Problem D: 16/15 minus 1. What's the answer? (Signal.) *1-fifteenth.*
- e. Raise your hand if you got everything right.

EXERCISE 4 TEST 1

Note: **Students are not to use calculators for any part of the test except part 7.**

- a. This is a test. You should only have your textbook, a sharpened pencil and lined paper on your desk.
- b. Find part 1 of test 1 in your textbook. √
- c. Do the test on your own. Raise your hand when you've completed part 6. (Pass out calculators to students as they complete part 6.)

EXERCISE 5 MARKING THE TEST

- a. (Collect the students' papers. Use the Answer Key to score tests. Award scores for test 1 as follows:)

Test 1 Percent Summary					
SCORE	%	SCORE	%	SCORE	%
57	100	51	89	45	79
56	98	50	87	44	77
55	96	49	86	43	75
54	95	48	84	42	74
53	93	47	82	41	72
52	91	46	81	40	70

- b. (Complete the Test 1 Remedy Summary to determine whether remedies are needed. Reproducible Summary Sheets are at the back of the Teacher's Guide.)
- (If more than 1/4 of the students did not pass a test part, present the remedy for that part before beginning lesson 11. Remedies appear at the end of the Test 1 Answer Key.)

Lesson 11

Objectives

- **Write a fraction for a whole number from a description.** (Exercise 1)
  Note: The term **numerator** is introduced. The description tells the number of times greater the numerator is than the denominator. For example: The denominator of the fraction is 6 and the fraction equals 5. Students write: $\frac{30}{6}$.

- **Use a number family to solve an addition or subtraction word problem that involves binary categories.** (Exercise 2)

- **Write a division problem as a fraction, and figure out the whole number it equals.** (Exercise 3)
  Note: Students say the division problem for the fraction, starting with the numerator.

  For example: $\frac{560}{8}$. Students say the problem: 560 divided by 8. They then work the problem on their calculator and write the complete equation: $\frac{560}{8} = 70$.

- **Compute the missing number in each row or column of a number-family table.** (Exercise 4)

- **Complete an equation that shows equivalent fractions by figuring out the fraction that equals 1.** (Exercise 5)
  Note: For $\frac{3}{8} \quad = \frac{12}{32}$ students write

  $\frac{3}{8} \left(\frac{4}{4} \right) = \frac{12}{32}$.

- Write a pair of equivalent fractions from pictures. (Exercise 6)

EXERCISE 1 FRACTIONS
From Descriptions

a. Open your textbook to lesson 11 and find part 1.
- I'll read what it says:
  You've learned a name for the bottom number of a fraction. It's the **denominator.** The top number of a fraction is called the **numerator.**
b. What's the new name for the top number of a fraction? (Signal.) *Numerator.*

- What's the new name for the bottom number of a fraction? (Signal.) *Denominator.*
  (Repeat step b until firm.)
c. Find part 2.
- You're going to write fractions from descriptions. We'll work the first two problems together. These descriptions refer to the denominator.
d. Problem A: The denominator of the fraction is 4.
- (Write on the board:)

$$a. \ \frac{}{4}$$

- The fraction equals 3 whole units, so the numerator must be 3 times the denominator.
- What's 4 times 3? (Signal.) *12.*
- Write the fraction for description A. √
- (Write to show:)

$$a. \ \frac{12}{4}$$

- Here's what you should have.
- What fraction has a denominator of 4 and equals 3 whole units? (Signal.) *12-fourths.*
e. Problem B says the denominator of the fraction is 8. The fraction equals 4 whole units.
- (Write on the board:)

$$b. \ \frac{}{8}$$

- The denominator is 8. How many whole units does the fraction equal? (Signal.) *4.*
- So the numerator is 4 times the denominator. What's the numerator? (Signal.) *32.*
- Write the fraction for description B. √
- (Write to show:)

$$b. \ \frac{32}{8}$$

- 32/8 is the fraction that equals 4 and has a denominator of 8.
f. Problem C: The denominator is 6 and the fraction equals 5. Write the fraction. Raise your hand when you've written the fraction for description C. √
- (Write on the board:)

$$c. \ \frac{30}{6}$$

- 30/6 is the fraction that has a denominator of 6 and that equals 5 whole units. The numerator is 5 times the denominator.
g. Problem D: The denominator is 3 and the fraction equals 5 whole units. Write the denominator. Then figure out 5 times that number and write the numerator of the fraction. Raise your hand when you're finished. √
- Everybody, what's the fraction that has a denominator of 3 and equals 5 whole units? (Signal.) *15-thirds.*
h. Problem E: The denominator is 7. The fraction equals 3. Write the fraction. Raise your hand when you're finished. √
- The fraction equals 3, so the numerator is 3 times the denominator. What's the numerator? (Signal.) *21.*
- The fraction that equals 3 and has a denominator of 7 is 21/7.
i. Problem F: The denominator is 9. The fraction equals 10. Write the fraction. Raise your hand when you're finished. √
- The fraction equals 10, so the numerator is 10 times the denominator. What's the numerator? (Signal.) *90.*
- The fraction that equals 10 and has a denominator of 9 is 90/9.

EXERCISE 2 PROBLEM SOLVING
Addition/Subtraction

a. Find part 3.
- These are problems that give information for making a number family. The problems tell about the parts that make up a whole. Remember, the whole is the big number in your family.
b. You can see the steps for working each problem in the box above the problems.
- 1. Make a family with three names and two numbers.
- 2. Calculate the missing number.
- 3. Write the unit name for the answer and box the answer.
c. Problem A: There were sick sheep and sheep that were not sick. If there were 68 sheep in all and 44 of them were not sick, how many were sick?
- The first sentence tells how to make the number family. The second sentence gives two numbers for the family.
- Make the family with three names. Raise your hand when you've done that much. (Observe students and give feedback.)

- (Write on the board:)

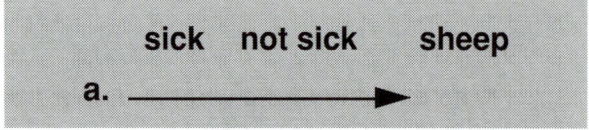

- The name for the big number is **sheep.** The names for the small numbers are **sick** and **not sick.** The problem gives two numbers. Write each number in the family below the name it tells about. Raise your hand when you've done that much.
(Observe students and give feedback.)
- (Write to show:)

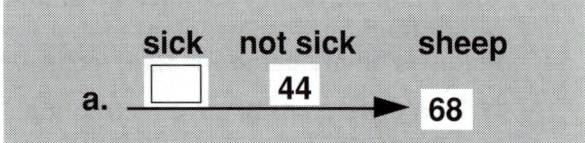

- Here's what you should have. The problem gives a number for all the sheep. That's 68. The problem gives a number for the sheep that were not sick. That's 44. The problem asks about the sheep that were sick. To figure out the number of sheep that were sick, you work the problem 68 minus 44.
- Work the problem. Write the unit name for the answer and box the answer. Raise your hand when you're finished.
(Observe students and give feedback.)
- The answer is 24 sick sheep.
d. Problem B: Make the family with three names and two numbers. Raise your hand when you've done that much.
(Observe students and give feedback.)
- (Write on the board:)

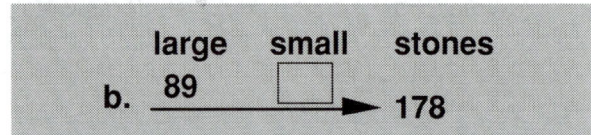

- Here's what you should have. The name for the big number is **stones.** The names for the small numbers are **large** and **small.** The problem gives the number 89 for large stones and 178 for all the stones. The problem asks about small stones. That's the problem for the missing number. Say that problem. (Signal.) *178 minus 89.*
- Work the problem. Write the unit name for the answer and box the answer. Raise your hand when you're finished.
(Observe students and give feedback.)

- (Write to show:)

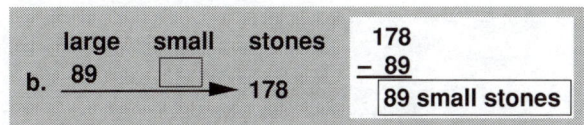

large   small   stones

b. 89   ☐ → 178

178
− 89
89 small stones

- Here's what you should have. There are 89 small stones.
- e. Work problem C. Follow the steps in the box. Raise your hand when you're finished. (Observe students and give feedback.)
- (Write on the board:)

new   used   books

c. 239   ☐ → 345

345
− 239
106 used books

- Here's what you should have. Problem C says that there were 345 books in a room. 239 of these books were new.
- You figured out how many were used. The family has the names **new, used** or **not new** and **books.** You worked the problem: 345 minus 239. The answer is 106. There were **106 used books** in the room.
- f. Work problem D. Raise your hand when you're finished. (Observe students and give feedback.)
- (Write on the board:)

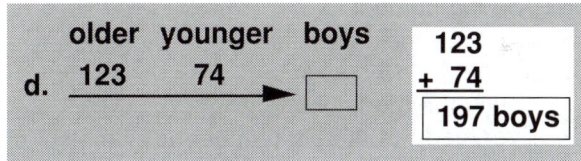

older   younger   boys

d. 123   74 → ☐

123
+ 74
197 boys

- Here's what you should have. Problem D says that there were 123 older boys and 74 younger boys on a camping trip. You figured out how many boys there were in all.
- The number family has names for **older** boys, **younger** boys and **boys.** You worked the problem: 123 plus 74. The answer is 197 boys. There were 197 boys on the camping trip.
- Raise your hand if you got everything right.

EXERCISE 3 FRACTIONS
For Division

a. Find part 4.
- I'll read what it says. Follow along: You can write any division problem as a fraction.
- When you read the division problem, the first number you say is the top number of the fraction. That's the numerator.

- You can see the division problem: 266 divided by 19. When you work division problems on a calculator, the division sign is like a diagram of a fraction.
- You can see the keys you'd press to work 266 divided by 19 on your calculator. 266 is the numerator. The division sign is like a fraction bar. 19 is the denominator. You can see the fraction after the arrow. 266/19.
- Remember, when you say the division problem, the first number you say is the numerator of the fraction.
b. Find part 5.
c. Problem A. Say the division problem. (Signal.) *560 divided by 8.*
- Write the fraction for that division problem. √
- (Write on the board:)

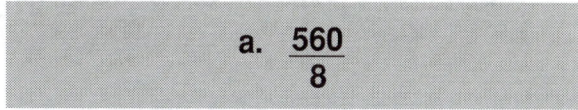

a. $\dfrac{560}{8}$

- Here's the fraction for 560 divided by 8.
d. Write the fraction for division problem B. Raise your hand when you're finished. (Observe students and give feedback.)
- (Write to show:)

a. $\dfrac{560}{8}$ b. $\dfrac{891}{3}$

- Here's the fraction for 891 divided by 3.
e. Your turn: Write fractions for the rest of the items in part 5. Raise your hand when you're finished. (Observe students and give feedback.)
f. (Write on the board:)

c. $\dfrac{1125}{25}$ d. $\dfrac{3780}{18}$

- Here are the fractions you should have for division problems C and D.
g. Listen: Each of the fractions you wrote equals a whole number. You can figure out the whole number by working the division problem.
- Fraction A. Start with 560 and say the division problem. (Signal.) *560 divided by 8.*
- Fraction B. Start with 891 and say the division problem. (Signal.) *891 divided by 3.*
h. Work the division problem for fraction A on your calculator. Write an equal sign after 560/8. Complete the equation to show the whole number that equals 560/8. Raise your hand when you're finished. (Observe students and give feedback.)

- (Write to show:)

a. $\dfrac{560}{8} = 70$	b. $\dfrac{891}{3}$

- Here's the equation you should have for fraction A. 560/8 equals 70.
i. Work the division problem for each of the other fractions, and complete the equations. Raise your hand when you're finished.
(Observe students and give feedback.)
j. (Write to show:)

a. $\dfrac{560}{8} = 70$	b. $\dfrac{891}{3} = 297$
c. $\dfrac{1125}{25} = 45$	d. $\dfrac{3780}{18} = 210$

- Here's what you should have for fractions B, C and D.
- Fraction B: 891/3 equals 297.
- Fraction C: 1125/25 equals 45.
- Fraction D: 3780/18 equals 210.
k. Raise your hand if you got everything right.

EXERCISE 4 NUMBER FAMILIES
Tables

a. Find part 6.
- Table A has rows that work like number families. Remember, the big number is at the end of the row. It's the total. The other numbers are small numbers.
- Touch the top row. √
- Is the big number or a small number missing in that row? (Signal.) *A small number.*
- Say the problem to figure out the missing number. (Signal.) *350 minus 202.*
- Touch the middle row.
- Is the big number or a small number missing in that row? (Signal.) *The big number.*
- Say the problem to figure out the missing number. (Signal.) *96 plus 207.*
b. Your turn: Copy table A. Write the missing numbers for each row. Raise your hand when you've finished.
(Observe students and give feedback.)

c. (Write on the board:)

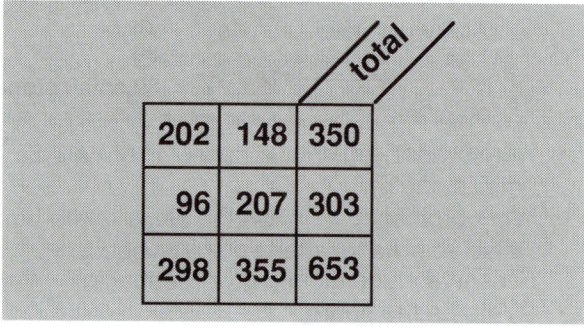

- Here's the table you should have. Raise your hand if you got everything right.
d. Touch table B.
- The **columns** of this table work just like number families. The arrows show the big number is at the bottom of each column. The numbers above the big number are small numbers.
e. Touch the first column. √
- Everybody, is the big number or a small number missing in that column? (Signal.) *A small number.*
- Say the problem to find the missing number. (Signal.) *112 minus 59.*
- Touch the second column.
- Everybody, is the big number or a small number missing in that column? (Signal.) *The big number.*
- Say the problem to find the missing number. (Signal.) *148 plus 90.*
- Touch the third column.
- Everybody, is the big number or a small number missing in that column? (Signal.) *A small number.*
- Say the problem to find the missing number. (Signal.) *350 minus 149.*
- (Repeat step e until firm.)
f. Your turn: Copy table B. Write the missing number for each column. Raise your hand when you're finished.
(Observe students and give feedback.)
- (Write on the board:)

	53	148	201
	59	90	149
total	112	238	350

g. Check your work.
- The problem for the first column is 112 minus 59. Everybody, what's the missing number in the first column? (Signal.) *53.*

- The problem for the second column is 148 plus 90. What's the missing number in the second column? (Signal.) *238.*
- The problem for the third column is 350 minus 149. What's the missing number in the third column? (Signal.) *201.*

h. Raise your hand if you got everything right.

EXERCISE 5 FRACTIONS
Equivalence

a. Find part 7.
- Each equation shows a pair of equivalent fractions. You're going to show that the fractions are equivalent.
- You'll copy each equation, write parentheses and the missing fraction that equals 1.

b. Problem A. Copy the problem and work it. Show the parentheses and the fraction that equals 1. Raise your hand when you're finished. (Observe students and give feedback.)
- (Write on the board:)

$$\text{a.} \quad \frac{1}{8} \left(\frac{6}{6} \right) = \frac{6}{48}$$

- Here's what you should have. The fraction you multiply by is 6/6.

c. Work problem B. Raise your hand when you're finished. (Observe students and give feedback.)
- (Write on the board:)

$$\text{b.} \quad \frac{7}{9} \left(\frac{5}{5} \right) = \frac{35}{45}$$

- Here's what you should have. The fraction you multiply by is 5/5.

d. Your turn: Work the rest of the problems in part 7. Raise your hand when you're finished. (Observe students and give feedback.)

e. (Write on the board:)

$$\text{c.} \quad \frac{9}{1} \left(\frac{7}{7} \right) = \frac{63}{7}$$

$$\text{d.} \quad \frac{2}{9} \left(\frac{8}{8} \right) = \frac{16}{72}$$

$$\text{e.} \quad \frac{3}{4} \left(\frac{4}{4} \right) = \frac{12}{16}$$

- Check your work. Here's what you should have for problems C, D and E.
f. Raise your hand if you got everything right.

EXERCISE 6 FRACTIONS
Equivalence

a. Find part 8.
- Each item shows pictures of fractions.

b. Write the fraction for the first picture and for the other fraction with the same shaded area. Then complete the equation to show that you multiply the first fraction by 1 to get the other fraction. Remember, if the fractions are equivalent, you multiply by 1.
- Raise your hand when you've written the complete equation for each item. (Observe students and give feedback.)

c. (Write on the board:)

$$\text{a.} \quad \frac{3}{4} \left(\frac{2}{2} \right) = \frac{6}{8}$$

$$\text{b.} \quad \frac{2}{5} \left(\frac{3}{3} \right) = \frac{6}{15}$$

- Here's what you should have.
d. For problem A, the fraction for the first picture is 3/4. The fraction for the picture with the same shaded area is 6/8. 3/4 times 2/2 equals 6/8.
- For problem B, the fraction for the first picture is 2/5. The fraction for the picture with the same shaded area is 6/15. 2/5 times 3/3 equals 6/15.

EXERCISE 7 INDEPENDENT WORK

a. Do the independent work for lesson 11. You'll use a calculator for part 13.
b. (Before beginning the next lesson, check students' independent work.)

Lesson 12

Objectives

- **Complete a pair of equivalent fractions.** (Exercise 1)
  Note: Problems are of the form:

  $\frac{3}{7}\left(\ \ \right)=\frac{15}{\blacksquare}$. Students identify whether

  they can work the problem on top or on the bottom. They work the problem and write the missing number. They then complete the

  fraction that equals 1. $\frac{3}{7}\left(\frac{5}{5}\right)=\frac{15}{\blacksquare}$. Then

  they work the problem for the box: **7 (5).**

- **Rewrite an equation of the form:**
  15 − 12 = 3 so that the equation begins with the number that is alone:
  3 = 15 − 12. (Exercise 2)
  Note: Students follow the rule that they read the side with two numbers and write that side in its original order: **15 − 12** not **12 − 15.**

- **Complete a table to show division problems and the corresponding fraction equations.**

  For example: $15\overline{)495}\ \dfrac{33}{\ }\ \left|\ \dfrac{495}{15}=33\right.$.
  (Exercise 3)

- Use a number family to solve an addition or subtraction word problem that involves binary categories. (Exercise 4)

- Write a fraction for a whole number from a description. (Exercise 5)

- **Write names and a fraction for a sentence that gives ratio information.** For example: In a room, there are 3 children for every 5

  adults. Students write: $\dfrac{\text{children}}{\text{adults}}=\dfrac{3}{5}$.

  (Exercise 6)

- Compute the missing number in each column of a number-family table. (Exercise 7)

EXERCISE 1 FRACTIONS
Equivalence

a. Open your textbook to lesson 12 and find part 1.
- I'll read what it says. Follow along: You've worked multiplication problems that have a missing fraction.
- You work the problem for the numerators and the problem for the denominators.

- You can see the problem 3/2 times some fraction equals 12/10. The missing fraction is 4/5.
- You can see a different kind of problem below.
- Part of the last fraction is missing. The equal sign tells you that the two fractions are equivalent. 4/7 equals 12 over some number. You multiply the first fraction by 1 to get the other fraction.
- To figure out the fraction that equals 1, you work the problem for either the numerators or the denominators.
- You can't work the problem for the denominators. That problem would be 7 times some value equals some value. You don't have enough numbers to work that problem.
- But you can work the problem for the numerators. That problem is: 4 times some value equals 12.
- The missing value is 3. So the fraction that equals 1 is 3/3.
- Now you have two numbers on the bottom, so you can work that problem: 7 times 3.
- The number that goes in the box is 21.
- 4/7 is equal to 12/21.
  Remember the steps:
  1. Work the problem for either the numerators or the denominators.
  2. Complete the fraction that equals 1.
  3. Work the problem for the number that goes in the box.
b. Find part 2.
c. (Write on the board:)

$$\frac{10}{2}\ =\ \frac{\square}{18}$$

- We'll work the sample problem together. The fractions are equivalent. So we multiply the first fraction by 1. We'll figure out the fraction that equals 1.
- (Write to show:)

$$\frac{10}{2}\left(\frac{\ }{\ }\right)=\frac{\square}{18}$$

- The first thing we do is work the problem for either the numerators or the denominators. Can you work the problem for the numerators? (Signal.) *No.*
- Can you work the problem for the denominators? (Signal.) *Yes.*
- The problem for the denominators is 2 times some value equals 18. What's the missing value? (Signal.) *9.*

- (Write to show:)

$$\frac{10}{2}\left(\frac{}{9}\right) = \frac{\boxed{}}{18}$$

- We've done step 1. We've worked the problem for the denominators.
- Next step: Complete the fraction that equals 1. The denominator is 9, so what's the numerator? (Signal.) *9.*
- (Write to show:)

$$\frac{10}{2}\left(\frac{9}{9}\right) = \frac{\boxed{}}{18}$$

- We've done step 2.
- Now we do step 3: Work the problem for the number that goes in the box. That problem is 10 times 9. What's the answer? (Signal.) *90.*
- (Write to show:)

$$\frac{10}{2}\left(\frac{9}{9}\right) = \frac{\boxed{90}}{18}$$

- The number in the box is 90. 10/2 is equivalent to 90/18.
d. Problem A. You're going to copy the problem and do step 1. Figure out whether you can work the problem for the numerators or the denominators. Work that problem. Then stop. √
- (Write on the board:)

$$\text{a.} \quad \frac{7}{3}\left(\frac{5}{}\right) = \frac{35}{\boxed{}}$$

- Here's what you should have for step 1. The missing number for the top is 5.
- Now do step 2. Complete the fraction that equals 1. √
- (Write to show:)

$$\text{a.} \quad \frac{7}{3}\left(\frac{5}{5}\right) = \frac{35}{\boxed{}}$$

- Here's the fraction: 5/5.
- Now do step 3. Work the problem for the number that goes in the box. √
- (Write to show:)

$$\text{a.} \quad \frac{7}{3}\left(\frac{5}{5}\right) = \frac{35}{\boxed{15}}$$

- Here it is. The answer is 15. So 7/3 equals 35/15. Raise your hand if you got everything right.
e. Problem B. Copy the problem and do step 1. Work the problem for either the numerators or the denominators. Raise your hand when you've done that much. √
- (Write on the board:)

$$\text{b.} \quad \frac{2}{5}\left(\frac{6}{}\right) = \frac{12}{\boxed{}}$$

- Now do steps 2 and 3. Complete the fraction that equals 1. Then work the problem for the denominators and write the value that goes in the box. Raise your hand when you're finished. (Observe students and give feedback.)
- (Write to show:)

$$\text{b.} \quad \frac{2}{5}\left(\frac{6}{6}\right) = \frac{12}{\boxed{30}}$$

- Here's what you should have for problem B. 2/5 equals 12/30. Raise your hand if you got everything right.
f. Problem C. Copy the problem and do step 1. Work the problem for either the numerators or the denominators. Raise your hand when you've done that much. √
- (Write on the board:)

$$\text{c.} \quad \frac{4}{9}\left(\frac{6}{}\right) = \frac{\boxed{}}{54}$$

- Now do steps 2 and 3. Raise your hand when you're finished. (Observe students and give feedback.)
- (Write to show:)

$$\text{c.} \quad \frac{4}{9}\left(\frac{6}{6}\right) = \frac{\boxed{24}}{54}$$

- Here's what you should have for problem C.
g. Raise your hand if you got everything right.

EXERCISE 2 EQUATIONS
Rewriting

a. Find part 3.
- These are equations that have two values on one side of the equal sign and one value on the other side. You're going to write them so the value that is **alone** comes first.

b. Look at the sample equation: 12 plus 15 equals 27.
• The value that is alone is 27. Here's how you rewrite the equation.
• (Write on the board:)

$$12 + 15 = 27$$
$$27 = 12 + 15$$

• Don't write 27 equals 15 plus 12 even though that's true. Keep the numbers for each side in the same order they are written. The **12** is first on one of the sides, so the **12** remains first on the rewritten side.
c. Equation A. One value is all alone on a side. Which value is that? (Signal.) *107.*
• What's on the other side? (Signal.) *115 minus 8.*
• Rewrite the equation so that **107** is first. Remember to keep **115** and **8** in the same order they are now. Raise your hand when you're finished. √
• Here's the equation you started with: 115 minus 8 equals 107. What's the rewritten equation? (Signal.) *107 equals 115 minus 8.*
• Both the original equation and the one you wrote say the same thing. They tell you that if you want to end up with 107, you start with 115 and subtract 8.
d. Your turn: Rewrite the rest of the equations in part 3. Raise your hand when you're finished.
(Observe students and give feedback.)
Key:
    b. 14 = 2 × 7
    c. 208 = 96 + 112
    d. 153 = 9 × 17

e. Check your work. I'll read the equations you started with. You say the rewritten equations.
• Equation B: 2 times 7 equals 14. (Signal.) *14 equals 2 times 7.*
• Equation C: 96 plus 112 equals 208. (Signal.) *208 equals 96 plus 112.*
• Equation D: 9 times 17 equals 153. (Signal.) *153 equals 9 times 17.*

EXERCISE 3 FRACTIONS
For Division

a. Find part 4.
• This table has division problems in the first column. You'll write fraction equations in the second column.
b. The sample row is completed. The **answer** to the division problem is the same number the fraction equals.

• 462 divided by 77 is 6. The equation shows 462 divided by 77 as a fraction. That fraction equals 6.
c. Your turn: Use your calculator. Copy row A and complete it. Remember to show the equal sign in your fraction equation. Raise your hand when you're finished.
(Observe students and give feedback.)
• (Write on the board:)

	Division	Fraction equation
a.	$15\overline{)495}$ (33)	$\dfrac{495}{15} = 33$

• Here's what you should have for row A. The division problem is: 495 divided by 15 is 33. The fraction equation shows the same thing: 495 divided by 15 equals 33.
e. Your turn: Copy row B and complete it. Raise your hand when you're finished.
(Observe students and give feedback.)
• (Write to show:)

	Division	Fraction equation
a.	$15\overline{)495}$ (33)	$\dfrac{495}{15} = 33$
b.	$61\overline{)793}$ (13)	$\dfrac{793}{61} = 13$

• Here's what you should have for row B.
f. Copy row C and complete it. Raise your hand when you're finished.
(Observe students and give feedback.)
• (Write to show:)

	Division	Fraction equation
a.	$15\overline{)495}$ (33)	$\dfrac{495}{15} = 33$
b.	$61\overline{)793}$ (13)	$\dfrac{793}{61} = 13$
c.	$13\overline{)585}$ (45)	$\dfrac{585}{13} = 45$

• This is what you should have for row C. 585/13 equals 45. Raise your hand if you got it right.

g. Remember, you can read any fraction as a division problem. You just start with the top number and say **divided by** for the fraction bar.

EXERCISE 4 PROBLEM SOLVING
Addition/Subtraction

a. Find part 5.
• These problems give information for making number families.
b. Work problem A. Remember to show the names in your family. Raise your hand when you're finished. (Observe students and give feedback.)
• (Write on the board:)

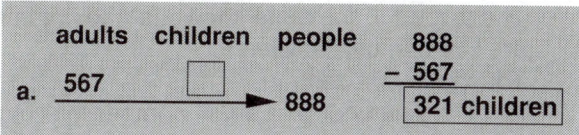

• Here's what you should have. The problem says that there were 567 adults at a picnic. The rest of the people were children. There were 888 people in all. You figured out the number of children.
• The family has the names **adults, children** and **people.** You worked the problem 888 minus 567. The answer is **321 children.** Raise your hand if you got it right.
c. Work the rest of the items in part 5. Raise your hand when you're finished.
(Observe students and give feedback.)
Key:

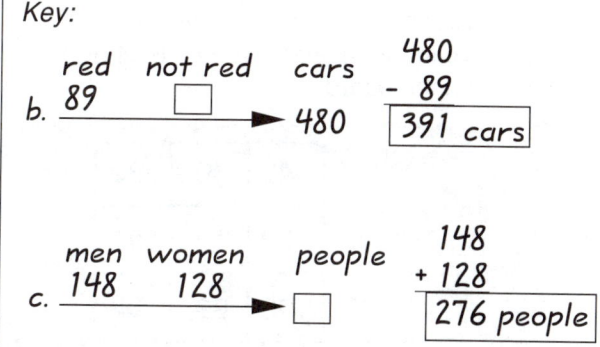

d. Find part J on page 45.
That shows what you should have.
• Problem B: The names are **red, not red** and **cars.** There were 391 cars that were not red.
• Problem C: The names are **men, women** and **people.** 276 people worked in the office.
• Raise your hand if you got everything right.

EXERCISE 5 FRACTIONS
From Descriptions

a. Find part 6.
• Each item describes a fraction.

b. Fraction A has a denominator of 2. The fraction equals 13. Write the fraction. Raise your hand when you're finished. √
• (Write on the board:)

$$\text{a. } \frac{26}{2}$$

• Here's what you should have. The fraction equals 13, so the numerator is 13 times the denominator.
c. Fraction B has a denominator of 8. The fraction equals 6. Write the fraction. Raise your hand when you're finished. √
• (Write on the board:)

$$\text{b. } \frac{48}{8}$$

• Here's what you should have.
d. Your turn: Write the fractions for the rest of the items in part 6. Raise your hand when you're finished. √
e. (Write on the board:)

$$\text{c. } \frac{6}{1} \qquad \text{d. } \frac{30}{5}$$

• Check your work. Here's what you should have for fractions C and D.
• Raise your hand if you got everything right.

EXERCISE 6 PROBLEM SOLVING
Ratios and Proportions

a. Find part 7.
• I'll read what it says. Follow along: Some problems tell about **ratios.** When you work ratio problems, you have **two unit names.** You write fractions so that one unit name tells about the numerator and the other unit name tells about the denominator.
• You can see part of a problem: In a mixture, there are 3 parts of sand for every 5 parts of water.
• The unit names are **parts of sand** and **parts of water. Parts of sand** is mentioned first, so you write it as the top name for your fraction. **Parts of water** goes on the bottom.
• Now you put the numbers where they go. There are 3 parts of sand. 3 goes in the numerator. There are 5 parts of water. 5 goes in the denominator.
b. Find part 8.

- For each item, you'll write the names and one fraction. Each number in the fraction will go with the name that is next to it.
c. Sentence A: There are 8 dogs for every 300 fleas.
- The names are **dogs** and **fleas.** Which is named first? (Signal.) *Dogs.*
- So **dogs** is the top name and **fleas** is the bottom name.
- Write the names. Then write the numbers. The sentence gives a number for dogs and a number for fleas. Remember to show the fraction bar in your fraction. Raise your hand when you've written the names and the fraction.
  (Observe students and give feedback.)
- (Write on the board:)

> **a.** $\dfrac{\textbf{dogs}}{\textbf{fleas}}\ \dfrac{8}{300}$

- Here's what you should have.
d. Your turn: Work problem B. Raise your hand when you're finished.
  (Observe students and give feedback.)
- (Write on the board:)

> **b.** $\dfrac{\textbf{pounds}}{\textbf{cartons}}\ \dfrac{7}{3}$

- Problem B says: There are 7 pounds in every 3 cartons. The names are **pounds** and **cartons.** The numbers are **7** for pounds and **3** for cartons.
- Make sure you have exactly what is on the board.
e. Your turn: Write names and fractions for the rest of the items in part 8. Raise your hand when you're finished.
  (Observe students and give feedback.)
f. (Write on the board:)

> **c.** $\dfrac{\textbf{plates}}{\textbf{pots}}\ \dfrac{100}{13}$
>
> **d.** $\dfrac{\textbf{containers}}{\textbf{gallons}}\ \dfrac{4}{11}$
>
> **e.** $\dfrac{\textbf{children}}{\textbf{adults}}\ \dfrac{9}{5}$

- Here's what you should have for items C, D and E.
- Item C says: There were 100 plates for every 13 pots. The names are **plates** and **pots.** The numbers are **100** for plates and **13** for pots.

- Item D says: Every 4 containers hold 11 gallons. The names are **containers** and **gallons.** The numbers are **4** for containers and **11** for gallons.
- Item E says: There was a ratio of 9 children to every 5 adults. The names are **children** and **adults.** The numbers are **9** for children and **5** for adults.
g. Raise your hand if you got everything right.

EXERCISE 7 NUMBER FAMILIES
Tables

a. Find part 9.
- The arrows in the table show that the columns are like number families.
- Touch the first column.
- Is the big number or a small number missing in that column? (Signal.) *A small number.*
- Say the problem for figuring out the missing number. (Signal.) *100 minus 42.*
- Touch the middle column.
- Is the big number or a small number missing in that column? (Signal.) *A small number.*
- Say the problem. (Signal.) *89 minus 49.*
- Touch the last column.
- Is the big number or a small number missing in that column? (Signal.) *The big number.*
- Say the problem. (Signal.) *82 plus 107.*
b. Your turn: Copy the table. Figure out the missing numbers and write them in the table. Raise your hand when you're finished.
  (Observe students and give feedback.)
c. (Write on the board:)

	42	40	82
	58	49	107
total	100	89	189

- Here's what you should have.
- Raise your hand if you got everything right.

EXERCISE 8 INDEPENDENT WORK

a. Do the independent work for lesson 12. You'll use a calculator for part 15.
b. (Before beginning the next lesson, check students' independent work.)

Lesson 13

Objectives

- **Work a mixed set of equivalent-fraction problems.** (Exercise 1)
  Note: Some problems have a missing number in the last fraction. $\frac{3}{5}\left(\quad\right)=\frac{12}{\blacksquare}$

  Some have the entire last fraction missing.
  $\frac{3}{5}\left(\frac{6}{6}\right)=\frac{\blacksquare}{\blacksquare}$

- Rewrite an equation of the form: 15 - 12 = 3 so that the equation begins with the number that is alone: 3 = 15 − 12. (Exercise 2)

- Write names and a fraction for a sentence that gives ratio information. (Exercise 3)

- **Compute perimeters of various polygons.** (Exercise 4)
  Note: Students add the length of each side and write the perimeter as a number and a unit name.

- **Complete a number-family table by first working rows with two numbers and then columns with two numbers.** (Exercise 5)
  Note: Problems are of the form:

  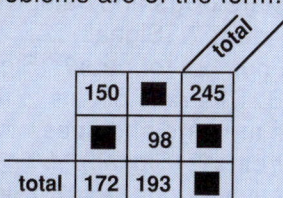

  Students first complete the top row and the bottom row. Then they work the first column and the last column.

- **Complete a table to show a multiplication problem, the corresponding division problem, and the fraction equation.**

Multiplication	Division	Fraction equation
For example: 12 (96) = 1152	$\frac{\boxed{96}}{12\overline{)1152}}$	$\frac{1152}{12}=\boxed{96}$

(Exercise 6)

EXERCISE 1 FRACTIONS
Equivalence

a. Open your textbook to lesson 13 and find part 1.
- These are different types of equivalent-fraction problems.

- Some of them show the first fraction and the fraction that equals 1. That's the fraction you multiply by when you work these problems. You figure out **the entire** equivalent fraction.
- Other problems show the first fraction and **part** of the equivalent fraction. You have to figure out **the missing part** of the equivalent fraction.

b. Problem A: You'll figure out the entire fraction that equals 4/5. Copy the problem and work it. Raise your hand when you're finished. √
- (Write on the board:)

$$\text{a. } \frac{4}{5}\left(\frac{8}{8}\right)=\boxed{\frac{32}{40}}$$

- Here's what you should have. 4/5 equals 32/40.
c. Problem B: You'll figure out the missing **part** of the fraction that equals 4/5. Show the fraction that equals 1. Raise your hand when you're finished.
- (Write on the board:)

$$\text{b. } \frac{4}{5}\left(\frac{7}{7}\right)=\frac{28}{\boxed{35}}$$

- Here's what you should have. The missing value is **35**.
e. Work the rest of the problems in part 1. Raise your hand when you're finished.
  (Observe students and give feedback.)
f. (Write on the board:)

$$\text{c. } \frac{10}{9}\left(\frac{2}{2}\right)=\frac{\boxed{20}}{18}$$

$$\text{d. } \frac{7}{3}\left(\frac{10}{10}\right)=\boxed{\frac{70}{30}}$$

- Here's what you should have for problems C and D.
- Problem C: What's the missing top number? (Signal.) 20.
- Problem D: What fraction equals 7/3? (Signal.) 70-thirtieths.
f. Raise your hand if you got everything right.

EXERCISE 2   EQUATIONS
Rewriting

a. Find part 2.
- You can rewrite equations so the value that is alone on one side of the equation comes first.
b. Equation A. What's the value that is alone? (Signal.) *400.*
- Say the equation that begins with 400. (Signal.) *400 equals 900 minus 500.*
- Equation B. What's the value that is alone? (Signal.) *99.*
- Say the equation that begins with 99. (Signal.) *99 equals 11 times 9.*
(Repeat step b until firm.)
c. Your turn: Rewrite each equation so it begins with the value that is alone on one side of the equation. Raise your hand when you're finished. (Observe students and give feedback.)

  Key:

  a. $400 = 900 - 500$

  b. $99 = 11 \times 9$

  c. $1058 = 358 + 700$

  d. $2120 = 53 \times 40$

d. Check your work. I'll read the equations you started with. You say the rewritten equations.
- Equation A: 900 minus 500 equals 400. What's the rewritten equation? (Signal.) *400 equals 900 minus 500.*
- Equation B: 11 times 9 equals 99. What's the rewritten equation? (Signal.) *99 equals 11 times 9.*
- Equation C: 358 plus 700 equals 1058. What's the rewritten equation? (Signal.) *1058 equals 358 plus 700.*
- Equation D: 53 times 40 equals 2120. What's the rewritten equation? (Signal.) *2120 equals 53 times 40.*

EXERCISE 3   PROBLEM SOLVING
Ratios and Proportions

a. Find part 3.
- These are sentences that tell about ratios.
b. For each sentence, write the names and the fraction. Remember, the first name in the problem is the name for the numerator of the fraction. Raise your hand when you're finished. (Observe students and give feedback.)

c. (Write on the board:)

a.	square feet	15
	boxes	4
b.	miles	85
	hours	3
c.	books	5
	students	9

- Here's what you should have.
- Sentence A says: 15 square feet of cardboard are used for every 4 boxes. The names are **square feet** and **boxes.** The numbers are 15 and 4.
- Sentence B says: The truck moved at the steady rate of 85 miles every 3 hours. The names are **miles** and **hours.** The numbers are 85 and 3.
- Sentence C says: The classroom needed 5 books for every 9 students. The names are **books** and **students.** The numbers are 5 and 9.
d. Find part 4.
- These are sentences of a different type. The name does not come right after the number.
e. Sentence A: The ratio of dogs to cats is 2 to 7. The names are **dogs** and **cats.** What's the number for **dogs?** (Signal.) *2.*
- What's the number for **cats?** (Signal.) *7.*
- Sentence B: The ratio of cats to fleas is 4 to 81. The names are **cats** and **fleas.** What's the number for **cats?** (Signal.) *4.*
- What's the number for **fleas?** (Signal.) *81.*
f. Your turn: Write the names and the fraction for each item in part 4. Raise your hand when you're finished.
(Observe students and give feedback.)
g. (Write on the board:)

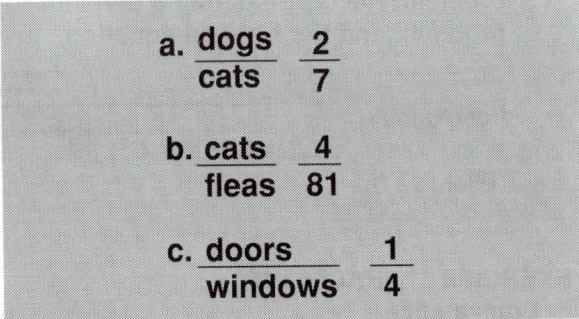

a.	dogs	2
	cats	7
b.	cats	4
	fleas	81
c.	doors	1
	windows	4

- Here's what you should have for each item. Make sure you have exactly what is on the board.

EXERCISE 4 PERIMETER
Polygons

a. Find part 5.
- I'll read what it says. Follow along: The perimeter is the distance around a figure. To find the perimeter, you add the length of each side.
- You can see the addition for the triangle. The sides are 8, 7 and 10 feet. So the perimeter is 25 feet.
- Below, you can see a funny 4-sided figure. The perimeter of that figure is 39 yards.
- Remember, to figure out the perimeter, you just add the length of each side.
- The answer has a number and a unit name.

b. Find part 6.

c. Figure out the perimeter for A. Write the answer as a number and unit name. Raise your hand when you're finished.
   (Observe students and give feedback.)
- (Write on the board:)

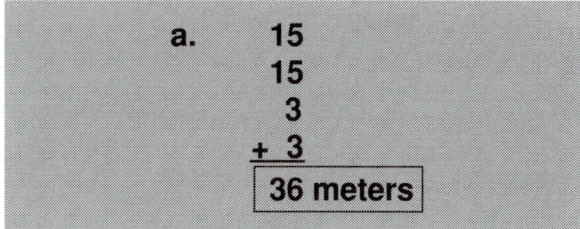

- Here's what you should have. You can have these numbers in any order, but you add these four values to get the total distance around the figure. You must have a plus sign, the four numbers and the answer **36 meters.**

d. Your turn: Figure out the perimeter for B. Raise your hand when you're finished.
   (Observe students and give feedback.)
- (Write on the board:)

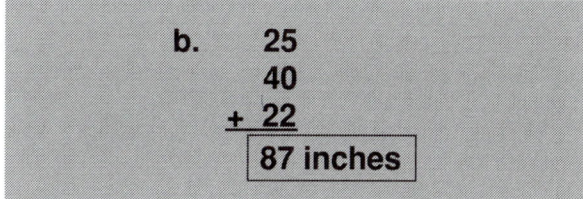

- Here's what you should have. You add three numbers. You must have a plus sign and the answer **87 inches.**

e. Figure out the perimeter for figure C. Raise your hand when you're finished.
   (Observe students and give feedback.)
- Check your work. Figure C: The sides are 46, 19, 28, 16, 20 and 14 centimeters long. What's the perimeter? (Signal.) *143 centimeters.*

EXERCISE 5 NUMBER FAMILIES
Tables

a. Find part 7.
   You're going to figure out all the missing numbers in this table.
- Here's the rule: If there are two numbers in a row, you can figure out the missing number in that row. If there's only one number in a row, you can't figure out the missing numbers.
- The same rule holds for columns: If there are two numbers in a column, you can figure out the missing number in a column. If there's only one number in a column, you can't find the missing numbers.

b. Look at the table.
- You **can't** work one of the rows.
- Can you work the top row? (Signal.) *Yes.*
- Can you work the middle row? (Signal.) *No.*
- Can you work the bottom row? (Signal.) *Yes.*

c. Copy the table. Figure out the missing numbers for each **row** that you can work. Write the missing numbers for those rows in the table. Raise your hand when you've done that much.
   (Observe students and give feedback.)

d. (Write on the board:)

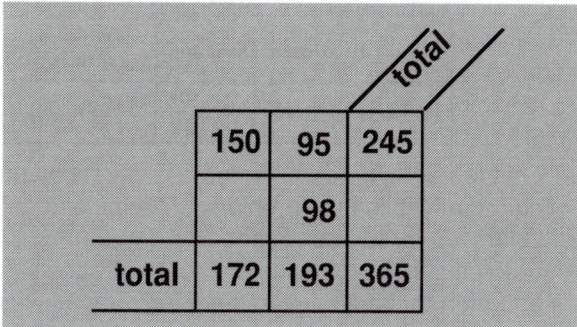

e. Check your rows. The problem for the top row is 245 minus 150. What's the missing number? (Signal.) *95.*
- The problem for the bottom row is 172 plus 193. What's the missing number? (Signal.) *365.*

f. Now you can work all the columns that have two numbers. Can you work the first column? (Signal.) *Yes.*
- The middle column already has three numbers.
- Can you work the third column? (Signal.) *Yes.*
- Work the first and third columns. Raise your hand when you're finished.
   (Observe students and give feedback.)

g. (Write to show:)

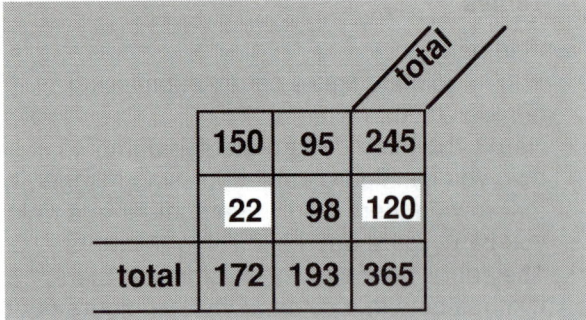

- Check your columns.
- The problem for the first column is 172 minus 150. What's the missing number? (Signal.) *22.*
- The problem for the last column is 365 minus 245. What's the missing number? (Signal.) *120.*

h. Remember how we do that. First, work all the rows you can work. Then work all the columns that have two numbers.

EXERCISE 6 FRACTIONS
For Division

a. Find part 8.
- (Teacher reference:)

Multiplication	Division	Fraction equation
12 (96) = 1152	$\dfrac{96}{12\overline{)1152}}$	$\dfrac{1152}{12}= 96$

- This table has three columns. The first column has a multiplication problem. The middle column is for the corresponding division problem. The last column is for the fraction equation that says the same thing as the multiplication problem and the division problem.
- The sample row is completed. The multiplication equation is 12 times 96 equals 1152. The division problem and answer is 1152 divided by 12 equals 96. The fraction equation is 1152/12 equals 96.

b. Your turn: Copy row A and complete it. Use your calculator. Raise your hand when you're finished. (Observe students and give feedback.)
- (Write on the board:)

	Multiplication	Division	Fraction equation
a.	37 (25) = 925	$\dfrac{25}{37\overline{)925}}$	$\dfrac{925}{37} = 25$

- Here's what you should have for row A. 37 times 25 equals 925. 925 divided by 37 equals 25. The fraction equation is 925/37 equals 25.

c. Your turn: Copy and complete row B. Remember the equal sign in your fraction equation. Raise your hand when you're finished. (Observe students and give feedback.)
- (Write to show:)

	Multiplication	Division	Fraction equation
a.	37 (25) = 925	$\dfrac{25}{37\overline{)925}}$	$\dfrac{925}{37} = 25$
b.	52 (13) = 676	$\dfrac{13}{52\overline{)676}}$	$\dfrac{676}{52} = 13$

- Here's what you should have for row B. 52 times 13 equals 676. 676 divided by 52 equals 13. The fraction equation is 676/52 equals 13.

d. Raise your hand if you got everything right.

EXERCISE 7 INDEPENDENT WORK

a. Do the independent work for lesson 13.
b. (Before beginning the next lesson, check students' independent work.)

Lesson 14

Objectives

- **Complete a fraction that equals a whole number.** (Exercise 1)

  Note: Problems are of the form: $\dfrac{\blacksquare}{7} = 6$.

- **Rewrite an equation so it begins with the unknown.** (Exercise 2)
  Note: Problems are of the form: $50 - 7 = \blacksquare$.
  Students write: $\quad = 50 - 7$, then work the problem and complete the equation:

  $\boxed{43} = 50 - 7$.

- **Refer to a word problem to write a ratio equation.** (Exercise 3)
  Note: Equations are of the form:

  $\dfrac{\text{people}}{\text{benches}} \ \dfrac{4}{3} = \dfrac{24}{\blacksquare}$. Students figure out

  the value that goes in the box.

- Compute perimeters of various polygons. (Exercise 4)

- Complete a number-family table by first working rows with two numbers and then columns with two numbers. (Exercise 5)

- **Treat a fraction as a division problem.** (Exercise 6)
  Note: Students write the fraction then divide to figure out the whole number it

  equals, e.g.: $\dfrac{2444}{52} = \boxed{47}$.

EXERCISE 1 FRACTIONS
For Whole Numbers

a. Open your textbook to lesson 14 and find part 1.
- These are equations with part of a fraction and the whole number it equals.
- The whole number tells how many times bigger the numerator of the fraction is than the denominator.
- To work the problems, start with the denominator, multiply to figure out the numerator, and write the **complete equation.**
b. Look at the sample problem. What's the denominator? (Signal.) *5.*
- How many whole units does the fraction equal? (Signal.) *3.*
- So the numerator is 5 times 3. What's the number? (Signal.) *15.*
- You'd write the equation: 3 equals 15/5.

c. Write the complete equation for problem A. Remember the whole number tells how many times bigger the numerator is than the denominator. Raise your hand when you're finished. √
- (Write on the board:)

$$\text{a.} \quad 5 = \dfrac{\boxed{40}}{8}$$

- Here's what you should have. The numerator is 8 times 5. That's 40. You should have written the equation: 5 equals 40/8.
d. Work problem B. Raise your hand when you're finished. √
- (Write on the board:)

$$\text{b.} \quad \dfrac{\boxed{42}}{7} = 6$$

- Here's what you should have. The numerator is 7 times 6. That's 42. You should have the equation: 42/7 equals 6.
e. Work the rest of the problems in part 1. Remember to write the complete equations. Raise your hand when you're finished. (Observe students and give feedback.)
f. (Write on the board:)

$$\text{c.} \quad 9 = \dfrac{\boxed{36}}{4} \qquad \text{d.} \quad \dfrac{\boxed{30}}{10} = 3$$

$$\text{e.} \quad 10 = \dfrac{\boxed{90}}{9}$$

- Here's what you should have for problems C, D and E. Equation C: 9 equals 36/4. Equation D: 30/10 equals 3. Equation E: 10 equals 90/9.
g. Raise your hand if you got everything right.

EXERCISE 2 EQUATIONS
Rewriting

a. Find part 2.
b. I'll read what it says. Follow along: You've worked with equations that have an unknown after the equal sign.
- You can see the equation. 50 minus 7 equals an unknown.
- We'll write the same equation so the unknown comes first. The side with 50 minus 7 comes after the equal sign.

- You can see the equation with the box. It says that the unknown equals 50 minus 7.
- You figure out the answer the same way you would if the box were on the other side. You figure out what 50 minus 7 equals. That's the number that goes in the box.
- You can see the box with 43 in it. That's the answer. 43 equals 50 minus 7.

c. Find part 3.
- The box is alone on one side of each equation.
- Rewrite each equation so the box comes first. Raise your hand when you've done that much.
  (Observe students and give feedback.)

d. (Write on the board:)

a.	$\square$	= 5 x 4
b.	$\square$	= 13 − 11
c.	$\square$	= 50 + 70
d.	$\square$	= 30 − 12

- Here's what you should have.
- For equation A: The unknown equals 5 times 4.
- For equation B: The unknown equals 13 minus 11.
- For equation C: The unknown equals 50 plus 70.
- For equation D: The unknown equals 30 minus 12.

e. You can figure out what the box equals in each equation. Read the side with two values. Start with the value that comes right after the equal sign.
- Problem A. Here's the problem you work: 5 times 4.
- Problem B. Say the problem you'll work. (Signal.) *13 minus 11.*
- Problem C. Say the problem you'll work. (Signal.) *50 plus 70.*

f. Your turn: Figure out the answer to each problem and write it in the box. Raise your hand when you're finished.
  (Observe students and give feedback.)

g. (Write to show:)

a.	20	= 5 x 4
b.	2	= 13 − 11
c.	120	= 50 + 70
d.	18	= 30 − 12

- Here's what you should have. Raise your hand if you got everything right.

EXERCISE 3 PROBLEM SOLVING
Ratios and Proportions

a. Find part 4.
- You'll complete equations by writing a fraction that has a box and a number.
- Each problem starts with the same names and numbers. The names are **bottles** and **pounds.** 5 bottles weigh 2 pounds. What's the top name? (Signal.) *Bottles.*
- All numbers for bottles go in the numerator. What's the bottom name? (Signal.) *Pounds.*
- All numbers for pounds go in the denominator.
- (Write on the board:)

$$\frac{\text{bottles}}{\text{pounds}}$$

- Here are the names for all the problems in part 4.
- 5 bottles weigh 2 pounds. What's the number for bottles? (Signal.) *5.*
- What's the number for pounds? (Signal.) *2.*
- (Write to show:)

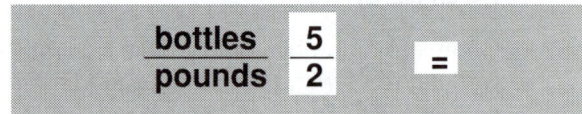

- Here's the first fraction. You're going to write an equivalent fraction with a box and a number.

b. If the problem asks about pounds, you write a box in the denominator.
- If the problem asks about bottles, you write a box in the numerator.

c. Problem A: 5 bottles weigh 2 pounds. How many bottles weigh 20 pounds? The question asks about bottles. So does the box go on top or on the bottom? (Signal.) *On top.*
- The question gives a number for pounds. Where does that number go? (Signal.) *On the bottom.*

d. Your turn: Copy the first part of the equation and write the box and the 20 where they go in the last fraction. √
- (Write to show:)

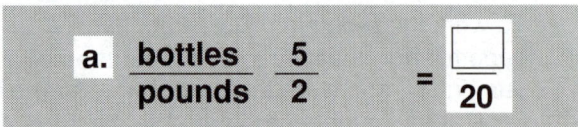

- Here's what you should have.

e. Problem B: 5 bottles weigh 2 pounds. How many pounds do 15 bottles weigh?

- The problem asks about **pounds** and gives a number for bottles. Write the complete equation with a box and a number in the last fraction. Raise your hand when you're finished. √
- (Write on the board:)

$$\text{b.} \quad \frac{\text{bottles}}{\text{pounds}} \quad \frac{5}{2} \quad = \quad \frac{15}{\boxed{}}$$

- Here's what you should have.
f. Problem C: The ratio of bottles to pounds is 5 to 2. How many pounds do 30 bottles weigh?
- Write the complete equation. Raise your hand when you're finished. √
- (Write on the board:)

$$\text{c.} \quad \frac{\text{bottles}}{\text{pounds}} \quad \frac{5}{2} \quad = \quad \frac{30}{\boxed{}}$$

- Here's what you should have.
g. Problem D: 5 bottles weigh 2 pounds. How many bottles weigh 30 pounds?
- Write the complete equation. Raise your hand when you're finished. √
- (Write on the board:)

$$\text{d.} \quad \frac{\text{bottles}}{\text{pounds}} \quad \frac{5}{2} \quad = \quad \frac{\boxed{}}{30}$$

- Here's what you should have.
h. Your turn: Work each problem and write the answer as a number and a **unit name.** Remember, figure out the fraction that equals 1. Figure out the number that goes in the box. Write that number and the unit name. Box the answer. Raise your hand when you've worked all the problems.
  (Observe students and give feedback.)
i. (Write to show:)

$$\text{a.} \quad \frac{\text{bottles}}{\text{pounds}} \quad \frac{5}{2} \left(\frac{10}{10}\right) = \frac{50}{20} \quad \boxed{50 \text{ bottles}}$$

$$\text{b.} \quad \frac{\text{bottles}}{\text{pounds}} \quad \frac{5}{2} \left(\frac{3}{3}\right) = \frac{15}{\boxed{6}} \quad \boxed{6 \text{ pounds}}$$

$$\text{c.} \quad \frac{\text{bottles}}{\text{pounds}} \quad \frac{5}{2} \left(\frac{6}{6}\right) = \frac{30}{\boxed{12}} \quad \boxed{12 \text{ pounds}}$$

$$\text{d.} \quad \frac{\text{bottles}}{\text{pounds}} \quad \frac{5}{2} \left(\frac{15}{15}\right) = \frac{\boxed{75}}{30} \quad \boxed{75 \text{ bottles}}$$

- Here's what you should have for each problem.
- Problem A: How many bottles weigh 20 pounds? The answer is 50 bottles.
- Problem B: How many pounds do 15 bottles weigh? The answer is 6 pounds.
- Problem C: How many pounds do 30 bottles weigh? The answer is 12 pounds.
- Problem D: How many bottles weigh 30 pounds? The answer is 75 bottles.
j. Raise your hand if you got everything right.

EXERCISE 4 PERIMETER
Polygons

a. Find part 5.
- Remember how to find the perimeter of figures with straight sides. You add the length of each side. The answer tells the distance around the figure. The answer has a number and a unit name.
b. Your turn: Find the perimeter of the figures in part 5. Raise your hand when you're finished. (Observe students and give feedback.)

Key:

a.
$$\begin{array}{r} 18 \\ 18 \\ 12 \\ + 12 \\ \hline \boxed{60 \ meters} \end{array}$$

b.
$$\begin{array}{r} 2 \\ 2 \\ 2 \\ 2 \\ + 2 \\ \hline \boxed{10 \ miles} \end{array}$$

c.
$$\begin{array}{r} 8 \\ 4 \\ 12 \\ + 10 \\ \hline \boxed{34 \ feet} \end{array}$$

d.
$$\begin{array}{r} 16 \\ 16 \\ 38 \\ + 38 \\ \hline \boxed{108 \ yards} \end{array}$$

c. Check your work.
- Find part J on page 53. That shows what you should have.
- Figure A. The units are meters. The sides are 18, 18, 12 and 12 meters. What's the perimeter of the figure? (Signal.) *60 meters.*
- Make sure you have 60 meters.
- Figure B. The sides are 2, 2, 2, 2 and 2 miles. What's the perimeter? (Signal.) *10 miles.*
- Figure C: The sides are 8, 4, 12 and 10 feet. What's the perimeter? (Signal.) *34 feet.*
- Figure D: The sides are 16, 16, 38 and 38 yards. What's the perimeter? (Signal.) *108 yards.*
- Make sure all your problems have a plus sign and the correct unit name.
d. Raise your hand if you got all of them right.

EXERCISE 5 NUMBER FAMILIES
Tables

a. Find part 6.
- You're going to figure out all the missing numbers in the tables. Remember, first work all the rows you can work.

b. Touch table A.
- Can you figure out the missing number in the top row? (Signal.) *Yes.*
- Can you figure out the missing number in the middle row? (Signal.) *No.*
- After you work all the **rows** you can work, work all the **columns** that have two numbers.

c. Copy table A. Work the problems carefully. Copy your answers in the table. Remember, first the rows; then the columns. Raise your hand when you're finished.
   (Observe students and give feedback.)
- (Teacher reference:)

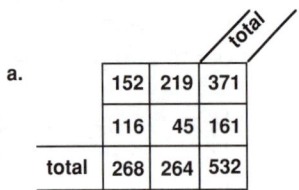

a.

	total	
152	219	371
116	45	161
total 268	264	532

d. (Write on the board:)

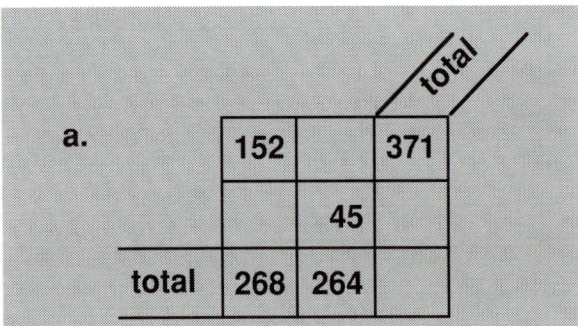

a.

		total
152		371
	45	
total 268	264	

d. Here's the table you started with. You worked the rows first.
- The problem for the top row is 371 minus 152. That's 219.
- (Write: **219.**)
- The problem for the bottom row is 268 plus 264. That's 532.
- (Write: **532.**)
- Then you worked the columns.
  The problem for the first column is 268 minus 152. That's 116.
- (Write: **116.**)
- The problem for the last column is 532 minus 371. That's 161.
- (Write: **161.**)

e. Raise your hand if you figured out all the missing numbers in the table.

f. Work table B. First work the rows, then the columns. Raise your hand when you're finished.
   (Observe students and give feedback.)
   Key:

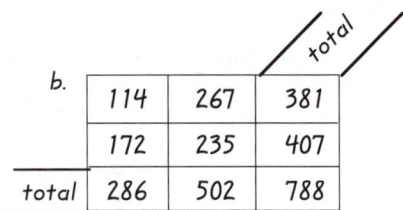

b.

		total
114	267	381
172	235	407
total 286	502	788

- Check your work. Find part K. That shows table B with all the missing numbers.
- Raise your hand if you figured out all the missing numbers in the table.

EXERCISE 6 FRACTIONS
As Division

a. Find part 7.
- These are fractions that equal whole numbers.

b. You can figure out the whole number **by reading the fraction as a division problem.**

c. Read fraction A as a division problem. (Signal.) *2444 divided by 52.*
- Read fraction B as a division problem. (Signal.) *282 divided by 47.*
- Read fraction C as a division problem. (Signal.) *532 divided by 28.*
- Read fraction D as a division problem. (Signal.) *1848 divided by 56.*
   (Repeat step C until firm.)

d. Work fraction A on your calculator and write the equation that shows the fraction and the whole number it equals. Raise your hand when you're finished. (Observe students and give feedback.)
- (Write on the board:)

$$\text{a.} \quad \frac{2444}{52} = \boxed{47}$$

- Here's what you should have for fraction A: 2444 over 52 equals 47.

e. Your turn: Work the division problems for the rest of the items. Write the complete equations that show the fraction and the number it equals. Raise your hand when you're finished.
   (Observe students and give feedback.)

f. (Write to show:)

a. $\dfrac{2444}{52}$ = $\boxed{47}$	b. $\dfrac{282}{47}$ = $\boxed{6}$
c. $\dfrac{532}{28}$ = $\boxed{19}$	d. $\dfrac{1848}{56}$ = $\boxed{33}$

- Here's what you should have for problems B, C and D.
- Equation B: 282/47 equals 6.
- Equation C: 532/28 equals 19.
- Equation D: 1848/56 equals 33.

g. Raise your hand if you got everything right.

EXERCISE 7 INDEPENDENT WORK

a. Do the independent work for lesson 14.

b. (Before beginning the next lesson, check students' independent work.)

Lesson 15

Objectives

- Refer to a word problem to write a ratio equation. (Exercise 1)

- Rewrite an equation so it begins with the unknown. (Exercise 2)

- **Write an equation that shows an improper fraction and the mixed number it equals.** (Exercise 3)
  Note: Students refer to a number line such as:

  $$\begin{array}{c} 0 \quad 1 \quad 2 \quad 3 \quad 4 \end{array}$$

  Students write the equation: $\dfrac{11}{4} = 2\dfrac{3}{4}$.

- Complete a number-family table by first working rows with two numbers and then columns with two numbers. (Exercise 4)

- Complete a fraction that equals a whole number. (Exercise 5)

- Compute perimeters of various polygons. (Exercise 6)

EXERCISE 1 PROBLEM SOLVING
Ratios and Proportions

a. Open your textbook to lesson 15 and find part 1.
- (Teacher reference:)

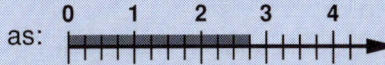

a. During part of last summer, there were 8 sunny days for every 7 cloudy days. There were 56 sunny days. How many cloudy days were there?

- These problems are partially worked.
- The word problem asks about one of the names. You write a box for that name in the last fraction.
- The problem also gives a number for the last fraction.

b. Problem A: During part of last summer, there were 8 sunny days for every 7 cloudy days. There were 56 sunny days. How many cloudy days were there?
- The problem gives a number for the last fraction. Is that the number for sunny days or cloudy days? (Signal.) *Sunny days.*
- Does that number go on top or on the bottom of the last fraction? (Signal.) *On top.*

- Copy the equation. Show a number and a box for the last fraction. Raise your hand when you've done that much.
  (Observe students and give feedback.)
- (Write on the board:)

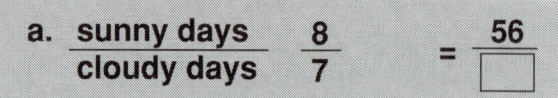

a.	sunny days	8	=	56
	cloudy days	7		☐

- Here's what you should have.
c. Item B: Write the equation with a box and a number in the last fraction. Don't work the problem. Raise your hand when you've done that much.
  (Observe students and give feedback.)
- (Write on the board:)

b.	inches	9	=	639
	seconds	7		☐

- Here is the equation you should have. The problem gives a number for inches. 639 is the numerator of the last fraction and a box is in the denominator.
d. Your turn: Write equations with a box and a number for items C and D. Raise your hand when you've done that much.
  (Observe students and give feedback.)
- (Write on the board:)

c.	cups of flour	2	=	☐
	cups of sugar	7		28

d.	hours	3	=	☐
	cars	8		48

- Here's what you should have for each equation.
e. Fix up any mistakes. Then work the problems and answer the questions. Remember, you answer the question by writing the number that goes in the box and the unit name. Raise your hand when you're finished.
  (Observe students and give feedback.)
f. (Write to show:)

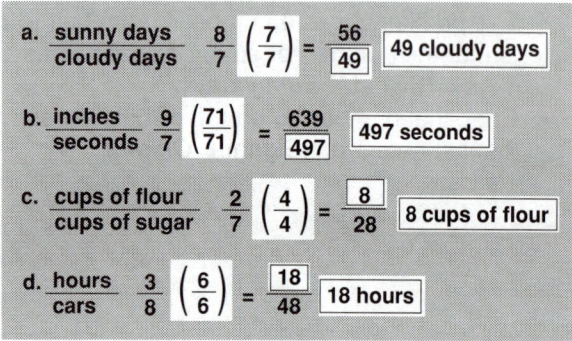

a.	sunny days	$\frac{8}{7}$	$\left(\frac{7}{7}\right)$	=	$\frac{56}{49}$	49 cloudy days
	cloudy days					

b.	inches	$\frac{9}{7}$	$\left(\frac{71}{71}\right)$	=	$\frac{639}{497}$	497 seconds
	seconds					

c.	cups of flour	$\frac{2}{7}$	$\left(\frac{4}{4}\right)$	=	$\frac{8}{28}$	8 cups of flour
	cups of sugar					

d.	hours	$\frac{3}{8}$	$\left(\frac{6}{6}\right)$	=	$\frac{18}{48}$	18 hours
	cars					

- Here's what you should have for each problem.
- Problem A: How many cloudy days were there? The answer is 49 cloudy days.
- Problem B: How many seconds would it take the ant to move 639 inches? The answer is 497 seconds.
- Problem C: How many cups of flour are needed for 28 cups of sugar? The answer is 8 cups of flour.
- Problem D: How many hours does it take the shop to fix 48 cars? The answer is 18 hours.
- Raise your hand if you got everything right.

EXERCISE 2 EQUATIONS
Rewriting

a. Find part 2.
b. You're going to rewrite each equation so it begins with the box. Then you'll use your calculator to work each problem. Raise your hand when you've rewritten all the equations, figured out the answers and written them in the boxes. (Observe students and give feedback.)
c. (Write on the board:)

a.	☐ 407	= 456 − 49
b.	☐ 117	= 13 × 9
c.	☐ 650	= 562 + 88
d.	☐ 226	= 501 − 275

- Check your work. Here's what you should have.
- Raise your hand if you got everything right.

EXERCISE 3 MIXED NUMBERS
On the Number Line

a. Find part 3.
- (Teacher reference:)

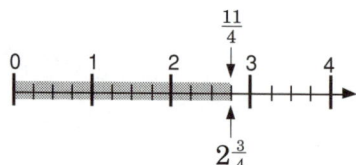

b. I'll read what it says. Follow along: Some fractions that are more than 1 do not equal whole numbers. These fractions can be written as mixed numbers.
- The mixed number shows the number of **whole units** and shows the number of **parts that are leftover.** The fraction in the mixed number is always less than 1 whole unit.

- Look at the number line. The bar on the number line goes past 2 whole units. The parts after 2 are leftover parts.
- The fraction for the whole bar is 11/4. Each unit is divided into 4 parts. 11 parts are shaded.
- The mixed number that equals 11/4 is 2 and 3/4. That's 2 full units, and 3/4 of the next unit.
c. Find part 4.
- For each item, you'll write an equation that shows the fraction and the mixed number it equals.
d. Bar A. You write the number of parts in each unit as the denominator of the fraction. You write the number of parts from the **beginning of the number line** as the numerator.
- Write the fraction for bar A. Raise your hand when you've done that much.
  (Observe students and give feedback.)
- (Write on the board:)

> a. $\dfrac{9}{2}$

- Here's what you should have. The fraction is 9/2.
e. Now write an equal sign and the mixed number that equals 9/2. That number shows how many units are shaded and the fraction of the next unit that is shaded.
- The denominator of that fraction is the same as the denominator of the first fraction. Write the mixed number. Raise your hand when you're finished. (Observe students and give feedback.)
- (Write to show:)

> a. $\dfrac{9}{2} = 4\dfrac{1}{2}$

- Here's what you should have. 9/2 equals 4 and 1/2.
f. Write the complete equation for item B. Write the fraction first, then an equal sign and the mixed number the fraction equals. Raise your hand when you're finished.
  (Observe students and give feedback.)
- (Write on the board:)

> b. $\dfrac{5}{3} = 1\dfrac{2}{3}$

- Here's what you should have. The fraction is 5/3. The mixed number is 1 and 2/3. Raise your hand if you got it right.

g. Write the equation for item C. Raise your hand when you're finished.
  (Observe students and give feedback.)
- (Write on the board:)

> c. $\dfrac{13}{4} = 3\dfrac{1}{4}$

- Here's what you should have. The fraction for the picture is 13/4. 13/4 equals 3 and 1/4.
h. Write an equation for item D. Raise your hand when you're finished.
  (Observe students and give feedback.)
- (Write on the board:)

> d. $\dfrac{8}{5} = 1\dfrac{3}{5}$

- Here's what you should have. The fraction for the picture is 8/5. 8/5 equals 1 and 3/5.
i. Raise your hand if you got everything right.

EXERCISE 4 NUMBER FAMILIES
Tables

a. Find part 5.
- These are tables that have missing numbers. Remember how to work the tables. First you work any rows that have two numbers. Then you work the columns.
b. Copy table A and figure out the missing numbers. Raise your hand when you're finished. (Observe students and give feedback.)
c. (Write on the board:)

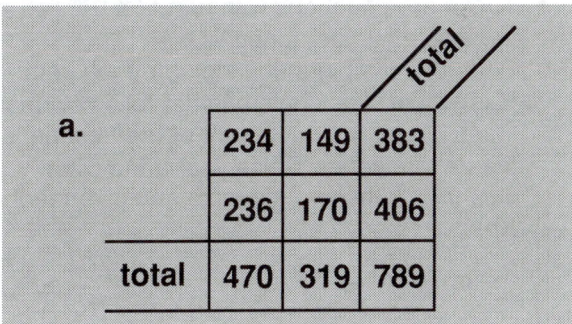

a.

		total
234	149	383
236	170	406
total 470	319	789

- Here's what you should have for table A. In the top row, you subtracted 149 from 383. The answer is 234.
- In the bottom row, you added 470 and 319. The answer is 789.
- In the first column, you subtracted 234 from 470. The answer is 236.
- In the middle column, you subtracted 149 from 319. The answer is 170.

- In the last column, you subtracted 383 from 789. The answer is 406.
- Raise your hand if you got everything right.
d. Copy and work table B. Raise your hand when you're finished.
   (Observe students and give feedback.)
e. (Write on the board:)

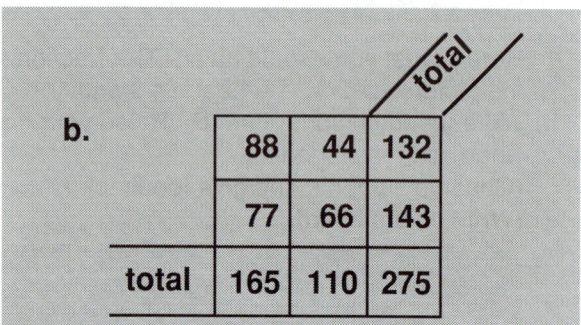

			total
b.	88	44	132
	77	66	143
total	165	110	275

- Here's what you should have for table B.
- In the top row, you subtracted 88 from 132. The answer is 44.
- In the middle row, you added 77 and 66. The answer is 143.
- In the first column, you added 88 and 77. The answer is 165.
- In the second column you added 44 and 66. The answer is 110.
- In the last column, you added 132 and 143. The answer is 275.
- Raise your hand if you got everything right.

EXERCISE 5 FRACTIONS
For Whole Numbers

a. Find part 6.
- These are equations with part of a fraction and the whole number it equals.
b. Write the complete equation for each item. Remember, the whole number tells how many times bigger the numerator of the fraction is than the denominator. Raise your hand when you're finished. (Observe students and give feedback.)

   Key:

   a. $\dfrac{\boxed{20}}{5} = 4$ b. $\dfrac{\boxed{20}}{4} = 5$

   c. $6 = \dfrac{\boxed{54}}{9}$ d. $\dfrac{\boxed{60}}{20} = 3$

c. Find part J on page 57. That shows what you should have for each item.
- Equation A: 20/5 equals 4.
- Equation B: 20/4 equals 5.
- Equation C: 6 equals 54/9.
- Equation D: 60/20 equals 3.
- Raise your hand if you got everything right.

EXERCISE 6 PERIMETER
Polygons

a. Find part 7.
b. Find the perimeter of these figures. Remember to write the plus sign. Show the correct unit name in your answer. Raise your hand when you're finished.
   (Observe students and give feedback.)
c. Check your work.
- Figure A: The units are inches. You added 15 plus 15 plus 11 plus 11. The answer is **52 inches.**
- Figure B: The units are meters. You added 12 plus 10 plus 5. The answer is **27 meters.**
- Figure C: The units are feet. You added 10 plus 10 plus 8 plus 8. The answer is **36 feet.**
- Figure D: The units are centimeters. You added 6 plus 6 plus 20 plus 20. The answer is **52 centimeters.**

EXERCISE 7 INDEPENDENT WORK

a. Do the independent work for lesson 15. You'll use a calculator for parts 10 and 11.
b. (Before beginning the next lesson, check students' independent work.)

Lesson 16

Objectives

- **Solve a problem that illustrates two values and the difference between them.** (Exercise 1)

  Note: Problems are of the form:

dif	21 B
40 N	

  The larger value is the big number in the family. The smaller value is a small number. The difference number is the other small number. Students make a number family with the abbreviation **dif**, two letters and the two values the problem gives:

  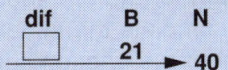

- Write an equation that shows an improper fraction and the mixed number it equals. (Exercise 2)

- **Write and solve a ratio equation for a word problem.** (Exercise 3)

  Note: Problems are of the form: **There are 20 fleas for every 3 cats. If there are 21 cats, how many fleas are there?** Students write the equation with the names and the three numbers the problem gives:

  $$\frac{fleas}{cats} \quad \frac{20}{3} = \frac{\blacksquare}{21}$$

  Students then figure out the missing number and write the answer as a number and a unit name: 140 fleas.

- Rewrite an equation so it begins with the unknown. (Exercise 4)

- **Write an equation that shows a mixed number and the fraction it equals.** (Exercise 5)

  Note: Students copy the denominator of the fraction in the mixed number. Then they compute the numerator by multiplying the denominator by the whole number, then adding the numerator. Students write:

  $$3\frac{1}{4} = \frac{13}{4}$$

EXERCISE 1 NUMBER FAMILIES
Comparison

a. Open your textbook to lesson 16 and find part 1. (Teacher reference:)

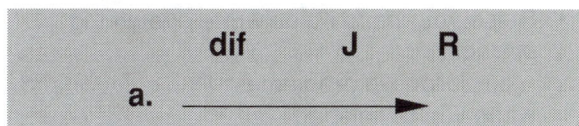

b. I'll read what is says. Follow along: You can make number families that compare two things. A comparison number family shows the **bigger thing** as the **big number.** The **smaller thing** is one of the **small numbers.** The **difference** between the two things is the other **small number.** It's always the first small number.

- You can see two boards, N and B. The dotted part next to board N is the difference.
- Below you can see the comparison number family.
- B is larger, so it is the name for the big number.
- N is smaller; it's the name for one of the small numbers.
- The name for the other small number is the **difference.**
- The difference shows how much you'd have to subtract from B to make it the same size as N.

c. Find part 2.

- Each diagram shows two things and a dotted part for the difference between the two things.

d. Problem A: The two things shown are bars.

- Which bar is bigger? (Signal.) *R.*
- Which bar is smaller? (Signal.) *J.*
- Make the number family with R as the name for the big number, J as the name for the second small number and the letters **d-i-f** for the **difference.** Raise your hand when you're finished. (Observe students and give feedback.)

e. (Write on the board:)

> dif   J   R
>
> a. ⟶

- Here's what you should have.

f. Your turn. Work the rest of the items. Make number families with two letters and **dif** for difference. Remember, the difference is always the first small number. Raise your hand when you're finished.

(Observe students and give feedback.)

g. (Write on the board:)

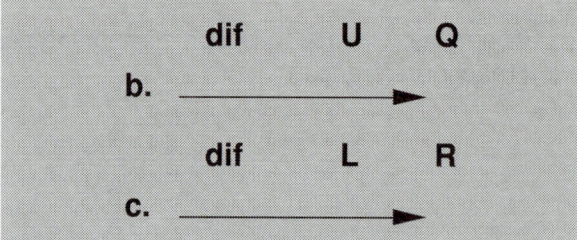

- Here's what you should have for items B and C.
- Raise your hand if you got everything right.
h. Find part 3.
- These are like the problems you just worked except that each diagram shows two numbers. You'll make the number family with three names. Below two of the names, you'll show the numbers.
i. Look at the sample problem. M is bigger, so it's the name for the big number of the family. The small numbers are for the **difference** and K.
- The diagram shows that the number for K is 37. The number for **difference** is 82.
- To figure out M, you just add 82 and 37. M is 119.
j. Remember, the family has to have the three names and two numbers.
k. Work problem A. Write the three names in the family and put in the two numbers. Then figure out the missing number. Raise your hand when you're finished.
(Observe students and give feedback.)
l. (Write on the board:)

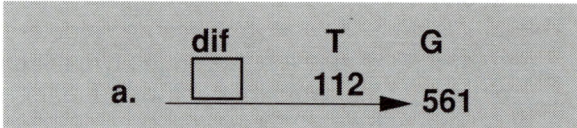

- Here's the family you should have. The big number is for G. The small numbers are for the difference and for T.
- You figured out the difference number by working the problem 561 minus 112. Everybody, what's the answer? (Signal.) *449.*
- You should have written 449.
- Raise your hand if you wrote everything correctly.
m. Your turn. Do problem B. Raise your hand when you're finished.
(Observe students and give feedback.)
n. (Write on the board:)

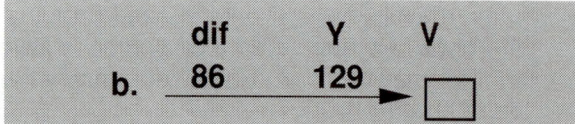

- Here's the family you should have. The names are Y, V and the difference. V is the big number. The problem gives values for the difference and for Y. To find the big number, you work the problem 86 plus 129. Everybody, what's the answer? (Signal.) *215.*
- Yes, V is 215.
o. Remember how to work problems that refer to the difference. You'll do a lot of work with those problems.

EXERCISE 2 MIXED NUMBERS
On the Number Line

a. Find part 4.
b. Item A: Write the **fraction** for the bar shown on the number line. Then complete the equation by showing the mixed number the fraction equals. Raise your hand when you're finished.
(Observe students and give feedback.)
- (Write on the board:)

$$\text{a.} \quad \frac{7}{3} = 2\frac{1}{3}$$

- Here's what you should have for item A. The fraction for A is 7/3. 7/3 equals 2 and 1/3.
c. Your turn: Write equations for the rest of the items in part 4. Raise your hand when you're finished. (Observe students and give feedback.)
- (Write on the board:)

$$\text{b.} \quad \frac{7}{5} = 1\frac{2}{5}$$
$$\text{c.} \quad \frac{7}{2} = 3\frac{1}{2}$$
$$\text{d.} \quad \frac{9}{4} = 2\frac{1}{4}$$
$$\text{e.} \quad \frac{13}{5} = 2\frac{3}{5}$$

- Here's what you should have for items B through E.
d. Raise your hand if you got everything right.

EXERCISE 3 PROBLEM SOLVING
Ratios and Proportions

a. Find part 5.

- For each problem, you'll write the names and a pair of equivalent fractions. You'll write the information from the **second sentence** in the fraction **after** the equal sign.
b. Problem A. First sentence: There are 21 fleas for every 3 cats.
- Write the names and the **first fraction**. Raise your hand when you've done that much. (Observe students and give feedback.)
- (Write on the board:)

> a. $\dfrac{\text{fleas}}{\text{cats}}$ $\dfrac{21}{3}$

- Here's what you should have so far.
- Now the second sentence: If there are 21 cats, how many fleas are there?
- Write the fraction that comes after the equal sign. Raise your hand when you've done that much. (Observe students and give feedback.)
- (Write to show:)

> a. $\dfrac{\text{fleas}}{\text{cats}}$ $\dfrac{21}{3}$ $= \dfrac{\boxed{}}{21}$

- Here's what you should have. The second sentence gives a number for cats and asks about fleas.
- Now figure out the number of fleas. Raise your hand when you're finished. (Observe students and give feedback.)
- (Write to show:)

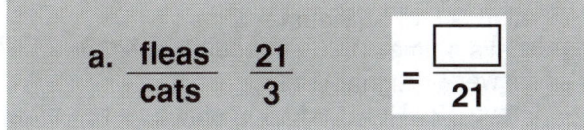

> a. $\dfrac{\text{fleas}}{\text{cats}}$ $\dfrac{21}{3}$ $\left(\dfrac{7}{7}\right)$ $= \dfrac{\boxed{147}}{21}$ $\boxed{147 \text{ fleas}}$

- Check your work. Here's what you should have. The fraction that equals 1 is 7/7. If there are 21 cats, there are **147 fleas.**
c. Write the names and the fractions for problem B. The fraction after the equal sign should have a box and a number. Raise your hand when you've done that much. (Observe students and give feedback.)
- (Write on the board:)

> b. $\dfrac{\text{trees}}{\text{cones}}$ $\dfrac{4}{7}$ $= \dfrac{20}{\boxed{}}$

- Here's what you should have. The names are **trees** and **cones.** The number for trees is 4 and the number for cones is 7. The problem asks about cones, so the denominator for the fraction after the equal sign is a box.
- Work the rest of problem B. Raise your hand when you've answered the question. (Observe students and give feedback.)
- (Write to show:)

> b. $\dfrac{\text{trees}}{\text{cones}}$ $\dfrac{4}{7}$ $\left(\dfrac{5}{5}\right)$ $= \dfrac{20}{\boxed{35}}$ $\boxed{35 \text{ cones}}$

- Check your work. Here's the equation you should have. If there are 20 trees, there are 35 cones. The answer is **35 cones.**
d. Work problem C. Raise your hand when you're finished. (Observe students and give feedback.)
- (Write on the board:)

> c. $\dfrac{\text{buses}}{\text{trucks}}$ $\dfrac{9}{4}$ $\left(\dfrac{8}{8}\right)$ $= \dfrac{72}{\boxed{32}}$ $\boxed{32 \text{ trucks}}$

- Check your work. Here's what you should have. If there are 72 buses, there are 32 trucks. The answer is **32 trucks.**
e. Raise your hand if you got everything right.

EXERCISE 4 EQUATIONS
Solving for an Unknown

a. Find part 6.
- Each equation has the letter **T** instead of a box. The **T** works just like a box. When you find the missing number, you know what **T** equals.
- To show the answer to the problem, you write a simple equation for **T**. We'll work the sample problem together: 543 minus 345 equals T.
- I'll rewrite the equation so it begins with T.
- (Write on the board:)

> $T = 543 - 345$

b. Now we figure out what **T** equals. Say the problem we'll work. (Signal.) *543 minus 345.*
- The answer is 198. Here's how I show the answer. I write **T** equals 198.
- (Write to show:)

> $T = 543 - 345$
> $\boxed{T = 198}$

- All done. Remember, when there's a letter, your last equation will show the letter and the number it equals.
c. Your turn: Work problem A. First rewrite it so it begins with **T.** Then use your calculator to figure out the answer. Remember to write a simple equation for **T.** Raise your hand when you're finished.
   (Observe students and give feedback.)
- (Write on the board:)

> **a. T = 978 + 269**
> **T = 1247**

- Here's what you should have. You rewrote the equation to say **T** equals 978 plus 269. The answer is 1247. You wrote **T** equals 1247. Raise your hand if you did all that exactly as it is on the board.
d. Work problem B. Use your calculator if you need it. Raise your hand when you're finished.
   (Observe students and give feedback.)
- (Write on the board:)

> **b. T = 120 x 8**
> **T = 960**

- Here's what you should have. You rewrote the equation to say **T** equals 120 times 8. The answer is 960. You wrote **T** equals 960.
e. Your turn: Work problem C. Raise your hand when you're finished.
   (Observe students and give feedback.)
- Problem C: Check your work. Read the first equation that starts with **T.** (Signal.) *T equals 470 minus 248.*
- Read the simple equation. (Signal.) *T equals 222.*
- Raise your hand if you got the answer of 222.

EXERCISE 5 MIXED NUMBERS
As Fractions

a. Find part 7.
b. I'll read what it says. Follow along: Here's how to write mixed numbers as fractions.
- First, you copy the denominator. You can see the mixed number 5 and 2/3. The denominator in the mixed number is 3. That's the denominator in the fraction.
- Next, you figure out the numerator of the fraction.
- You start with the denominator of the fraction and **multiply** by the whole number. Then you **add** the numerator of the fraction in the mixed number.

- The arrows show that you start with the denominator and **multiply** by the whole number. That's 3 times 5 equals 15.
- Then you **add** the numerator. That's 15 plus 2 equals 17. So, the fraction that equals 5 and 2/3 is 17/3.
- Remember, first **multiply.** Then **add.**
c. We'll work the sample problem together.
- (Write on the board:)

$$3\frac{1}{4} =$$

- I start with the denominator of the mixed number. What's the denominator? (Signal.) *4.*
- (Write to show:)

$$3\frac{1}{4} = \frac{}{4}$$

- I'm going to multiply 4 by another value. What value is that? (Signal.) *3.*
- That's 4 times 3. Everybody, what's the answer? (Signal.) *12.*
- I have 12. Now I **add** something to 12. What do I add? (Signal.) *1.*
- I add 1 to 12. What's the answer? (Signal.) *13.*
- (Write to show:)

$$3\frac{1}{4} = \frac{13}{4}$$

- 3 and 1/4 equals 13/4.
- Remember, start with the denominator, multiply by the whole number, then add the numerator of the fraction.
d. (Write on the board:)

$$a. \quad 4\frac{2}{9} = \frac{}{9}$$
$$b. \quad 4\frac{3}{5} = \frac{}{5}$$
$$c. \quad 1\frac{5}{9} = \frac{}{9}$$

- Here are some mixed numbers we'll work the fast way. Remember, you'll start with the denominator and multiply by the whole number. Then you'll add the top number of the fraction.

e. Problem A: 4 and 2/9. The fraction will be ninths. Start with 9. Say the multiplication problem and the answer. (Signal.) *9 times 4 equals 36.*
- How many do you add to 36? (Signal.) *2.*
- What's the answer? (Signal.) *38.*
- (Write to show:)

$$\text{a.} \quad 4\frac{2}{9} = \frac{38}{9}$$

- So 4 and 2/9 equals 38/9.
f. Problem B: 4 and 3/5. The fraction will be fifths. Say the multiplication problem and the answer. (Signal.) *5 times 4 equals 20.*
- How many do you add to 20? (Signal.) *3.*
- What's the answer? (Signal.) *23.*
- (Write to show:)

$$\text{b.} \quad 4\frac{3}{5} = \frac{23}{5}$$

- So 4 and 3/5 equals 23/5.
g. Problem C: 1 and 5/9. The fraction will be ninths. Say the multiplication problem and the answer. (Signal.) *9 times 1 equals 9.*
- How many do you add to 9? (Signal.) *5.*
- What's the answer? (Signal.) *14.*
- (Write to show:)

$$\text{c.} \quad 1\frac{5}{9} = \frac{14}{9}$$

- So 1 and 5/9 equals 14/9.
h. Find part 8.
i. Mixed number A is: 6 and 1/3. You'll copy the mixed number and write the fraction it equals. What's the denominator of that fraction? (Signal.) *3.*
- You'll write **3** in the denominator. Then multiply and add to figure out the top number. Raise your hand when you've finished the equation for A. √
- (Write on the board:)

$$\text{a.} \quad 6\frac{1}{3} = \boxed{\frac{19}{3}}$$

- Here's what you should have. 3 times 6 equals 18, plus 1 equals 19. So 6 and 1/3 equals 19/3. Raise your hand if you got it right.

j. Mixed number B: 2 and 4/5. You'll copy the mixed number and write the fraction it equals. What's the denominator of that fraction? (Signal.) *5.*
- You'll write **5** in the denominator. Then multiply and add to figure out the top number. Raise your hand when you've finished equation B. √
- (Write on the board:)

$$\text{b.} \quad 2\frac{4}{5} = \boxed{\frac{14}{5}}$$

- Here's what you should have. 5 times 2 equals 10, plus 4 equals 14. So 2 and 4/5 equals 14/5. Raise your hand if you got it right.
k. Mixed number C: 3 and 2/7. What's the denominator of that fraction? (Signal.) *7.*
- You'll write **7** in the denominator. Then multiply and add to figure out the top number. Raise your hand when you've finished the equation. √
- (Write on the board:)

$$\text{c.} \quad 3\frac{2}{7} = \boxed{\frac{23}{7}}$$

- Here's what you should have. 7 times 3 equals 21, plus 2 equals 23. So 3 and 2/7 equals 23/7. Raise your hand if you got it right.

EXERCISE 6 INDEPENDENT WORK

a. Find part 9 of your independent work.
- The table has missing numbers. Remember how to work the table. First you work any rows that have two numbers. Then you work the columns.
b. Do the independent work for lesson 16.
c. (Before beginning the next lesson, check students' independent work.)

Lesson 17

Objectives

- **Work an area-of-rectangle problem from a diagram.** (Exercise 1)
  Note: Students use the equation
  Area = base x height. Students write three equations, the last of which indicates the number and the unit name:

  A = b x h
  A = 12 x 10
  A = 120 square inches

- Write an equation that shows a mixed number and the fraction it equals. (Exercise 2)

- Rewrite an equation so it begins with the unknown. (Exercise 3)

- Solve a problem that illustrates two values and the difference between them. (Exercise 4)

- **Complete a number-family table.** (Exercise 5)
  Note: To figure all the numbers, students first work all the rows that have two numbers, then all the columns that have two numbers, then any rows or columns that now have two numbers.

- **Answer a set of questions by referring to a number-family table.** (Exercise 6)
  Note: Tables are of the form:

	brown	not brown	total for both colors
horses	18	47	65
cows	115	63	178
total for both animals	133	110	243

Students use the column and row headings to identify the cell that is referred to in each item. For example: What is the number of cows that are not brown?

EXERCISE 1 AREA
Rectangles

a. Open your textbook to lesson 17 and find part 1. I'll read what it says. Follow along: You've found the perimeter of rectangles. You can also find the area. To find the area of rectangles, you multiply the **base** of the rectangle by the **height.** The base is the side that is on the bottom.

- You can make any side the base.
- The second rectangle is turned so that a longer side is the base.
- Everybody, what's the length of the base? (Signal.) *7 feet.*
- Below, the same rectangle is turned so that a shorter side is the base.
- Everybody, what's the length of the base? (Signal.) *5 feet.*
- You can see the equation for finding the area of a rectangle: **Area equals base times height.**
- Below the equation you can see the multiplication with numbers for the base and height.
- The answer is **square** units, not regular units. The area of the figure is **35 square feet.**
- The bottom rectangle has the square feet visible.
- If you count them, you'll see that there are 35 square units inside the figure.
- Remember, the area of figures is always measured in **square** units.

b. Find part 2.
   You're going to find the area of each figure.

c. Figure A. What's the base? (Signal.) *9 centimeters.*

- What's the height? (Signal.) *2 centimeters.*
- Write the equation for the area of a rectangle. You can use initials for **area, base** and **height.** Raise your hand when you've written the equation. √
- (Write on the board:)

  a. A = b x h

- Here's what you should have. Area equals base times height.
- (Write to show:)

  a. A = b x h
  A =

- Below, write the equation with numbers for the **base** and **height.** Remember to show the **times** sign. √
- (Write to show:)

  a. A = b x h
  A = 9 x 2

- Here's the equation.

- (Write to show:)

a. $A = b \times h$
$A = 9 \times 2$
$A =$

- The units for the base and the height are **centimeters** so the units for the area are **square centimeters.** Write the answer and the unit name. Raise your hand when you've completed problem A. √
- (Write to show:)

a. $A = b \times h$
$A = 9 \times 2$
$A = $ 18 square centimeters

- Here's what you should have. The area is 18 square centimeters.
d. Your turn: Find the area for rectangle B. Show **three** equations. Start with the equation for the area. Then put in the numbers you multiply. Then show the answer. Remember the unit in the answer is always a **square** unit. Raise your hand when you're finished.
  (Observe students and give feedback.)
- (Write on the board:)

b. $A = b \times h$
$A = 5 \times 9$
$A = $ 45 square yards

- Here's what you should have. The area of rectangle B is 45 square yards.
e. Your turn: Work problem C. Remember to show three equations. Raise your hand when you're finished.
  (Observe students and give feedback.)
- (Write on the board:)

c. $A = b \times h$
$A = 12 \times 10$
$A = $ 120 square inches

- Here's what you should have. The area for rectangle C is 120 square inches.

EXERCISE 2 MIXED NUMBERS
As Fractions

a. Find part 3.
- Remember how to figure out fractions for mixed numbers. You start with the denominator, **multiply** by the whole number, then **add** the numerator of the fraction.
b. Your turn: Copy mixed number A. Write an equal sign and write the fraction 3 and 7/8 equals. Raise your hand when you're finished. √
- (Write on the board:)

a. $3 \frac{7}{8} = \frac{31}{8}$

- Here's what you should have: The denominator is 8. 8 times 3 equals 24, plus 7 equals 31. 3 and 7/8 equals 31/8.
c. Your turn: Copy mixed number B. Write the fraction it equals. Raise your hand when you're finished. √
- (Write on the board:)

b. $1 \frac{7}{9} = \frac{16}{9}$

- Here's what you should have: 9 times 1 equals 9, plus 7 equals 16. 1 and 7/9 equals 16/9.
d. Your turn: Write fractions for the rest of the mixed numbers in part 3. Raise your hand when you're finished.
  (Observe students and give feedback.)
  (Write on the board:)

c. $6 \frac{2}{3} = \frac{20}{3}$

d. $5 \frac{3}{7} = \frac{38}{7}$

e. Check your work.
- Mixed number C: 6 and 2/3. What fraction does it equal? (Signal.) *20-thirds.*
- Mixed number D: 5 and 3/7. What fraction does it equal? (Signal.) *38-sevenths.*
f. Raise your hand if you got all those fractions right.

EXERCISE 3 EQUATIONS
Solving for an Unknown

a. Find part 4.
- One of the values in each equation is a letter. You'll rewrite the equation so it begins with the letter. You'll work the problem and show the answer by writing a simple equation with the letter and the number it equals.

b. Your turn: Work problem A. Show two equations. Raise your hand when you're finished. (Observe students and give feedback.)
- Read the first equation you wrote. (Signal.) *R equals 18 times 9.*
- Read the simple equation. (Signal.) *R equals 162.*

c. Your turn: Work the rest of the problems in part 4. Raise your hand when you're finished. (Observe students and give feedback.)

Key:

b. $D = 850 - 196$
$$\boxed{D = 654}$$

c. $M = 117 + 685$
$$\boxed{M = 802}$$

d. Check your work.
- Problem B. Read the first equation you wrote. (Signal.) *D equals 850 minus 196.*
- Read the simple equation. (Signal.) *D equals 654.*
- Problem C. Read the first equation you wrote. (Signal.) *M equals 117 plus 685.*
- Read the simple equation. (Signal.) *M equals 802.*

EXERCISE 4 NUMBER FAMILIES
Comparison

a. Find part 5.
- You're going to make a number family for each diagram. You'll show two letters and **d-i-f** for the difference. Then you'll put in the two numbers the problem gives and figure out the third number.

b. Work problem A. Raise your hand when you're finished. (Observe students and give feedback.)

c. (Write on the board:)

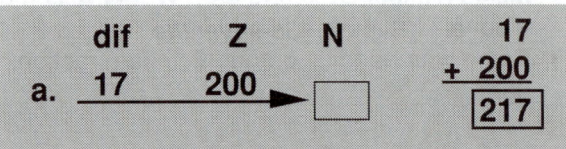

- Here's what you should have. The names are: N, Z and **difference.** N is the big number. The problem gives numbers for Z and the difference. You figured out N by working the problem: 200 plus 17. The answer is 217.

d. Work the rest of the problems in part 5. Raise your hand when you're finished. (Observe students and give feedback.)

e. (Write on the board:)

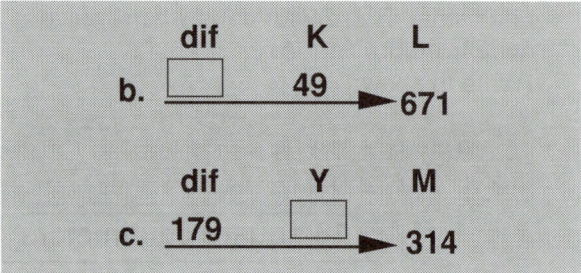

- Here are the families you should have for problems B and C.
- For problem B you figured out the difference number by working the problem: 671 minus 49. The answer is 622.
- For problem C, you figured out small number Y by working the problem: 314 minus 179. The answer is 135.
- Raise your hand if you got everything right.

EXERCISE 5 NUMBER FAMILIES
Tables

a. Find part 6.
- This table is harder than the ones you've worked. After you've worked all the rows and all the columns that you can work, you'll find that there are still missing numbers in the table.

b. Work all the rows you can work. Then work all the columns you can work. Raise your hand when you've done that much. (Observe students and give feedback.)
- (Write on the board:)

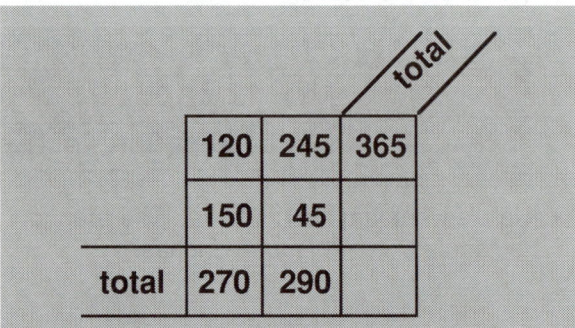

- Here's what you should have so far.
- There are still two missing numbers, one in the second row and one in the third row.

- Can you work the problem for the middle row now? (Signal.) *Yes.*
- Can you work the problem for the bottom row now? (Signal.) *Yes.*

c. Work those problems. Raise your hand when you're finished.
(Observe students and give feedback.)
- (Write to show:)

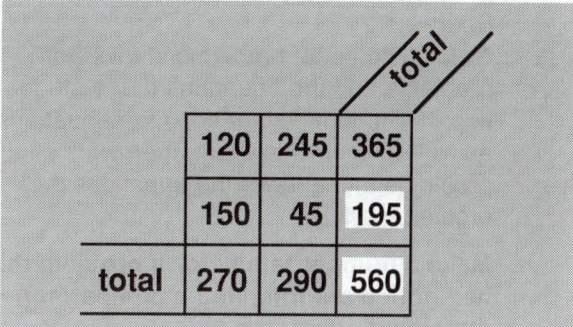

		total
120	245	365
150	45	195
total	270	290	560

- Here's what you should have.
- Check your work.
- For the middle row, you added 150 and 45. The missing number is 195. Raise your hand if you did that.
- You could have added 270 and 290 for the **bottom row.** Or you could have added the numbers in the **last column**—365 and 195 to figure out the **total** total of 560. Raise your hand if you got it right.

d. Remember how you worked this table. You worked all the rows you could work. Then you worked all the columns you could work. Then you looked for a row or a column with two numbers. You worked each row or column that had two numbers.

EXERCISE 6 DATA ANALYSIS
Tables

a. Find part 7.
(Teacher reference:)

	brown	not brown	total for both colors
horses	18	47	65
cows	115	63	178
total for both animals	133	110	243

- You've worked with tables that have totals for the rows and totals for the columns. There are similar tables that have names for each row and column.

b. Look at table A. This table shows the number of horses and cows on a ranch.
- Touch the name for the top row. √
The name is **horses.** What's the name for the middle row? (Signal.) *Cows.*
- What's the name for the bottom row? (Signal.) *Total for both animals.*
- Now touch the first **column.** What's the name? (Signal.) *Brown.*
- What's the name for the next column? (Signal.) *Not brown.*
- Listen: You can find the number for cows that are not brown. You go to the **row** for **cows,** then you go to the column for **not brown.** The cell where the row and column meet shows the number for cows that are not brown. √
- Everybody, touch the number for cows that are not brown.
Everybody, what's the number? (Signal.) *63.*
- Listen: You can find the number for horses that are brown. You start with the row for horses and you go down the column for brown. The cell where that row and that column meet shows the number for horses that are brown.
- Find the number for brown horses. √
- Everybody, what's the number? (Signal.) *18.*

c. Touch table B. √
- (Teacher reference:)

	lot A	lot B	total for both lots
cars	132	125	257
trucks	16	64	80
total for both vehicles	148	189	337

- This table shows the number of cars and trucks on two different lots—lot A and lot B.

d. Touch the first **column.** √
- That shows the number of cars and trucks on lot A.
- How many **cars** are on lot A? (Signal.) *132.*
- How many **trucks** are on lot A? (Signal.) *16.*
- How many total **vehicles** are on lot A? (Signal.) *148.*
- Touch the second column. √
- That shows the number of cars and trucks on lot B.
- How many **vehicles** are on lot B? (Signal.) *189.*
- Look at the rows. The top **row** shows **cars.** There are 132 cars on lot A and 125 cars on lot B.
- What's the total number of cars for both lots? (Signal.) *257.*
- The next row shows **trucks** on lot A and lot B.

- How many trucks are on lot A? (Signal.) *16.*
- How many trucks are on lot B? (Signal.) *64.*
- What's the total number of trucks for both lots? (Signal.) *80.*
- (Repeat step d until firm.)

e. Your turn: Write answers to items A through E. Raise your hand when you're finished. (Observe students and give feedback.)

f. Check your work. Tell me the answer to each question.
- Item A: What's the total number of cars for both lots? (Signal.) *257.*
- Item B: How many vehicles are on lot B? (Signal.) *189.*
- Item C: How many trucks are on lot B? (Signal.) *64.*
- Item D: How many trucks are on lot A? (Signal.) *16.*
- Item E: What's the total number of vehicles for both lots? (Signal.) *337.*

EXERCISE 7 INDEPENDENT WORK

a. Do the independent work for lesson 17.
b. (Before beginning the next lesson, check students' independent work.)

Lesson 18

Objectives

- Answer a set of questions by referring to a number-family table. (Exercise 1)

- **Identify whether equations for ratio problems are appropriately written.** (Exercise 2)
  Note: The set of problems shows word problems and the equations that a student wrote. Students either copy the equations as written or rewrite them, then work all the problems and answer the questions the problems ask.

- **Make a number family for a problem that has both a diagram and a comparison statement.** (Exercise 3)
  Note: Problems are of the form:

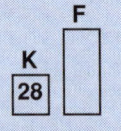

F is 31 more than K.

Students write: <u>31</u> <u>28</u> ⟶ ☐, and

figure out the missing value.

- Complete a number-family table. (Exercise 4)

- Work an area-of-rectangle problem from a diagram. (Exercise 5)

- Write an equation that shows a mixed number and the fraction it equals. (Exercise 6)

EXERCISE 1 DATA ANALYSIS
Tables

a. Open your textbook to lesson 18 and find part 1.
- This table shows the number of two different kinds of animals that were observed on two beaches. First, animals. What kinds of animals were they? (Signal.) *Penguins and seals.*
- Those are the column names—**penguins** and **seals.**
- Now, the names for the beaches. What were the two beaches? (Signal.) *Black Beach and Surf Beach.*
- Those are the names for the rows.
- Touch the number 47. √
- That number tells about one of the animals. Which animal? (Signal.) *Penguins.*
- That number also tells about one of the beaches. Which beach? (Signal.) *Black Beach.*

- Yes, the number 47 tells about the number of penguins on Black Beach.
- Touch the number 203. √
- That's the total for one of the **beaches.** Which beach? (Signal.) *Surf Beach.*
- Yes, the total number of penguins **and** seals on Surf Beach was 203.
- Touch the number 380. √
- That's the total for one of the **animals.** Which animal? (Signal.) *Seals.*

b. Your turn: Write answers to questions A through D. Raise your hand when you've finished question D.
(Observe students and give feedback.)

c. Check your answers.
- Question A asks about the number of seals for both beaches. What's the number? (Signal.) *380.*
- There were **380 seals** for both beaches.
- Question B asks about the number of penguins on Surf Beach. What's the number? (Signal.) *108.*
- **108 penguins** were observed on Surf Beach.
- Question C asks about the total number of both animals on Black Beach. What number? (Signal.) *332.*
- There were **332 animals** on Black Beach.
- Question D asks about the beach with fewer seals. Which beach? (Signal.) *Surf Beach.* There were 95 seals on Surf Beach and 285 seals on Black Beach.

d. Items E through G tell about a number. You have to **name the animal and name the place.**

e. Item E says: 108 tells about the number of BLANK on BLANK. Find the cell with 108 in it. Write the name that goes in the first blank. Then write a comma and the name that goes in the second blank. Raise your hand when you've completed item E. √
- You should have written the names **penguins** and **Surf Beach.** 108 tells about the number of **penguins** on **Surf Beach.** Raise your hand if you got it right.

f. Do item F. Raise your hand when you've written the names for the blanks.
- Item F says that 95 tells about the number of BLANK on BLANK. The two names are **seals** and **Surf Beach.** 95 tells the number of **seals** on **Surf Beach.**

g. Do item G. Raise your hand when you're finished. √
- Item G says: 332 tells about the number of BLANK on BLANK. For the first blank, you should have written **both animals.** For the second blank, you should have written **Black Beach.** 332 tells about the number of **both animals** on **Black Beach.**

EXERCISE 2 PROBLEM SOLVING
Ratio Equation Workcheck

a. Find part 2.
- Each item shows a word problem and the ratio equation a student wrote.
- The student did not write all the equations correctly.

b. Don't work the problems. Read each problem. Check the student's equation. If the equation is correct, copy it. If the equation is not correct, write the correct equation.
- Be careful. The student may have made a mistake with either the first fraction or the last fraction. Raise your hand when you're finished.
(Observe students and give feedback.)

Key:

a. $\dfrac{women}{men} \dfrac{2}{5} = \dfrac{48}{\blacksquare}$

b. $\dfrac{cows}{bulls} \dfrac{3}{1} = \dfrac{\blacksquare}{15}$

c. $\dfrac{buttonholes}{seconds} \dfrac{6}{17} = \dfrac{\blacksquare}{85}$

d. $\dfrac{leaves\ on\ the\ ground}{leaves\ in\ the\ tree} \dfrac{2}{7} = \dfrac{\blacksquare}{308}$

c. You should have copied equations A and D. You should have rewritten equations B and C.
- Here's what you should have for B and C.

d. Problem B: The ratio of cows to bulls on the farm was 3 to 1. If there were 15 bulls, how many cows were there?
- The student wrote the first fraction incorrectly. The 3 goes on top, not on the bottom.
- Raise your hand if you caught that mistake.
- Your equation should be: 3 over 1 equals box over 15.

e. Problem C: The machine made 6 buttonholes every 17 seconds. How many buttonholes did the machine make in 85 seconds?
- The student wrote the second fraction incorrectly. The box goes on top; 85 goes on the bottom.

f. Make sure you have the correct equations. Then, work all the problems in part 2. Remember, the answer is a number and a unit name. Raise your hand when you're finished.
(Observe students and give feedback.)

Key:

a. $\dfrac{\text{women}}{\text{men}}\ \dfrac{2}{5}\left(\dfrac{24}{24}\right)=\dfrac{48}{\boxed{120}}\ \boxed{120\ men}$

b. $\dfrac{\text{cows}}{\text{bulls}}\ \dfrac{3}{1}\left(\dfrac{15}{15}\right)=\dfrac{\boxed{45}}{15}\ \boxed{45\ cows}$

c. $\dfrac{\text{buttonholes}}{\text{seconds}}\ \dfrac{6}{17}\left(\dfrac{5}{5}\right)=\dfrac{\boxed{30}}{85}\ \boxed{30\ buttonholes}$

d. $\dfrac{\substack{\text{leaves on}\\ \text{the ground}}}{\substack{\text{leaves in}\\ \text{the tree}}}\ \dfrac{2}{7}\left(\dfrac{44}{44}\right)=\dfrac{\boxed{88}}{308}\ \boxed{88\ leaves}$

g. Check your work.
- Find part J on page 69. That shows what you should have for each problem.

h. Raise your hand if you got everything right.

EXERCISE 3 NUMBER FAMILIES
Comparison

a. Find part 3.
- These are problems that show two things. The difference is not shown. The statement under each problem tells the difference number or asks about the difference. You're going to make number families with two letters and the name **difference.**

b. Problem A. Which is the big number? (Signal.) *F.*
- Make the family with the names F, K and **difference.** Don't write any numbers. Raise your hand when you've done that much.
  (Observe students and give feedback.)
- (Write on the board:)

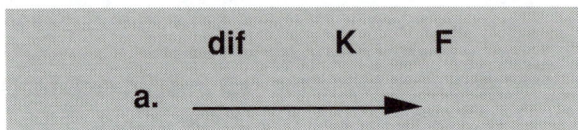

- Here's what you should have so far. F is more, so F is the name for the big number. K is the name for a small number.

c. The diagram shows a number for K. The statement under the diagram says: F is 31 more than K. The sentence names both F and K. That sentence tells about the **difference.** The difference is 31.
- Put the two numbers in your family. Then figure out the value for F. Raise your hand when you're finished.
  (Observe students and give feedback.)

d. (Write to show:)

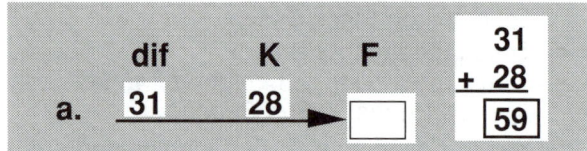

- Here's what you should have. The difference is 31. K is 28. You figure out the big number by adding 31 and 28. The answer is 59. That's F.

e. Problem B. The question under the diagram asks: How much less is M than R? That question asks about the difference, so you need a number family with the name **difference.** Make the family. Put in the numbers you know. Figure out the missing number. Raise your hand when you're finished.
  (Observe students and give feedback.)

f. (Write on the board:)

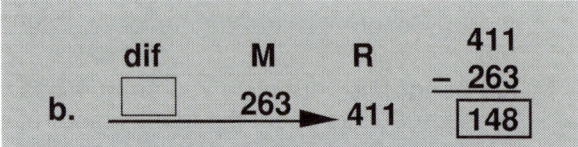

- Here's what you should have. The big number is R. The problem gives a number for M and a number for R. You figure out the difference number by working the problem: 411 minus 263. The answer is 148.

g. Work the rest of the problems in part 3. Remember each family needs the name **difference.** Raise your hand when you're finished. (Observe students and give feedback.)

h. Check your work.
- (Write on the board:)

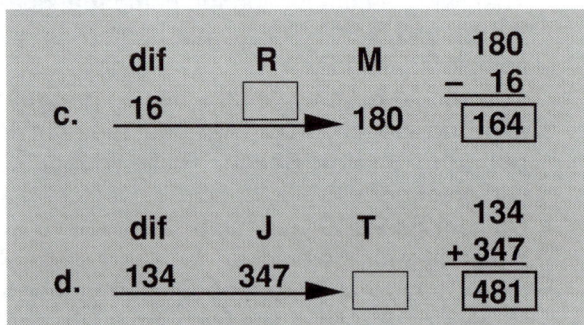

- Here are the number families you should have for items C and D.
- Item C: How much is R? (Signal.) *164.*
- Item D: How much is T? (Signal.) *481.*

i. Raise your hand if you got everything right.

EXERCISE 4 NUMBER FAMILIES
Tables

a. Find part 4.
- This is a hard table. Remember, first complete all the rows that have two numbers. Then complete all the columns that have two numbers. Then go back to the rows or columns and complete any that have two numbers. Keep going until you've figured out all the missing numbers.

b. Work the table. Raise your hand when you're finished. (Observe students and give feedback.)

Key:

		total
126	22	148
	53	101	154
total	179	123	302

- (Write on the board:)

		total
126		
		154
total	179	123	302

- Here's what you get when you work the bottom row. There are still lots of missing numbers. Everybody, what's the missing middle number in the first column? (Signal.) *53.*
- (Write **53.**)
- What's the missing top number in the last column? (Signal.) *148.*
- (Write **148.**)
- Now, there are two numbers in the top row. So you can figure out the missing middle number. Everybody, what's the number? (Signal.) *22.*
- (Write **22.**)
- Now there are two numbers in the middle row. So you can figure out the missing middle number. Everybody, what number? (Signal.) *101.*
- (Write **101.**)

c. All done. Raise your hand if you got everything right.
- That's pretty good work!

EXERCISE 5 AREA
Rectangles

a. Find part 5.
- Remember the equation for figuring out the area of rectangles. The **area** equals the **base** times the **height.** The units in the answer are **square** units.
- You'll write three equations. You'll start with the equation: **area equals base times height.** Use letters for the words. Below, write an equation with numbers. Below that equation, write the equation with the answer. Remember, the answer has a number and a unit name.

b. Figure out the area of rectangle A. Raise your hand when you've finished item A.
   (Observe students and give feedback.)
- (Write on the board:)

> **a. A = b x h**
> **A = 18 x 10**
> **A = 180 square inches**

- Here's what you should have. Check your work. Rectangle A has a base of 18 inches and a height of 10 inches. The area is 180 **square inches.**

c. Your turn: Find the area for problems B and C in part 5. Raise your hand when you're finished. (Observe students and give feedback.)
- (Write on the board:)

> **b. A = b x h**
> **A = 9 x 30**
> **A = 270 square yards**
>
> **c. A = b x h**
> **A = 12 x 14**
> **A = 168 square centimeters**

- Here's what you should have.

EXERCISE 6 MIXED NUMBERS
As Fractions

a. Find part 6.
- You're going to write equations that show each mixed number and the fraction it equals.

b. Write the equation for item A. Raise your hand when you're finished.
   (Observe students and give feedback.)

- (Write on the board:)

$$\text{a.} \quad 1\frac{6}{7} = \boxed{\frac{13}{7}}$$

- Here's what you should have. 1 and 6/7 equals 13/7.
c. Write equations for the rest of the items in part 6. Raise your hand when you're finished. (Observe students and give feedback.)
- (Write on the board:)

$$\text{a.} \quad 1\frac{6}{7} = \boxed{\frac{13}{7}} \qquad \text{d.} \quad 3\frac{1}{4} = \boxed{\frac{13}{4}}$$

$$\text{b.} \quad 2\frac{1}{6} = \boxed{\frac{13}{6}} \qquad \text{e.} \quad 4\frac{1}{3} = \boxed{\frac{13}{3}}$$

$$\text{c.} \quad 2\frac{3}{5} = \boxed{\frac{13}{5}}$$

- Here are the equations you should have.
- Equation B: 2 and 1/6 equals 13/6.
- Equation C: 2 and 3/5 equals 13/5.
- Equation D: 3 and 1/4 equals 13/4.
- Equation E: 4 and 1/3 equals 13/3.
d. Raise your hand if you got everything right.

EXERCISE 7 INDEPENDENT WORK

a. Do the independent work for lesson 18.
b. (Before beginning the next lesson, check students' independent work.)

Lesson 19

Objectives

- Answer a set of questions by referring to a number-family table. (Exercise 1)
- Complete a pair of equivalent fractions. (Exercise 2)
- Make a number family for a problem that has both a diagram and a comparison statement. (Exercise 3)
- Write and solve a ratio equation for a word problem. (Exercise 4)
- **Compute both the area and the perimeter of a rectangle.** (Exercise 5)

EXERCISE 1 DATA ANALYSIS
Tables

a. Open your textbook to lesson 19 and find part 1. (Teacher reference:)

	cows	bulls	total for both animals
Blue Ranch	618	189	807
Russell Ranch	285	376	661
total for both ranches	903	565	1468

- This table shows the number of two kinds of animals at two different places. What are the two types of animals? (Signal.) *Cows and bulls.*
- What are the two places? (Signal.) *Blue Ranch and Russell Ranch.*
b. Answer questions A through C in part 1. Raise your hand when you've finished question C. (Observe students and give feedback.)
c. Check your answers.
- Question A asks about the number of cows on Russell Ranch. What's the number? (Signal.) *285.*
- Question B asks about the ranch that has more bulls. Which ranch? (Signal.) *Russell Ranch.* You should have written the name **Russell** or **Russell Ranch.**
- Question C asks about the total number of bulls for both ranches. What's the number? (Signal.) *565.*

d. Items D through F ask about the names for a number. For each item, write the names that go in each blank. You don't have to copy the whole item. Just write the missing names. Raise your hand when you're finished.
(Observe students and give feedback.)
e. Check your work.
• Item D says: 376 tells about the number of BLANK on BLANK. You should have written **bulls** and **Russell Ranch.** There are 376 bulls on Russell Ranch.
• Item E: 903 tells about the number of BLANK on BLANK. You should have written **cows** and **both ranches.** The number 903 tells about the number of cows for both ranches.
• Item F: 189 tells about the number of BLANK on BLANK. You should have written **bulls** and **Blue Ranch.** There are 189 bulls on Blue Ranch.
f. Raise your hand if you got everything right.

EXERCISE 2 FRACTIONS
Equivalence

a. Find part 2.
• These are equivalent-fraction problems that you probably can't figure out in your head.
• You work the problem by saying a division problem for either the numerators or the denominators. The answer to that problem tells you about the fraction that equals 1. If the answer to the division problem is 100, the fraction that equals 1 is 100/100.
• What if the answer to the division problem is 192? What's the fraction that equals 1? (Signal.) *192 over 192.*
b. Problem A: 2/5 equals some value over 940. You can work a division problem for the denominators. Say that problem. (Signal.) *940 divided by 5.*
• When you work that problem, the answer is 188. So, the fraction that equals 1 is 188/188.
• Your turn: Copy the problem and work it. Write the fraction that equals 1 and figure out the missing value in the last fraction. You can use your calculator. Raise your hand when you're finished. (Observe students and give feedback.)
• (Write on the board:)

$$\text{a.} \quad \frac{2}{5} \left(\frac{188}{188} \right) = \frac{\boxed{376}}{940}$$

• Here's what you should have. The fraction that equals 1 is 188/188. When you multiply on top, you get 376. So 2/5 equals 376/940.

c. Problem B: Say the division problem for the denominators. (Signal.) *1022 divided by 14.*
• Copy the problem and work it. Raise your hand when you're finished.
(Observe students and give feedback.)
• (Write on the board:)

$$\text{b.} \quad \frac{3}{14} \left(\frac{73}{73} \right) = \frac{\boxed{219}}{1022}$$

• Here's what you should have. 1022 divided by 14 is 73. So the fraction that equals 1 is 73/73. When you multiply on top, you get 219. 3/14 equals 219/1022.
d. Your turn: Work the rest of the problems in part 2. Raise your hand when you're finished.
(Observe students and give feedback.)
e. (Write on the board:)

$$\text{c.} \quad \frac{8}{7} \left(\frac{91}{91} \right) = \frac{728}{\boxed{637}}$$

$$\text{d.} \quad \frac{15}{4} \left(\frac{31}{31} \right) = \frac{465}{\boxed{124}}$$

• Here's what you should have for problems C and D.
• Problem C: 637 is the missing number.
• Problem D: 124 is the missing number.
f. Raise your hand if you got everything right.

EXERCISE 3 NUMBER FAMILIES
Comparison

a. Find part 3.
• These are problems that tell about the difference number or ask about the difference number. Remember, your number family must have the name for **difference.**
b. Make families for the problems in part 3. Figure out the missing numbers and answer the questions. Raise your hand when you're finished. (Observe students and give feedback.)

Key:

a. dif □ — Z 103 → 188   K: $\begin{array}{r} 188 \\ -\ 103 \\ \hline \boxed{85} \end{array}$

b. dif 68 — J 511 → □   M: $\begin{array}{r} 68 \\ +\ 511 \\ \hline \boxed{579} \end{array}$

c. dif 17 — F □ → 318   P: $\begin{array}{r} 318 \\ -\ 17 \\ \hline \boxed{301} \end{array}$

d. dif 56 — H □ → 106   K: $\begin{array}{r} 106 \\ -\ 56 \\ \hline \boxed{50} \end{array}$

c. Find part J on page 72. That shows what you should have for each family.

d. Raise your hand if you got everything right.

EXERCISE 4 PROBLEM SOLVING
Ratios and Proportions

a. Find part 4.

• You're going to work ratio problems. You write the names and a pair of equivalent fractions. You figure out the missing number and answer the question the problem asks.

b. Work problem A. Remember, first write the complete ratio equation with names. Then figure out the missing number and answer the question. Raise your hand when you're finished. (Observe students and give feedback.)

• (Write on the board:)

a. $\dfrac{\text{tall trees}}{\text{short trees}}\ \dfrac{2}{7}\left(\dfrac{60}{60}\right)=\dfrac{120}{\boxed{420}}$ $\boxed{\text{420 short trees}}$

• Here's what you should have. The answer to the question is **420 short trees.**

• Raise your hand if you had the correct equation and the answer, 420 short trees.

c. Your turn: Work problem B. Raise your hand when you're finished. (Observe students and give feedback.)

• (Write on the board:)

b. $\dfrac{\text{yards}}{\text{turns}}\ \dfrac{8}{3}\left(\dfrac{16}{16}\right)=\dfrac{\boxed{128}}{48}$ $\boxed{\text{128 yards}}$

• Here's what you should have. The names are **yards** and **turns.** The tire travels 128 yards in 48 turns.

• Raise your hand if you had the correct equation and the answer, 128 yards.

d. Work problem C. Raise your hand when you're finished. (Observe students and give feedback.)

• (Write on the board:)

c. $\dfrac{\text{girls}}{\text{students}}\ \dfrac{4}{9}\left(\dfrac{21}{21}\right)=\dfrac{\boxed{84}}{189}$ $\boxed{\text{84 girls}}$

• Here's what you should have. The names are **girls** and **students.** There are 84 girls in Newton School.

• Raise your hand if you had the correct equation and the answer, 84 girls.

EXERCISE 5 PERIMETER AND AREA
Rectangles

a. Find part 5.

• For each rectangle, you're going to find the **perimeter and the area.** Remember, the perimeter is just the distance around. You add up the lengths of the sides. The unit in the answer is a regular unit of length.

• The area is different. You multiply to find the number of **square units.** If the sides are measured in inches, the name in the answer is **square inches.**

b. (Write on the board:)

P =

• When you find the perimeter, write **P equals.** Show the number and the unit name.

c. Rectangle A. First figure out the perimeter. Raise your hand when you've done that much. (Observe students and give feedback.)

• (Write on the board:)

a. $\begin{array}{r} 7 \\ 7 \\ 5 \\ +\ 5 \\ \hline \boxed{P = 24 \text{ inches}} \end{array}$

• The perimeter is 24 inches.

• Now figure out the area. Remember to show the three equations for finding the area. Raise your hand when you're finished. (Observe students and give feedback.)

- (Write to show:)

a.
```
   7       A = b x h
   7       A = 5 x 7
   5       A = 35 square inches
 + 5
 P = 24 inches
```

- Here's what you should have. The area of rectangle A is 35 square inches.
d. Find the perimeter and the area for rectangle B. Remember, your answer shows what P equals and what A equals. Raise your hand when you're finished.
  (Observe students and give feedback.)
- (Write on the board:)

b.
```
   12      A = b x h
   12      A = 12 x 12
   12      A = 144 square yards
 + 12
 P =  48 yards
```

- Here's what you should have for rectangle B. The perimeter is 48 yards. The area is 144 square yards.
e. Your turn: Work item C. Raise your hand when you're finished.
  (Observe students and give feedback.)
- (Write on the board:)

c.
```
   9       A = b x h
   9       A = 20 x 9
   20      A = 180 square centimeters
 + 20
 P =  58 centimeters
```

- Here's what you should have. The perimeter of rectangle C is 58 centimeters and the area is 180 square centimeters.
f. Raise your hand if you got all of them right.

EXERCISE 6 INDEPENDENT WORK

a. Do the independent work for lesson 19.
b. (Before beginning the next lesson, check students' independent work.)

Lesson 20 – Test 2

Objectives

- Write and solve a ratio equation for a word problem. (Exercise 1)

- Identify whether equations for ratio problems are appropriately written. (Exercise 2)

- Perform on a mastery test of skills presented in lessons 11 through 19. (Exercise 3)

Note: Exercise 4 provides instructions for marking the test.

EXERCISE 1 PROBLEM SOLVING
Ratios and Proportions

a. Open your textbook to lesson 20 and find part 1.
- These are ratio problems. For each problem, you write the names and a pair of equivalent fractions. You figure out the missing number and write the answer to the question.
b. Work all the problems in part 1. You can use your calculator. Raise your hand when you're finished. (Observe students and give feedback.)

Key:

a. $\dfrac{red\ books}{blue\ books}\ \dfrac{6}{7}\left(\dfrac{71}{71}\right)=\dfrac{\boxed{426}}{497}$ 426 red books

b. $\dfrac{ducks}{geese}\ \dfrac{5}{4}\left(\dfrac{18}{18}\right)=\dfrac{\boxed{90}}{72}$ 90 ducks

c. $\dfrac{yards}{rods}\ \dfrac{11}{2}\left(\dfrac{22}{22}\right)=\dfrac{\boxed{242}}{44}$ 242 yards

d. $\dfrac{rabbits}{carrots}\ \dfrac{8}{3}\left(\dfrac{6}{6}\right)=\dfrac{48}{\boxed{18}}$ 18 carrots

c. Check your work.
- Find part J on page 74. That shows what you should have for each problem.
- Problem A: How many red books did the library have? (Signal.) *426.*
- Problem B: How many ducks lived in the marsh? (Signal.) *90.*
- Problem C: If something is 44 rods long, how many yards long is that object? (Signal.) *242.*
- Problem D: If there are 48 rabbits, how many carrots are there? (Signal.) *18.*
d. Raise your hand if you got everything right.

EXERCISE 2 PROBLEM SOLVING
Ratio Equation Workcheck

a. Find part 2.
- Each item shows a word problem and the ratio equation a student wrote.

- The student did not write all the equations correctly.
b. Don't work the problems. Read each problem. Check the student's equation. If the equation is correct, copy it. If the equation is not correct, write the correct equation.
- Be careful. The student may have made a mistake with either the first fraction or the last fraction. Raise your hand when you've written the correct equation for all the problems.
  (Observe students and give feedback.)
c. (Write on board:)

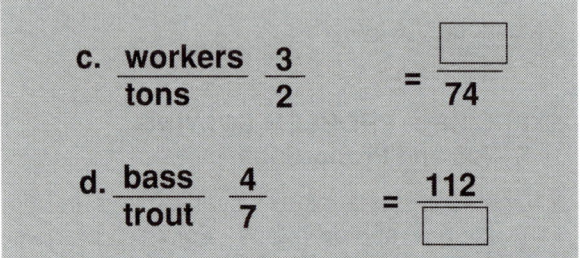

c. $\dfrac{\text{workers}}{\text{tons}} \quad \dfrac{3}{2} \quad = \quad \dfrac{\boxed{}}{74}$

d. $\dfrac{\text{bass}}{\text{trout}} \quad \dfrac{4}{7} \quad = \quad \dfrac{112}{\boxed{}}$

You should have copied equations A and B. You should have rewritten equations C and D.
d. Problem C: If it takes 3 workers to move 2 tons, how many workers are needed to move 74 tons? The student wrote the last fraction incorrectly. 74 is the number for tons, so 74 goes on the bottom.
- Raise your hand if you caught that mistake.
e. Problem D: In a lake, the ratio of bass to trout is 4 to 7. There are 112 bass. How many trout are in the lake?
- The student wrote the first fraction incorrectly. The ratio is 4 to 7. The 4 goes on top and the 7 goes on the bottom.
f. Work all the problems in part 2. Remember, the answer is a number and a unit name. Raise your hand when you're finished.
  (Observe students and give feedback.)
  Key:

  a. $\dfrac{\text{fish}}{\text{birds}} \dfrac{3}{4} \left(\dfrac{200}{200}\right) = \dfrac{600}{\boxed{800}}$ $\boxed{800 \text{ birds}}$

  b. $\dfrac{\text{bricks}}{\text{pounds}} \dfrac{2}{9} \left(\dfrac{28}{28}\right) = \dfrac{\boxed{56}}{252}$ $\boxed{56 \text{ bricks}}$

  c. $\dfrac{\text{workers}}{\text{tons}} \dfrac{3}{2} \left(\dfrac{37}{37}\right) = \dfrac{\boxed{111}}{74}$ $\boxed{111 \text{ workers}}$

  d. $\dfrac{\text{bass}}{\text{trout}} \dfrac{4}{7} \left(\dfrac{28}{28}\right) = \dfrac{112}{\boxed{196}}$ $\boxed{196 \text{ trout}}$

g. Check your work.
- Find part K. That shows what you should have for each problem.

- Problem A: How many birds are in the swamp? (Signal.) *800.*
- Problem B: How many bricks weigh 252 pounds? (Signal.) *56.*
- Problem C: How many workers are needed to move 74 tons? (Signal.) *111.*
- Problem D: How many trout are in the lake? (Signal.) *196.*
h. Raise your hand if you got everything right.

EXERCISE 3 TEST 2

> *Note:* **Students are not to use calculators for any part of the test.**

a. The rest of the lesson is a test. You should only have your textbook, a sharpened pencil and lined paper on your desk.
b. Find part 1 of test 2 in your textbook. √
c. Do the test on your own. Raise your hand when you're finished.

EXERCISE 4 MARKING THE TEST

a. (Collect the students' papers. Use the Answer Key to score tests. Award scores for test 2 as follows:)

Test 2 Percent Summary					
SCORE	%	SCORE	%	SCORE	%
67	100	60	90	53	79
66	99	59	88	52	78
65	97	58	87	51	76
64	96	57	85	50	75
63	94	56	84	49	73
62	93	55	82	48	72
61	91	54	81	47	70

b. (Complete the Test 2 Remedy Summary to determine whether remedies are needed. Reproducible Summary Sheets are at the back of the Teacher's Guide.)
- (If more than 1/4 of the students did not pass a test part, present the remedy for that part before beginning lesson 21. Remedies appear at the end of the Test 2 Answer Key.)

Lesson 21

EXERCISE 1 FRACTIONS
Equivalence

a. Open your textbook to lesson 21 and find part 1.
- These are equivalent-fraction problems that have big numbers.
b. Work problem A. Raise your hand when you're finished. (Observe students and give feedback.)
- (Write on the board:)

$$\text{a.} \quad \frac{7}{15} \left(\frac{121}{121} \right) = \frac{847}{1815}$$

- Here's what you should have. You worked the division problem 847 divided by 7. That's 121. So the fraction that equals 1 is 121/121. The missing number for the equivalent fraction is 1815.
c. Work problem B. Raise your hand when you're finished. (Observe students and give feedback.)
- (Write on the board:)

$$\text{b.} \quad \frac{10}{3} \left(\frac{223}{223} \right) = \frac{2230}{669}$$

- Here's what you should have. You worked the division problem 669 divided by 3. That's 223. So the fraction that equals 1 is 223/223. The missing number for the equivalent fraction is 2230.

d. Work the rest of the problems in part 1. Raise your hand when you're finished. (Observe students and give feedback.)
e. (Write on the board:)

$$\text{c.} \quad \frac{7}{8} \left(\frac{80}{80} \right) = \frac{560}{640}$$

$$\text{d.} \quad \frac{5}{17} \left(\frac{125}{125} \right) = \frac{625}{2125}$$

- Here's what you should have for problems C and D.
  Problem C: The missing number for the equivalent fraction is 560.
  Problem D: The missing number for the equivalent fraction is 2125.
- Raise your hand if you got everything right.

EXERCISE 2 NUMBER FAMILIES
Comparison

a. Find part 2.
- The information under each diagram tells about two numbers and gives the unit name. One of the sentences asks or tells about the difference number.
b. Problem A: The information under the diagram says: Bar T is 143 meters long. Bar M is 112 meters shorter than bar T. That statement tells about the difference number.
- Make the number family. Put in the numbers you know. Figure out the missing number. Remember to write the unit name in your answer. Raise your hand when you're finished. (Observe students and give feedback.)
- (Write on the board:)

$$\text{a.} \quad \xrightarrow[\quad 112 \quad]{\text{dif} \quad \text{M} \quad \text{T}} 143 \qquad \begin{array}{r} 143 \\ -\ 112 \\ \hline \boxed{31 \text{ meters}} \end{array}$$

- Here's what you should have for problem A. The difference is 112. T is the big number. You figured out M by working the problem: 143 minus 112. The answer is 31 meters.
c. Your turn. Work the rest of the problems in part 2. Raise your hand when you're finished. (Observe students and give feedback.)

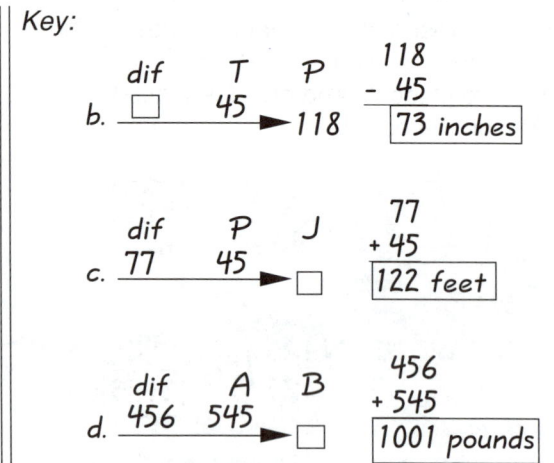

b. dif ☐ 45 → 118 | T 45 → 118 | P 118 − 45 = **73 inches**

c. dif 77 45 → ☐ | P 45 | J 77 + 45 = **122 feet**

d. dif 456 545 → ☐ | A 545 | B 456 + 545 = **1001 pounds**

d. Find part J on page 81. That shows the families you should have for problems B, C and D.
- Problem B: The difference number is missing. You worked the problem 118 minus 45. The answer is 73 inches.
- Problem C: The number for J is missing. You worked the problem 77 plus 45. The answer is 122 feet.
- Problem D: The number for pile B is missing. You worked the problem 456 plus 545. The answer is 1001 pounds.
- Raise your hand if you got everything right.

EXERCISE 3 GEOMETRY
Parallel/Intersecting Lines

a. Find part 3.
- (Teacher reference:)

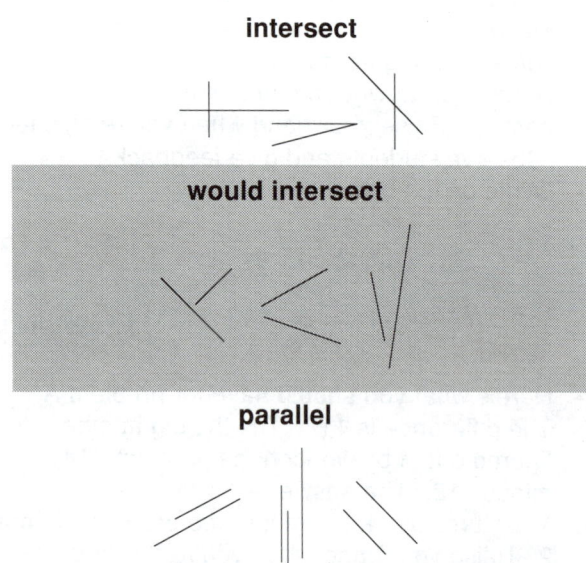

intersect

would intersect

parallel

- I'll read what it says. Follow along: Some lines **intersect.** Those are lines that touch each other or cross each other.
- You can see pairs of lines that intersect.

- Some lines do not intersect, but they would intersect if the lines were extended.
- You can see pairs of lines that would intersect if they were extended.
- Some lines are **parallel.** Those lines would never intersect no matter how long you made them. Parallel lines are always the same distance from each other.
- You can see groups of lines that are parallel. So they do not intersect.

b. Find part 4.
Each picture shows groups of lines. Some of the lines are parallel and some are not parallel. The parallel lines would never intersect no matter how far you followed them.

c. Item A. Are the lines parallel? (Signal.) *No.*
- They are not parallel. Are they intersecting in the picture? (Signal.) *No.*
- You can see that they **would** intersect if you made them longer.
- Item B. Are the lines parallel? (Signal.) *Yes.*
- Those are parallel lines. They would not intersect no matter how long you made them.
- Item C. Are the lines parallel? (Signal.) *No.*
- They are not parallel. Are the lines intersecting in the picture? (Signal.) *Yes.*
- Item D. Are the lines parallel? (Signal.) *No.*
- Are they intersecting in the picture? (Signal.) *No.*
- Would they intersect if you made them longer? (Signal.) *Yes.*

d. Your turn: Write **parallel** or **not parallel** for each item in part 4. Raise your hand when you're finished.
(Observe students and give feedback.)

e. Check your work. For each item, tell me what you wrote.
- Item A. (Signal.) *Not parallel.*
- Item B. (Signal.) *Parallel.*
- Item C. (Signal.) *Not parallel.*
- Item D. (Signal.) *Not parallel.*
- Item E. (Signal.) *Parallel.*
- Item F. (Signal.) *Parallel.*
- Remember, lines intersect if they touch each other or cross each other. Lines are parallel if they would never cross each other no matter how long you made them.

EXERCISE 4 PERIMETER AND AREA
Rectangles

a. Find the list of abbreviations on the last page of your textbook. √
- The unit names are in the first column. The abbreviations are in the second column. You can use these abbreviations when you write the unit name in your answers.

- The abbreviation for inches is **i-n.** The abbreviation for feet is **f-t.** You can see the abbreviation for yards, miles, miles per hour, seconds, minutes, hours, days, weeks, months, and years. The abbreviation for pounds is **l-b.**
- You can see the rest of the abbreviations. Look at the bottom of the list. The abbreviation for square is **s-q.** So the abbreviation for square feet is **s-q f-t.** The abbreviation for inches is **i-n.**
- So what's the abbreviation for square inches? (Signal.) *S-q i-n.*

b. Find part 5 on page 79. √
- For each rectangle, you'll find the perimeter and the area. What's the equation for area? (Signal.) *Area equals base times height.*
- The answer always tells about **square** units. Remember, you'll use the abbreviations.

c. Your turn: Figure the perimeter and the area for rectangle A. Write what P equals and what A equals. Raise your hand when you're finished. (Observe students and give feedback.)
- (Write on the board:)

$$
\begin{array}{ll}
16 & A = b \times h \\
16 & A = 34 \times 16 \\
34 & \boxed{A = 544 \text{ sq ft}} \\
\underline{+\ 34} & \\
\boxed{P = 100 \text{ ft}} &
\end{array}
$$

- Check your work. Here's what you should have. Perimeter equals 100 **feet.** Area equals 544 **square feet.** Raise your hand if you wrote the right answer with the correct abbreviations.

d. Your turn. Work the rest of the problems in part 5. Raise your hand when you're finished. (Observe students and give feedback.)

Key:

$$
\begin{array}{lll}
b. & 21 & A = b \times h \\
& 21 & A = 12 \times 21 \\
& 12 & \boxed{A = 252 \text{ sq m}} \\
& \underline{+\ 12} & \\
& \boxed{P = 66 \text{ m}} &
\end{array}
$$

$$
\begin{array}{lll}
c. & 15 & A = b \times h \\
& 15 & A = 15 \times 15 \\
& 15 & \boxed{A = 225 \text{ sq in}} \\
& \underline{+\ 15} & \\
& \boxed{P = 60 \text{ in}} &
\end{array}
$$

e. Find part K. That shows you what you should have for rectangles B and C.
- Rectangle B: What's the perimeter? (Signal.) *66 meters.*
- What's the area? (Signal.) *252 square meters.*
- Rectangle C: What's the perimeter? (Signal.) *60 inches.*
- What's the area? (Signal.) *225 square inches.*
- Raise your hand if you got all of them right.

EXERCISE 5 NUMBER FAMILIES
Tables

a. Find part 6.
b. Figure out all the missing numbers in the table. Remember, first work all the rows you can work. Then work all the columns you can work. Then find any row or column with two numbers and work it. Keep going until you've filled in all the numbers. Raise your hand when you've finished the table.
(Observe students and give feedback.)
- (Write on the board:)

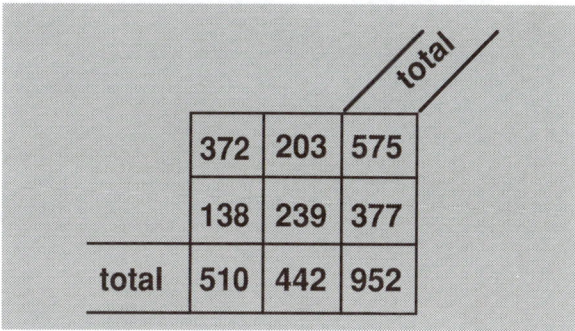

		total
372	203	575
138	239	377
total 510	442	952

c. Check your work. Here's what you should have for the table.
- Raise your hand if you figured out all the missing numbers.

EXERCISE 6 DATA ANALYSIS
Tables

a. Find part 7.
- This table shows the number of ripe grapes and unripe grapes in two gardens.
b. Answer the questions in part 7. Raise your hand when you're finished.
(Observe students and give feedback.)
c. Check your work.
- Question A: Everybody, which garden has more grapes that are ripe? (Signal.) *Brandon's garden.*
- Question B: What is the total number of grapes for both gardens? (Signal.) *2212.*

- Question C: Which is greater, the total for ripe grapes or the total for grapes that are unripe? (Signal.) *Ripe grapes.*
- Item D: 1158 tells about the number of BLANK in BLANK. You should have written **grapes** or **total grapes** for the first blank and **Eden's garden** for the other blank. 1158 tells about the number of total grapes in Eden's garden.
- Item E: 104 tells about the number of BLANK in BLANK. You should have written **unripe** for the first blank and **Brandon's garden** for the second blank. 104 tells about the number of unripe grapes in Brandon's garden.

d. Raise your hand if you answered all the questions correctly.

EXERCISE 7 INDEPENDENT WORK

- (In addition to independent work in the textbook, assign **Bridge to Connecting Math Concepts** *Independent Worksheet* 1 as classwork or homework. Before beginning the next lesson, check the students' independent work.)

Lesson 22

Objectives

- **Indicate whether non-parallel lines intersect.** (Exercise 1)

- Make a number family for a problem that has both a diagram and a comparison statement. (Exercise 2)

- **Figure out the perimeter of a parallelogram.** (Exercise 3)
  Note: Students first identify which figures are parallelograms. For those figures, both pairs of opposite sides are parallel.

- **Complete a number-family table and answer questions by referring to the table.** (Exercise 4)

- **Make a number family for a comparison sentence.** (Exercise 5)
  Note: Sentences are of the form: Jim was 18 pounds lighter than Tim. Students write:

dif	Jim	Tim
18 →		

- **Complete a table to show mixed numbers and the corresponding fractions.** (Exercise 6)

EXERCISE 1 GEOMETRY
Parallel/Intersecting Lines

a. Open your textbook to lesson 22 and find part 1.
- Each item shows lines. The lines for some items are parallel. The lines for other items are not parallel. Write **parallel** or **not parallel** for each item in part 1. Raise your hand when you're finished.
  (Observe students and give feedback.)
b. Check your work. For each item tell me what you wrote.
- Item A. (Signal.) *Not parallel.*
- Item B. (Signal.) *Not parallel.*
- Item C. (Signal.) *Parallel.*
- Item D. (Signal.) *Not parallel.*
c. You should have written **not parallel** for items A, B, and D.
- Some of the lines that are not parallel intersect. Your turn: Write **intersect** or **do not intersect** for each pair of lines that are not parallel.
  (Observe students and give feedback.)
d. Check your work. For each item tell me what you wrote.
- Item A. (Signal.) *Do not intersect.*

- Item B. (Signal.) *Do not intersect.*
- Item D. (Signal.) *Intersect.*

EXERCISE 2 NUMBER FAMILIES
Comparison

a. Find part 2.
- These are problems that tell about the difference number or ask about the difference number.
b. Make a number family for each problem. Figure out the missing number. Show your answer with a number and a unit name. Use abbreviations for the unit names. Raise your hand when you're finished.

(Observe students and give feedback.)
Key:

a. $\dfrac{dif \quad B \quad R}{171 \quad 98} \blacktriangleright \square \quad \begin{array}{r} 171 \\ +\ 98 \\ \hline \boxed{269\ in} \end{array}$ b. $\dfrac{dif \quad B \quad D}{\square \quad 345} \blacktriangleright 459 \quad \begin{array}{r} 459 \\ -\ 345 \\ \hline \boxed{114\ oz} \end{array}$

c. $\dfrac{dif \quad K \quad F}{37 \quad \square} \blacktriangleright 199 \quad \begin{array}{r} 199 \\ -\ 37 \\ \hline \boxed{162\ m} \end{array}$ d. $\dfrac{dif \quad G \quad V}{62 \quad 102} \blacktriangleright \square \quad \begin{array}{r} 62 \\ +\ 102 \\ \hline \boxed{164\ lb} \end{array}$

c. Find part J on page 86. That shows what you should have for each problem.
- Problem A asks: What is the length of pole R? The answer is 269 inches. That's 269 **i - n.**
- Problem B asks: How much lighter is pile B than pile D? The answer is 114 ounces. That's 114 **o - z.**
- Problem C asks: How long is bar K? The answer is 162 meters. That's 162 **m.**
- Problem D asks: How much does box V weigh? The answer is 164 pounds. That's 164 **l - b.**
d. Raise your hand if you got everything right.

EXERCISE 3 PERIMETER
Parallelograms

a. Find part 3.
(Teacher reference:)

I'll read what it says. Follow along: On some 4-sided figures, **both pairs of opposite sides are parallel.**
- You can see some of those figures. One of the figures is a rectangle. The others are not rectangles. The left and right sides are parallel to each other. The top and bottom sides are parallel to each other.
- Look at figure A below.
- (Teacher reference:)

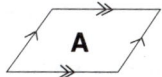

- The arrows show which sides are parallel. The pair of sides with single arrowheads are parallel. The left side is parallel to the right side. The pair of sides with double arrowheads are parallel. The top side is parallel to the bottom side.
- Here's the rule: **When both pairs of opposite sides are parallel, the two sides in each pair are the same length.**
- You can see the figure that shows the length of two sides.
- The left side is 12 units long. So what's the length of the opposite side? (Signal.) *12 units.*
- The bottom side is 14 units long. So what's the length of the opposite side? (Signal.) *14 units.*
- Remember, if **both pairs** of opposite sides are parallel, the two sides in each pair are the same length.
b. Find part 4.
- The length of two sides is shown for all the figures. In some of these figures, **one pair** of opposite sides is parallel. For other figures, **both pairs** of opposite sides are parallel.
c. Figure A. Is the top side parallel to the bottom side? (Signal.) *Yes.*
- Is the left side parallel to the right side? (Signal.) *No.*
- The figure does not have two pairs of parallel sides so you can't figure out the length of any unmarked side. You can see that the top side is not the same length as the bottom side. And the left side is not the same length as the right side.
d. Figure B. Is the top side parallel to the bottom side? (Signal.) *Yes.*
- Is the left side parallel to the right side? (Signal.) *Yes.*
- The figure has two pairs of parallel sides, so you can figure out the length of the sides that aren't marked.
e. Figure C. Is the top side parallel to the bottom side? (Signal.) *Yes.*
- Is the left side parallel to the right side? *Yes.*
- So you can figure out the length of both sides that aren't marked.
- Figure D: Are both pairs of opposite sides parallel? (Signal.) *No.*
- So can you figure out the length of the sides that aren't marked? (Signal.) *No.*
f. For two of the figures in part 4, both pairs of opposite sides are parallel.
- Your turn: Put your paper on top of those two figures and trace them. Then write the number and unit name for all four sides. Don't trace figures that do not have **two pairs** of parallel sides. Raise your hand when you're finished.
(Observe students and give feedback.)

g. Check your work. You should have traced figures B and C.
- Figure B. What did you write for the bottom side? (Signal.) *10 feet.*
- That's 10 f - t.
- What did you write for the right side? (Signal.) *6 feet.*
- That's 6 f - t.
- Figure C. What did you write for the top side? (Signal.) *9 inches.*
- That's 9 i - n.
- What did you write for the left side? (Signal.) *18 inches.*
- That's 18 i - n.

h. You can figure out the perimeter of figures B and C because you know the length of each side. You can't figure out the perimeter of the other figures.
- Find the perimeter of B and C. Show the answer as: P equals. Remember the unit name in the answer. Raise your hand when you're finished. (Observe students and give feedback.)
- (Write on the board:)

b.		c.	
	6		18
	6		18
	10		9
	+ 10		+ 9
	P = 32 ft		**P = 54 in**

- Here's what you should have for B and C.
- Raise your hand if you got everything right.

EXERCISE 4 DATA ANALYSIS
Tables

a. Find part 5.
- The table is supposed to show the number of inches of snow that fell in two cities during 1986 and 1987, but some of the numbers are missing.

b. Your turn: Copy the table and fill in all the missing numbers. Raise your hand when you've done that much.
(Observe students and give feedback.)

c. (Write on the board:)

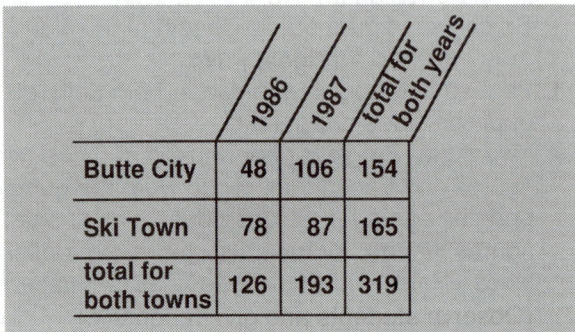

	1986	1987	total for both years
Butte City	48	106	154
Ski Town	78	87	165
total for both towns	126	193	319

- Here are the numbers you should have.
- The missing number in the top row is 106. The missing numbers in the middle row are 87 and 165. The missing numbers in the bottom row are 126 and 193.
- Raise your hand if you got all of the numbers right.
- Make sure your table has all the right numbers. √

d. Now answer the questions in part 5. Raise your hand when you're finished.
(Observe students and give feedback.)

e. Check your work. I'll read each question; you tell me the answer.
- Item A: In which year did more snow fall in Ski Town? (Signal.) *1987.*
- Item B: How many inches of snow fell in Butte City in 1986? (Signal.) *48.*
- Item C: Which city had a higher total snowfall for both years? (Signal.) *Ski Town.*
- Item D: What was the total number of inches of snow for both cities during both years? (Signal.) *319.*
- Item E: 106 tells about the number of inches of snow that fell in Butte City during 1987. You should have written **Butte City** and **1987.**

f. Raise your hand if you answered all the questions correctly.

EXERCISE 5 NUMBER FAMILIES
Comparison

a. Find part 6.
- Each sentence in part 6 compares two things. The sentences name things and give a number to show how much more or how much less one of those things is.
- For each sentence, you'll make a number family with three names. The name for the first small number is **difference.** The name for the big number is the name for the value that is larger.

b. Look at the sample sentences.
- Sample sentence 1: Building M is 121 feet shorter than building C. The taller building is the big number. If M is **shorter than** C, which is taller? (Signal.) *C.*
- The sentence gives the difference number. Everybody, what's the difference number? (Signal.) *121.*

c. (Write on the board:)

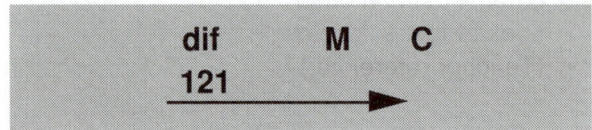

- Here's the family for sample sentence 1.

d. Sample sentence 2: The pile of salt weighed 16 ounces more than the pile of baking powder. Which has the larger number of ounces, the pile of salt or the pile of baking powder? (Signal.) *The pile of salt.*
- So, what's the name for the big number? (Signal.) *Salt.*
- And baking powder is the name for a small number. The other small number is the difference number. Everybody, what number is the difference number? (Signal.) *16.*
e. (Write on the board:)

- Here's the number family for sample sentence 2.
f. Find items A through E in part 6.
- Each sentence gives you enough information to make a family with three names and the difference number.
g. Sentence A: James was 23 centimeters taller than Dan. Make the number family with two names and a difference number. Raise your hand when you're finished.
  (Observe students and give feedback.)
- (Write on the board:)

- Here's what you should have for sentence A. James is the big number. The difference number is 23.
h. Sentence B says: There were 89 more red trucks than yellow trucks. Make the family for sentence B. Raise your hand when you're finished. (Observe students and give feedback.)
- (Write on the board:)

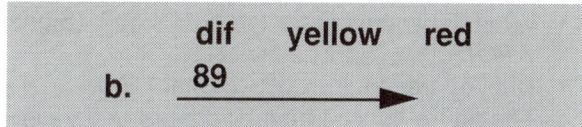

- Here's what you should have for sentence B. The big number is for red trucks. The difference number is 89.
i. Make families for the rest of the items in part 6. Raise your hand when you're finished.
  (Observe students and give feedback.)

- (Write on the board:)

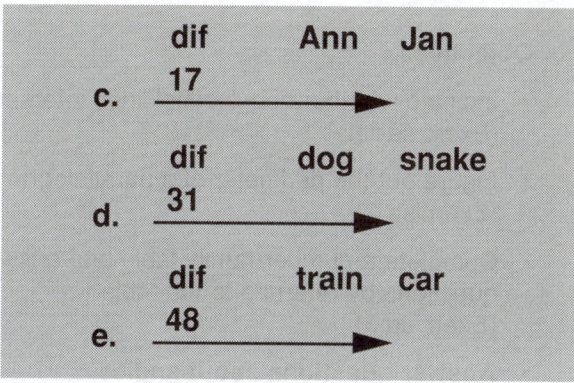

- Here's what you should have for sentences C, D and E.
j. Raise your hand if you got everything right.

EXERCISE 6 MIXED NUMBERS
As Fractions

a. Find part 7.
- This is a table. The first column shows mixed numbers. The second column will have fractions that equal mixed numbers.
b. Copy the table and complete it. Raise your hand when you're finished.
  (Observe students and give feedback.)
- (Write on the board:)

	Mixed number	Fraction
a.	$4\frac{1}{2}$	$\frac{9}{2}$
b.	$2\frac{4}{9}$	$\frac{22}{9}$
c.	$3\frac{5}{8}$	$\frac{29}{8}$
d.	$4\frac{2}{8}$	$\frac{34}{8}$

- Here's what you should have.
c. Raise your hand if you got everything right.

EXERCISE 7 INDEPENDENT WORK

- (In addition to independent work in the textbook, assign **Bridge to Connecting Math Concepts** *Independent Worksheet* 2 as classwork or homework. Before beginning the next lesson, check the students' independent work.)

Lesson 23

EXERCISE 1 GEOMETRY
Parallel/Intersecting Lines

a. Open your textbook to lesson 23 and find part 1.
- Each item shows lines. The lines for some items are parallel. The lines for other items are not parallel. Write **parallel** or **not parallel** for each item in part 1. Raise your hand when you're finished.
  (Observe students and give feedback.)
b. Check your work. For each item tell me what you wrote.
- Item A. (Signal.) *Parallel.*
- Item B. (Signal.) *Not parallel.*
- Item C. (Signal.) *Not parallel.*
- Item D. (Signal.) *Parallel.*
c. You should have written **not parallel** for B and C. Some lines that are not parallel intersect. Your turn: Write **intersect** or **do not intersect** for each pair of lines that are not parallel.
  (Observe students and give feedback.)
d. Check your work. For items B and C tell me what you wrote.

- Item B. (Signal.) *Intersect.*
- Item C. (Signal.) *Intersect.*

EXERCISE 2 PERIMETER
Parallelograms

a. Find part 2.
  Remember the rule about four-sided figures with parallel sides. If **both** pairs of opposite sides are parallel, the opposite sides are the same length. If both pairs are not parallel, you can't figure out the length of any opposite sides.
b. Your turn. Put your paper over the figures and trace the figures that have two pairs of parallel sides. Write the length of the sides you can figure out. Write the number and unit name for each side. Don't trace the figures that don't have two pairs of parallel sides. Raise your hand when your finished.
  (Observe students and give feedback.)

Key:

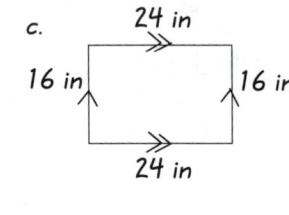

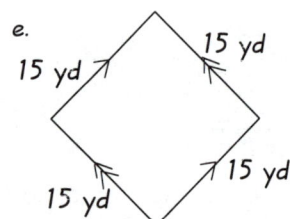

c. Check your work.
- Did you trace figure A? (Signal.) *Yes.*
  Both pairs of opposite sides are parallel.
- What did you write for the bottom side? (Signal.) *3 feet.*
  That's 3 f - t.
- What did you write for the right side? (Signal.) *7 feet.*
- Did you trace figure B? (Signal.) *No.*
  The figure doesn't have two pairs of parallel sides.
- Did you trace figure C? (Signal.) *Yes.*
  Both pairs of opposite sides are parallel.
- What did you write for the bottom side? (Signal.) *24 inches.*
  That's 24 i - n.
- What did you write for the left side? (Signal.) *16 inches.*
- Did you trace figure D? (Signal.) *No.*

- Did you trace figure E? (Signal.) *Yes.*
- What did you write for the upper left side? (Signal.) *15 yards.*
  That's 15 y-d.
- What did you write for the lower left side? (Signal.) *15 yards.*

d. Your turn: Find the perimeter of A, C and E. Show what P equals for each figure. Raise your hand when you're finished.
(Observe students and give feedback.)

Key:

a.	c.	e.
7	16	15
7	16	15
3	24	15
+ 3	+ 24	+ 15
$P = 20 \ ft$	$P = 80 \ in$	$P = 60 \ yd$

- Find part J on page 92. That shows what you should have for each figure.
- Figure A: The perimeter equals 20 feet.
- Figure C: The perimeter equals 80 inches.
- Figure E: The perimeter equals 60 yards.
- Raise your hand if you got everything right.

EXERCISE 3 DATA ANALYSIS
Tables

a. Find part 3.
- This table is supposed to show the number of cars and trucks parked on two lots.
- What are the names of the lots? (Signal.) *A and B.*
- Some of the numbers in the table are missing.

b. Your turn. Copy the table and fill in all the missing numbers. Raise your hand when you've done that much.
- (Write on the board:)

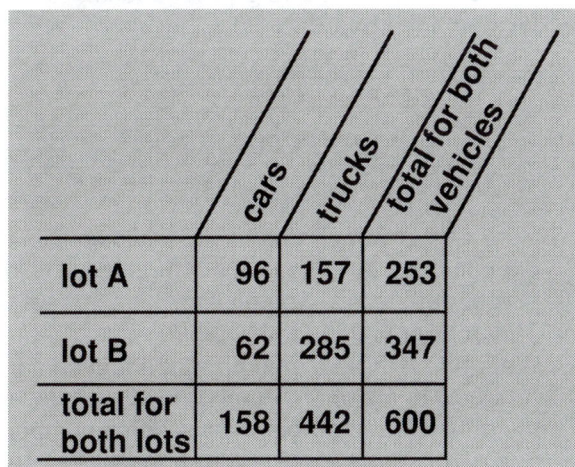

	cars	trucks	total for both vehicles
lot A	96	157	253
lot B	62	285	347
total for both lots	158	442	600

- Here are the numbers you should have. The missing number in the top row is 253. The missing numbers in the middle row are 62 and 285. The missing numbers in the bottom row are 442 and 600. Make sure your table has all the right numbers. √

c. Now write the answers to all the items in part 3. Raise your hand when you're finished.
(Observe students and give feedback.)

d. Check your work. I'll read each item. You tell me the answer.
- Item A: Are there more cars or more trucks on lot A? (Signal.) *Trucks.*
- Item B: What is the total number of cars on both lots? (Signal.) *158.*
- Item C: Which lot has more vehicles in all, lot A or lot B? (Signal.) *Lot B.*
- Item D: How many cars are parked on lot B? (Signal.) *62.*
- Item E: 285 tells the number of trucks on lot B. You should have written **trucks** and **B.**
- Raise your hand if you got everything right.

EXERCISE 4 GEOMETRY
Angles

a. Find part 4.
- (Teacher reference:)

- I'll read what it says. Follow along: An angle is formed when two lines **come together.**
- The angle is marked as part of a circle between the lines.
- We can measure angles in units called **degrees.**
- You can see an angle of 5 degrees.
- Touch the angle of 5 degrees. √
- The two lines are very close together.
- You can see how to write 5 degrees.
- The symbol for degrees is a little circle.
- An angle with more than 5 degrees shows a bigger part of the circle.
- Touch the first angle of 45 degrees. √
- That angle is a lot larger than the 5-degree angle.
- Below are other angles that are 45 degrees.
- Touch angle A. √
- The arrow is close to the place where the lines intersect.
- Touch angle B. √

- The arrow is farther from the place where the lines intersect. Both angles are the same size— 45 degrees. It doesn't matter how far the arrow is from the place where the lines intersect.
- The same angle can be shown in different positions. All the angles at the bottom of part 4 are also 45 degrees.

b. Find part 5.
(Teacher reference:)

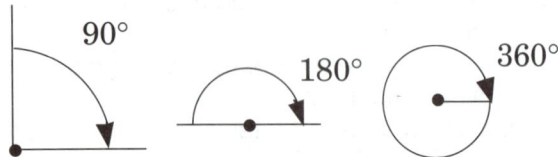

This part gives important facts about angles.
- If you remember these facts, you'll be able to work a lot of very difficult angle problems.
- The first fact is: The angle for the corner of a room or the corner of a rectangle is 90 degrees.
- Touch the 90-degree angle. √
- Next fact: The angle for half a circle is 180 degrees.
- Touch the angle for 180 degrees. √
- Remember, a straight line forms a 180-degree angle.
- Last fact: The angle for a complete circle is 360 degrees. That's the largest angle you can show.
- Touch the angle for 360 degrees. √

c. Let's review: How many degrees are in the corner of a room or the corner of a rectangle? (Signal.) *90 degrees.*
- How many degrees are in half of a circle? (Signal.) *180 degrees.*
- How many degrees are in a whole circle? (Signal.) *360 degrees.*
(Repeat step c until firm.)

d. Find part 6.
(Teacher reference:)

  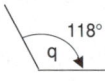 

- Four angles are shown. Each angle has a letter that identifies it.

e. Item A says: Write the degrees for the largest angle.
- Find the largest angle. Write the number shown in the angle. Use a small circle for the degree symbol. Raise your hand when you're finished. √
- (Write on the board:)

a. 118°

- The largest angle is 118 degrees.
f. Your turn: Write answers to the rest of the items in part 6. Remember to use the symbol for degrees. Raise your hand when you're finished. (Observe students and give feedback.)
g. Check your work.
- Item B: Write the letter of each angle that is smaller than angle P. What are the letters? (Signal.) *R and V.*
- Yes, angle P is 80 degrees. Angle R is 30 degrees. Angle V is 25 degrees.
- Item C: How many degrees are in a whole circle? (Signal.) *360 degrees.*
- Item D: How many degrees are in half a circle? (Signal.) *180 degrees.*
- Item E: What's the letter of the smallest angle? (Signal.) *V.*

EXERCISE 5 PROBLEM SOLVING
Comparison

a. Find part 7.
- These are word problems that compare two things. Remember, if a sentence compares, you make a number family with a difference number.
b. Problem A: Building C is 56 feet taller than building D. Building D is 194 feet tall.
- The first sentence gives the information you need to make a number family with three names and a difference **number.** The second sentence gives a number for building D.
- Make the family with three names and two numbers. Figure out the missing number and answer the question the problem asks. Remember, the answer has a number and a unit name. Use abbreviations. Raise your hand when you're finished.
(Observe students and give feedback.)
- (Write on the board:)

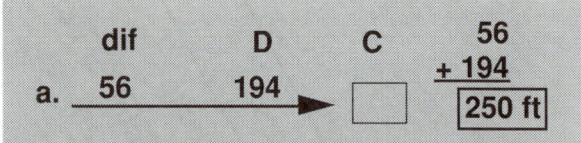

- Here's what you should have. Building C is the big number. Building D is 194 feet tall. Building C is 56 feet taller. So building C is 250 feet tall. The answer is 250 feet.
c. Work problem B. Remember, the first sentence tells how to make the number family. Answer the question the problem asks with a number and a unit name. Raise your hand when you're finished. (Observe students and give feedback.)

- (Write on the board:)

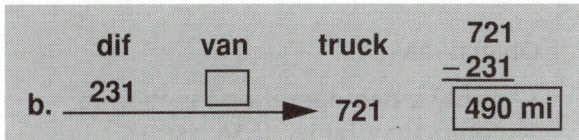

	dif	van	truck	721 − 231
b.	231	□ → 721		490 mi

- Here's what you should have. The truck is the big number. That's 721. The difference is 231. You figured out the distance the van went. The answer is 490 miles.
d. Work the rest of the problems in part 7. Make the number family. Figure out the answer to each problem. Remember to answer the question with a number and a unit name. Raise your hand when you're finished.
  (Observe students and give feedback.)
- (Write on the board:)

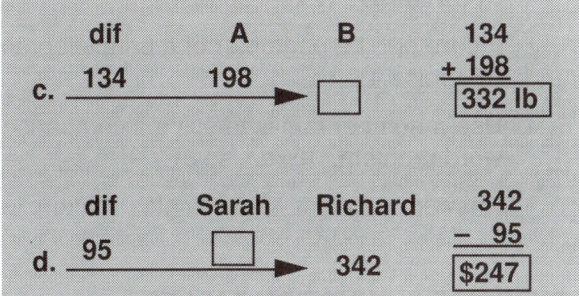

	dif	A	B	134 + 198
c.	134	198	□	332 lb

	dif	Sarah	Richard	342 − 95
d.	95	□ → 342		$247

- Here's what you should have.
- Problem C: Pile B is the big number. The difference number is 134. Pile A is 198. You figured out what pile B weighed. The answer is 332 pounds.
- Problem D: Richard is the big number — 342. The difference number is 95. You figured out how much money Sarah saved. The answer is 247 dollars.

EXERCISE 6 DIVISION
Mixed Number Answers

a. Find part 8.
- I'll read what it says. Follow along: You have learned to write the remainder for division problems. The remainder can also be written as a fraction. When you write the remainder as a fraction, you show the answer as a mixed number.
- **You change the remainder into a fraction by dividing.**
- The first problem is 34 divided by 7.
- The answer is 4 with a remainder of 6.
- The problem divides by 7, so you can divide the remainder by 7. That's 6/7.

- You can see the answer: 4 and 6/7.
- Next problem: 31 divided by 9. The answer has a remainder of 4.
- The problem divides by 9, so you can divide the remainder by 9. That's 4/9.
- You can see the answer: 3 and 4/9.
- Next problem: 16 divided by 5. The answer to this problem has a remainder of 1.
- The problem divides by 5, so you can divide the remainder by 5. That's 1/5.
- You can see the answer: 3 and 1/5.
- Remember, when you have a remainder, you can write it as a fraction. **The number you divide by tells the denominator for the fraction.**
b. Find part 9.
  You're going to rewrite each answer as a mixed number. You'll change the remainder into a fraction.
c. Sample problem. 25 divided by 7. The answer is 3 with a remainder of 4. We'll change 4 into a fraction. What fraction? (Signal.) *4-sevenths.*
- Yes, we're dividing by 7. So the fraction is 4 divided by 7.
- (Write on the board:)

$$3\frac{4}{7}$$

- Here's the answer: 3 and 4/7. There's no R before the fraction 4/7.
d. Your turn: Write the answer to problem A as a mixed number. Raise your hand when you're finished. √
  (Write on the board:)

$$\text{a.} \quad 6\frac{3}{5}$$

- Here's what you should have for problem A. You're dividing by 5, so the fraction is 3 divided by 5.
e. Your turn: Write the answer to problem B as a mixed number. Raise your hand when you're finished. √
- (Write on the board:)

$$\text{b.} \quad 2\frac{3}{4}$$

- Here's what you should have for problem B. You're dividing by 4, so the fraction is 3 divided by 4.

f. Your turn: Write mixed numbers for the rest of the answers in part 9. Raise your hand when you're finished.
(Observe students and give feedback.)

g. (Write on the board:)

c. $7\frac{1}{3}$ d. $6\frac{5}{6}$

- Check your work. Here's what you should have for problems C and D.
- Problem C: 22 divided by 3 equals 7 and 1/3.
- Problem D: 41 divided by 6 equals 6 and 5/6.

EXERCISE 7 INDEPENDENT WORK

- (In addition to independent work in the textbook, assign **Bridge to Connecting Math Concepts** *Independent Worksheet* 3 as classwork or homework. Before beginning the next lesson, check the students' independent work.)

Lesson 24

Objectives

- **Apply a two-operation function to complete a table.** (Exercise 1)
  Note: Functions are stated as a rule, such as: Multiply by 4 and add 5. Students apply the rule to each row of a table:

3	17
7	■
9	■
5	■

- Rewrite the answer to a division problem as a mixed number. (Exercise 2)

- **Use facts to complete a number-family table and answer questions by referring to the table.** (Exercise 3)

- Figure out the perimeter of a parallelogram. (Exercise 4)

- Use a number family to solve a comparison word problem. (Exercise 5)

- Answer questions about angles. (Exercise 6)

EXERCISE 1 FUNCTION TABLES
Two-Step Rules

a. Open your textbook to lesson 24 and find part 1.
- For each item there are numbers and a rule. Each rule is shown in a box. You'll follow the rule to work problems in your head.
b. Table A. The rule is: Multiply each number by 5 and add 4. For the first number, you multiply 3 times 5. That's 15. Then you add 4. That's 19.
- The next number is 7. What's the first thing you do to 7? (Signal.) *Multiply by 5*.
- Then what do you do to the answer? (Signal.) *Add 4*.
- Copy the table and the numbers you'll start with. Write **19** as the first number in the second column. Then write the answers for the rest of the numbers in the table. Raise your hand when you're finished.
(Observe students and give feedback.)
- (Write on the board:)

a.

3	19
7	39
9	49
5	29

- Here's what you should have.
- 7 times 5 is 35, plus 4 is 39.
- 9 times 5 is 45, plus 4 is 49.
- 5 times 5 is 25, plus 4 is 29.

c. For table B there's a different rule. First multiply each number by 9. Then subtract 2.
- Copy the table, follow the rule and write the numbers. Raise your hand when you're finished. (Observe students and give feedback.)
- (Write on the board:)

b.	6	52
	2	16
	8	70
	1	7
	9	79

- Here's what you should have for table B.
- 6 times 9 is 54, minus 2 is 52.
- 2 times 9 is 18, minus 2 is 16.
- 8 times 9 is 72, minus 2 is 70.
- 1 times 9 is 9, minus 2 is 7.
- 9 times 9 is 81, minus 2 is 79.

EXERCISE 2 DIVISION
Mixed Number Answers

a. Find part 2.
- You're going to rewrite each answer as a mixed number. Remember, the number you divide by tells the denominator for the fraction.

b. Item A. Write the answer as a mixed number. Raise your hand when you're finished. (Observe students and give feedback.)
- (Write on the board:)

$$a.\ 9\frac{1}{3}$$

- Check your work. Here's what you should have for item A. You're dividing by 3. So the fraction is 1 divided by 3. 28 divided by 3 equals 9 and 1/3.

c. Your turn: Work the rest of the items in part 2. Raise your hand when you're finished. (Observe students and give feedback.)

d. (Write on the board:)

$$b.\ 4\frac{3}{7} \quad c.\ 6\frac{1}{2} \quad d.\ 6\frac{3}{4}$$

$$e.\ 4\frac{4}{5} \quad f.\ 2\frac{5}{9}$$

- Check your work. Here's what you should have for items B through F.
- Raise your hand if you wrote all the mixed numbers correctly.

EXERCISE 3 DATA ANALYSIS
Tables

a. Find part 3.
- The table will show the number of visitors to two parks during two months.
- What are the names of the parks? (Signal.) *Valley Park and Mountain Park.*
- Which months does the table tell about? (Signal.) *July and August.*

b. All of the numbers in the table are missing. The facts give information about four numbers. Copy the headings. Read the facts. Put the numbers in the table. Raise your hand when you've done that much.
- (Write on the board:)

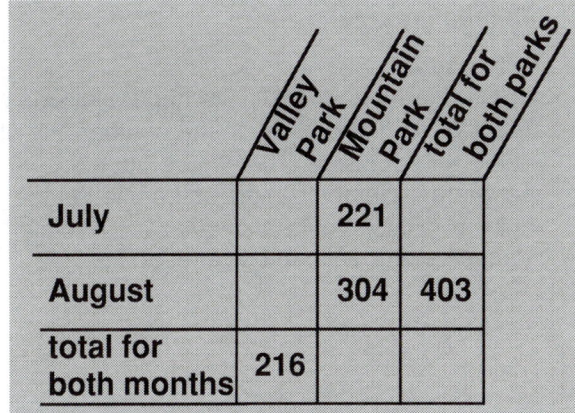

	Valley Park	Mountain Park	total for both parks
July		221	
August		304	403
total for both months	216		

- Here's what you should have.

c. Now, fill in all the missing numbers. Raise your hand when you've done that much. (Observe students and give feedback.)
- (Write to show:)

	Valley Park	Mountain Park	total for both parks
July	117	221	338
August	99	304	403
total for both months	216	525	741

- Here's what you should have. Make sure your table has all the right numbers. √

d. Now figure out the answers to all the items in part 3. Raise your hand when you're finished. (Observe students and give feedback.)

e. Check your work.
• Item A: Everybody, did Mountain Park have more visitors in July or August? (Signal.) *August.*
• Item B: 99 tells about the number of visitors in Valley Park during August. You should have written: **Valley Park** and **August.**
• Item C: What was the total number of visitors for July? (Signal.) *338.*
• Item D: What was the total number of visitors for both months? (Signal.) *741.*
• Item E : 525 tells about the number of visitors in Mountain Park during July and August. You should have written: **Mountain Park** and **July** and **August** or **both months.**

f. Raise your hand if you got everything right.

EXERCISE 4 PERIMETER
Parallelograms

a. Find part 4.
• Some figures have two pairs of parallel sides. Trace those figures. Write the number and unit name for each side. Then figure out the perimeter. Remember the unit name in the answer. Raise your hand when you're finished. (Observe students and give feedback.)

Key:

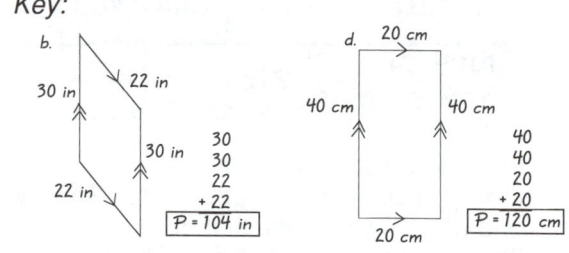

• Find part J on page 96. That shows what you should have for each figure that you traced. You should have traced B and D.
• Figure B: What is the perimeter? (Signal.) *104 inches.*
• Figure D: What is the perimeter? (Signal.) *120 centimeters.*
• Raise your hand if you got everything right.

EXERCISE 5 PROBLEM SOLVING
Comparison

a. Find part 5.
You're going to make the number family for each comparison problem. Then you'll write the number problem and the answer.

b. Problem A: Blaze Lake is 54 miles shorter than Scott Lake. If Scott Lake is 143 miles long, how long is Blaze Lake?
• Make the complete number family for that problem. Then write the number problem and figure out the answer. Use an abbreviation. Raise your hand when you're finished. (Observe students and give feedback.)
• (Write on the board:)

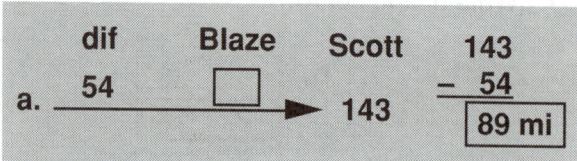

• Here's what you should have. Blaze Lake is 89 miles long. The answer is **89 miles.**

c. Problem B: There are 15 fewer vacation days in the winter than in the summer. If there are 41 vacation days in the winter, how many vacation days are there in the summer?
• Make the complete number family for that problem. Then write the number problem and figure out the answer. Raise your hand when you're finished. (Observe students and give feedback.)
• (Write on the board:)

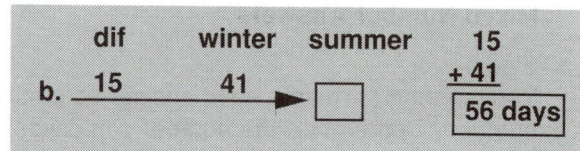

• Here's what you should have. The answer is **56 days.** That's how many vacation days there are in the summer.

d. Your turn: Work the rest of the problems in part 5. Use an abbreviation if you can. Raise your hand when you're finished. (Observe students and give feedback.)

Key:

$$c. \quad \frac{dif \quad Nov \quad Feb}{32 \quad 120 \quad \square} \qquad \begin{array}{r} 32 \\ +120 \\ \hline 152\ lb \end{array}$$

$$d. \quad \frac{dif \quad Peter \quad Simon}{88 \quad \square \quad 205} \qquad \begin{array}{r} 205 \\ -88 \\ \hline 117\ cards \end{array}$$

e. Check your work.
• Find part K. That shows what you should have for problems C and D.
• Problem C: Everybody, what did Sally weigh in February? (Signal.) *152 pounds.*

- Problem D: How many cards does Peter have? (Signal.) *117 cards*.

EXERCISE 6 GEOMETRY
Angles

a. Find part 6.
- (Teacher reference:)

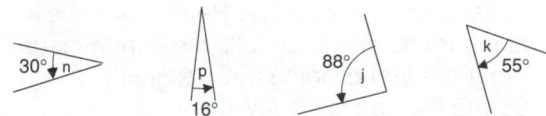

- These items ask about the angles that are shown.
b. Work item A. Raise your hand when you've written the answers to both questions in item A. √
- (Write on the board:)

a.	**j**	**88°**

- Item A asks: Which angle is the largest? That's angle J. Item A also asks how large that angle is. The answer is 88 degrees.
c. Item B asks: What is the sum of angles N and P? You find the sum by adding.
- Work item B. Raise your hand when you're finished. (Observe students and give feedback.)
- (Write on the board:)

b.	**46°**

- Angle N is 30 degrees. Angle P is 16 degrees. The sum of those angles is 46 degrees.
d. Work the rest of the items in part 6. Raise your hand when you're finished.
  (Observe students and give feedback.)
- (Write on the board:)

c.	**16°**	**p**
d.	**33°**	

- Item C asks two questions. The first is: How many degrees are in the smallest angle? The answer is 16 degrees. Item C also asks: What's the letter of the smallest angle? The answer is P.
- Item D asks: How much larger is angle J than angle K? Angle J is 88 degrees. Angle K is 55 degrees. So, the answer is 33 degrees.

EXERCISE 7 INDEPENDENT WORK

- (In addition to independent work in the textbook, assign *Bridge to Connecting Math Concepts Independent Worksheet* 4 as classwork or homework. Before beginning the next lesson, check the students' independent work.)

Lesson 25

Objectives

- Use a number family to solve a comparison word problem. (Exercise 1)

- **Solve a division problem and write the answer as a mixed number.** (Exercise 2)

- **Make a number family to solve a problem involving an angle that is divided into two smaller angles.** (Exercise 3)
  Note: Problems are of the form:

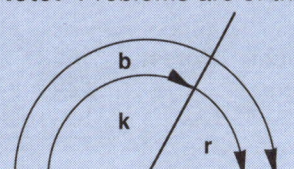

  Angle k equals 120°.
  Angle b is 1/2 a circle.
  How large is angle r? Students make the number family:

k	r	b
120	☐	180

- Compute both the area and the perimeter of a rectangle. (Exercise 4)

- Apply a two-operation function to complete a table. (Exercise 5)

- Use facts to complete a number-family table and answer questions by referring to the table. (Exercise 6)

EXERCISE 1 PROBLEM SOLVING
Comparison

a. Open your textbook to lesson 25 and find part 1. You're going to make a number family for each comparison problem. Then you'll write the number problem and the answer.
b. Problem A: There are 128 more students in school A than in school B. There are 439 students in school B. How many students are in school A?
- Make the complete number family for that problem. Then write the number problem and figure out the answer. Raise your hand when you're finished.
  (Observe students and give feedback.)
- (Write on the board:)

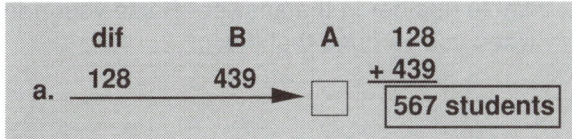

- Here's what you should have. The answer is **567 students.** So there are 567 students in school A.
- c. Problem B: There were 165 fewer residents in Bloom Town than in Forest Town. There were 829 residents in Bloom Town. How many residents were in Forest Town?
- Make the complete number family for that problem. Then write the number problem and figure out the answer. Raise your hand when you're finished.
  (Observe students and give feedback.)
- (Write on the board:)

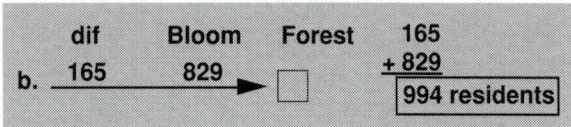

- Here's what you should have. The answer is **994 residents.** That's how many residents there were in Forest Town.
- d. Your turn: Work the rest of the problems in part 1. Raise your hand when you're finished.
  (Observe students and give feedback.)
  Key:

  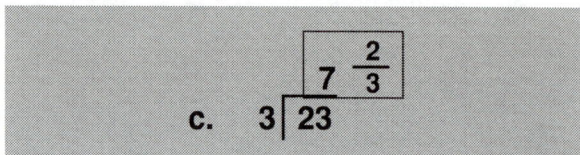

- e. Check your work.
- Find part J on page 100. That shows what you should have for each problem.
- Problem C: How much did the skateboard weigh? (Signal.) *8 pounds.*
- Problem D: How far is it to Cool Lake? (Signal.) *424 miles.*
- Raise your hand if you got everything right.

EXERCISE 2 DIVISION
Mixed Number Answers

a. Find part 2.
- All these problems will have a remainder in the answer. You're going to write the remainder as a fraction.
b. Problem A. Copy the problem. Figure out the whole number in the answer. Raise your hand when you've done that much. √

- (Write on the board:)

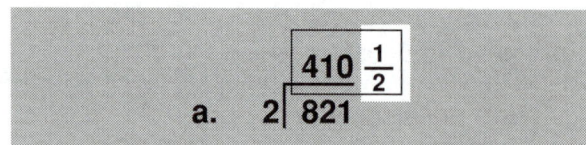

- Here's what you should have. 821 divided by 2 equals 410. The remainder is **1.** You'll write the remainder as a fraction. Remember, the number you divide by tells the denominator. What are you dividing by? (Signal.) *2.*
- So the fraction is 1/2. Write it. √
- (Write to show:)

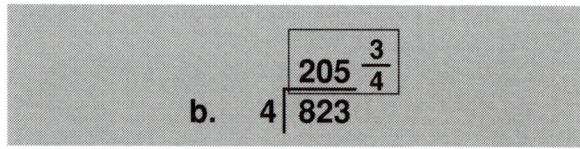

- Here's what you should have. 821 divided by 2 is 410 and 1/2.
- c. Problem B. What are you dividing by? (Signal.) *4.*
- That's the denominator of the fraction. Work the problem. Remember, write the remainder as a fraction. Raise your hand when you're finished.
  (Observe students and give feedback.)
- (Write on the board:)

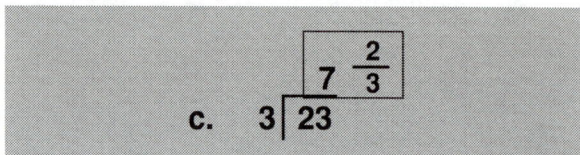

- Here's what you should have. You're dividing by 4. So the fraction is 3/4.
- d. Work problem C. Raise your hand when you're finished. (Observe students and give feedback.)
- (Write on the board:)

- Here's what you should have. You're dividing by 3. So the fraction is 2/3.
- e. Your turn: Work the rest of the problems in part 2. Raise your hand when you're finished.
  (Observe students and give feedback.)
- f. Check your work.
- Problem D: 43 divided by 5. What's the answer? (Signal.) *8 and 3-fifths.*
- Problem E: 121 divided by 6. What's the answer? (Signal.) *20 and 1-sixth.*
- Problem F: 79 divided by 7. What's the answer? (Signal.) *11 and 2-sevenths.*

EXERCISE 3 GEOMETRY
Angles

a. Find part 3.
• You're going to make number families to work angle problems.
• Each problem shows two smaller angles that equal a whole angle. Each angle has a letter. The problems tell about two of the angles. You're going to make the number family that shows the two smaller angles as small numbers and the whole angle as the big number.
b. Problem A. Raise your hand when you know the letters for the two smaller angles that add up to the whole angle.
• Everybody, what are the letters for the two smaller angles? (Signal.) *F and G.*
• What's the letter for the whole angle? (Signal.) *H.*
• (Write on the board:)

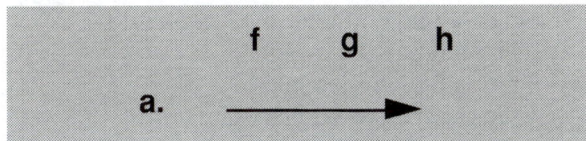

• Here's the number family with three letters.
• Your turn: Copy the number family. Then read problem A. Put in numbers for two of the angles. For angle H, you'll write the degrees for half a circle. Figure out the missing number. Raise your hand when you're finished. (Observe students and give feedback.)
• (Write to show:)

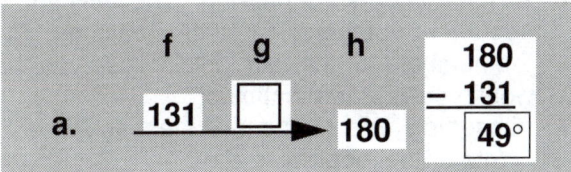

• The problem tells that angle F is 131 degrees and angle H is half a circle. That's 180 degrees. The third angle is G. That angle is 180 degrees minus 131 degrees. How big is angle G? (Signal.) *49 degrees.*
c. Your turn: Make the number family with three letters for problem B. Remember, the letters should show the two small angles that are combined and the big angle that's the sum of the small angles. Raise your hand when you have a number family with three letters. √
• (Write on the board:)

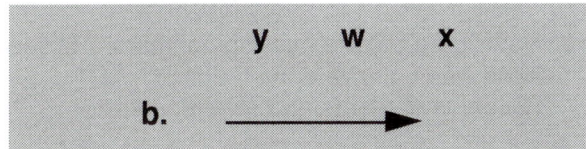

• Here's the family with three letters. Now put in numbers for two of the angles and figure out the missing number. Raise your hand when you're finished. (Observe students and give feedback.)
• (Write to show:)

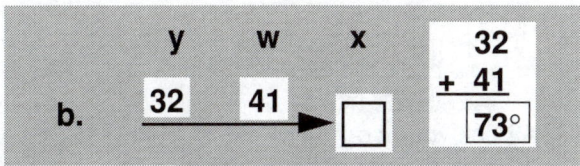

• Angle Y is 32 degrees and angle W is 41 degrees. Everybody, how large is angle X? (Signal.) *73 degrees.*
d. Your turn: Work the rest of the problems in part 3. Raise your hand when you're finished. (Observe students and give feedback.)
e. (Write on the board:)

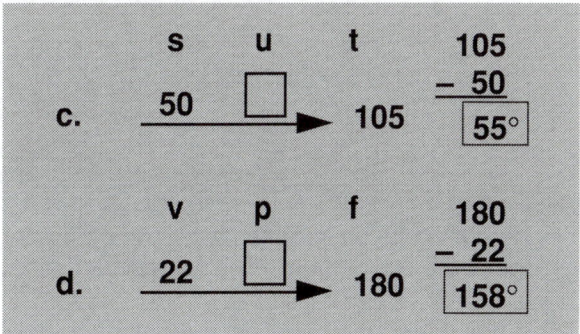

f. Check your work. Here's what you should have for each item.
• Item C: Angle T is 105 degrees. Angle S is 50 degrees. Everybody, what's angle U? (Signal.) *55 degrees.*
• Item D: Angle V is 22°. Angle F is half a circle. What's angle F? (Signal.) *180 degrees.*
• What's angle P? (Signal.) *158 degrees.*

EXERCISE 4 AREA AND PERIMETER
Rectangles

a. Find part 4.
• These are rectangles. You're going to find the area and the perimeter. To find the perimeter, you have to use what you know about figures with two pairs of sides that are parallel.
b. Work problem A: Figure out the perimeter and the area. Raise your hand when you're finished. (Observe students and give feedback.)

- (Write on the board:)

a.	12	A = b x h
	12	A = 11 x 12
	11	**A = 132 sq cm**
	+ 11	
	P = 46 cm	

- Check your work. The sides are 12, 12, 11 and 11 centimeters. The perimeter is 46 centimeters. The area is 11 times 12. That's 132 square centimeters.
- c. Your turn: Work problem B. Raise your hand when you're finished.
  (Observe students and give feedback.)
- (Write on the board:)

b.	38	A = b x h
	38	A = 10 x 38
	10	**A = 380 sq ft**
	+ 10	
	P = 96 ft	

- Here's what you should have. The perimeter of rectangle B is 96 feet. The area is 380 square feet.
- d. Raise your hand if you got everything right.
- Remember, you can figure out the opposite sides, so you can find the perimeter.

EXERCISE 5 FUNCTION TABLES
Two-Step Rules

a. Find part 5.
   Each item has a rule in the box and numbers for the first column of your table. The rule tells what to do with the numbers.
b. Item A: What's the rule? (Signal.) *Multiply each number by 8. Then add 7.*
- The numbers for the first column of your table are 3, 9, 7 and 6.
- Make a table and follow the rule for item A. Raise your hand when you've completed the table. (Observe students and give feedback.)
- (Write on the board:)

a.	3	31
	9	79
	7	63
	6	55

- Here's what you should have. The rule is multiply each number by 8, then add 7.
- 3 times 8 plus 7 equals 31.
- 9 times 8 plus 7 equals 79.
- 7 times 8 plus 7 equals 63.
- 6 times 8 plus 7 equals 55.
c. Make a table and follow the rule for item B. Raise your hand when you've completed the table. (Observe students and give feedback.)
- (Write on the board:)

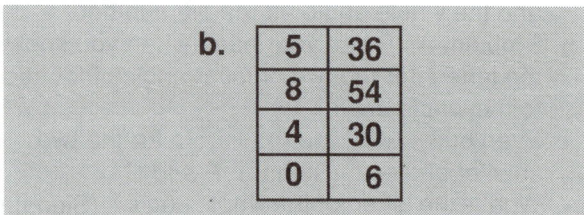

b.	5	36
	8	54
	4	30
	0	6

- Here's what you should have. The rule is: Multiply each number by 6, then add 6.
- 5 times 6 plus 6 equals 36.
- 8 times 6 plus 6 equals 54.
- 4 times 6 plus 6 equals 30.
- Zero times 6 plus 6 equals 6.
d. Raise your hand if you got everything right.

EXERCISE 6 DATA ANALYSIS
Tables

a. Find part 6.
- The table will show the number of children and adults at two different camps.
b. All the numbers are missing. The facts tell about four numbers. Copy the headings. Put in the numbers for the four facts. Raise your hand when you've done that much.
  (Observe students and give feedback.)
- (Write on the board:)

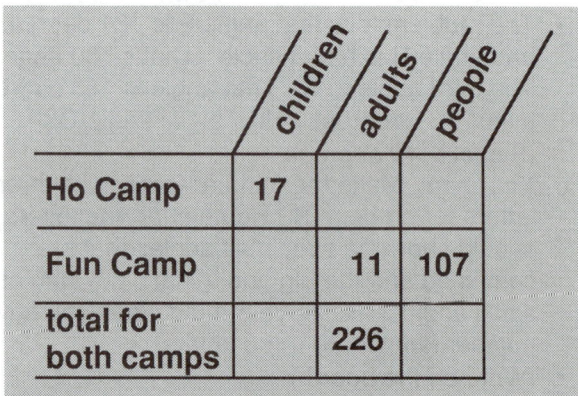

	children	adults	people
Ho Camp	17		
Fun Camp		11	107
total for both camps		226	

c. Now figure out the missing numbers. Raise your hand when you've done that much.
  (Observe students and give feedback.)

- (Write to show:)

	children	adults	people
Ho Camp	17	215	232
Fun Camp	96	11	107
total for both camps	113	226	339

- Here's what you should have. Make sure your table has all the right numbers. Then answer the questions in part 6. Raise your hand when you're finished.
  (Observe students and give feedback.)
d. Check your work.
- Item A: Did Fun Camp have more children or adults? (Signal.) *Children.*
- Item B: What was the total number of people for both the camps? (Signal.) *339.*
- Item C: Which camp had fewer children, Ho Camp or Fun Camp? (Signal.) *Ho Camp.*
- Item D: What was the total number of people in Ho Camp? (Signal.) *232.*
- Item E: 11 tells about the number of adults in Fun Camp. You should have written **adults** for the first blank and **Fun Camp** for the second blank.
- Raise your hand if you got everything right.

EXERCISE 7 INDEPENDENT WORK

- (In addition to independent work in the textbook, assign *Bridge to Connecting Math Concepts Independent Worksheet* 5 as classwork or homework. Before beginning the next lesson, check the students' independent work.)

Lesson 26

Objectives

- **Work a word problem by constructing a number-family table with column and row headings.** (Exercise 1)

- **Rewrite a fraction as a mixed number.** (Exercise 2)
  Note: Students read the fraction as a division problem, work it and show the answer as a mixed number.

- Make a number family to solve a problem involving an angle that is divided into two smaller angles. (Exercise 3)

- Compute both the area and the perimeter of a rectangle. (Exercise 4)

- Use a number family to solve a comparison word problem. (Exercise 5)

- Apply a two-operation function to complete a table. (Exercise 6)

EXERCISE 1 PROBLEM SOLVING
Number-Family Tables

a. Open your textbook to lesson 26 and find part 1.
- I'll read what it says. Follow along: The simplest way to solve some problems is to make a table.
- You can see a big problem in the box.
- Below the problem are rules for making the table.
b. Listen: The first names in the problem are **column** headings. Those are the names in the first sentence.
- Listen: There were red cars and black cars on two lots.
- The column headings are **red cars** and **black cars.** You also need a heading for **total** cars.
c. The next names in the problem are **row** headings. Those are the names in the second sentence.
- Listen: The lots are Al's lot and Lisa's lot. The row headings are **Al's lot** and **Lisa's lot.** You also need the heading **total** for both lots.
- Remember, the **first** names in the problem are the **column** headings. The **next** names are the **row** headings.
- Now you can read the rest of the problem, fill in the missing numbers and answer the questions.

d. Find part 2.
This problem is just like the problem in part 1. Remember, the **first** names in the problem tell about the **column headings.**
- The first sentence says: Game wardens put trout and bass in two different lakes. What are the first two column headings? (Signal.) *Trout and bass.*
- The next sentence tells about the row headings. Your turn: Make the table with headings for the columns and rows. Remember the **totals.** Raise your hand when you've done that much. (Observe students and give feedback.)
- (Write on the board:)

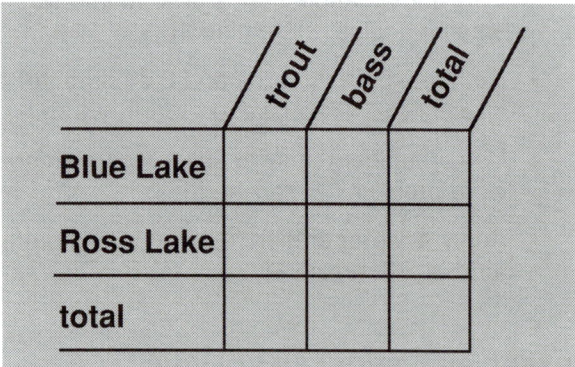

	trout	bass	total
Blue Lake			
Ross Lake			
total			

- Here's what you should have. Now put in all the numbers the facts tell about. Raise your hand when you have four numbers in your table. (Observe students and give feedback.)
- (Write to show:)

	trout	bass	total
Blue Lake	125	74	
Ross Lake		128	231
total			

- Here's what you should have.
e. Now figure out the missing numbers for the table and answer the questions. Raise your hand when you're finished.
(Observe students and give feedback.)

- (Write to show:)

	trout	bass	total
Blue Lake	125	74	199
Ross Lake	103	128	231
total	228	202	430

- Here's what you should have. Raise your hand if you got everything right.
f. Everybody, how many trout were put in Ross Lake? (Signal.) *103.*
- How many total fish were put in Blue Lake? (Signal.) *199.*
- What's the total number of bass for both lakes? (Signal.) *202.*
- That was a hard problem.

EXERCISE 2 FRACTIONS
As Mixed Numbers

a. Find part 3.
- You're going to write mixed numbers that equal fractions. To do that, you read the fraction as a division problem, write it and show the remainder as a fraction. Remember, you can read a fraction as a division problem by starting with the top number.
b. Fraction A. Read it as a division problem. (Signal.) *18 divided by 7.*
- Fraction B. Read it as a division problem. (Signal.) *25 divided by 3.*
- Fraction C. Read it as a division problem. (Signal.) *44 divided by 6.* (Repeat step b until firm.)
c. Your turn: Don't copy the fractions. For each fraction, write the division problem and the answer. Show the remainder as a fraction. The number you divide by is the bottom number of the fraction. Raise your hand when you're finished. (Observe students and give feedback.)
d. Check your work.
- Fraction A: 18/7. Read it as a division problem. (Signal.) *18 divided by 7.*
- What's the answer? (Signal.) *2 and 4-sevenths.*
- Fraction B: 25/3. Read it as a division problem. (Signal.) *25 divided by 3.*
- What's the answer? (Signal.) *8 and 1-third.*
- Fraction C: 44/6. Read it as a division problem. (Signal.) *44 divided by 6.*

- What's the answer? (Signal.) *7 and 2-sixths.*
- Fraction D: 37/4. Read it as a division problem. (Signal.) *37 divided by 4.*
- What's the answer? (Signal.) *9 and 1-fourth.*
- Fraction E: 23/7. Read it as a division problem. (Signal.) *23 divided by 7.*
- What's the answer? (Signal.) *3 and 2-sevenths.*
- Raise your hand if you figured out the mixed number that each fraction equals.

EXERCISE 3 GEOMETRY
Angles

a. Find part 4.
   These are problems that deal with degrees. Remember, you can add or subtract degrees to find missing angles.
b. Problem A. The smaller angles are M and N. The whole angle is angle K. You have to figure out how many degrees are in angle N. Make the number family. The written information names angle M first. Show angle M as the first small number. Figure out the missing angle. Remember to show the degree symbol in your answer. Raise your hand when you're finished.
- (Write on the board:)

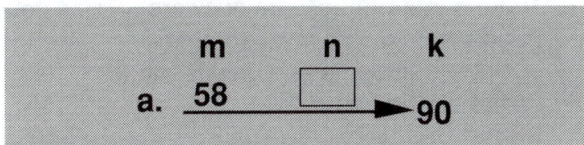

- Here's what you should have. Angle M is 58 degrees. Angle K is 90 degrees. Everybody, how many degrees are in angle N? (Signal.) *32.*
c. Your turn: Work the rest of the problems in part 4. Raise your hand when you're finished.
   (Observe students and give feedback.)
d. (Write on the board:)

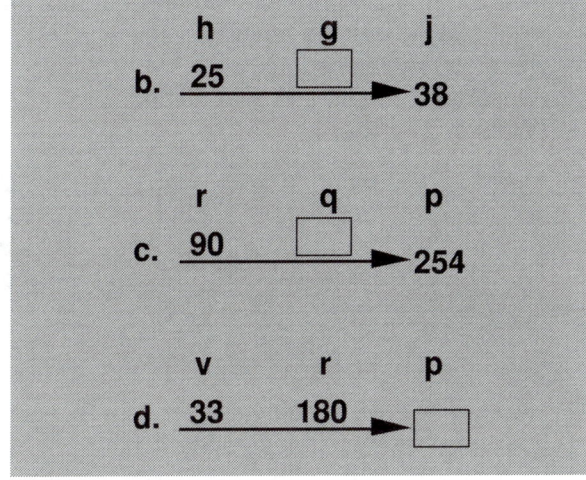

e. Check your work. Here's the number family you should have for each item.
- Item B: Everybody, what's angle G? (Signal.) *13 degrees.*
- Item C: What's angle Q? (Signal.) *164 degrees.*
- Item D: What's angle P? (Signal.) *213 degrees.*

EXERCISE 4 PERIMETER AND AREA
Rectangles

a. Find part 5.
- Lengths are shown for only two sides in each rectangle. You can figure out the lengths of the missing sides. Then you can figure out the perimeter.
b. Your turn: Find the perimeter and the area for each rectangle in part 5. Don't trace the figures. Just find the perimeters and areas. Raise your hand when you're finished.
   (Observe students and give feedback.)
   Key:

   a.   13
        13
         3
      + 3
      P = 32 mi   A = b x h
                         A = 3 x 13
                         A = 39 sq mi

   b.   28
        28
        45
      + 45
      P = 146 cm   A = b x h
                         A = 45 x 28
                         A = 1260 sq cm

   c.   6
         6
        15
      + 15
      P = 42 ft   A = b x h
                         A = 15 x 6
                         A = 90 sq ft

- Find part J on page 104. That shows what you should have for each figure.
- Rectangle A: What's the perimeter? (Signal.) *32 miles.*
  What's the area? (Signal.) *39 square miles.*
- Rectangle B: What's the perimeter? (Signal.) *146 centimeters.*
- What's the area? (Signal.) *1260 square centimeters.*
- Rectangle C: What's the perimeter? (Signal.) *42 feet.*
  What's the area? (Signal.) *90 square feet.*
- Raise your hand if you got everything right.

EXERCISE 5 PROBLEM SOLVING
Comparison

a. Find part 6.
   You're going to make the number family for each comparison problem. Then you'll write the number problem and the answer.

b. Problem A: 145 more students attend the middle school than the high school. There are 540 students at the middle school. How many students attend the high school?

• Make the complete number family for that problem. Then write the number problem and figure out the answer. Raise your hand when you're finished.
   (Observe students and give feedback.)

• (Write on the board:)

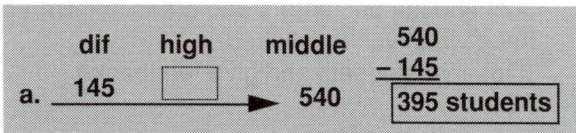

• Here's what you should have. The answer is **395 students.** 395 students attend the high school.

c. Problem B: There are 155 fewer cherry trees in the orchard than apple trees. There are 49 cherry trees. How many apple trees are there?

• Make the complete number family for that problem. Then write the number problem and figure out the answer. Raise your hand when you're finished.
   (Observe students and give feedback.)

• (Write on the board:)

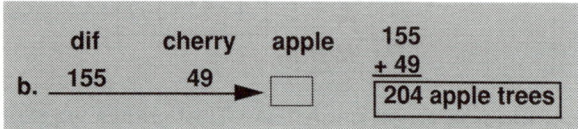

• Here's what you should have. The answer is **204 apple trees.** 204 apple trees are in the orchard.

d. Your turn: Work problem C. Raise your hand when you're finished.
   (Observe students and give feedback.)

• (Write on the board:)

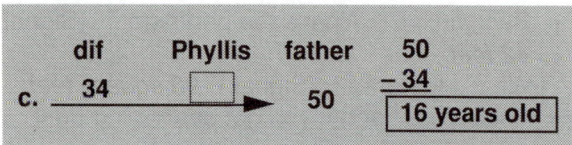

• Here's what you should have for problem C. The answer is 16 years old. Phyllis is 16 years old.

EXERCISE 6 FUNCTION TABLES
Two-Step Rules

a. Find part 7.
   Each item has a rule and some numbers.

b. Item A: The rule says: Multiply each number by 4. Then subtract the product **from 37.**

• The first number is 6. You multiply that by 4. 6 times 4 is 24, so the **product** is 24. You **subtract** 24 **from** 37. The answer is 13.

• The next number is 5. You multiply that by 4. 5 times 4 is 20, so the **product** is 20. Then you subtract 20 **from** 37. The answer is 17.

• What's the next number? (Signal.) *3.*

• You multiply 3 times 4. What's the product? (Signal.) *12.*

• Then you subtract 12 from 37. The answer is 25.

• What's the last number? (Signal.) *2.*

• What do you multiply 2 by? (Signal.) *4.*

• What's the product? (Signal.) *8.*

• Then you subtract 8 from 37.

c. Go back to the first number for item A. What's the first number? (Signal.) *6.*

• What do you multiply by? (Signal.) *4.*

• What's the **product?** (Signal.) *24.*

• Then you subtract 24 from 37.

d. Make a table; then write answers. Raise your hand when you've finished item A.
   (Observe students and give feedback.)

• (Write on the board:)

a.	6	13
	5	17
	3	25
	2	29

• Here's the table you should have for item A.

e. Item B has a different rule. Follow the rule and write a table with the answers. Raise your hand when you're finished.
   (Observe students and give feedback.)

• (Write on the board:)

b.	6	5
	4	19
	5	12
	2	33

• Here's what you should have for item B. Raise your hand if you got everything right.

EXERCISE 7 INDEPENDENT WORK

- (In addition to independent work in the textbook, assign **Bridge to Connecting Math Concepts** *Independent Worksheet* 6 as classwork or homework. Before beginning the next lesson, check the students' independent work.)

Lesson 27

Objectives

- **Figure out the degrees in the four angles that are formed by two perpendicular intersecting lines.** (Exercise 1)
  Note: The term **perpendicular** and the 90° symbol (⌐) are introduced.

- **Rewrite a fraction as a whole number or a mixed number.** (Exercise 2)

- **Figure out the degrees in corresponding angles formed by a line that intersects parallel lines.** (Exercise 3)

- Write and solve a ratio equation for a word problem. (Exercise 4)

- Work a word problem by constructing a number-family table with column and row headings. (Exercise 5)

- Apply a two-operation function to complete a table. (Exercise 6)

EXERCISE 1 GEOMETRY
Perpendicular Lines

a. Open your textbook to lesson 27 and find part 1. (Teacher reference:)

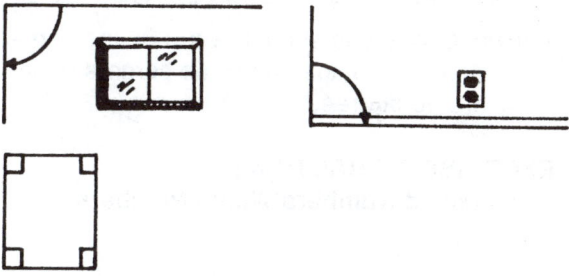

- I'll read what it says. Follow along: If two lines form a 90-degree angle, the lines are **perpendicular.** Perpendicular edges are found in lots of things that people build.
- The walls of a room are perpendicular to the ceiling.
- You can see the 90-degree angle.
- The floor of a room is perpendicular to the walls.
- The top and bottom of your paper are perpendicular to the sides.
- Perpendicular lines are marked with a special angle marker. The marker is shaped like a square corner to show that the angle is 90 degrees.

- Remember, if an angle marker is shown as a corner, the angle is 90 degrees.
b. Find part 2.
- (Teacher reference:)

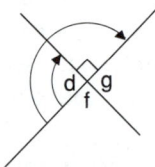

c. Trace the figure. √
- The figure has a 90-degree angle marker. Together the 90-degree angle and angle D make up half a circle. So you can figure out angle D. You can also figure out G because the 90-degree angle and angle **G** make up half a circle. Write 90 degrees for the angle with a 90-degree marker. Then write the degrees for angles D, F and G. Remember, if an angle is 90 degrees, also make the appropriate angle marker. Raise your hand when you're finished.
(Observe students and give feedback.)
- (Write on the board:)

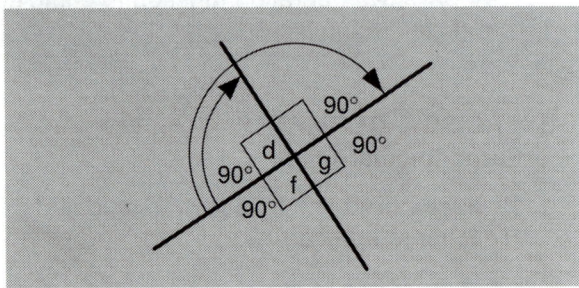

- Here's what you should have. All the angle markers you made should be corners because all are 90 degrees.

EXERCISE 2 FRACTIONS
As Mixed Numbers/Whole Numbers

a. Find part 3.
These are fractions. You can write them as division problems and figure out the answer.
b. Fraction A. Say the division problem. (Signal.) *31 divided by 3.*
- Fraction B. Say the division problem. (Signal.) *127 divided by 4.*
- Fraction C. Say the division problem. (Signal.) *95 divided by 5.*
- (Repeat step b until firm.)
c. Your turn: Write each fraction as a division problem and figure out the answer. Show the answer as a mixed number or a whole number. Raise your hand when you're finished.
(Observe students and give feedback.)

d. Check your work. I'll say each fraction. You say what it equals.
- Problem A: 31/3. What's the answer? (Signal.) *10 and 1-third.*
- Problem B: 127/4. What's the answer? (Signal.) *31 and 3-fourths.*
- Problem C: 95/5. What's the answer? (Signal.) *19.*
- Problem D: 163/6. What's the answer? (Signal.) *27 and 1-sixth.*
- Problem E: 462/7. What's the answer? (Signal.) *66.*
e. You figured out the whole number or mixed number that equals each fraction.

EXERCISE 3 GEOMETRY
Corresponding Angles

a. Find part 4.
I'll read what it says. Follow along: A line that intersects parallel lines creates the same pair of angles at each parallel line.
- You can see a pair of parallel lines.
- There's a line intersecting both parallel lines.
- The intersecting line creates the same angle at the top line—and at the bottom line.
- Angle A equals angle B. Those angles are called **corresponding angles** because they are in the same position. They are just above each parallel line and on the left side of the intersecting line.
- You can see angle **M** at the top line and corresponding angle **N** at the bottom line.
- They're equal to each other.
- You can see another pair of parallel lines and an intersecting line.
- Look at the next page.
- The angle at the top line is 50 degrees. So the corresponding angle at the bottom line is 50 degrees. That's angle **R.**
- If you know that angle **R** is 50 degrees, you can figure out angle **T.** Angles **R** and **T** make half a circle. That's 180 degrees. So angle T is: 180 minus 50. That's 130 degrees.
b. Find part 5.
- For each figure, you're going to identify the corresponding angles and figure out some of the angles.
c. Item A: Trace the figure. Make sure you show the lines and angle markers. These are the arrows for each angle. Raise your hand when you're finished. √
d. Find angle G and shade in the angle. Then find the corresponding angle to G and shade it in. Raise your hand when you're finished. √

- (Write on the board:)

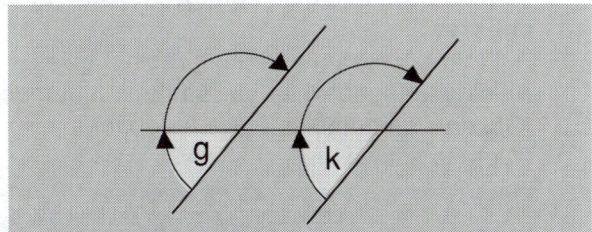

- Here are the corresponding angles. They are to the left of the parallel lines and below the intersecting line. Angles G and K are corresponding angles.
e. Listen: Angle G is 50 degrees. Write 50 degrees for angle G on the diagram.
- Then write 50 degrees for the angle that corresponds to G. Raise your hand when you're finished. √
- (Write to show:)

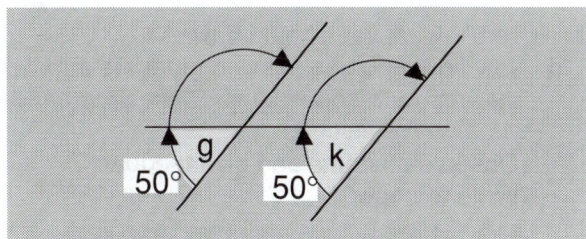

- Here's what you should have.
f. Now you're going to figure out angle H. You know that angle G and angle H together make half a circle.
- You know how many degrees are in half a circle. You also know how many degrees are in angle G. So you can figure out the degrees in angle H. Do it. Write the degrees for angle H. Raise your hand when you're finished. √
- Everybody, how many degrees are in angle H? (Signal.) *130.*
- That's also the number of degrees in the corresponding angle for H. Write the number of degrees for that angle. Raise your hand when you're finished. √
- Everybody, which angle corresponds to angle H? (Signal.) *Angle J.*
- Yes, both are to the left of the parallel lines and above the intersecting line.
g. Trace item B. Raise your hand when you're finished. √
- I'll give you information about one of the angles. You'll write the number of degrees for that angle.
- Find angle F. It's to the right of the intersecting line and above the bottom parallel line. Angle F. Listen: Angle F is 70 degrees. Write 70

degrees for angle F and for the corresponding angle. Raise your hand when you're finished. √
- Everybody, which angle corresponds to angle F? (Signal.) *Angle P.*
- Yes, both are on the top of the parallel lines and to the right of the intersecting line. You should have written 70 degrees for angle P.
- Angle P and angle M make up half a circle. Figure out the number of degrees in angle M and write it. Then write the number of degrees for the angle that corresponds to M. Raise your hand when you're finished. √
- (Write on the board:)

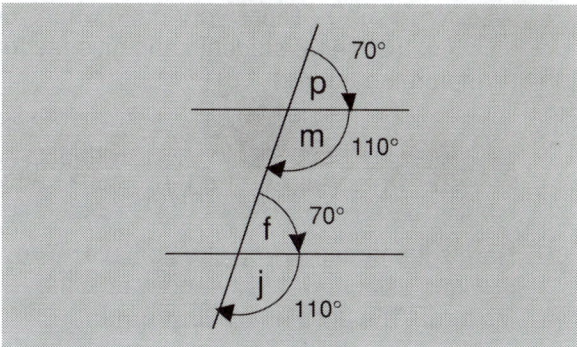

- Here's what you should have. Angle M is 110 degrees. The corresponding angle on the other line is angle J. It is also 110 degrees.
- Raise your hand if you got everything right.

EXERCISE 4   PROBLEM SOLVING
Ratios and Proportions

a. Find part 6.
- These are ratio problems. For each problem, you'll write an equation with the names. You'll figure out the missing number and answer the question the problem asks.
b. Work problem A. Raise your hand when you're finished. (Observe students and give feedback.)
- (Write on the board:)

a. $\dfrac{\text{red flowers}}{\text{yellow flowers}}$ $\dfrac{2}{7}\left(\dfrac{5}{5}\right)=\dfrac{10}{\boxed{35}}$ $\boxed{\text{35 yellow flowers}}$

- Here's the equation you should have. The problem asks about the number of yellow flowers. The answer is **35 yellow flowers.**
c. Work problem B. Raise your hand when you're finished. (Observe students and give feedback.)
- (Write on the board:)

b. $\dfrac{\text{bees}}{\text{flowers}}$ $\dfrac{8}{3}\left(\dfrac{33}{33}\right)=\dfrac{\boxed{264}}{99}$ $\boxed{\text{264 bees}}$

- Here's the equation you should have. The problem asks about the number of bees. The answer is **264 bees.**
d. Work problem C. Raise your hand when you're finished. (Observe students and give feedback.)
- (Write on the board:)

$$\text{c.} \quad \frac{\text{Tom}}{\text{sister}} \quad \frac{3}{2} \left(\frac{23}{23} \right) = \frac{69}{\boxed{46}} \quad \boxed{46 \text{ oz}}$$

- Here's the equation you should have. The problem asks how much Tom's sister eats. The answer is **46 ounces.**
e. Work problem D. Raise your hand when you're finished. (Observe students and give feedback.)
- (Write on the board:)

$$\text{d.} \quad \frac{\text{lb of sand}}{\text{lb of gravel}} \quad \frac{5}{6} \left(\frac{40}{40} \right) = \frac{\boxed{200}}{240} \quad \boxed{200 \text{ lb}}$$

- Here's the equation you should have. The problem asks how much sand was in the mixture. The answer is **200 pounds.**

EXERCISE 5 PROBLEM SOLVING
Number-Family Tables

a. Find part 7.
- You're going to make a table for the problem in part 7.
b. The problem says: There were both boys and girls at two different camps. The camps were Rainier Camp and Maxwell Camp.
- The rest of the problem tells about the numbers and asks the questions. Make the table with the headings. Don't put in any numbers. Remember, the first names are column headings. Raise your hand when you've written the headings.
  (Observe students and give feedback.)
- (Write on the board:)

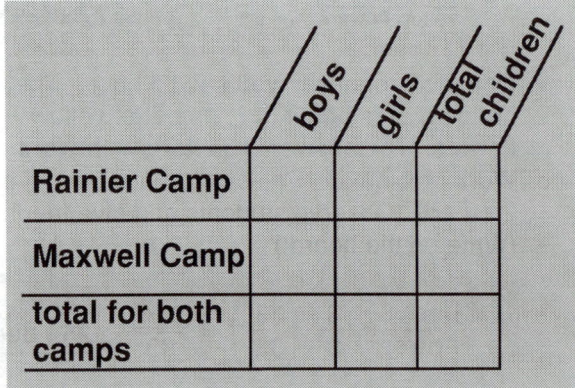

- Here's what you should have. You can call the total for boys and girls: **children, total,** or **total children.**
c. Put in the four numbers the problem gives. Raise your hand when you've done that much. (Observe students and give feedback.)
- (Write to show:)

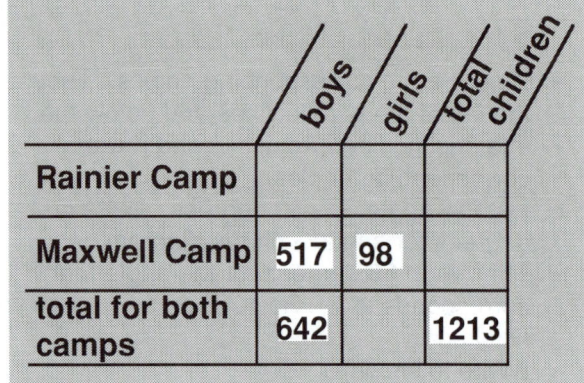

	boys	girls	total children
Rainier Camp			
Maxwell Camp	517	98	
total for both camps	642		1213

- Here's what you should have.
d. Now figure out the missing numbers and write answers to the questions. Raise your hand when you're finished.
  (Observe students and give feedback.)
- (Write to show:)

	boys	girls	total children
Rainier Camp	125	473	598
Maxwell Camp	517	98	615
total for both camps	642	571	1213

e. Check your work.
- Question A: Everybody, what was the total number of children at Maxwell Camp? (Signal.) *615.*
- Question B: Were there more boys or girls at Rainier Camp? (Signal.) *Girls.*
- Question C: Which of the two camps had fewer girls? (Signal.) *Maxwell Camp.*
- Question D: 125 tells about the number of boys at Rainier Camp. You should have written **boys** and **Rainier Camp** for question D.
- Raise your hand if you got everything right.

EXERCISE 6 FUNCTION TABLES
Two-Step Rules

a. Find part 8.
b. Each item has a rule and some numbers. Make two tables. Follow the rules and write the answers. Raise your hand when you're finished.
• (Write on the board:)

a.		b.	
10	8	7	31
4	6	2	191
1	5	5	95
7	7		

• Here are the tables you should have for each item. Raise your hand if you got everything right.

EXERCISE 7 INDEPENDENT WORK

• (In addition to independent work in the textbook, assign **Bridge to Connecting Math Concepts** *Independent Worksheet* 7 as classwork or homework. Before beginning the next lesson, check the students' independent work.)

Lesson 28

Objectives

• **Indicate whether lines are parallel or perpendicular.** (Exercise 1)

• Work a word problem by constructing a number-family table with column and row headings. (Exercise 2)

• Figure out the degrees in corresponding angles formed by a line that intersects parallel lines. (Exercise 3)

• **Compute the area of a parallelogram.** (Exercise 4)
  Note: Students learn the rule that the height of the figure is shown by a line that is perpendicular to the base.

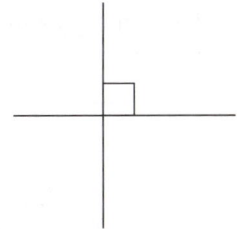

• **Complete a ratio table.** (Exercise 5)
  Note: These tables have columns that work like those in number-family tables. The total is at the bottom of the table. The rows are governed by a multiplication rule, such as times 5. Students multiply the values in the first column by 5 to find the values in the second column.

	x 5	
	8	40
	4	20
total	12	60

• Rewrite a fraction as a mixed number. (Exercise 6)

EXERCISE 1 GEOMETRY
Parallel/Perpendicular Lines

a. (Draw on the board:)

• This angle mark is used for a special angle. How many degrees are in the angle? (Signal.) *90 degrees.*
• And what do we call lines that form a 90-degree angle? (Signal.) *Perpendicular.* (Repeat step a until firm.)

b. Open your textbook to lesson 28 and find part 1.
- Some pairs of lines in part 1 are parallel. Some intersect. They are not parallel. Some of the lines that intersect have a square angle marker to indicate that the lines are perpendicular. Remember, perpendicular lines form an angle of 90 degrees.
c. For each item write **parallel, not parallel,** or **intersect** to tell about the lines. Raise your hand when you've done that much.
  (Observe students and give feedback.)
- Tell me what you wrote for each item.
- Item A: What did you write? (Signal.) *Intersect.*
- Item B: What did you write? (Signal.) *Not parallel.*
- Item C: What did you write? (Signal.) *Parallel.*
- Item D: What did you write? (Signal.) *Intersect.*
- Item E: What did you write? (Signal.) *Parallel.*
- Item F: What did you write? (Signal.) *Intersect.*
- Item G: What did you write? (Signal.) *Not parallel.*
d. Circle the letter of each item that shows a pair of perpendicular lines. Remember, those are lines that form a 90-degree angle. Raise your hand when you're finished. √
- You should have circled **A** and **D**. Both those items show lines that are perpendicular.
- Raise your hand if you got everything right.

EXERCISE 2 PROBLEM SOLVING
Number-Family Tables

a. Find part 2.
- You're going to make a table for the problem in part 2.
b. The problem says: In 1980 and 1981, babies were born in Queen's Hospital and Marist Hospital.
- The rest of the problem tells about the numbers and asks the questions. Make the table with the headings. Don't put in any numbers. Raise your hand when you've written the headings.
  (Observe students and give feedback.)
- (Write on the board:)

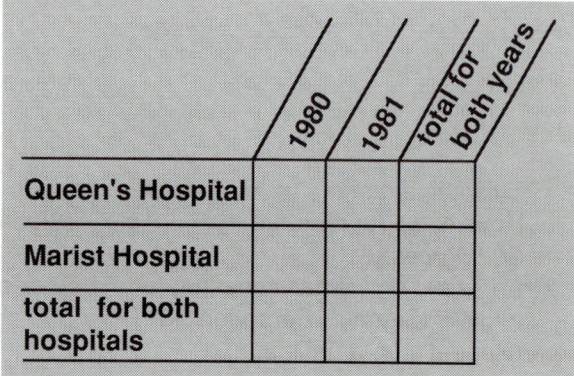

	1980	1981	total for both years
Queen's Hospital			
Marist Hospital			
total for both hospitals			

- Here's what you should have.
c. Put in the four numbers the problem gives. Raise your hand when you've done that much.
  (Observe students and give feedback.)
- (Write to show:)

	1980	1981	total for both years
Queen's Hospital	173		
Marist Hospital			463
total for both hospitals	237	575	

- Here's what you should have.
d. Now figure out the missing numbers and write answers to the questions. Raise your hand when you're finished.
  (Observe students and give feedback.)
- (Write to show:)

	1980	1981	total for both years
Queen's Hospital	173	176	349
Marist Hospital	64	399	463
total for both hospitals	237	575	812

e. Check your work.
- Everybody, how many babies were born in Queen's Hospital in 1981? (Signal.) *176.*
- In 1981, were more babies born in Queen's Hospital or Marist Hospital? (Signal.) *Marist Hospital.*
- In which year were fewer babies born in Marist Hospital? (Signal.) *1980.*
- 349 tells about the number of babies born in Queen's Hospital during both years. You should have written **Queen's Hospital** and **both years** or **1980 and 1981.**
- Raise your hand if you got everything right.

EXERCISE 3 GEOMETRY
Corresponding Angles

a. Find part 3.
- Each item has parallel lines and an intersecting line. Remember, corresponding angles formed at parallel lines are equal. Each item gives you information about one of the angles. You'll figure out the corresponding angle that's equal to that angle.
b. Trace figure A. Raise your hand when you're finished. √
- Listen: Angle P is 72 degrees. Write the number of degrees for angle P and for the angle that corresponds to angle P. Raise your hand when you're finished.
  (Observe students and give feedback.)
- (Write on the board:)

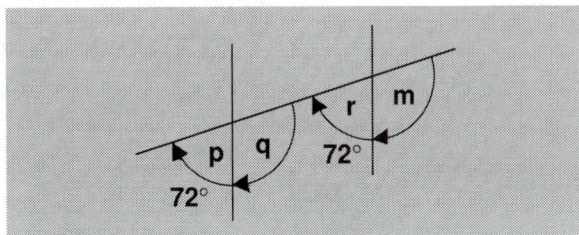

- What's the angle that corresponds to angle P? (Signal.) *Angle R.*
- How many degrees is angle R? (Signal.) *72.*
c. If you know what angle R equals, you can figure out what angle M equals because R and M together make half a circle. Your turn: Figure out angle M. Write the number of degrees. Then write the degrees for the corresponding angle. Raise your hand when you're finished.
  (Observe students and give feedback.)
- (Write to show:)

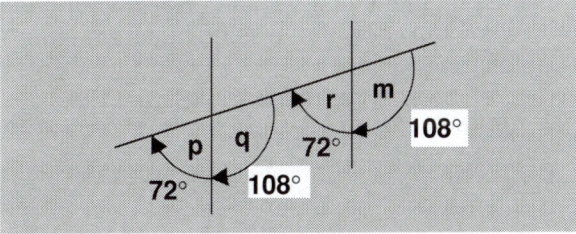

- Check your work. Everybody, how many degrees is angle M? (Signal.) *108.*
- What angle corresponds to M? (Signal.) *Angle Q.*
- How many degrees is angle Q? (Signal.) *108.*
d. Trace figure B. Raise your hand when you're finished.
- Item B gives you information about one of the angles. You'll figure out what the degrees are for all of the marked angles.

- Listen: Angle G is 125 degrees. Figure out all the other marked angles and write the degrees. Raise your hand when you're finished.
  (Observe students and give feedback.)
- (Write on the board:)

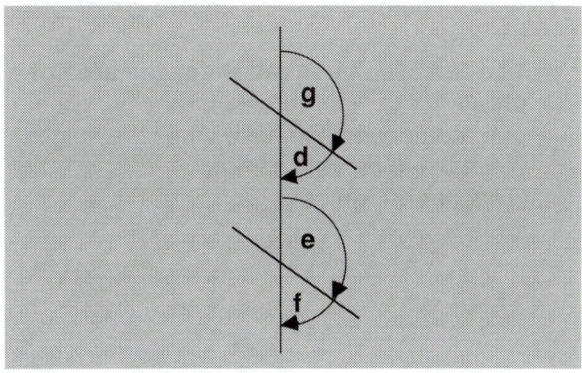

- Check your work. Angle G: Everybody, how many degrees is it? (Signal.) *125.*
- What's the other angle that equals 125 degrees? (Signal.) *Angle E.*
- Yes, angle E corresponds to angle G.
- Angle D: How many degrees does it equal? (Signal.) *55.*
- What other angle equals 55 degrees? (Signal.) *Angle F.*
- (Write to show:)

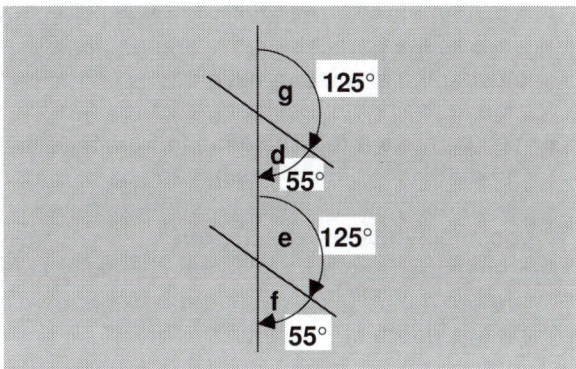

- Here's what you should have.
- Raise your hand if you got everything right.

EXERCISE 4 AREA
Parallelograms

a. Find part 4.
  (Teacher reference:)

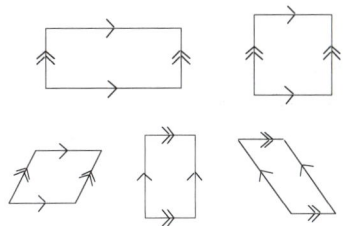

b. I'll read what it says. Follow along: You've worked with figures that have pairs of parallel sides. These figures are called **parallelograms.**
- A **rectangle** is a special kind of parallelogram.
- A **square** is a parallelogram.
- All the figures below the square are also parallelograms.
- You can use the **equation for the area of a rectangle** to find the area of any parallelogram. The area equals the base times the height. But you have to be very careful.
(Teacher reference:)

- For parallelograms that are not rectangles, the **height** of the figure is **not** the length of a side.
- You can see a parallelogram with numbers for two sides. The left side does not show the height. The **height** is a dotted line that is perpendicular to the base of the figure. The height is 8.
- To find the area for the parallelogram, you work the problem 20 times 8. The problem actually tells about a rectangle. You can see the parallelogram with part of it shaded.
- You can make a rectangle that has the same area as the parallelogram. You do that by moving the shaded triangle that is on the left side to the right side.
- You can see a rectangle made from the parts of the parallelogram. Parts haven't been added or taken away, so the parallelogram and the rectangle have the same area.
- Remember, the base times the height gives the area of a parallelogram, but the height may not be the length of a side.
c. Find part 5.
Find the area of each figure. Remember to show three equations. Raise your hand when you're finished.
(Observe students and give feedback.)
Key:

a. $A = b \times h$
    $A = 16 \times 22$
    $\boxed{A = 352 \text{ sq in}}$

b. $A = b \times h$
    $A = 12 \times 8$
    $\boxed{A = 96 \text{ sq mi}}$

c. $A = b \times h$
    $A = 18 \times 18$
    $\boxed{A = 324 \text{ sq m}}$

d. $A = b \times h$
    $A = 20 \times 14$
    $\boxed{A = 280 \text{ sq ft}}$

d. Check your work.

- Figure A. What's the base? (Signal.) *16 inches.*
- What's the height? (Signal.) *22 inches.*
- What's the area? (Signal.) *352 square inches.*
- Figure B. What's the base? (Signal.) *12 miles.*
- What's the height? (Signal.) *8 miles.*
- What's the area? (Signal.) *96 square miles.*
- Figure C. What's the base? (Signal.) *18 meters.*
- What's the height? (Signal.) *18 meters.*
- What's the area? (Signal.) *324 square meters.*
- Figure D. What's the base? (Signal.) *20 feet.*
- What's the height? (Signal.) *14 feet.*
- What's the area? (Signal.) *280 square feet.*
- Raise your hand if you got everything right.

EXERCISE 5 RATIOS AND PROPORTIONS
Tables

a. (Write on the board:)

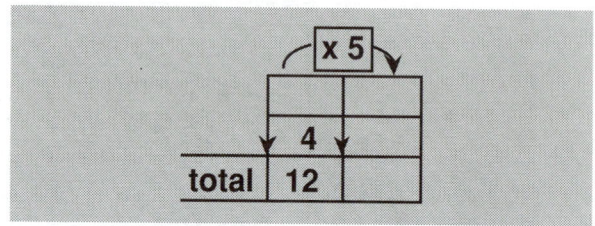

- This is a new kind of table. The total is at the bottom of each column, so if you have two numbers in a column, you can figure out the missing number.
- **You can't add for the rows.** You multiply to go from the first number in the row to the second number in that row. At the top of the table is **times 5.** That means you multiply the first number in each row by 5 to get the other number in the row.
- First work the column. The top number is missing in the first column. Raise your hand when you know the missing number in that column. √
- Everybody, what number? (Signal.) *8.*
b. (Write to show:)

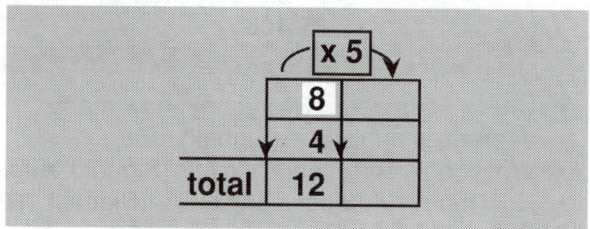

- 8 plus 4 equals 12.
- Now we can multiply to complete each row.
- For the top row, we multiply 8 by 5. What's the answer? (Signal.) *40.*
- For the next row, we multiply 4 by 5. What's the answer? (Signal.) *20.*

- For the bottom row, we multiply 12 by 5. What's the answer? (Signal.) *60.*
c. (Write to show:)

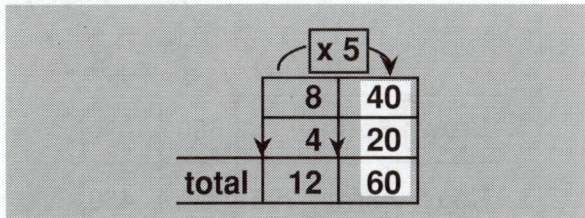

d. Find part 6.
- Copy the table and figure out the missing numbers. Raise your hand when you're finished. (Observe students and give feedback.)
e. (Write on the board:)

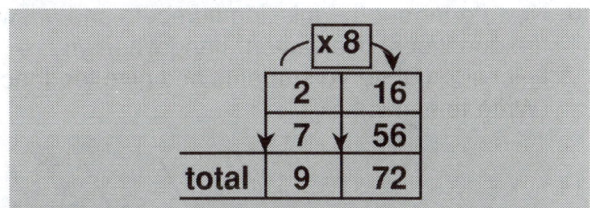

- Here's what you should have. The missing number in the first column is 7. The missing numbers in the last column are 16, 56, 72. 16 plus 56 equals 72.

EXERCISE 6 FRACTIONS
As Mixed Numbers

a. Find part 7.
   You're going to change fractions into mixed numbers. To do that, work the division problem and write the remainder as a fraction.
b. Fraction A. Read it as a division problem. (Signal.) *38 divided by 5.*
- Fraction B. Read it as a division problem. (Signal.) *146 divided by 7.*
- Fraction C. Read it as a division problem. (Signal.) *27 divided by 2.*
- (Repeat step b until firm.)
c. Your turn: Write the division problem and the answer for each fraction in part 7. Remember, write the remainder as a fraction. Raise your hand when you're finished.
   (Observe students and give feedback.)
   Key:

$$a.\ 5\overline{)38}\,{}^{7\frac{3}{5}} \qquad b.\ 7\overline{)146}\,{}^{20\frac{6}{7}}$$

$$c.\ 2\overline{)27}\,{}^{13\frac{1}{2}} \qquad d.\ 9\overline{)352}\,{}^{39\frac{1}{9}}$$

d. Check your work.
- Fraction A: 38/5. Say the division problem and the whole answer. (Signal.) *38 divided by 5 equals 7 and 3-fifths.*
- So 7 and 3-fifths equals the fraction you started with—38/5.
- Fraction B: 146/7. Say the division problem and the whole answer. (Signal.) *146 divided by 7 equals 20 and 6-sevenths.*
- So 20 and 6-sevenths equals the fraction you started with—146/7.
- Fraction C: 27/2. Say the division problem and the whole answer. (Signal.) *27 divided by 2 equals 13 and 1-half.*
- So 13 and 1/2 equals the fraction you started with—27/2.
- Fraction D: 352/9. Say the division problem and the whole answer. (Signal.) *352 divided by 9 equals 39 and 1-ninth.*
- So 39 and 1/9 equals the fraction you started with—352/9.
- Raise your hand if you wrote all the right mixed numbers.

EXERCISE 7 INDEPENDENT WORK

- (In addition to independent work in the textbook, assign *Bridge to Connecting Math Concepts Independent Worksheet* 8 as classwork or homework. Before beginning the next lesson, check the students' independent work.)

Lesson 29

EXERCISE 1 PROBLEM SOLVING
Number-Family Tables

a. Open your textbook to lesson 29 and find part 1.
- This is a table problem.
b. Read the first two sentences of the problem and make the table with the headings. Raise your hand when you've done that much. √
- (Write on the board:)

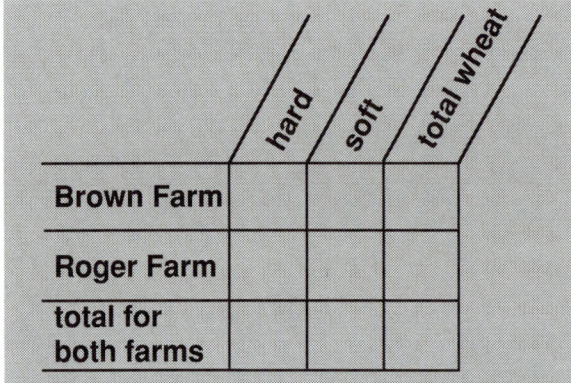

- Here's what you should have for the headings.
c. Put in the four numbers the problem gives. Raise your hand when you've done that much. (Observe students and give feedback.)

- (Write to show:)

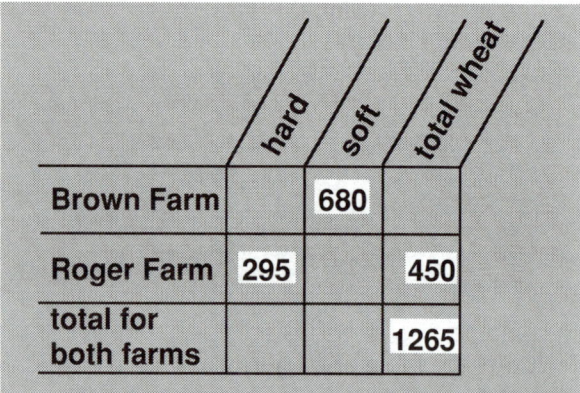

- Here's what you should have. Make sure each number is in the right place.
d. Now figure out the missing numbers and answer the questions. Raise your hand when you're finished. (Observe students and give feedback.)
- (Write to show:)

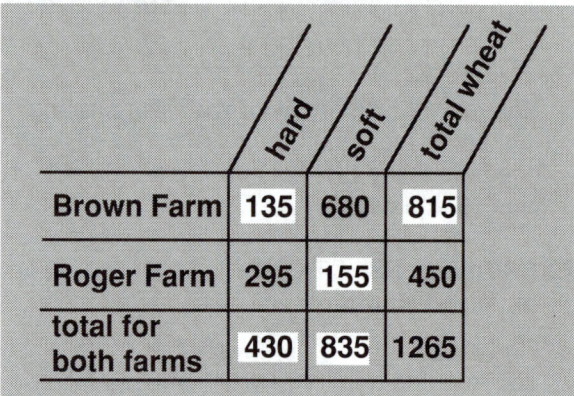

- Here's what you should have.
e. Check your work.
- Item A: Everybody, what is the total amount of hard wheat grown on both farms? (Signal.) *430 tons.*
- Item B: Was more hard wheat grown on Brown Farm or Roger Farm? (Signal.) *Roger Farm.*
- Item C: Was there less hard wheat or soft wheat on Roger Farm? (Signal.) *Soft wheat.*
- Raise your hand if you got everything right.

EXERCISE 2 FRACTIONS
As Mixed Numbers

a. Find part 2.
   You're going to change fractions into mixed numbers. To do that, work the division problem and write the remainder as a fraction.
b. Fraction A. Read it as a division problem. (Signal.) *191 divided by 4.*
- Fraction B. Read it as a division problem. (Signal.) *35 divided by 8.*

- Fraction C. Read it as a division problem.
  (Signal.) *141 divided by 5.*
  (Repeat step b until firm.)
c. Your turn: Write the division problem and the
  answer for each fraction in part 2. Remember,
  write the remainder as a fraction. Raise your
  hand when you're finished.
  (Observe students and give feedback.)
d. Check your work.
- Fraction A: 191/4. Say the division problem
  and the whole answer. (Signal.)
  191 divided by 4 equals 47 and 3-fourths.
- Fraction B: 35/8. Say the division problem and
  the whole answer. (Signal.)
  35 divided by 8 equals 4 and 3-eighths.
- Fraction C: 141/5. Say the division problem
  and the whole answer. (Signal.)
  141 divided by 5 equals 28 and 1-fifth.
- Fraction D: 26/3. Say the division problem and
  the whole answer. (Signal.)
  26 divided by 3 equals 8 and 2-thirds.

EXERCISE 3 AREA
Rectangles and Parallelograms

a. Find part 3.
  You're going to find the area of rectangles and
  parallelograms that are not rectangles.
  Remember, for those figures, the height is
  measured perpendicular to the base. The base
  times the height gives the area.
b. Find the area of the figures in part 3. Raise your
  hand when you're finished.
  (Observe students and give feedback.)
  Key:

  *a. A = b x h
      A = 15 x 10
      A = 150 sq ft*

  *b. A = b x h
      A = 16 x 21
      A = 336 sq cm*

  *c. A = b x h
      A = 23 x 14
      A = 322 sq yd*

c. Check your work.
- Figure A. What's the base? (Signal.) *15 feet.*
- What's the height? (Signal.) *10 feet.*
- What's the area? (Signal.) *150 square feet.*
- Figure B. What's the base? (Signal.) *16
  centimeters.*
- What's the height? (Signal.) *21 centimeters.*
- What's the area? (Signal.) *336 square
  centimeters.*
- Figure C. What's the base? (Signal.) *23 yards.*
- What's the height? (Signal.) *14 yards.*
- What's the area? *322 square yards.*

- Raise your hand if you wrote the correct area for
  all the parallelograms.

EXERCISE 4 RATIOS AND PROPORTIONS
Tables

a. Find part 4.
- You worked tables like these last time.
  Remember, you multiply for each **row.** You add
  for each **column.**
b. Copy table A and work it. Raise your hand
  when you're finished.
  (Observe students and give feedback.)
- (Write on the board:)

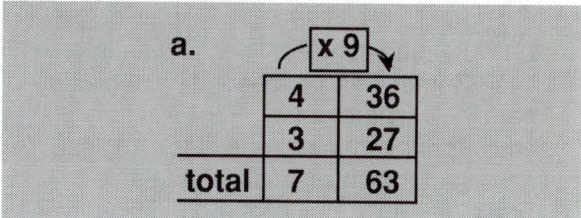

- Here's what you should have. The missing
  number in the first column is 3. The numbers in
  the second column are 36, 27, and 63. 36 plus
  27 equals 63.
c. Table B doesn't show what you multiply by.
- Figure out the multiplication rule for the top row
  of the table. Then multiply each of the other
  rows by the same number. Raise your hand
  when you've completed the table.
  (Observe students and give feedback.)
- (Write on the board:)

b.		
	7	56
	9	72
total	16	128

- Here's what you should have. You multiply by 8.
  The missing number for the middle row is 72.
  The missing number in the bottom row is 128.
  Raise your hand if you got it right.
- You can test the numbers in the second column
  by seeing if the top two equal the total. You add
  56 and 72 and see if you get 128. Raise your
  hand when you know the answer. √
- Everybody, what's 56 plus 72? (Signal.) *128.*
d. Table C. There's a row with two numbers.
  That's the middle row.
- Figure out what you multiply by in that row.
  Complete the other rows.
- Check the numbers in the last column to make
  sure the top numbers add up to the total. Raise
  your hand when you're finished.
  (Observe students and give feedback.)

- (Write on the board:)

	3	12
	10	40
total	13	52

- Here's what you should have. You multiply by 4 in each row. The missing number for the top row is 12. The total for the second column is 52.

EXERCISE 5 GEOMETRY
Corresponding Angles

a. Find part 5.
   Each item has parallel lines and an intersecting line. Remember, corresponding angles at parallel lines are equal. For each item, I'll give you information about one of the angles. You'll figure out the corresponding angle.
b. Trace figure A. Raise your hand when you've done that much. √
- Listen: Angle W is 28 degrees. Write the number of degrees for angle W and for the angle that corresponds to angle W. Raise your hand when you're finished.
  (Observe students and give feedback.)
- (Write on the board:)

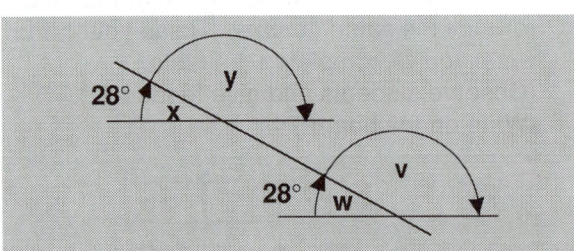

- What's the angle that corresponds to angle W? (Signal.) Angle X.
- How many degrees is angle X? (Signal.) 28.
c. If you know what angle X equals, you can figure out what angle Y equals because X and Y together make half a circle. Your turn: Figure out angle Y. Write the number of degrees. Then write the degrees for the corresponding angle. Raise your hand when you're finished.
  (Observe students and give feedback.)
- (Write to show:)

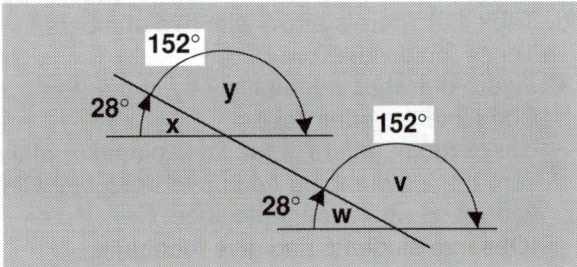

- Check your work. Here's what you should have. Angle Y is 152 degrees.
- Which angle corresponds to Y? (Signal.) Angle V.
- How many degrees is angle V? (Signal.) 152.
d. Trace figure B. Raise your hand when you're finished.
- The item gives you information about one of the angles.
- Angle N is 147 degrees. Figure out all the other marked angles and write the degrees. Raise your hand when you're finished.
  (Observe students and give feedback.)
- (Write on the board:)

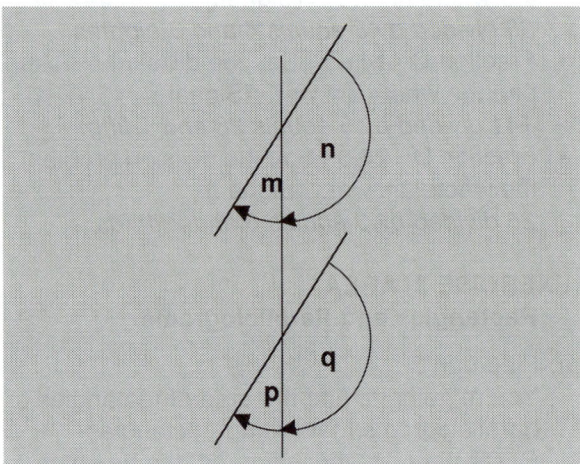

- Check your work. Everybody, how many degrees is angle N? (Signal.) 147.
- What's the other angle that equals 147 degrees? (Signal.) Angle Q.
- Q and N are corresponding angles.
- How many degrees does angle P equal? (Signal.) 33.
- What other angle equals 33 degrees? (Signal.) Angle M.
- M and P are corresponding angles.
- (Write to show:)

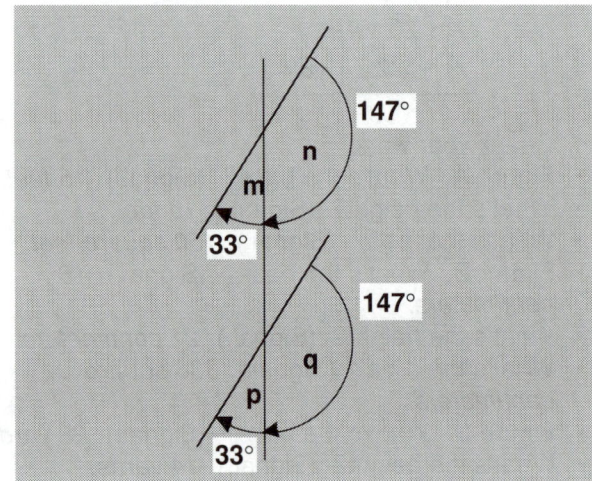

- Here's what you should have.
- Raise your hand if you got everything right.

EXERCISE 6 FUNCTION TABLES
Two-Step Rules

a. Find part 6.
b. Each item has a rule and some numbers. Follow the rule and write the answers. Raise your hand when you've completed a table for each item.
- (Write on the board:)

a.

2	4
8	4
5	4
9	4

b.

3	35
1	25
8	60

c.

5	0
2	60
3	40

- Here's what you should have for each item.
- Raise your hand if you got everything right.

EXERCISE 7 INDEPENDENT WORK

- (In addition to independent work in the textbook, assign **Bridge to Connecting Math Concepts** *Independent Worksheet* 9 as classwork or homework. Before beginning the next lesson, check the students' independent work.)

Lesson 30 – Test 3

Objectives

- **Complete a table to show a division problem and a corresponding fraction-to-mixed number equation.** (Exercise 1)

  Note: For example: $\frac{4\frac{3}{5}}{5\overline{)23}}$ $\quad \frac{23}{5} = 4\frac{3}{5}$

- Compute the area of a parallelogram. (Exercise 2)

- Perform on a mastery test of skills presented in lessons 21 through 29. (Exercise 3)

Note: Exercise 4 provides instructions for marking the test.

EXERCISE 1 FRACTIONS
As Mixed Numbers

a. Open your textbook to lesson 30 and find part 1.
- This is a table that should show division problems in the first column and fraction equations in the second column.
- The sample row shows 23 divided by 5. The answer is 4 and 3/5. The fraction equation shows that 23/5 equals 4 and 3/5.
b. Row A has 39 divided by 9.
- Copy the division problem and write the mixed-number answer. Then write the fraction equation with the same numbers. Raise your hand when you've completed row A. (Observe students and give feedback.)
- (Write on the board:)

a. $\quad \frac{4\frac{3}{9}}{9\overline{)39}} \quad \Bigg| \quad \frac{39}{9} = 4\frac{3}{9}$

- Here's what you should have for row A.
c. Your turn: Complete the rows for the rest of the table. Raise your hand when you're finished. (Observe students and give feedback.)

- (Write to show:)

a.	$9\overline{)39}^{\,4\frac{3}{9}}$	$\dfrac{39}{9} = 4\frac{3}{9}$
b.	$8\overline{)71}^{\,8\frac{7}{8}}$	$\dfrac{71}{8} = 8\frac{7}{8}$
c.	$3\overline{)25}^{\,8\frac{1}{3}}$	$\dfrac{25}{3} = 8\frac{1}{3}$
d.	$4\overline{)14}^{\,3\frac{2}{4}}$	$\dfrac{14}{4} = 3\frac{2}{4}$

- Here's what you should have for rows B, C and D.
- Raise your hand if you got everything right.

EXERCISE 2 AREA
Rectangles and Parallelograms

a. Find part 2.
- Two of these figures are parallelograms that are not rectangles. Remember, for those figures, the height is not the length of a side. The height is a line that is **perpendicular** to the base.
b. Find the area of each figure in part 2. Remember to show three equations for each problem. Raise your hand when you're finished. (Observe students and give feedback.)
  Key:

a. $A = b \times h$
   $A = 30 \times 36$
   $\boxed{A = 1080 \text{ sq ft}}$

b. $A = b \times h$
   $A = 20 \times 12$
   $\boxed{A = 240 \text{ sq in}}$

c. $A = b \times h$
   $A = 17 \times 9$
   $\boxed{A = 153 \text{ sq km}}$

- Find part J on page 119. That shows what you should have for each figure.
- Figure A: The area is 1080 square feet.
- Figure B: The area is 240 square inches.
- Figure C: The area is 153 square kilometers.
- Raise your hand if you got everything right.

EXERCISE 3 TEST 3

Note: **Students are not to use calculators for any part of the test.**

a. This is a test. You should only have your textbook, a sharpened pencil and lined paper on your desk.
b. Find part 1 of test 3 in your textbook. √
c. Do the test on your own. Raise your hand when you've completed the test.
   (Observe students but do not give feedback.)

EXERCISE 4 MARKING THE TEST

a. (Collect the students' papers. Use the Answer Key to score tests. Award scores for test 3 as follows:)

Test 3 Percent Summary					
SCORE	%	SCORE	%	SCORE	%
59	100	53	90	47	80
58	98	52	88	46	78
57	97	51	86	45	76
56	95	50	85	44	75
55	93	49	83	43	73
54	92	48	81	42	71

b. (Complete the Test 3 Remedy Summary to determine whether remedies are needed. Reproducible Summary Sheets are at the back of the Teacher's Guide.)
- (If more than 1/4 of the students did not pass a test part, present the remedy for that part before beginning lesson 31. Remedies appear at the end of the Test 3 Answer Key.)

EXERCISE 5 INDEPENDENT WORK

- (Assign **Bridge to Connecting Math Concepts** *Independent Worksheet* 10 as classwork or homework. Before beginning the next lesson, check the students' independent work.)

CUMULATIVE TEST REMINDER

Before presenting lesson 31, present Cumulative Test 1 that appears in Appendix A of the Teacher's Guide. Also, provide remedies for students who do not pass the test.

Lesson 31

EXERCISE 1 LONG DIVISION
Quotient Given

a. Open your textbook to lesson 31 and find part 1. You're going to work problems that divide by a two-digit value. For these problems, the number is shown under the division sign. Part of the answer is also shown, but there's a remainder for each answer. You're going to multiply and subtract to figure out the remainder.

b. I'll work problem A.

- (Write on the board:)

$$\text{a. } 47\overline{\smash{\big)}289} \quad 6$$

- We'll multiply 47 by 6 and write the answer below.

- (Write to show:)

$$\text{a. } 47\overline{\smash{\big)}289} \quad 6 \\ \phantom{\text{a. } 47)}-282$$

- If the original problem had 282 under the division sign, there wouldn't be a remainder. To find the remainder, we just subtract.

- (Write to show:)

$$\text{a. } 47\overline{\smash{\big)}289} \quad 6 \\ \phantom{\text{a. } 47)}\underline{-282} \\ \phantom{\text{a. } 47)2}7$$

- 289 minus 282. The remainder is 7. We can write that remainder as a fraction. Everybody, what fraction? (Signal.) *7-forty-sevenths.*

- (Write to show:)

$$\text{a. } 47\overline{\smash{\big)}289} \quad 6\frac{7}{47} \\ \phantom{\text{a. } 47)}\underline{-282} \\ \phantom{\text{a. } 47)2}7$$

- Your turn: Copy problem A. Raise your hand when you're finished.
  (Observe students and give feedback.)

c. Your turn: Use your calculator to work problem B. Multiply and subtract to find the remainder. Write the remainder as a fraction. Raise your hand when you're finished.
  (Observe students and give feedback.)

- (Write on the board:)

$$\text{b. } 36\overline{\smash{\big)}205} \quad 5\frac{25}{36} \\ \phantom{\text{b. } 36)}\underline{-180} \\ \phantom{\text{b. } 36)}25$$

- Here's what you should have for problem B. 5 times 36 is **180.** When you subtract from **205,** you end up with a remainder of **25.** That's 25/36. 205 divided by 36 equals 5 and 25/36. Raise your hand if you got it right.

d. Your turn: Work the rest of the problems in part 1. Remember, first multiply. Then subtract to figure out the remainder.

- Write each remainder as a fraction. Raise your hand when you're finished.
  (Observe students and give feedback.)

  Key:

  c. $23\overline{)227}$ $9\frac{20}{23}$
  -207
  20

  d. $16\overline{)139}$ $8\frac{11}{16}$
  -128
  11

e. Check your work.
  Find part J on page 123. That shows the work you should have for problems C and D.
- Problem C: 227 divided by 23 equals 9 and 20/23.
- Problem D: 139 divided by 16 equals 8 and 11/16.
- Check your work carefully and fix any mistakes.

EXERCISE 2 RATIOS AND PROPORTIONS
Tables

a. Find part 2.
- These are tables with missing numbers in both columns. You can figure out the missing number in the first column by adding or subtracting. Then you find the row with two numbers and figure out what you multiply by.
b. Copy table A and write the missing number in the first column. Raise your hand when you've done that much.
  (Observe students and give feedback.)
- Now you should have **9** as the total for the first column. Figure out what you multiply by in the top row. Then complete the other rows of the table. Raise your hand when you're finished.
  (Observe students and give feedback.)
- (Write on the board:)

a.	4	28
	5	35
total	9	63

- Here's what you should have. You multiply by 7 in each row. The missing number in the middle row is 35. The missing total number is 63.
- You can check those numbers by adding. 28 plus 35 equals 63. So the numbers in the second column are correct.
c. Copy table B. Write the missing number in the first column. Then use the row with two numbers to figure out what you multiply by.
- Check your answers by adding the two top numbers in the second column. Raise your hand when you've finished table B.
  (Observe students and give feedback.)

- (Write on the board:)

b.	3	33
	1	11
total	4	44

- Here's what you should have. The total for the first column is 4. You multiply by 11 in each row. The numbers in the second column add up. 33 plus 11 equals 44.

EXERCISE 3 PLACE VALUE
Hundreds

a. Find part 3.
- You can read thousands numbers as hundreds. You just read to the last digits that are not underlined. Then say **hundred.**
b. Number A: 4 thousand 500. The last two digits are underlined. I'll read the digits that are not underlined and then say **hundred: 45 hundred.**
- Number B: 3 thousand 600. I'll read the digits that are not underlined and say **hundred: 36 hundred.**
c. Number C. Your turn: Read it as hundreds. (Signal.) *11 hundred.*
- Number D. Read it as hundreds. (Signal.) *18 hundred.*
- Number E. Read it as hundreds. (Signal.) *92 hundred.*
- (Repeat step c until firm.)

EXERCISE 4 GEOMETRY
Supplementary Angles

a. Find part 4.
- You're going to figure out all the angles in the circle. The circle is formed by two intersecting lines.
- (Write on the board:)

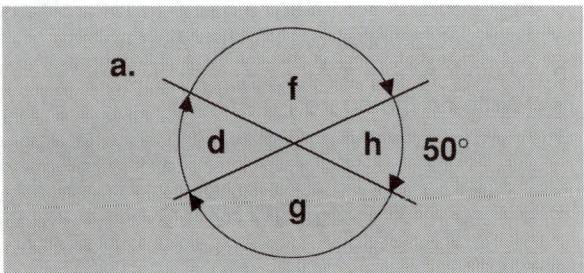

- Here's item A. Here's information about one of the angles: Angle H is 50 degrees. Trace figure A. You don't have to copy the arrow heads. Write 50 degrees for angle H. Raise your hand when you've done that much. √

- Angle H can be added to angle F to form half a circle.
- (Shade to show:)

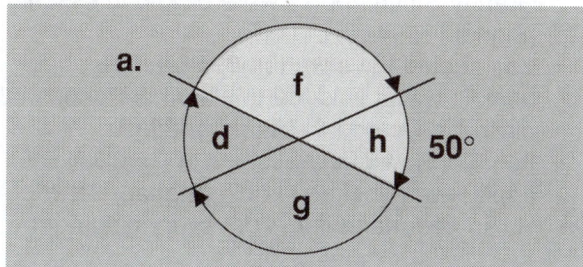

b. Your turn: Figure out angle F. Raise your hand when you've done that much.
  (Observe students and give feedback.)
- Angle F is 180 minus 50. Everybody, how many degrees are in angle F? (Signal.) *130.*
- (Write to show:)

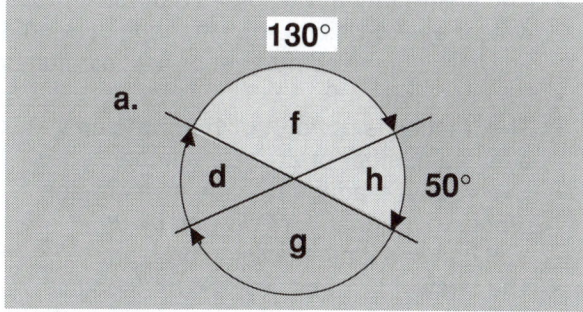

- Angle F is 130 degrees. Look at angle H again. Angle H and angle F form half a circle. Angle H and a **different angle form another half circle.** Everybody, which angle is that? (Signal.) *Angle G.*
- (Change shading to show:)

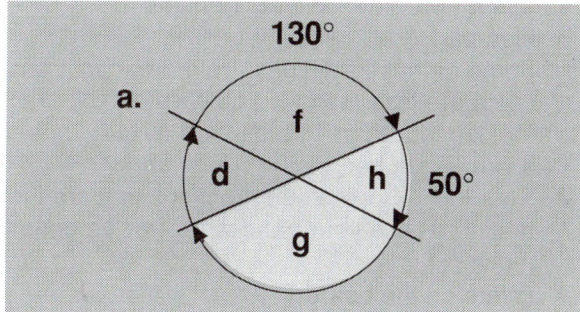

- Figure out angle G. Raise your hand when you're finished. √
- Angle G is 180 minus 50. That's **130.** (Write **130°**.)
- Angle G is the same size as angle F.

c. How could we figure out how many degrees are in angle D?
  (Call on a student. Ideas: *Combine D and F to form half a circle; combine D and G to form half a circle. Subtract 130° from 180°.*)
- Do it. Figure out angle D. Raise your hand when you're finished.
- Everybody, how many degrees is angle D? (Signal.) *50.*
- Angle D is 180 minus 130. That's **50.** Angle D is the same size as angle H.

d. Item B. You're going to figure out every angle in the circle. Listen: Angle M is 115 degrees. Trace figure B. Write 115 degrees for angle M. Then figure out what angle F equals. Raise your hand when you're finished.
- (Write on the board:)

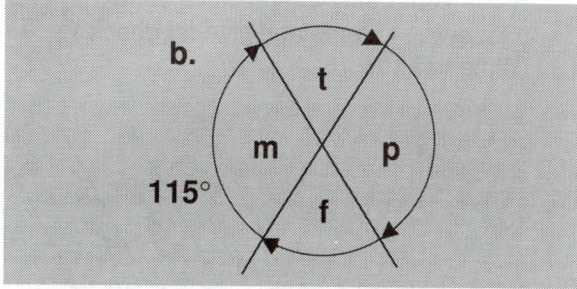

- Angle M and angle F form half a circle. Angle M equals 115 degrees. So what does angle F equal? (Signal.) *65 degrees.*
- (Write **65°**.)
- Angle M and angle F form half a circle. Angle M and another angle form a different half circle. Everybody, which angle is that? (Signal.) *Angle T.*
- Figure out angle T. Then figure out the last angle. Raise your hand when you're finished. (Observe students and give feedback.)
- Everybody, what does angle T equal? (Signal.) *65 degrees.*
- What does angle P equal? (Signal.) *115 degrees.*
- Angle M is the same size as angle P. Angle T is the same size as angle F. Think about that. See if you can figure out a rule for angles in a circle.

EXERCISE 5 PROBLEM SOLVING
Number-Family Tables

a. Find part 5.
- This is a table problem.
b. Read the first two sentences of the problem and make the table with the headings. Raise your hand when you've done that much.

- (Write on the board:)

	September	October	total for both months
Granny Smith			
Red Delicious			
total for both apples			

- Here's what you should have for the headings.
- c. Put in the four numbers the problem gives. Raise your hand when you've done that much. (Observe students and give feedback.)
- (Write to show:)

	September	October	total for both months
Granny Smith	276		1170
Red Delicious	191		
total for both apples			3356

- Here's what you should have. Make sure you have each number in the right place.
- d. Now figure out the missing numbers and answer the questions. Remember the unit name **bushels** in your answers. Raise your hand when you're finished. (Observe students and give feedback.)
- (Write to show:)

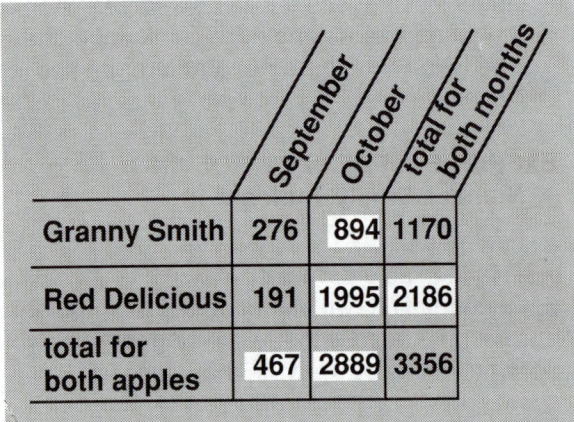

	September	October	total for both months
Granny Smith	276	894	1170
Red Delicious	191	1995	2186
total for both apples	467	2889	3356

- e. Check your work.
- Here's what you should have.
- Item A: Everybody, how many bushels of Granny Smith apples were picked in October? (Signal.) *894.*
- You should have written **894 bushels.**
- Item B: What was the total number of bushels picked in October? (Signal.) *2889.*
- Item C: How many bushels of Red Delicious apples were picked in all? (Signal.) *2186.*
- Raise your hand if you got everything right.

EXERCISE 6 FRACTIONS
As Mixed Numbers/Whole Numbers

- a. Find part 6.
- I'll read what it says. Follow along: Any fraction that is more than 1 can be written as a whole number or a mixed number.
- You just read the fraction as a division problem, write it as a division problem, and figure out the answer as a mixed number or a whole number.
- You can see 14/9.
- The fraction is more than 1, so we can write it as a mixed number or a whole number.
- Everybody, read the fraction as a division problem. (Signal.) *14 divided by 9.*
- You can see it as a division problem.
- The answer is 1 and 5/9. So 14/9 equals 1 and 5/9.
- The next fraction is 36/9.
- The fraction is more than 1, so we can write it as a mixed number or a whole number.
- Read the fraction as a division problem. (Signal.) *36 divided by 9.*
- You can see the division problem.
- The answer is 4. So 36/9 equals 4.
- b. Find part 7.
  Some of these fractions are more than 1 and some are less than 1. You'll write each fraction that is more than 1 as either a mixed number or a whole number.
- c. Your turn: Copy all the fractions that are more than 1. Raise your hand when you've done that much. √
- (Write on the board:)

a. $\dfrac{17}{8}$	d. $\dfrac{29}{26}$	e. $\dfrac{63}{7}$	g. $\dfrac{67}{9}$

- The fractions that are more than 1 are: A, D, E and G.

d. Your turn: Below each fraction, write the division problem and figure out the mixed number or whole number each fraction equals. Remember to box your answer. Raise your hand when you're finished.
(Observe students and give feedback.)

• (Write on the board:)

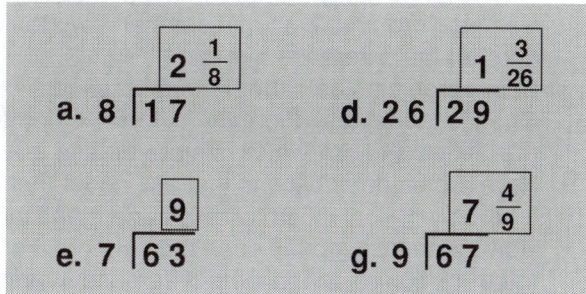

• Check your work. Here are the answers you should have for each division problem. You rewrote each fraction as a mixed number or a whole number.
• Fraction A: 17/8 equals 2 and 1/8.
• Fraction D: 29/26 equals 1 and 3/26.
• Fraction E: 63/7 equals 9.
• Fraction G: 67/9 equals 7 and 4/9.

EXERCISE 7 INDEPENDENT WORK

a. From now on, you'll be able to use an answer key to check whether some of your answers are correct.
b. Find page K1. The K pages start near the end of your textbook. Raise your hand when you've found page K1. √
• Find the row for 31. √
It shows the answers for every other problem in parts 8 through 15 of your independent work. Notice that the answer key does not show you how to work the problems. It just shows the answers.
• When you work problems, you'll have to show the work. If your answer is not correct, you can try working the problem again or working it a different way. Remember, when you work your problems, check the answer key. If your answer is not correct, see if you can find your mistake.
• (In addition to independent work in the textbook, assign **Bridge to Connecting Math Concepts** Independent Worksheet 11 as classwork or homework. Before beginning the next lesson, check the students' independent work.)

Lesson 32

Objectives

• Complete a division problem that has a two-digit divisor. (Exercise 1)

• **Find the area and perimeter of a parallelogram.** (Exercise 2)

• **Solve a comparison word problem that asks about the difference.** (Exercise 3)

• **Complete a ratio table that has headings for the rows.** (Exercise 4)

• Read a thousands numeral as a hundreds numeral. (Exercise 5)

• Figure out the degrees in the four angles that are formed by two intersecting lines. (Exercise 6)

• Write a fraction as a division problem and show the answer as a whole number or a mixed number. (Exercise 7)

EXERCISE 1 LONG DIVISION
Quotient Given

a. Open your textbook to lesson 32 and find part 1. These are division problems. Part of the answer is shown, but there's a remainder for each answer. You'll multiply and subtract to find the remainder. You'll write the remainder as a fraction.
b. Use your calculator and work all the problems in part 1. Raise your hand when you're finished.
(Observe students and give feedback.)
c. (Write on the board:)

$$\begin{array}{ll} \text{a.} \quad 63\overline{)337} \quad 5\frac{22}{63} & \text{b.} \quad 91\overline{)809} \quad 8\frac{81}{91} \\ \quad\quad -315 & \quad\quad -728 \\ \quad\quad\;\; 22 & \quad\quad\;\; 81 \\[2mm] \text{c.} \quad 49\overline{)358} \quad 7\frac{15}{49} & \text{d.} \quad 17\overline{)160} \quad 9\frac{7}{17} \\ \quad\quad -343 & \quad\quad -153 \\ \quad\quad\;\; 15 & \quad\quad\;\;\; 7 \end{array}$$

d. Here's what you should have. Check your work over and fix any mistakes.
e. Raise your hand if you didn't make any mistakes.
• Good work!

EXERCISE 2 AREA AND PERIMETER
Parallelograms

a. Find part 2.
* (Teacher reference:)

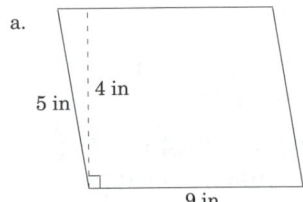 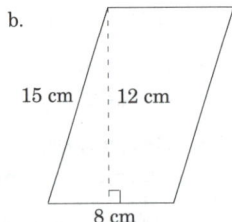

* You're going to find the area and perimeter of these parallelograms. The height is shown with a dotted line. Remember, that's **not** the length of any side.
b. Figure A. The length of the left side is shown. What's the length of that side? (Signal.) *5 inches*.
* The height is shown. What's the height? (Signal.) *4 inches*.
* Figure B: The length of the left side is shown. What's the length of that side? (Signal.) *15 centimeters*.
* The height is shown. What's the height? (Signal.) *12 centimeters*.
* Remember, you use the height when you find the area, not the perimeter.
c. Go back to figure A. Find the **area** and the **perimeter.** Raise your hand when you're finished. (Observe students and give feedback.)
* (Write on the board:)

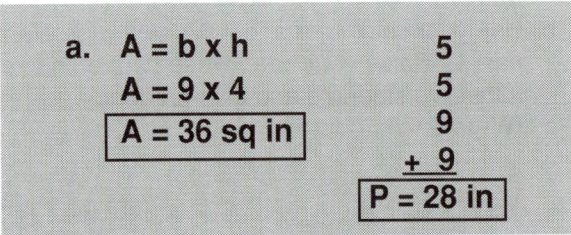

* Check your work. Here's what you should have. Raise your hand if you got everything right.
d. Find the area and the perimeter for figure B. Raise your hand when you're finished. (Observe students and give feedback.)
* (Write on the board:)

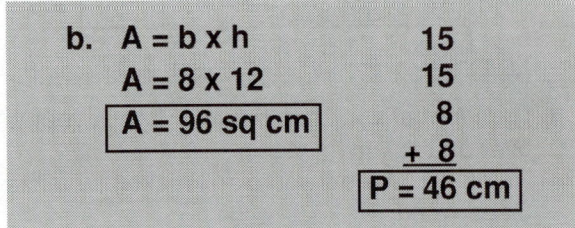

* Here's what you should have. Raise your hand if you got everything right.

EXERCISE 3 PROBLEM SOLVING
Comparison

a. Find part 3.
* I'll read what it says. Follow along: You've worked word problems that compare two things and **tell** the difference number.
* Some problems **ask** about the difference. Questions that ask about the difference name two things and ask which is more or less.
* Here are some questions that ask about the difference: How much taller is Jane than Dan? How much less money does the car cost than the truck costs? What's the difference in price between the basket and the board?
* If a problem **asks** about the difference, make a number family with the name **difference.**
* The next page shows a problem: A TV costs $295. A radio costs $48. What is the difference in the price of these two items?
* The question asks about the difference. So you start with a number family that compares.
* You can figure out the name for the big number by comparing the cost of the TV and the radio. The TV costs more. It's the big number.
* You can see the number family.
* To find the difference number, **you always subtract.** The difference is $247. That's how much more the TV costs than the radio costs.
b. Find part 4.
  Each problem asks about the difference.
c. Problem A: Jenny is 58 inches tall. Tony is 39 inches tall. What is the difference in height?
* The question asks about the difference. So one of the names in the family will be **difference.** What are the other names? (Signal.) *Jenny and Tony.*
* The one with the bigger number is the big number in the family. Make the family with three names and two numbers. Raise your hand when you've done that much. √
* (Write on the board:)

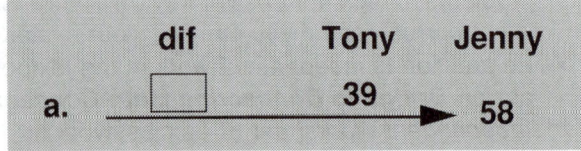

* Here's what you should have. To find the difference, you always subtract. You'll do that later.
d. Problem B: The cable car went 15 miles. The bus went 60 miles. What is the difference in distance?

- The name for the big number is the name of the vehicle that went farther. Write three names and two numbers. Make sure you put the big number in the right place. Raise your hand when you've written your number family. √
- (Write on the board:)

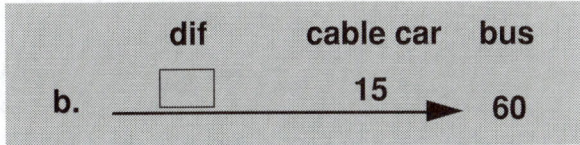

- Here's what you should have for problem B.
e. Problem C. Read the problem. Make the number family. Raise your hand when you're finished. √

Key:

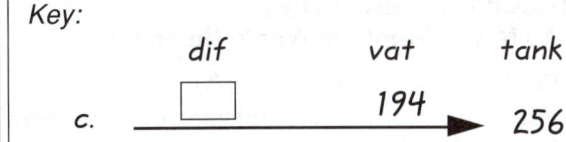

- A vat holds 194 gallons. A tank holds 256 gallons. Everybody, what's the name for the big number? (Signal.) *Tank.*
- The names for the small numbers are **difference** and **vat.**
f. Your turn: Work the number problems for the items in part 4. Box each answer; write the unit name. Raise your hand when you're finished.
  (Observe students and give feedback.)
g. Check your work.
- Problem A: 58 minus 39. The difference is 19 inches. Jenny is **19 inches** taller than Tony.
- Problem B: 60 minus 15. The difference is 45 miles. The bus went **45 miles** farther than the cable car.
- Problem C: 256 minus 194. The difference is **62 gallons.** The vat holds 62 fewer gallons than the tank.
h. Raise your hand if you got everything right.

EXERCISE 4 RATIOS AND PROPORTIONS
Tables

a. Find part 5.
- These are like tables you've worked before, but they have names. You'll copy the table and figure out the missing numbers.
b. Table A. Copy it and write the missing number in the first column. Raise your hand when you've done that much.
  (Observe students and give feedback.)

- (Write on the board:)

a.	perch	2	
	bass	3	
	fish	5	40

- Here's what you should have so far. All the rows follow the same multiplication rule.
c. One of the rows has two numbers. What's the name for that row? (Signal.) *Fish.*
- Figure out the number you multiply by in that row. Then complete the table. Raise your hand when you're finished.
  (Observe students and give feedback.)
- (Write on the board:)

a.	perch	2	16
	bass	3	24
	fish	5	40

- Here's what you should have. You multiply by 8 for each row. In the second column, the number of perch is 16. The number of bass is 24. 16 plus 24 equals 40 fish.
- Raise your hand if you got everything right.
d. Table B. Copy it. Complete the first column. Then complete the second column. Raise your hand when you're finished.
  (Observe students and give feedback.)
e. (Write on the board:)

b.	red	5	35
	not red	3	21
	cars	8	56

- Here's what you should have.
- The ratio of red cars to total cars is 5 to 8. If there are 21 cars that are not red, there are 35 red cars and 56 total cars.
- Raise your hand if you got everything right.

EXERCISE 5 PLACE VALUE
Hundreds

a. Find part 6.
  These are thousands numbers. You can read them as hundreds numbers by reading the part that is not underlined and then saying **hundred.**
b. Number A. Everybody, read it as hundreds. (Signal.) *13 hundred.*

c. My turn to read number B. First I read the part that is not underlined and say **hundred.** Then I read the underlined part. Here I go: **13 hundred 24.**

d. Your turn: Read number B as hundreds. (Signal.) *13 hundred 24.*

• Read number C as hundreds. (Signal.) *56 hundred 6.*

• Number D. Read it as hundreds. (Signal.) *11 hundred 28.*

• Number E. Read it as hundreds. (Signal.) *72 hundred 10.*

e. (Repeat step d until firm.)

EXERCISE 6 GEOMETRY
Supplementary Angles

a. Find part 7.

• For each item, you'll figure out all the angles. Remember, the way you do that is to combine angles that form half a circle. You can combine an angle with two different neighbors to form two different half circles.

b. Item A. Here's the fact: Angle Y is 60 degrees. Your turn: Trace figure A. Figure out all the angles in the circle. Raise your hand when you're finished.
(Observe students and give feedback.)

• (Draw on the board:)

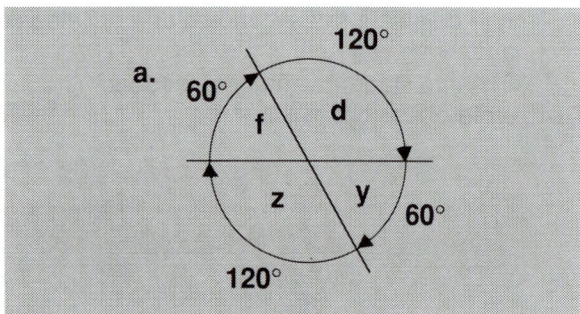

• Here's what you should have. Angle Y is 60 degrees. Angle Z is 120 degrees. Together, those angles form half a circle. Angle Y and angle D form a different half circle. So angle D is 120 degrees. Everybody, what does angle F equal? (Signal.) *60 degrees.*

c. The fact is shown for item B. Trace the figure. Figure out all the angles. Raise your hand when you're finished.
(Observe students and give feedback.)

• (Write on the board:)

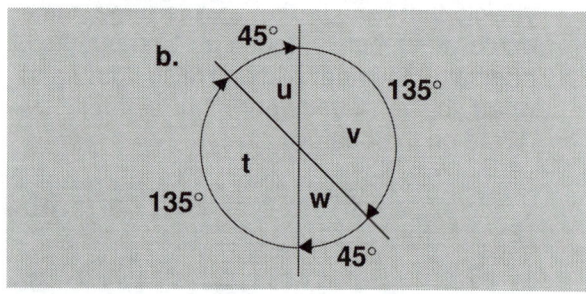

• Here's what you should have for the angles in item B. As part of the independent work, you'll write the rule for the angles that are on opposite sides of a circle.

EXERCISE 7 FRACTIONS
As Mixed Numbers/Whole Numbers

a. Find part 8.

b. Your turn: Copy all the fractions that are more than 1. Raise your hand when you've done that much. √

c. (Write on the board:)

| b. $\frac{77}{6}$ | d. $\frac{580}{5}$ | f. $\frac{18}{17}$ | g. $\frac{47}{3}$ |

• The fractions that are more than 1 are : B, D, F and G.

d. Your turn: Below each fraction, write the division problem and figure out the mixed number or whole number each fraction equals. Raise your hand when you're finished.
(Observe students and give feedback.)

e. (Write on the board:)

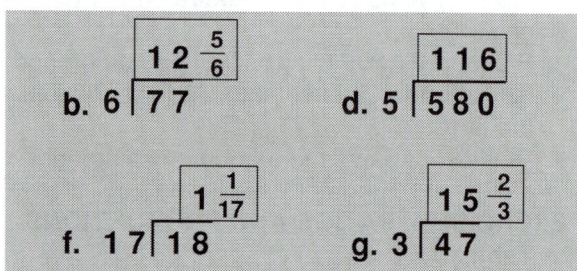

f. Check your work. Here are the answers you should have.

• Item B: 77/6 equals 12 and 5/6.

• Item D: 580/5 equals 116.

• Item F: 18/17 equals 1 and 1/17.

• Item G: 47/3 equals 15 and 2/3.

EXERCISE 8 INDEPENDENT WORK

• (In addition to independent work in the textbook, assign ***Bridge to Connecting Math Concepts*** *Independent Worksheet* 12 as classwork or homework. Before beginning the next lesson, check the students' independent work.)

Lesson 33

Objectives

- **Workcheck: lesson 32, part 9.** (Exercise 1)
  Note: Students confirm the rule that vertically opposite angles are equal.

- **Complete a fraction number family that has a big number of 1.** (Exercise 2)
  Note: Problems are of the form:

  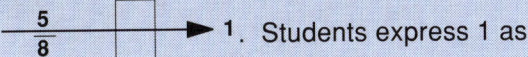

  $$\frac{5}{8} \longrightarrow \boxed{} \longrightarrow 1.$$ Students express 1 as a fraction with a denominator of 8. They then figure out the missing small number by subtracting: $\frac{8}{8} - \frac{5}{8} = \frac{3}{8}$.

- Complete a ratio table that has headings for the rows. (Exercise 3)

- **Estimate the answer to a problem by rounding values to hundreds.** (Exercise 4)
  For example: $\begin{array}{r} 612 \\ + 880 \\ \hline \end{array}$. Students say the problem for the closest hundreds numbers: "600 plus 900." They write the estimated answer: **1500.**

- **Complete a division problem with an answer that is not too big.** (Exercise 5)
  Note: Problems are in pairs. One problem has an answer that is too big:

  $\begin{array}{r} 8 \\ 22\overline{)169} \end{array}$ $\begin{array}{r} 7 \\ 22\overline{)169} \end{array}$. Students work the problem with the appropriate answer and cross out the other problem.

- **Compute the area of a triangle.** (Exercise 6)
  Note: The triangle is shown as half of a rectangle with the same base and height.
  Students use the equation: $A \triangle = \frac{b \times h}{2}$.

- **Solve a set of comparison word problems, some of which ask about the difference.** (Exercise 7)

EXERCISE 1 OPPOSITE ANGLES
Workcheck

a. Open your textbook to lesson 33 and find part 1. For your lesson 32 independent work, you figured out the rule for opposite angles in a circle.

b. Here's the rule: Opposite angles are equal.
- Raise your hand if you figured out that rule. You could have said that the opposite angles are the same size or have the same number of degrees.
- That's an important rule for working hard problems. If you know the degrees in an angle formed by intersecting lines, you know the degrees in the opposite angle. You can see figures with the opposite angles marked.

EXERCISE 2 NUMBER FAMILIES
Fractions

a. Find part 2.
- I'll read what it says. Follow along: You can use number families to show fractions.
- The small numbers are fractions along the arrow. The big number is the fraction at the end of the arrow.
- You can see two number families. Both of them have a big number that equals 1.
- In family A, the small numbers are 2/5 and 3/5. What's the big number? (Signal.) *5-fifths.*
- In family B, the small numbers are 7/9 and 2/9. What's the big number? (Signal.) *9-ninths.*
- For each family, you can make addition statements and subtraction statements.
- You can see an addition statement and a subtraction statement for family A. The addition statement is 2/5 plus 3/5 equals 5/5. The subtraction statement is 5/5 minus 3/5 equals 2/5.
- You can make another subtraction statement for that family. Everybody, what's that statement? (Signal.) *5-fifths minus 2-fifths equals 3-fifths.*
- Fraction number families are important for working difficult word problems that have a big number of **1.**
- You can see a family with a big number of **1** and a missing small number.
- To solve the problem, you first change **1** into a fraction. That fraction must have the same denominator as 5/9. The fraction is **9/9.**
- On the next page, you can see the missing small number. You work the problem 9/9 minus 5/9.
- The answer is 4/9.
- Remember, the small numbers add up to **1.** All the denominators are the same.

b. Find part 3.
- In all these problems, the big number is 1. The small numbers are fractions that are less than 1. Remember, all the fractions must have the same denominator.

- We'll do the sample problem together. First we change 1 into a fraction with the same denominator as the fraction shown in the number family. What denominator is that? (Signal.) *10.*
- So what's the fraction for the big number? (Signal.) *10-tenths.*
- (Write on the board:)

- All the fractions in the family must have the same denominator. So we write that denominator for the missing fraction. What denominator is that? (Signal.) *10.*
- (Write to show:)

- The top numbers work just like the numbers in any other number family. A small number is missing. Everybody, what's that number? (Signal.) *3.*
- (Write to show.)

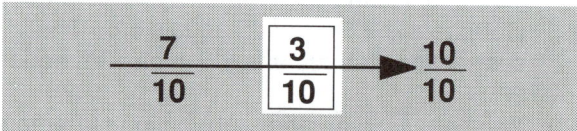

- Remember, all the denominators must be the same. The top numbers work just like the numbers in a regular number family.
c. Your turn: Write family A with the correct fraction that equals 1, then figure out the missing fraction and box it. Raise your hand when you're finished.
   (Observe students and give feedback.)
- (Write on the board:)

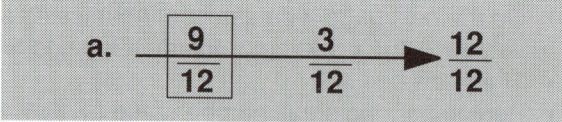

- Check your work. Here's what you should have. The fraction that equals 1 is 12/12. The missing fraction is 9/12.
- The fractions for the small numbers, 9/12 and 3/12, add up to 1.
d. Your turn: Work the rest of the problems in part 3. Make a box around the missing fraction. Raise your hand when you're finished.
   (Observe students and give feedback.)

Key:

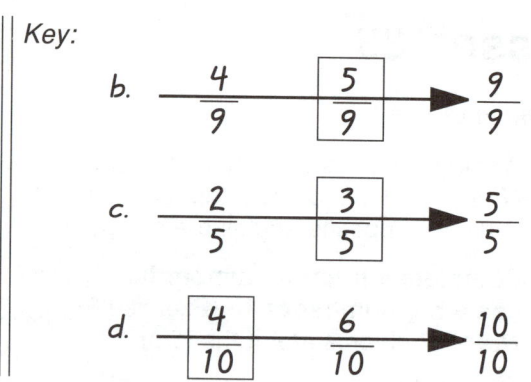

e. Check your work.
- Problem B. Tell me the fraction that equals 1. (Signal.) *9-ninths.*
- Tell me the missing fraction. (Signal.) *5-ninths.*
- Problem C. Tell me the fraction that equals 1. (Signal.) *5-fifths.*
- Tell me the missing fraction. (Signal.) *3-fifths.*
- Problem D. Tell me the fraction that equals 1. (Signal.) *10-tenths.*
- Tell me the missing fraction. (Signal.) *4-tenths.*
f. Raise your hand if you got everything right.

EXERCISE 3 RATIOS AND PROPORTIONS
Tables

a. Find part 4.
- This table requires multiplication to get from the numbers in the first column to the numbers in the second column.
b. The number for girls is missing in the first column. Copy the table and write that number. Raise your hand when you're finished.
- Everybody, what's the number for girls in the first column? (Signal.) *4.*
- (Write on the board:)

girls	4	
boys	3	18
children	7	

c. Figure out the missing numbers in the second column. Remember, use the row that has two numbers. Raise your hand when you're finished. (Observe students and give feedback.)
- (Write to show:)

girls	4	24
boys	3	18
children	7	42

- The second column should show 24 girls, 18 boys and 42 children. The numbers add up: 24 plus 18 equals 42.

EXERCISE 4 ESTIMATION
Hundreds

a. I'll show you how to write estimates for problems. The estimate will give an answer that is close to the real answer.
b. (Write on the board:)

$$612$$
$$+ 880$$

- The estimation problem is a simple hundreds problem: 612 is close to 600.
- Say the hundreds number that is closest to 880. (Signal.) *900.*
c. So I can work this problem: 600 plus 900. What's 6 plus 9? (Signal.) *15.*
- So 600 plus 900 is 15 hundred.
- (Write to show:)

$$612$$
$$+ 880$$
$$1500$$

- That's pretty close to the answer for 612 plus 880. The answer to that problem is 14 hundred 92.
- (Write to show:)

$$612$$
$$+ 880$$
$$\overline{1492} \qquad 1500$$

d. Find part 5.
For each problem, you'll say the problem for the closest hundreds number.
e. Problem A: 320 plus 706. What's the hundreds number that's closest to 320? (Signal.) *300.*
- What's the hundreds number that's closest to 706? (Signal.) *700.*
- The estimation problem for hundreds is: 3 plus 7. That's 10 hundred. So the answer to problem A is about 10 hundred. Write the estimation answer for A—10 and two zeros. √
- (Write on the board:)

a. 1000

- Here's what you should have.

f. Problem B: 13 hundred 10 plus 584. What's the hundreds number closest to 13 hundred 10? (Signal.) *13 hundred.*
- What's the hundreds number closest to 584? (Signal.) *600.*
- Say the estimation problem for hundreds. (Signal.) *13 plus 6.*
- What's the answer? (Signal.) *19.*
- So the answer to problem B is about **19 hundred.** Write that answer for B. √
- (Write on the board:)

b. 1900

- Here's 19 hundred. 19 and two zeros.
g. Problem C. Raise your hand when you can say the estimation problem for the closest hundreds.
- Everybody, say the problem. (Signal.) *10 plus 12.*
- What's the answer? (Signal.) *22.*
- So the answer to problem C is about 22 hundred. Write that answer. √
- (Write on the board:)

c. 2200

- Here's 22 hundred.
h. Your turn: Write answers to the rest of the problems in part 5. Say the estimation problem for hundreds to yourself. Then write the estimation answer with two zeros to show the hundreds. Raise your hand when you're finished. (Observe students and give feedback.)
i. (Write on the board:)

d. 800
e. 2200
f. 2500

j. Check your work. Here are the estimate answers you should have for problems D, E and F.
- Problem D. The estimation problem is 3 plus 5. So the estimation answer to D is 8 hundred.
- Problem E. The estimation problem is 15 plus 7. So the estimation answer to E is 22 hundred.
- Problem F. The estimation problem is 3 plus 22. So the estimation answer to F is 25 hundred.
k. Raise your hand if you got all of them right.

EXERCISE 5 LONG DIVISION
Quotient Too Large

a. Find part 6.
- These problems are in pairs. That's because two different people started to work them and they got different answers. You're going to figure out which answer is correct. For these problems, it's easy. The number you get when you multiply can't be any bigger than the number above it. If the number you get when you multiply is bigger, you can't subtract. So the answer is wrong. If you **can** subtract, the answer is right.

b. Look at the problems in A.
- The problem is: 283 divided by 34. One person thinks that the answer above the division sign is **8.** The other person thinks the answer above the division sign is **9.** You're going to figure out which answer is right. Copy the first problem in A and multiply. If the number you get is no bigger than **283,** subtract and figure out the remainder and write it as a fraction. Raise your hand when you've done that much.
(Observe students and give feedback.)
- Everybody, what did you get when you multiplied? (Signal.) *272.*
- When you subtract from **283,** what's the remainder? (Signal.) *11.*
So the fraction in the answer is 11/34.
- I'll work the other problem and show you why that answer is wrong.
- (Write on the board:)

$$
\begin{array}{r}
9 \\
34\overline{\smash)283} \\
-306
\end{array}
$$

a.

- This answer won't work. 34 times 9 equals 306. That's bigger than 283 so you can't subtract. The correct answer is **8,** not **9.**
- (Cross out problem.)

c. Your turn: Multiply for both problems in B. Don't do any subtracting. Just multiply and write the numbers. Raise your hand when you've done that much.
(Observe students and give feedback.)

Key:

b.
$$
\begin{array}{r}
7 \\
45\overline{\smash)275} \\
-315
\end{array}
\qquad
\begin{array}{r}
6 \\
45\overline{\smash)275} \\
-270
\end{array}
$$

- For one of those problems, the answer you get when you multiply is bigger than **275.** 315 is too big, so the answer above the division sign is too big. Everybody, which answer is too big? (Signal.) *7.*
- Which answer is right? (Signal.) *6.*
- Cross out the problem with the answer of **7.** Then figure out the remainder for the problem with the answer of **6.** Write the remainder as a fraction. Raise your hand when you're finished. √
- Everybody, what fraction did you write? (Signal.) *5-forty-fifths.*
- 275 divided by 45 equals 6 and 5/45.

d. Your turn: Multiply for both problems in C. If the number you get is too big, cross out the whole problem. Then work the problem with the right answer. Raise your hand when you're finished. (Observe students and give feedback.)
- (Write on the board:)

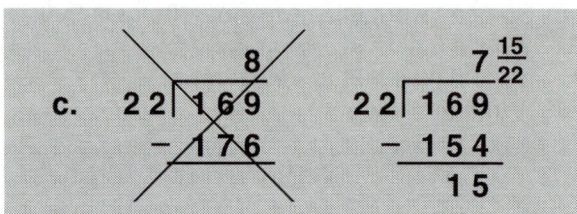

c.

- Here's what you should have for the problems in C. 169 divided by 22 equals 7 and 15/22.

e. Remember, if you can't subtract, the answer above the division sign is too big.

EXERCISE 6 AREA
Triangles

a. Find part 7.
- (Teacher reference:)

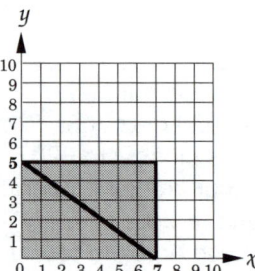

- I'll read what it says. Follow along: You can make a rectangle into triangles. You just start with a rectangle and draw a **diagonal** line. That line goes from one corner to the opposite corner.

b. Touch the diagonal line on the top coordinate grid. √

- The line divides the rectangle into two triangles. Each triangle has exactly half the area of the rectangle you started with. So the area of a triangle is half the area of a rectangle with the same base and same height.
- You can see the equation for the area of a triangle: **Area of a triangle equals base times height divided by 2.** Remember that equation. It's like the equation for the area of a rectangle except it divides the area in half. Notice that it has the symbol for a triangle.
c. Find part 8.
- Each grid shows a shaded triangle. You can see that each triangle is half of a rectangle that has the same base and the same height as the triangle.
d. We'll work the sample problem together.
- (Write on the board:)

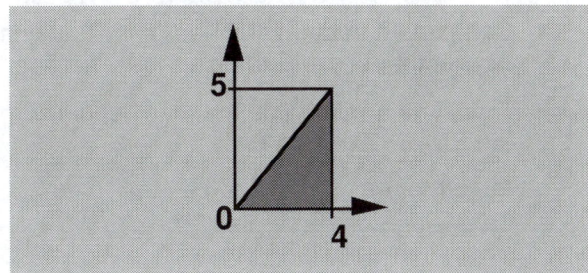

- To figure out the area of the shaded triangle, I figure the area of this **rectangle** and then divide by 2.
- (Write on the board:)

$$A \triangle = \frac{b \times h}{2}$$

$$A \triangle = \frac{4 \times 5}{2} = \frac{20}{2}$$

$$\boxed{A \triangle = 10 \text{ sq in}}$$

- The area of the rectangle is 20 square inches. The area of the triangle is 10 square inches.
- Copy the work for the sample problem. Raise your hand when you're finished. √
e. Problem A. Write the equation for the area of a triangle. Remember the symbol for the triangle. Work the problem and write the answer. Remember to write the unit name in the answer. Raise your hand when you're finished. (Observe students and give feedback.)

- (Write on the board:)

$$a. \quad A \triangle = \frac{b \times h}{2}$$

$$A \triangle = \frac{3 \times 8}{2} = \frac{24}{2}$$

$$\boxed{A \triangle = 12 \text{ sq m}}$$

- Here's what you should have. The area of the rectangle is 3 times 8. That's 24 square meters. The area of the triangle is half of 24. The answer is 12 square meters.
- Raise your hand if you got everything right.
f. Your turn: Work problem B. Raise your hand when you're finished.
(Observe students and give feedback.)
- (Write on the board:)

$$b. \quad A \triangle = \frac{b \times h}{2}$$

$$A \triangle = \frac{7 \times 6}{2} = \frac{42}{2}$$

$$\boxed{A \triangle = 21 \text{ sq yd}}$$

- Check your work. Here's what you should have.
- The area of the triangle is 21 square yards. That's half the area of the rectangle.

EXERCISE 7   PROBLEM SOLVING
Comparison

a. Find part 9.
   These problems compare. Some of them ask about the difference number. You'll make a number family for each problem.
b. Work problem A. Raise your hand when you're finished. (Observe students and give feedback.)
- (Write on the board:)

	dif	Rose	Jane
a.	☐	45 →	58

- Here's the family you should have.
- Everybody, what's the difference in the girls' height? (Signal.) *13 inches.*
c. Work problem B. Raise your hand when you're finished. (Observe students and give feedback.)

- (Write on the board:)

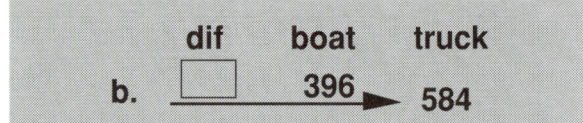

	dif	boat	truck
b.	☐	396 →	584

- Here's what you should have.
- Everybody, how much farther did the truck go? (Signal.) *188 miles*.
d. Work the rest of the problems in part 9. Raise your hand when you're finished.
   (Observe students and give feedback.)
   Key:

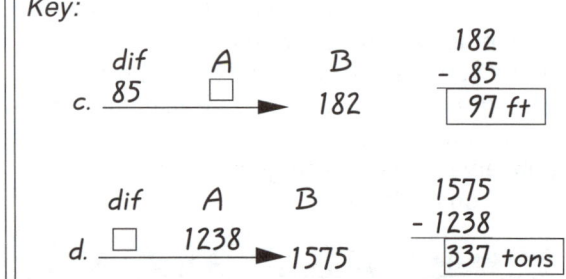

e. Find part J on page 132. That shows what you should have for problems C and D.
- Problem C gives the difference number. How tall was building A? (Signal.) *97 feet*.
- Problem D asks about the difference number. Everybody, how much lighter was load A than load B? (Signal.) *337 tons*.

EXERCISE 8 INDEPENDENT WORK

- (In addition to independent work in the textbook, assign **Bridge to Connecting Math Concepts** *Independent Worksheet* 13 as classwork or homework. Before beginning the next lesson, check the students' independent work.)

Lesson 34

Objectives

- Compute the area of a triangle. (Exercise 1)
- Complete a division problem with an answer that is not too big. (Exercise 2)
- Estimate the answer to a problem by rounding a value to hundreds. (Exercise 3)
- **Complete a division problem that has the correct whole-number answer.** (Exercise 4)
   Note: $47\overline{)367}$ (answer 6) $47\overline{)367}$ (answer 7). Problems are in pairs. One problem has an answer that is too small. Students multiply and subtract to determine which problem has the correct whole-number answer, then complete it.
- Complete a fraction number family that has a big number of 1. (Exercise 5)
- Complete a ratio table that has headings for the rows. (Exercise 6)
- Find the area and perimeter of a parallelogram. (Exercise 7)

EXERCISE 1 AREA
Triangles

a. Open your textbook to lesson 34 and find part 1.
- You're going to figure the area of the triangles. Remember, think of a rectangle that has the same base and height as the triangle. The equation for the **area of a triangle is base times height divided by 2.** Remember, all you're doing is figuring out the area of a rectangle and then dividing by 2.
b. Problem A. Figure out the area of the shaded triangle. Remember to show the equation for the area of a triangle. Raise your hand when you're finished.
   (Observe students and give feedback.)
- (Write on the board:)

a. $A \triangle = \dfrac{b \times h}{2}$

$A \triangle = \dfrac{4 \times 1}{2} = \dfrac{4}{2}$

$\boxed{A \triangle = 2 \text{ sq yd}}$

- Here's what you should have. The answer is 2

square yards. That's half the area of the rectangle 4 yards wide and 1 yard high.

c. Your turn: Work the rest of the problems in part 1. Raise your hand when you're finished.
(Observe students and give feedback.)

Key:

b. $A_\triangle = \dfrac{b \times h}{2}$

$A_\triangle = \dfrac{4 \times 5}{2} = \dfrac{20}{2}$

$\boxed{A_\triangle = 10 \text{ sq ft}}$

c. $A_\triangle = \dfrac{b \times h}{2}$

$A_\triangle = \dfrac{12 \times 10}{2} = \dfrac{120}{2}$

$\boxed{A_\triangle = 60 \text{ sq in}}$

d. Check your work.
• Find part J on page 135 in your textbook. That shows what you should have for problems B and C.
• Raise your hand if you got everything right.

EXERCISE 2 LONG DIVISION
Quotient Too Large

a. Find part 2.
• These problems are in pairs. You're going to figure out which answer above the division sign is correct.

b. Your turn: Multiply for both problems in A. Don't do any subtracting. Just multiply and write the numbers below. Raise your hand when you've done that much.
(Observe students and give feedback.)

Key:

$$
a. \quad 35\overline{)283} \;\; ^8 \qquad 35\overline{)283} \;\; ^9
$$
$$
 -280 \qquad\qquad -315
$$

• In one of the problems, the number you get when you multiply is bigger than 283. Which number is bigger than 283? (Signal.) *315.*
• So the answer above the division sign is too big. Which answer is too big? (Signal.) *9.*
• Which answer is right? (Signal.) *8.*

c. Cross out the problem with the answer of **9.** Then figure out the remainder for the problem with the answer of **8.** Write the remainder as a fraction. Raise your hand when you're finished.
(Observe students and give feedback.)
• Everybody, what fraction did you write? (Signal.) *3-thirty-fifths.*
• 283 divided by 35 equals 8 and 3/35.

d. Your turn: Multiply for both problems in B. If the number you get is too big, cross out the whole problem. Then work the problem with the right answer. Raise your hand when you're finished.
(Observe students and give feedback.)

• (Write on the board:)

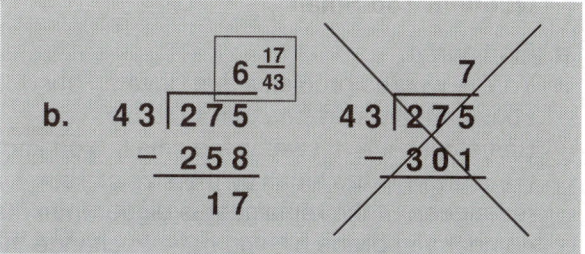

• Here's what you should have for the problems in B. 275 divided by 43 equals 6 and 17/43.

EXERCISE 3 ESTIMATION
Hundreds

a. Find part 3.
For each problem, you'll write the estimation answer. To do that, you'll just add or subtract the closest hundreds numbers.

b. Problem A: 291 plus 12 hundred 96. Raise your hand when you can say the estimation problem. √
• Everybody, what's the estimation problem? (Signal.) *3 plus 13.*
• What's the answer? (Signal.) *16.*
• Write the estimate for A as 16 and two zeros. √
• (Write on the board:)

a. 1600

c. Problem B: 15 hundred 18 minus 699. Say the estimation problem. (Signal.) *15 minus 7.*
• What's the answer? (Signal.) *8.*
• Write the estimation answer. √
• The estimation answer is 800.

d. Your turn: Write the estimation answers for the rest of the problems in part 3. Raise your hand when you're finished.
(Observe students and give feedback.)

e. (Write on the board:)

c. 2200
d. 2900

f. Check your work. Here are the estimation answers you should have for problems C and D.
• Problem C. The estimation problem is 26 minus 4. So the estimation answer to C is 22 hundred.
• Problem D. The estimation problem is 18 plus 11. So the estimation answer to D is 29 hundred.

EXERCISE 4 LONG DIVISION
Quotient Too Small

a. Find part 4.
• You've worked problems that have an answer
  above the division sign that's too big.
  Sometimes the answer is too small. You know
  the answer is too small by the size of your
  remainder. If the remainder is bigger than the
  number you divide by, you have to change the
  answer.
• The problems in part 4 are in pairs. One answer
  is right. The other answer is too small.
b. The sample problems have been worked. The
  first problem has an answer of **6.** What's the
  remainder? (Signal.) *37.*
• What number are you dividing by? (Signal.) *34.*
• The remainder is bigger than **34.** So you have
  to change the answer. You make the number in
  the answer bigger. You'd change it to 7.
• The second sample problem shows that answer.
  What's the remainder for an answer of **7?**
  (Signal.) *3.*
• Is that remainder bigger than the number you
  divide by? (Signal.) *No.*
• So 7 is the correct number. The fraction in the
  answer is 3/34. That's less than 1. 241 divided
  by 34 equals 7 and 3/34.
c. Your turn: Copy and work both problems in A.
  If the remainder is bigger than the number
  you're dividing by, circle the remainder. If the
  remainder is all right, write the fraction in the
  answer. Remember, the fraction in the answer
  can't be more than 1. Raise your hand when
  you're finished working both problems in A.
  (Observe students and give feedback.)
  Key:

$$a. \quad 6\,1\,\overline{)3\,4\,8} \quad \boxed{5\tfrac{43}{61}} \qquad 6\,1\,\overline{)3\,4\,8} \quad 4$$
$$\quad\quad -3\,0\,5 \qquad\qquad -2\,4\,4$$
$$\quad\quad\quad\quad 4\,3 \qquad\qquad \boxed{1\,0\,4}$$

• Check your work. Problem A: 348 divided by
  61. The correct answer is 5 and 43/61. The
  remainder for the second problem is 104. That's
  too big.
d. Your turn: Copy and work both problems in B.
  If the remainder is bigger than the number
  you're dividing by, circle the remainder. If the
  remainder is all right, write the fraction in the
  answer. Raise your hand when you're finished.
  (Observe students and give feedback.)

Key:

$$b. \quad 4\,7\,\overline{)3\,6\,7} \quad 6 \qquad 4\,7\,\overline{)3\,6\,7} \quad \boxed{7\tfrac{38}{47}}$$
$$\quad\quad -2\,8\,2 \qquad\qquad -3\,2\,9$$
$$\quad\quad \boxed{8\,5} \qquad\qquad 3\,8$$

• Check your work. Problem B: 367 divided by 47.
  The answer to the first problem in B is **6.** What's
  the remainder for that answer? (Signal.) *85.*
• Is that remainder bigger than **47?** (Signal.) *Yes.*
• So the answer is not big enough. The answer to
  the second problem is **7.** What's the remainder
  for that answer? (Signal.) *38.*
• Is that remainder bigger than **47?** (Signal.) *No.*
• What's the correct answer to 367 divided by 47?
  (Signal.) *7 and 38-forty-sevenths.*

EXERCISE 5 NUMBER FAMILIES
Fractions

a. Find part 5.
• All these fraction number families show 1 as the
  big number and have a missing fraction.
• You'll write the complete number family.
  Remember, the fraction that equals 1 has the
  same denominator as the other fraction shown.
b. Write the complete number family with three
  fractions for family A. Raise your hand when
  you're finished.
  (Observe students and give feedback.)
• (Write on the board:)

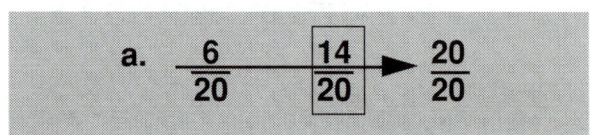

• Here's what you should have for family A.
• Raise your hand if you got everything right.
c. Your turn: Write complete families for B through
  D. Raise your hand when you're finished.
  (Observe students and give feedback.)

d. (Write on the board:)

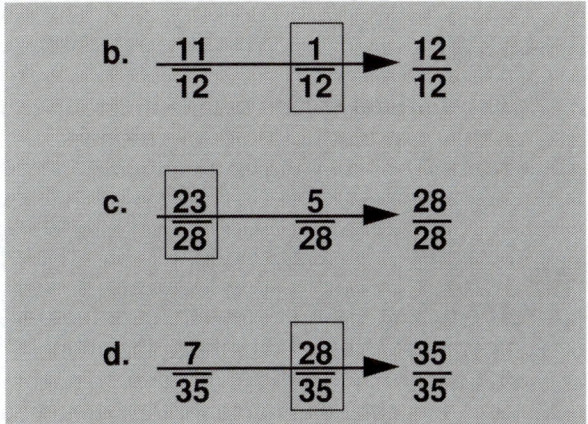

b. $\dfrac{11}{12}$ $\boxed{\dfrac{1}{12}}$ ➔ $\dfrac{12}{12}$

c. $\boxed{\dfrac{23}{28}}$ $\dfrac{5}{28}$ ➔ $\dfrac{28}{28}$

d. $\dfrac{7}{35}$ $\boxed{\dfrac{28}{35}}$ ➔ $\dfrac{35}{35}$

- Here's what you should have for families B through D.
e. Raise your hand if you got everything right.

EXERCISE 6 RATIOS AND PROPORTIONS
Tables

a. Find part 6.
- These are table problems.
- You'll copy each table, write the missing number in the first column, figure out the multiplication rule and complete the second column.
b. Copy table A and complete it. Raise your hand when you're finished.
  (Observe students and give feedback.)
- (Write on the board:)

a.			
	men	6	48
	women	1	8
	people	7	56

- Here's what you should have. The missing number in the first column is 7. You multiply by 8 for each row. The numbers in the second column add up: 48 men plus 8 women equals 56 people.
c. Copy table B and complete it. Raise your hand when you're finished.
  (Observe students and give feedback.)
- (Write on the board:)

b.			
	men	2	18
	women	3	27
	people	5	45

- Here's what you should have. The missing number in the first column is 2. You multiply by 9 for each row. The numbers in the second column add up: 18 men plus 27 women equals 45 people.
d. Copy table C and complete it. Raise your hand when you're finished.
  (Observe students and give feedback.)
- (Write on the board:)

c.			
	men	6	42
	women	3	21
	people	9	63

- Here's what you should have. The missing number in the first column is 6. You multiply by 7 for each row. The numbers in the second column add up: 42 men plus 21 women equals 63 people.
e. Raise your hand if you got everything right.

EXERCISE 7 AREA AND PERIMETER
Parallelograms

a. Find part 7.
- These are parallelograms. You'll find the perimeter and the area of each parallelogram. Remember, when you find the area, you use the height. That's the line perpendicular to the base. When you find the perimeter, you just add up the lengths of the sides.
b. Find the area and perimeter of figure A. Raise your hand when you're finished.
  (Observe students and give feedback.)
- (Write on the board:)

a. $A = b \times h$ 9
  $A = 15 \times 8$ 9
  $\boxed{A = 120 \text{ sq m}}$ 15
               + 15
           $\boxed{P = 48 \text{ m}}$

- Here's what you should have. The area is 120 square meters. The perimeter is 48 meters.
- Raise your hand if you got everything right.
c. Find the area and perimeter of figure B. Raise your hand when you're finished.
  (Observe students and give feedback.)

- (Write on the board:)

b.	A = b x h	21
	A = 13 x 18	21
	A = 234 sq ft	13
		+ 13
		P = 68 ft

- Here's what you should have. The area is 234 square feet. The perimeter is 68 feet.
- Raise your hand if you got everything right.

EXERCISE 8 INDEPENDENT WORK

- (In addition to independent work in the textbook, assign **Bridge to Connecting Math Concepts** Independent Worksheet 14 as classwork or homework. Before beginning the next lesson, check the students' independent work.)

Lesson 35

Objectives

- **Appropriately place names in a ratio table and complete the table.** (Exercise 1)
  Note: The table shows 3 numbers:

		4	68
		12	

The names are presented in an order that does not correspond to their placement in the table. For example: children, girls, boys. Students show the name for the big number (children) as the bottom name:

girls	4	68
boys		
children	12	

- **Work a division problem that shows an incorrect whole-number answer.** (Exercise 2)

- **Solve a word problem that involves a whole group by making a fraction number family.** (Exercise 3)
  Note: Problems are of the form: 2/5 of the children in the class are boys. What's the fraction for girls? Students make fraction

number families: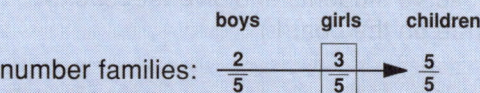

- **Use estimation to determine whether a problem has a wrong answer.** (Exercise 4)
  Note: Students round numbers to the nearest hundred. If the answer shown for each problem is not close to the estimation answer, students rework the problem.

- **Compute the area of a rectangle or a right triangle.** (Exercise 5)

EXERCISE 1 RATIOS AND PROPORTIONS
Tables

a. Open your textbook to lesson 35 and find part 1.
- These are tables without names. You're going to write the names where they are supposed to go.
- Here are the rules: The names for the smaller numbers are for the top two rows. The name for the total is the big number. It's the bottom name in the table.

b. Look at problem A. The names for the table are **children, girls** and **boys.** The number for boys in the first column is 4.

- Listen : The three names are **children, girls** and **boys.** One of those is the name for the big number. Which name is that? (Signal.) *Children.*
- The problem tells that the number for boys in the first column is 4. So you write **boys** for the row that has 4.
- Copy table A. Write the names where they go. Raise your hand when you've done that much.

(Observe students and give feedback.)
(Write on the board:)

a.	boys	4	68
	girls		
	children	12	

- Here's what you should have.
- Raise your hand if you got everything right.

c. Problem B: The names are **wet shirts, shirts** and **dry shirts.** The number of wet shirts in the first column is 7.

- Figure out which name is the total. Write the names where they belong and the numbers the table shows. Raise your hand when you've done that much.

(Observe students and give feedback.)
- (Write on the board:)

b.	dry		24
	wet	7	
	shirts	11	

- Here's what you should have. The names are **dry shirts, wet shirts** and **shirts. Wet shirts** is the name for the row with 7. The name for the bottom row has to be the name for the big number, **shirts.**

d. Your turn: Copy table C and put the names where they belong. Raise your hand when you've done that much.

(Observe students and give feedback.)
- (Write on the board:)

c.	tall	4	
	short	3	
	trees		42

- Here's what you should have. The names are **tall trees, short trees,** and **trees. Trees** is the name for the total. **Short trees** is the name for the row with 3.

e. Your turn: Complete each table. Raise your hand when you're finished.

(Observe students and give feedback.)
- (Write to show:)

a.	boys	4	68		b.	dry	4	24
	girls	8	136			wet	7	42
	children	12	204			shirts	11	66

c.	tall	4	24
	short	3	18
	trees	7	42

- Here's what you should have. Raise your hand if you figured out all the right numbers.

EXERCISE 2 LONG DIVISION
Incorrect Quotient

a. Find part 2.
- I'll read what it says. Follow along: The correct answer to a division problem may have a remainder. If the answer above the division sign is **too big,** the number you get when you multiply is too big. You can't subtract.
- 163 divided by 25 does not equal 7. 7 is **too big.** You can't subtract 175 from 163.
- If the answer above the division sign is **too small,** the number you get when you multiply is too small. So the remainder is bigger than the number you're dividing by.
- 163 divided by 25 does not equal 5. 5 is **too small.** 38 is larger than 25.
- If the answer above the division sign is **correct,** you **can subtract** and the remainder is smaller than the number you're dividing by.
- You can see the problem with the **correct** answer. 163 divided by 25 equals 6 and 13/25.
- Remember, if you have the right answer, you can subtract and the remainder is smaller than the number you're dividing by. That means the fraction in the answer is less than 1.

b. Find part 3.
Some of these problems have an answer that's too big. Some of them have an answer that's too small. You'll work the problem, figure out how to change the answer, then work the problem again with the correct answer.

c. Problem A: 82 divided by 18. The answer shown is 5. If you can't subtract, 5 is too big. You make the answer smaller. If you can subtract but the remainder is bigger than 18, 5 is not big enough.

- Work the problem. Figure out whether the answer is too big or too small. Then write the problem again with the correct answer and work it. Raise your hand when you've completed item A. (Observe students and give feedback.)

Key:

$$a. \quad 18\overline{|82} \quad \quad \boxed{4\frac{10}{18}} \\ \quad -90 \quad \quad 18\overline{|82} \\ \quad \quad \quad -72 \\ \quad \quad \quad \quad 10$$

- Everybody, what number did you get when you multiplied 5 times 18? (Signal.) *90.*
- Is that number too big? (Signal.) *Yes.*
- You couldn't subtract, so the answer is too big. The correct whole number is 4. When you multiply by 4, you get an answer of 4 and 10/18.

d. Work problem B. Raise your hand when you're finished. (Observe students and give feedback.)

Key:

$$b. \quad 33\overline{|145} \quad \quad \boxed{4\frac{13}{33}} \\ \quad -99 \quad \quad 33\overline{|145} \\ \quad \boxed{46} \quad \quad -132 \\ \quad \quad \quad \quad \quad 13$$

- Everybody, what number did you get when you multiplied 3 times 33? (Signal.) *99.*
- That number is not big enough because the remainder is bigger than 33.
- Everybody, what's the correct answer to 145 divided by 33? (Signal.) *4 and 13-thirty-thirds.*

e. Work problem C. Raise your hand when you're finished. (Observe students and give feedback.)

Key:

$$c. \quad 31\overline{|188} \quad \quad \boxed{6\frac{2}{31}} \\ \quad -217 \quad \quad 31\overline{|188} \\ \quad \quad \quad -186 \\ \quad \quad \quad \quad 2$$

- Everybody, what number did you get when you multiplied 7 times 31? (Signal.) *217.*
- Is that number too big? (Signal.) *Yes.*
- So the answer is too big. The correct answer is 6 and 2-thirty-firsts.

f. Work problem D. Raise your hand when you're finished. (Observe students and give feedback.)

Key:

$$d. \quad 59\overline{|417} \quad \quad \boxed{7\frac{4}{59}} \\ \quad -354 \quad \quad 59\overline{|417} \\ \quad \boxed{63} \quad \quad -413 \\ \quad \quad \quad \quad 4$$

- Everybody, what number did you get when you multiplied 6 times 59? (Signal.) *354.*
- That remainder is bigger than 59.
- Everybody, what's the correct answer to 417 divided by 59? (Signal.) *7 and 4-fifty-ninths.*

EXERCISE 3 PROBLEM SOLVING
Fraction Number Families

a. Find part 4.
These are word problems that involve fractions. For each problem, you'll make a number family and answer the question.

b. Sample problem: 2/5 of the children in a class are boys. What fraction of the children are girls?

- The names for this problem are **boys, girls** and **children.** I'll write those names above where the fractions go in the number family.
- (Write on the board:)

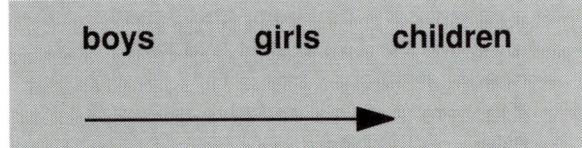

- The problem gives a fraction for boys. What fraction? (Signal.) *2-fifths.*
- (Write to show:)

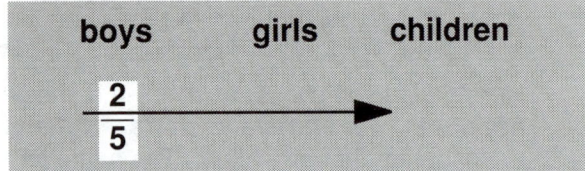

- All the children in the classroom equals 1 whole. It's the whole class.
- (Write to show:)

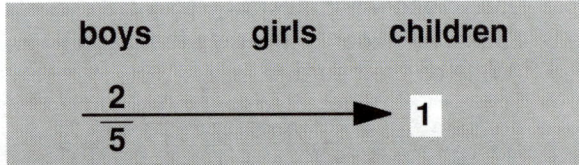

- I have to write 1 as a fraction. What fraction do I write for 1? (Signal.) *5-fifths.*

- (Change to show:)

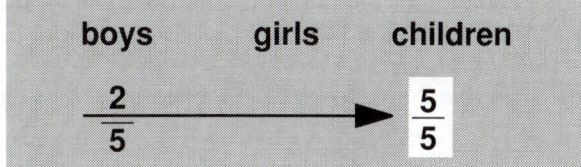

- What's the denominator of the missing fraction? (Signal.) *5.*
- (Write **5:**)

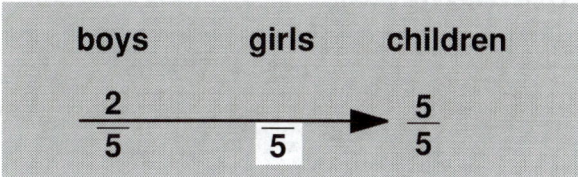

- What's the numerator of the missing fraction? (Signal.) *3.*
- (Write to show:)

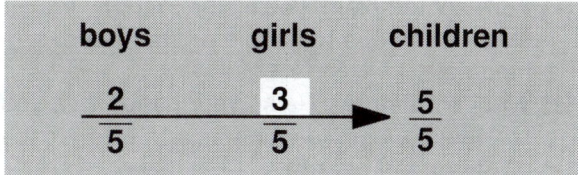

- So what fraction of the children are girls? (Signal.) *3-fifths.*
- Yes, the fraction for the boys plus the fraction for the girls equals the fraction for all the children in the class. 2/5 plus 3/5 equals the whole class.
- c. Problem A: 1/9 of the trees in a forest are pines. What's the fraction for the trees that are not pines?
- First I put the names on the arrow. Tell me the names. (Call on individual students. Idea: *Pines, not pines, trees.*)
- (Write on the board:)

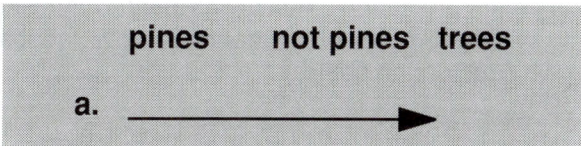

- I could have written **pine trees, not pine trees** and **total.** The important part is that we're adding the fraction for pine trees and the fraction for trees that are not pines to get the total for **all the trees.**
- Your turn: Make the arrow with the names. Write the fraction the problem gives you. Then write the other fractions in the number family. Remember, all the trees equals **1** whole group. Raise your hand when you're finished. (Observe students and give feedback.)

- (Write to show:)

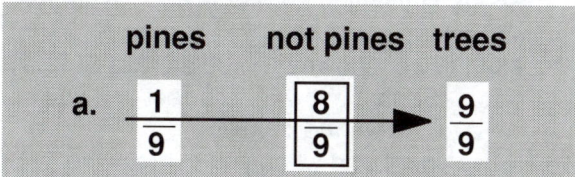

- Check your work. Here's what you should have.
- Everybody, what's the fraction for trees that are not pines? (Signal.) *8-ninths.*
- That's the answer. 8/9 of the trees are not pines.
- d. Your turn: Read problem B. Make an arrow and write the names for your number family. Raise your hand when you have three names. (Observe students and give feedback.)
- Problem B says: In Sage City, 8/10 of the days are dry. What's the fraction for days that are rainy?
- The names are **dry, rainy,** and **days.**
- (Write on the board:)

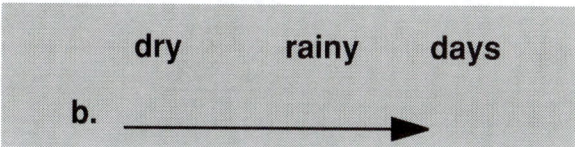

- You add the dry days and the rainy days to get all the days. Write the fractions for the number family. Raise your hand when you're finished. (Observe students and give feedback.)
- (Write to show:)

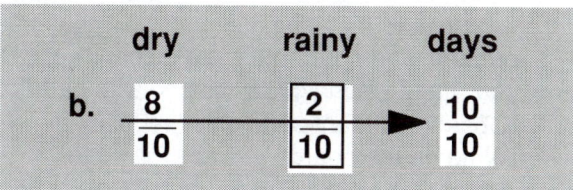

- 8/10 of the days are dry; 2/10 of the days are rainy. The fraction for all the days is 10/10.
- e. Raise your hand if you got everything right.

EXERCISE 4 ESTIMATION
Hundreds

- a. Find part 5.
  Some of these problems have answers that are very wrong. You can figure out which answers are wrong by working the estimation problem. If the answer shown is close to the estimation answer, the answer is correct. If the answer shown is not close to the estimation answer, the answer shown is wrong.
- b. Problem A. Raise your hand when you can say the estimation problem.
- Everybody, say the estimation problem. (Signal.) *15 minus 7.*

- What's the answer? (Signal.) *8.*
- So the estimation answer is 800. Write that answer.
- The answer shown for problem A is 837. That's close. So that answer is correct and you don't have to rework the problem.

c. Raise your hand when you can say the estimation problem for B.
- Everybody, say the problem. (Signal.) *20 minus 9.*
- What's the answer? (Signal.) *11.*
- So the estimation answer is 11 hundred. Is that close to the answer shown for B? (Signal.) *No.*
- B is wrong. Write the estimation answer. Next to that number, copy problem B and work it. You should get an answer close to 11 hundred. Raise your hand when you're finished.
(Observe students and give feedback.)
Key:

$$b. \quad 1100$$

$$\begin{array}{r} 2002 \\ -\ 878 \\ \hline \boxed{1124} \end{array}$$

- Check your work. The answer to B is 11 hundred 24.

d. Your turn: For the rest of the problems in part 5, say the estimation problems to yourself and write the answer. If the answer shown is close, go to the next problem. If the answer shown is not close to the estimation answer, copy the problem and figure out the exact answer. Raise your hand when you're finished with part 5.
(Observe students and give feedback.)
(Write on the board:)

c. 1400	d. 600	e. 2400
	$\begin{array}{r} 1441 \\ -\ 795 \\ \hline \boxed{646} \end{array}$	
f. 1300	g. 6700	h. 6500
	$\begin{array}{r} 4762 \\ +1943 \\ \hline \boxed{6705} \end{array}$	

e. Check your work. You should have reworked problems for D and G. Raise your hand if you did both those problems.

f. Problem D. The estimation problem is 14 minus 8. The estimation answer is 600.

- The exact answer is 6 hundred 46.
g. Problem G. The estimation problem is 48 plus 19. What's the estimation answer? (Signal.) *67 hundred.*
- The exact answer is 67 hundred 5. That's close to 67 hundred.

EXERCISE 5 AREA
Rectangles and Triangles

a. Find part 6.
b. (Write on the board:)

$$A\ \square = b \times h$$
$$A\ \triangle = \frac{b \times h}{2}$$

- These are the equations you'll use to work the problems in part 6. The first equation is for the area of a rectangle.
- For each problem, you'll start with the equation for the area of a rectangle or the area of a triangle.
c. Work problem A. Raise your hand when you're finished. (Observe students and give feedback.)
- (Write on the board:)

$$a. \quad A\ \triangle = \frac{b \times h}{2}$$
$$A\ \triangle = \frac{13 \times 6}{2} = \frac{78}{2}$$
$$\boxed{A\ \triangle = 39\ sq\ m}$$

- Here's what you should have. The area of the triangle is half the area of a rectangle with the same base and height. The area of the triangle is 39 square meters.
d. Work problem B. Raise your hand when you're finished. (Observe students and give feedback.)
- (Write on the board:)

$$b. \quad A\ \square = b \times h$$
$$A\ \square = 4 \times 14$$
$$\boxed{A\ \square = 56\ sq\ yd}$$

- Here's what you should have. You figured out the area of a rectangle. That's 56 square yards.
e. Your turn: Work the rest of the problems in part 6. Raise your hand when you're finished.
(Observe students and give feedback.)

f. (Write on the board:)

$$A \triangle = \frac{b \times h}{2}$$

$$A \triangle = \frac{12 \times 12}{2} = \frac{144}{2}$$

$$\boxed{A \triangle = 72 \text{ sq ft}}$$

$$A \square = b \times h$$

$$A \square = 20 \times 8$$

$$\boxed{A \square = 160 \text{ sq mi}}$$

- Here's what you should have for problems C and D. Raise your hand if you got everything right.

EXERCISE 6 INDEPENDENT WORK

- (In addition to independent work in the textbook, assign *Bridge to Connecting Math Concepts Independent Worksheet* 15 as classwork or homework. Before beginning the next lesson, check the students' independent work.)

Lesson 36

Objectives

- Appropriately place names in a ratio table and complete the table. (Exercise 1)
- Use estimation to determine whether a problem has a wrong answer. (Exercise 2)
- **Make number families for a mixed set of problems that involve classification or comparison.** (Exercise 3)
- Compute the area of a rectangle or a right triangle. (Exercise 4)
- Solve a word problem that involves a whole group by making a fraction number family. (Exercise 5)
- Work a division problem that shows an incorrect whole-number answer. (Exercise 6)

EXERCISE 1 RATIOS AND PROPORTIONS
Tables

a. Open your textbook to lesson 36 and find part 1.
- Each problem gives information about the names.
b. Problem A: The names are apples **with worms, apples** and apples **without worms.**
- Remember, the total is the big number at the bottom of the table.
- Copy the table. Write the three names where they go. Raise your hand when you've done that much.
(Observe students and give feedback.)
- (Write on the board:)

a.			
with worms	6	
without worms		
apples	9	36

- Here's what you should have.
- Listen: In the first column, the number of apples with worms is 6. The name for the total is **apples.**
c. Your turn: Copy the table for problem B. Write the names where they belong. Raise your hand when you've done that much.
(Observe students and give feedback.)

- (Write on the board:)

b.	clean		144
	dirty	8	
	shoes	10	

- Here's what you should have. The name for the total is **shoes.** The number for clean shoes is 2. 2 is the missing number in the top row. That's the row for clean shoes.

d. Copy the table for problem C. Write the names where they belong. Raise your hand when you've done that much.

   (Observe students and give feedback.)

- (Write on the board:)

c.	girls	5	
	boys	7	
	children		120

- Here's what you should have. The name for the total is **children.** The number for girls is 5.

e. Your turn: Copy the rest of the tables. Write the names where they belong. Raise your hand when you've done that much.

   (Observe students and give feedback.)

- (Write on the board:)

d.	closed	3	276
	open		
	doors	4	

e.	tall	5	
	short		44
	women	9	

- Here's what you should have for tables D and E.
- Raise your hand if you got everything right.
- Make sure all your tables are correct. As part of your independent work, you'll complete the tables.

EXERCISE 2 ESTIMATION
Hundreds

a. Find part 2.
   For some of these problems, the answers are wrong.

b. For each problem, quietly say the estimation problem and write the answer. Compare the estimation answer with the answer shown. If there's a big difference, the answer shown is wrong. For those problems, copy the problem, work it and write the correct answer. If the estimation answer is close, you don't have to rework the problem. Raise your hand when you're finished with part 2.

   (Observe students and give feedback.)

   Key:

a. 3 2 0 0   b. 4 0 0   c. 2 8 0 0   d. 9 0 0

   $$\begin{array}{r} 3587 \\ -\ 428 \\ \hline \boxed{3159} \end{array}$$
   $$\begin{array}{r} 1290 \\ -\ 405 \\ \hline \boxed{885} \end{array}$$

e. 3 9 0 0   f. 2 7 0 0   g. 2 0 0   h. 8 4 0 0

   $$\begin{array}{r} 1418 \\ +2475 \\ \hline \boxed{3893} \end{array}$$

c. Check your work. You should have worked problems A, D and E. Those are problems with the wrong answer shown.

d. Problem A. The estimation problem is 36 minus 4. The estimation answer is 32 hundred. What's the exact answer? (Signal.) *3159.*

- Problem D. The estimation problem is 13 minus 4. The estimation answer is 9 hundred. What's the exact answer? (Signal.) *885.*

- Problem E. The estimation problem is 14 plus 25. The estimation answer is 39 hundred. What's the exact answer? (Signal.) *3893.*

EXERCISE 3 PROBLEM SOLVING
Comparison/Classification

a. Find part 3.
- For some of these problems, you'll make a number family that has the name **difference.** For other problems, you won't. Remember, if the problem compares two things and refers to so many more or so many less, you'll make a family with a difference number. Otherwise, you won't.

- You're **not** going to work the problems in part 3. You're just going to make the number family with the names and the numbers the problem gives.

b. Make the number family for problem A. Raise your hand when you're finished.

   (Observe students and give feedback.)

- (Write on the board:)

	blue	not blue	shirts
a.	25	☐	300

- Here's the family you should have. Raise your hand if you got it right.
- The number family does not have the name **difference** because the problem does not compare two things.
- c. Make the number family for problem B. Raise your hand when you're finished.
  (Observe students and give feedback.)
- (Write on the board:)

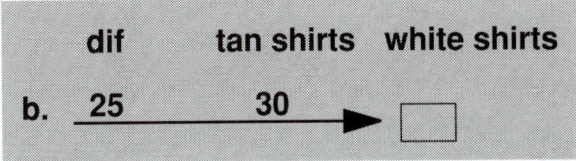

	dif	tan shirts	white shirts
b.	25	30	☐

- Here's what you should have. The problem asks about the difference. So your number family has the names **difference, tan shirts** and **white shirts.** Raise your hand if you got it right.
- d. Your turn. Make number families for the rest of the problems in part 3. Raise your hand when you're finished.
  (Observe students and give feedback.)
  Key:

c.
dif	cow	bull
☐	1658	2030

d.
paint	brushes	total
45	21	☐

e.
dif	cow	bull
307	789	☐

f.
hard-cover	soft-cover	books
808	☐	8003

- Find part J on page 143. That shows what you should have for problems C, D, E and F. You wrote number families with the name **difference** for items C and E.
- Raise your hand if you got everything right.

EXERCISE 4 AREA
Rectangles and Triangles

a. Find part 4.
- You can see the height of each figure. For two of the figures, the base is shown **at the top.**

- Figure B. What's the base? (Signal.) *36 miles.*
- Figure C. What's the base? (Signal.) *78 feet.*
- (Write on the board:)

$$A\ \square\ = b \times h$$

$$A\ \triangle\ = \frac{b \times h}{2}$$

- Here are the equations you'll use. Remember to make the symbols to show whether the equation is for the area of a rectangle or for the area of a triangle.
- b. Find the area of each figure in part 4. Remember to start each problem with the right equation. Raise your hand when you're finished. (Observe students and give feedback.)
  Key:

a. $A\triangle = \frac{b \times h}{2}$

$A\triangle = \frac{14 \times 29}{2} = \frac{406}{2}$

$\boxed{A\triangle = 203 \ sq \ in}$

b. $A\triangle = \frac{b \times h}{2}$

$A\triangle = \frac{36 \times 85}{2} = \frac{3060}{2}$

$\boxed{A\triangle = 1530 \ sq \ mi}$

c. $A\square = b \times h$

$A\square = 78 \times 40$

$\boxed{A\square = 3120 \ sq \ ft}$

- c. Check your work.
- Find part K. That shows what you should have for each problem.
- Raise your hand if you got everything right.

EXERCISE 5 PROBLEM SOLVING
Fraction Number Families

a. Find part 5.
- These are problems that you can solve with fraction number families. Remember, you need three names above the arrow. And you need three fractions with the same denominator. The fraction for the **big** number equals **1** whole because that's the fraction that tells about all the days, or all the children, or all the vehicles.

b. Your turn: Make the complete number family for problem A. Show three names and three fractions. Box the fraction that answers the question. Raise your hand when you've finished problem A.

(Observe students and give feedback.)

• (Write on the board:)

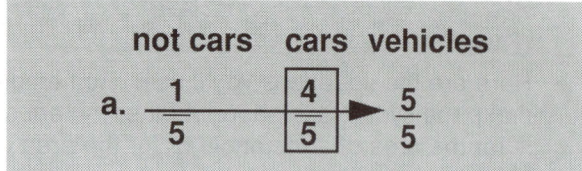

• Here's the family you should have. The names are **not cars, cars** and **vehicles. Vehicles** is the name for the big number. The fractions are 1/5, 4/5 and 5/5.

c. Your turn: Work problem B. Raise your hand when you're finished.

(Observe students and give feedback.)

• (Write on the board:)

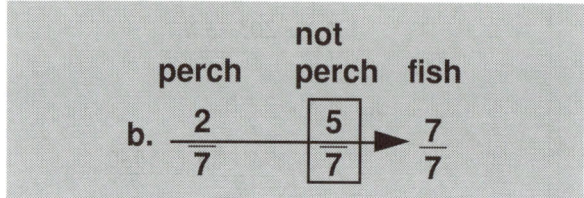

• Here's what you should have. The fraction for perch is 2/7. The fraction for fish that are not perch is 5/7. The fraction for all the fish is 7/7.

d. Your turn: Work problem C. Raise your hand when you're finished.

(Observe students and give feedback.)

• (Write on the board:)

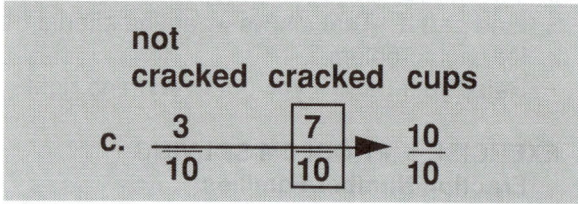

• Here's what you should have. The fraction for cups that are not cracked is 3/10. The fraction for cups that are cracked is 7/10. The fraction for all the cups is 10/10.

e. Raise your hand if you got all of the problems correct.

EXERCISE 6 LONG DIVISION
Incorrect Quotient

a. Find part 6.

• For these problems, the answer may be too big. Or the answer may be too small. Or the answer may be correct. Remember, if the number you get when you multiply is too big, the answer is too big. If the number you get when you multiply is too small, the answer is too small. You know the answer is too small when the remainder is at least as big as the number you divide by.

b. Work the problems in part 6. Raise your hand when you're finished.

(Observe students and give feedback.)

Key:

a. $64 \overline{)345}$ with quotient 6, -384 $64 \overline{)345}$ with quotient $5\frac{25}{64}$, -320, 25

b. $37 \overline{)227}$ with quotient 5, -185, $\boxed{42}$ $37 \overline{)227}$ with quotient $6\frac{5}{37}$, -222, 5

c. $42 \overline{)240}$ with quotient 4, -168, $\boxed{72}$ $42 \overline{)240}$ with quotient $5\frac{30}{42}$, -210, 30

c. Check your work.

• Find part L. That shows what you should have for the problems in part 6.

• Check your work carefully. Correct any mistakes.

EXERCISE 7 INDEPENDENT WORK

a. Remember, you're going to complete the ratio tables for part 1.

• (In addition to independent work in the textbook, assign **Bridge to Connecting Math Concepts** *Independent Worksheet* 16 as classwork or homework. Before beginning the next lesson, check the students' independent work.)

Lesson 37

Objectives

- **Determine whether a problem has a wrong answer by rounding each value to tens or hundreds.** (Exercise 1)

- **Find the area of a non-right triangle shown on the coordinate system.** (Exercise 2)

- **Work a division problem that divides by a two-digit value and has a one-digit quotient.** (Exercise 3)
  Note: Students estimate the answer by constructing an estimation problem for the tens. For example: $6\,2\,\overline{|\,3\,7\,8}$.
  The estimation problem is: $6\,\overline{|\,3\,7}$.

- **Solve a word problem by using a ratio table.** (Exercise 4)
  Note: Problems have three names. For example: **men, women,** and **adults.** Students make a ratio table with the ratio numbers in the first column and the other number the problem gives in the second column. They then figure out the multiplication rule for each row, complete the table, and answer the questions.

- Solve a word problem that involves a whole group by making a fraction number family. (Exercise 5)

EXERCISE 1 ESTIMATION
Tens and Hundreds

a. Open your textbook to lesson 37 and find part 1.
- You've made estimates for hundreds problems. You can do the same thing with tens problems. You just estimate to the closest tens number.

b. Problem A: 47 minus 18. 47 is closest to 50. That's 5 tens. 18 is closest to 20. That's 2 tens. The estimation problem is 5 minus 2. So the estimation answer is 3 tens. That's 30. The answer to problem A is correct.

c. For some of the other problems in part 1 you'll round to the nearest ten. For others, you'll round to the nearest hundred. Say the estimation problem to yourself and write the answer. If the answer shown is wrong, rework the problem and write the exact answer. Raise your hand when you're finished.
(Observe students and give feedback.)

Key:
a. 30   b. 40   c. 7200   d. 110

$$\begin{array}{r} 91 \\ -\ 47 \\ \hline \boxed{44} \end{array} \qquad \begin{array}{r} 6775 \\ +\ 438 \\ \hline \boxed{7213} \end{array}$$

e. 1200   f. 40   g. 170   h. 700

$$\begin{array}{r} 99 \\ +70 \\ \hline \boxed{169} \end{array}$$

d. Check your work. You should have reworked problems B, C and G.
- What's the estimation answer for problem B? (Signal.) *40.*
- What's the exact answer? (Signal.) *44.*
- What's the estimation answer for problem C? (Signal.) *72 hundred.*
- What's the exact answer? (Signal.) *7213.*
- What's the estimation answer for problem G? (Signal.) *170.*
- What's the exact answer? (Signal.) *169.*

EXERCISE 2 AREA
Non-Right Triangles

a. Find part 2.
- (Teacher reference:)

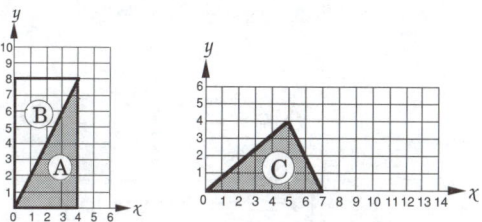

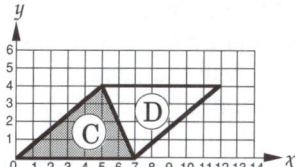

- I'll read what it says. Follow along: You've found the area of triangles that have a 90-degree angle.
- Triangle A has a 90-degree angle.
- You've used the equation: The area of a triangle equals half the base times height.
- Triangle A and triangle B are identical. Each is half the area of the rectangle with a base of 4 units and a height of 8 units.

- You can use the same equation to show that the area of **any triangle** is half the area of a **parallelogram** with the same base and same height.
- Touch triangle C on the next page. √
- Triangle C has no 90-degree angle.
- We can make a **parallelogram** by combining triangle C with another triangle that is exactly the same size and shape. That's triangle D.
- You can see triangle D.
- Triangle C is combined with triangle D to form a parallelogram.
- The base is 7 and the height is 4. So the whole parallelogram has an area of 28 square units.
- The area of triangle C is half that amount—14 square units.
- Remember, the area of any triangle is half the area of a parallelogram with the same base and the same height.
b. Find part 3. √
- The grid shows square units. Each shaded triangle is shown as part of a parallelogram. The triangle is half the area of the parallelogram.
c. Your turn: Start with the equation for the area of a triangle. Figure out the area of shaded triangle A. Raise your hand when you're finished. (Observe students and give feedback.)
- (Write on the board:)

a. $A_\triangle = \dfrac{b \times h}{2}$

$A_\triangle = \dfrac{5 \times 4}{2} = \dfrac{20}{2}$

$\boxed{A_\triangle = 10 \text{ sq in}}$

- Here's what you should have. The area of the parallelogram is 20 square inches. So the area of the triangle with the same base and the same height is 10 square inches. Raise your hand if you got it right.
d. Your turn: Find the area of shaded triangle B. Raise your hand when you're finished. (Observe students and give feedback.)
- (Write on the board:)

b. $A_\triangle = \dfrac{b \times h}{2}$

$A_\triangle = \dfrac{2 \times 5}{2} = \dfrac{10}{2}$

$\boxed{A_\triangle = 5 \text{ sq in}}$

- Here's what you should have. The area of the parallelogram is 10 square inches. So the area of the triangle with the same base and height is 5 square inches.
e. Your turn: Find the area of shaded triangle C. Raise your hand when you're finished. (Observe students and give feedback.)
- (Write on the board:)

c. $A_\triangle = \dfrac{b \times h}{2}$

$A_\triangle = \dfrac{6 \times 4}{2} = \dfrac{24}{2}$

$\boxed{A_\triangle = 12 \text{ sq in}}$

- Here's what you should have. The area of the parallelogram is 24 square inches. So the area of the triangle with the same base and height is 12 square inches.
- Raise your hand if you got everything right.

EXERCISE 3 LONG DIVISION
Estimated Quotient

a. (Write on the board:)

$$73 \overline{\smash{)}594}$$

- Here's 594 divided by 73. You're going to work problems like this one. To work the problem, you have to **estimate** the answer.
b. I'll show you a quick way to estimate the answer. You cover the last digit of the number you're dividing by and the last digit of the number under the division sign. Then you say the division problem for the tens. Those are the digits that are not covered.
- (Cover **3** and **4**.)
- The division problem for the tens is: 59 divided by 7. Everybody, what's the whole-number answer? (Signal.) *8.*
- We write **8** in the answer above the **last digit** of 594.
- (Write to show:)

$$73 \overline{\smash{)}594}^{8}$$

- Remember, we work a problem for the tens, but the answer goes above the last digit of 594, not above the tens digit.

- Now we test the answer by multiplying: 73 times 8. That's 584.
- (Write to show:)

$$73 \overline{\smash{\big)}594} \overset{8}{} \\ -584$$

- Now we subtract.
- (Write to show:)

$$73 \overline{\smash{\big)}594} 8\,\tfrac{10}{73} \\ -584 \\ 10$$

- We got the correct answer from the estimate.
c. (Write on the board:)

$$34 \overline{\smash{\big)}175}$$

- Your turn: Pretend the last digit of both values is covered.
d. Say the problem for the tens. Get ready. (Signal.) *17 divided by 3.*
- What's the whole-number answer? (Signal.) *5.*
- (Repeat step d until firm.)
e. You write the answer above the last digit.
- (Write to show:)

$$34 \overline{\smash{\big)}175} \overset{5}{}$$

f. Find part 4.
- The problem on the board is problem A. Copy the problem and see if your estimate is correct. Raise your hand when you've finished problem A. (Observe students and give feedback.)
Key:

$$a. \quad 34 \overline{\smash{\big)}175} 5\,\tfrac{5}{34} \\ -170 \\ 5$$

- Check your work. When you multiplied, you got **170.** You subtracted and got **5.** The fraction is 5/34. It's less than 1. So your answer is correct.

g. Problem B. Pretend the last digit of each number is covered. Say the problem for the tens. Get ready. (Signal.) *27 divided by 9.*
- What's the whole-number answer? (Signal.) *3.*
- Copy the problem and write that answer. Write it above the last digit of 278. Then complete the problem and see if your estimate is correct. Raise your hand when you're finished. (Observe students and give feedback.)
Key:

$$b. \quad 92 \overline{\smash{\big)}278} 3\,\tfrac{2}{92} \\ -276 \\ 2$$

- Check your work. When you multiplied, you got **276.** The fraction is 2/92.
h. Problem C. Say the estimation problem for the tens. Get ready. (Signal.) *27 divided by 4.*
- What's the whole-number answer? (Signal.) *6.*
- Copy the problem and write that answer. Then complete the problem and see if your estimate is correct. Raise your hand when you're finished. (Observe students and give feedback.)
Key:

$$c. \quad 43 \overline{\smash{\big)}271} 6\,\tfrac{13}{43} \\ -258 \\ 13$$

- Check your work. When you multiplied, you got **258.** You subtracted and got **13.** The fraction is 13/43.
i. Raise your hand if you got everything right.

EXERCISE 4 PROBLEM SOLVING
Ratio Tables

a. Find part 5.
- (Teacher reference:)

ratio		
rainy		
dry		
days		

ratio		
rainy	1	
dry	8	
days		54

ratio		
rainy	1	6
dry	8	48
days	9	54

b. I'll read what it says. Follow along: You've worked word problems that you solve by making ratio equations. The problems you've worked have **two names.**
- You can work problems that have **three names.** To work those problems, you first make a ratio table that shows the three names.

- You can see a problem that has three names. Those names are red.
- The problem says: The ratio of **rainy** days to **dry** days is 1 to 8. There was a **total** of 54 **days.** How many days were rainy? How many days were dry? The names are for **rainy** days, **dry** days, and **days.**
- The first sentence tells the ratio numbers for rainy days and dry days. The ratio numbers go in the first column. **1** is the number for **rainy** and **8** is the number for **dry.**
- The problem tells that there was a total of 54 days. That number goes in the **second** column.
- Now the table is just like the ones you've been working.
- You can figure out the total in the first column and the missing numbers in the second column.
- You can see the table with no missing numbers.
- Remember, the ratio numbers go in the first column.

c. Find part 6.
- Each problem has three names. The problem tells about ratio numbers and gives a number for the second column.

d. Problem A: There were **perch** and **bass** in a pond. For every 10 fish, 7 were perch. There were 900 bass in the pond. How many perch were in the pond? How many total fish were in the pond?
- What are the names for the table? (Call on an individual student.) *Perch, bass, fish.*
- Copy the table and write those names where they belong. Remember, **fish** is the big number. Raise your hand when you have the **names** in your table. √
- The problem says that for every 10 fish, 7 were perch. Put those numbers in the table. Put in the number you know for the second column. Then figure out the missing number in the first column. Raise your hand when you have four numbers in your table. √
- (Write on the board:)

		ratio	
a.	perch	7	
	bass	3	900
	fish	10	

- Here's what you should have so far. Now, complete the table and answer the questions. Raise your hand when you're finished. (Observe students and give feedback.)

- (Write to show:)

		ratio	
a.	perch	7	2100
	bass	3	900
	fish	10	3000

- You multiplied by 300 for each row. The numbers in the second column are correct because 2100 plus 900 is 3000.
- You answered two questions. There were **2100 perch.** There was a total of **3000 fish.**
- Raise your hand if you got everything right.

e. Problem B: There were **perch** and **bass** in a pond. The ratio of perch to bass was 4 to 5. There was a **total of 180 fish** in the pond.
- Make a ratio table with the names and four numbers. Raise your hand when you've done that much. Remember, the ratio numbers go in the first column. Fish is the big number. (Observe students and give feedback.)
- (Write on the board:)

		ratio	
b.	perch	4	
	bass	5	
	fish	9	180

- Here's what you should have. Now, complete the table and answer the questions. Raise your hand when you're finished. (Observe students and give feedback.)
- (Write to show:)

		ratio	
b.	perch	4	80
	bass	5	100
	fish	9	180

- You multiplied by 20 in each row. The numbers in the second column add up: 80 plus 100 equals 180.
- You answered two questions. There were **100 bass.** There were **80 perch.**

f. Problem C: There were parts in a junkyard. For every 8 **parts,** 7 were **rusty.** There were 60 parts that were **not rusty.**
- Which of the three names tells about the total or the big number? (Signal.) *Parts.*

- Make the table with the names and four numbers. Remember, the ratio numbers are in the first column. The other number goes in the second column. Then complete the table and answer the questions. Raise your hand when you're finished.
  (Observe students and give feedback.)
- (Write on the board:)

		ratio	
c.	rusty	7	
	not rusty	1	60
	parts	8	

- Check your work. The names are **rusty, not rusty** and **parts.** The ratio is 7 rusty parts to 8 parts. The missing ratio number is 1.
- How many parts were rusty? (Signal.) *420.*
- (Write **420.**)
- How many parts were there in all? (Signal.) *480.*
- (Write **480.**)

EXERCISE 5 PROBLEM SOLVING
Fraction Number Families

a. Find part 7.
- These are problems that you can work with a fraction number family.
b. Make the complete number family for problem A. Box the fraction that answers the question. Raise your hand when you're finished.
  (Observe students and give feedback.)
- (Write on the board:)

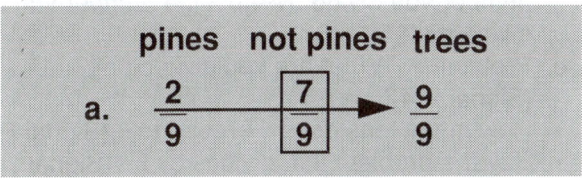

- The fractions are 2/9 for pines, 7/9 for trees that are not pines and 9/9 for all the trees.
c. Your turn: Work the rest of the problems in part 7. Remember to box the fraction that answers each question. Raise your hand when you're finished. (Observe students and give feedback.)

d. (Write on the board:)

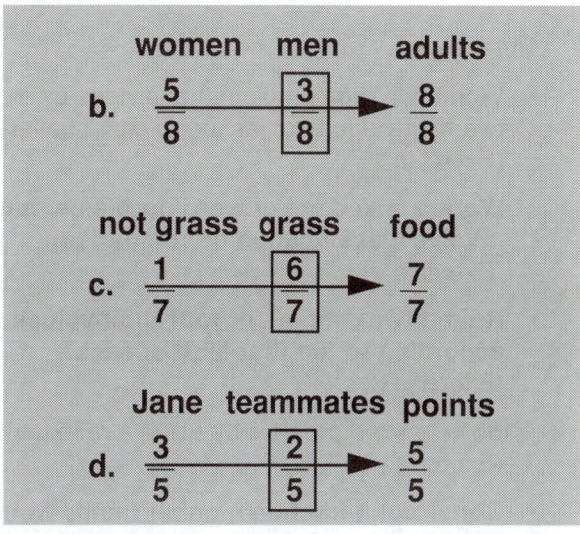

e. Check your work.
- Problem B. What fraction of the adults were men? (Signal.) *3-eighths.*
- Problem C. What's the fraction of food that was grass? (Signal.) *6-sevenths.*
- Problem D. What fraction of the points did Jane's teammates score? (Signal.) *2-fifths.*

EXERCISE 6 INDEPENDENT WORK

- (In addition to independent work in the textbook, assign **Bridge to Connecting Math Concepts** *Independent Worksheet* 17 as classwork or homework. Before beginning the next lesson, check the students' independent work.)

Lesson 38

Objectives

- Work a division problem that divides by a two-digit value and has a one-digit quotient. (Exercise 1)

- **Work a mixed set of word problems that involve classification or comparison.** (Exercise 2)

- **Round two-, three-, or four-digit values according to the first digit of each.** (Exercise 3)

- Solve a word problem by using a ratio table. (Exercise 4)

- **Construct a fraction number family from a diagram.** (Exercise 5)
  Note: Diagrams show a whole divided into equal parts. Students make a number family with a fraction for the whole, for the parts being removed and for the parts that remain. For example:

- **Find the area of a non-right triangle.** (Exercise 6)
  Note: To determine the height, students refer to a dotted line that is perpendicular to the base: ⟋|‾6‾⟍ /‾14‾ .

EXERCISE 1 LONG DIVISION
Estimated Quotient

a. Open your textbook to lesson 38 and find part 1.
- (Teacher reference:)

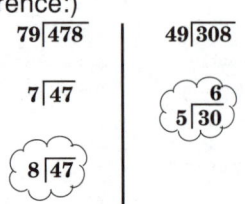

b. I'll read what it says. Follow along: You've worked problems that divide by two-digit numbers. You can estimate the answer by covering the last digit of each value in the problem.

- You can see the problem 478 divided by 79. If you cover the last digit of each number, you get the estimation problem, 47 divided by 7.
- You can get a more accurate estimation problem if you **round the number you divide by** to the **nearest ten.**
- Everybody, how many tens does 79 round to? (Signal.) *8.*
- In the cloud you can see the division problem based on the rounding: 47 divided by 8.
- Look at the next division problem: 308 divided by 49.
- You're dividing by 49. Raise your hand when you know how many tens 49 rounds to. √
- Everybody, how many tens does 49 round to? (Signal.) *5.*
- So the problem for the tens divides by 5.
- You can see the estimation problem in the cloud. It's 30 divided by 5.
- The estimated answer is **6.** That answer is written in the problem just after the cloud.
- Look at the next problem: 308 divided by 41.
- You're dividing by 41. Everybody, how many tens does that round to? (Signal.) *4.*
- The estimation problem divides by 4.
- You can see it in the cloud. It's 30 divided by 4.
- The estimated answer is **7.** You write that answer in the original problem.

c. Remember, round the number you divide by to the nearest ten. Then say the estimation problem using the rounded number.

d. Find part 2.
- For each problem, you'll say the estimation problem for the tens. You'll first round the number you divide by—just the number you divide by.

e. Problem A. What are you dividing by? (Signal.) *47.*
- How many tens does 47 round to? (Signal.) *5.*
- Say the problem that divides by 5. (Signal.) *20 divided by 5.*
- Problem B. What are you dividing by? (Signal.) *29.*
- How many tens does 29 round to? (Signal.) *3.*
- Say the problem that divides by 3. (Signal.) *24 divided by 3.*
- (Repeat step e until firm.)

f. Problem C. What are you dividing by? (Signal.) *52.*
- How many tens does 52 round to? (Signal.) *5.*
- Say the problem that divides by 5. (Signal.) *36 divided by 5.*

g. Work problem A. Say the estimation problem to yourself. Copy the problem and write the answer above the ones digit of 208. Then complete the problem. Raise your hand when you're finished.
(Observe students and give feedback.)
• (Write on the board:)

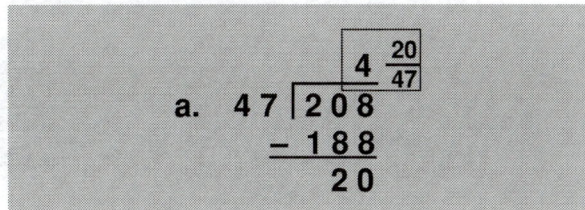

• Here's what you should have. The estimation problem is 20 divided by 5. The answer is 4. 208 divided by 47 equals 4 and 20/47. Raise your hand if you got it right.
• Remember, round the number you're dividing by to the nearest ten. Then work the estimation problem.

h. Copy and work problem B. Raise your hand when you're finished.
(Observe students and give feedback.)
• (Write on the board:)

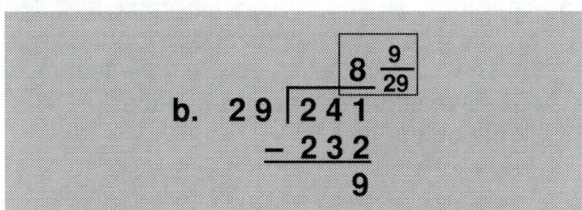

• Here's what you should have. The estimation problem is 24 divided by 3. The answer is 8. When you multiply, you get 232. The fraction is 9/29.

i. Copy and work problem C. Raise your hand when you're finished.
(Observe students and give feedback.)
• (Write on the board:)

c. $52\overline{)368}$ $7\frac{4}{52}$
-364
4

• Here's what you should have. The estimation problem is 36 divided by 5. The answer is 7. 52 times 7 is 364. The fraction is 4/52.

EXERCISE 2 PROBLEM SOLVING
Comparison/Classification

a. Find part 3.
• For some of the problems, you'll make a number family with the name **difference.** For others, you'll make a family that does not have the name **difference.**
• Remember, if the problem compares two things or asks about the difference between two things, your family needs the name **difference.** If the problem does not tell about the difference or ask about the difference, your family should not have the name **difference.**

b. Work problem A. Raise your hand when you're finished. (Observe students and give feedback.)
• (Write on the board:)

• Here's the family you should have. The problem does not tell about the difference or ask about the difference. There were 349 insects that could not fly.

c. Work problem B. Raise your hand when you're finished. (Observe students and give feedback.)
• (Write on the board:)

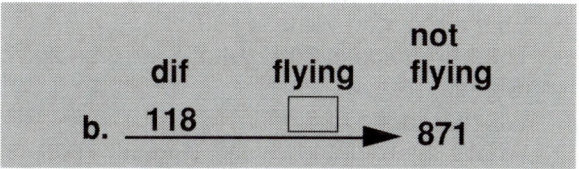

• Here's the family you should have. The problem tells about the difference. The answer is 753 flying insects.

d. Work problem C. Raise your hand when you're finished. (Observe students and give feedback.)
• (Write on the board:)

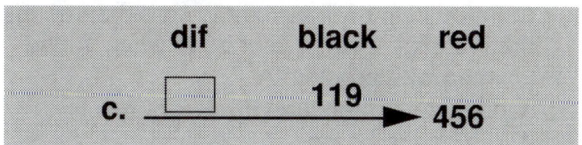

• Here's the family you should have. The problem asks about the difference. There were 337 more red ants than black ants.

e. Work problem D. Raise your hand when you're finished. (Observe students and give feedback.)

- (Write on the board:)

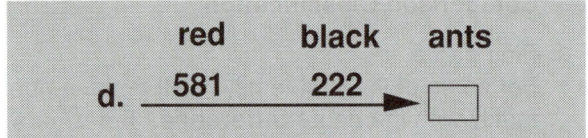

	red	black	ants
d.	581	222	▶ ☐

- Here's the family you should have. The problem does not tell about the difference or ask about the difference. The answer is 803 ants.

EXERCISE 3 ESTIMATION
Tens, Hundreds and Thousands

a. Find part 4.
(Teacher reference:)

4686 rounds to 5000

4329 rounds to 4000

379 rounds to ___

608 rounds to ___

- You can round most numbers so everything after the first digit is a zero. You do that by rounding the first digit.
- The first digit of 4 thousand 686 is 4. That rounds to 5. So the rounded value is 5000.
- The first digit of 4 thousand 329 is 4. That rounds to 4. So the rounded value is 4000.
- The next value is 379. What's the first digit? (Signal.) *3.*
- Will that round to 3 or 4? (Signal.) *4.*
- So the rounded value is 400.
- The next value is 608. What's the first digit? (Signal.) *6.*
- Will that round to 6 or 7? (Signal.) *6.*
- What's the rounded value? (Signal.) *600.*
- Remember, you can base the rounding on the first digit.
b. Each problem shows three values. You'll write a rounded value based on the first digit of each value. Write the estimation problem and the answer for problem A. You'll round 680, 5 thousand 290 and 390. Remember, each rounded value will have nothing but zeros after the first digit. Raise your hand when you're finished with problem A.
(Observe students and give feedback.)
- (Write on the board:)

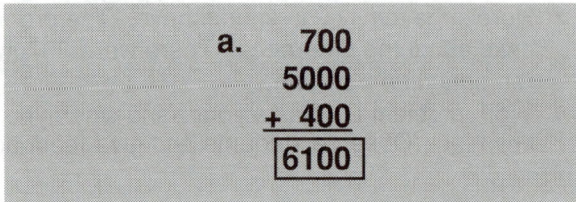

a.	700
	5000
	+ 400
	6100

- Here's what you should have. The rounded values are 700, 5000 and 400. The estimation answer is 6100.
c. Your turn: Work the rest of the estimation problems for part 4. Raise your hand when you're finished.
- (Write on the board:)

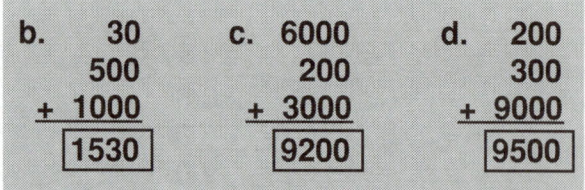

b.	30	c.	6000	d.	200
	500		200		300
	+ 1000		+ 3000		+ 9000
	1530		9200		9500

- Here's what you should have for problems B, C and D.
- Raise your hand if you got everything right.

EXERCISE 4 PROBLEM SOLVING
Ratio Tables

a. Find part 5.
- These are ratio problems that have three names. Remember, the name for the total or the big number goes in the bottom row. The ratio numbers go in the first column.
b. Your turn: Read the facts for problem A. Make a table with three names and four numbers. Raise your hand when you've done that much.
(Observe students and give feedback.)
- (Write on the board:)

a.	boys	2	402
	girls	3	
	children	5	

- Here's the table with four numbers.
- Your turn: Figure out all the missing numbers in the table and write answers to the questions. Raise your hand when you're finished.
(Observe students and give feedback.)
- Check your work. Everybody, how many children are in the school? (Signal.) *1005.*
- (Write **1005.**)
- How many girls are in the school? (Signal.) *603.*
- (Write **603.**)
c. Your turn: Read the facts for problem B. Fix up your table with three names and four numbers. One of the numbers is in the second column. Raise your hand when you've done that much.
(Observe students and give feedback.)

- (Write on the board:)

b.	**blue**	5	
	red	3	
	balloons	8	240

- Here's the table with four numbers.
- Your turn: Figure out all the missing numbers in the table and write answers to the questions. Raise your hand when you're finished.
  (Observe students and give feedback.)
- Check your work. Everybody, how many blue balloons are there? (Signal.) *150.*
- (Write **150.**)
- How many red balloons are there? (Signal.) *90.*
- (Write **90.**)

EXERCISE 5 NUMBER FAMILIES
Fraction Pictures

a. Find part 6.
- These are pictures of pies.
- For each pie, you'll make a fraction number family showing the fraction for the whole pie, the fraction for the pieces being removed, and the fraction for the remaining pieces.
- The sample item shows the pie divided into 5 pieces. One piece is being removed.
- The fraction number family shows 1/5 for removed, 4/5 for remaining and 5/5 for the whole pie.
b. Your turn: Make the fraction number family for pie A. Show the pieces being removed as the first fraction. Raise your hand when you're finished. (Observe students and give feedback.)
- (Write on the board:)

	removed	remaining	pie
a.	$\dfrac{3}{4}$	$\dfrac{1}{4}$ ▶	$\dfrac{4}{4}$

- Here's what you should have. The fractions are 3/4 for removed, 1/4 for remaining and 4/4 for the whole pie.
c. Write fractions for the other pies in part 6. Raise your hand when you're finished.
  (Observe students and give feedback.)

d. (Write to show:)

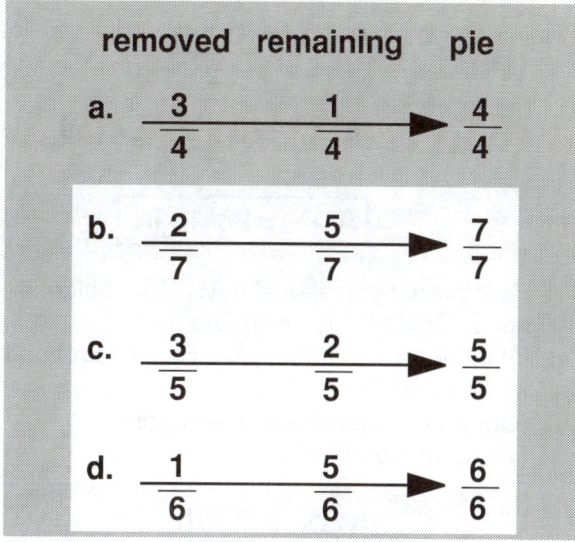

	removed	remaining	pie
a.	$\dfrac{3}{4}$	$\dfrac{1}{4}$ ▶	$\dfrac{4}{4}$
b.	$\dfrac{2}{7}$	$\dfrac{5}{7}$ ▶	$\dfrac{7}{7}$
c.	$\dfrac{3}{5}$	$\dfrac{2}{5}$ ▶	$\dfrac{5}{5}$
d.	$\dfrac{1}{6}$	$\dfrac{5}{6}$ ▶	$\dfrac{6}{6}$

- Here's what you should have for pies B, C and D.
e. Raise your hand if you got everything right.

EXERCISE 6 AREA
Non-Right Triangles

a. Find part 7.
- These triangles are not on a coordinate grid. The height of each triangle is shown with a dotted line that is perpendicular to the base. Remember, the length of the side on the left is not the height of the figure.
b. Find the area of triangle A. Raise your hand when you're finished.
  (Observe students and give feedback.)
- (Write on the board:)

$$a. \quad A \triangle = \frac{b \times h}{2}$$

$$A \triangle = \frac{14 \times 10}{2} = \frac{140}{2}$$

$$\boxed{A \triangle = 70 \text{ sq cm}}$$

- Here's what you should have. The height is 10, not 12. The area is 70 square centimeters.
c. Find the area of triangle B. Raise your hand when you're finished.
  (Observe students and give feedback.)

• (Write on the board:)

$$b. \quad A\triangle = \frac{b \times h}{2}$$

$$A\triangle = \frac{8 \times 7}{2} = \frac{56}{2}$$

$$\boxed{A\triangle = 28 \text{ sq in}}$$

• Here's what you should have. The height is 7, not 8. The area is 28 square inches.

d. Find the area of triangle C. Raise your hand when you're finished.

(Observe students and give feedback.)

• (Write on the board:)

$$c. \quad A\triangle = \frac{b \times h}{2}$$

$$A\triangle = \frac{20 \times 9}{2} = \frac{180}{2}$$

$$\boxed{A\triangle = 90 \text{ sq yd}}$$

• Here's what you should have. Raise your hand if you got everything right.

EXERCISE 7 INDEPENDENT WORK

• (In addition to independent work in the textbook, assign **Bridge to Connecting Math Concepts** *Independent Worksheet* 18 as classwork or homework. Before beginning the next lesson, check the students' independent work.)

Lesson 39

Objectives

• **Find the area and the perimeter of a rectangle, triangle or parallelogram.** (Exercise 1)

• Work a mixed set of word problems that involve classification or comparison. (Exercise 2)

• Work a division problem that divides by a two-digit value and has a one-digit quotient. (Exercise 3)

• Round two-, three-, four-digit values according to the first digit of each. (Exercise 4)

• Solve a word problem by using a ratio table. (Exercise 5)

• Construct a fraction number family from a diagram. (Exercise 6)

EXERCISE 1 AREA AND PERIMETER
Parallelograms/Rectangles/Triangles

a. Open your textbook to lesson 39 and find part 1.

• You'll find the area and the perimeter of each figure. Remember, for area, you use height. The height is marked with a dotted line. If there is no dotted line, one of the sides shows the height.

b. Work all the problems in part 1. Raise your hand when you're finished.

(Observe students and give feedback.)

Key:

a. $A\square = b \times h$

$A\square = 15 \times 8$

$\boxed{A\square = 120 \text{ sq m}}$

$$\begin{array}{r} 8 \\ 8 \\ 15 \\ + 15 \\ \hline \boxed{P = 46 \text{ m}} \end{array}$$

b. $A\triangle = \frac{b \times h}{2}$

$A\triangle = \frac{14 \times 12}{2} = \frac{168}{2}$

$\boxed{A\triangle = 84 \text{ sq ft}}$

$$\begin{array}{r} 15 \\ 13 \\ + 14 \\ \hline \boxed{P = 42 \text{ ft}} \end{array}$$

c. $A\square = b \times h$

$A\square = 7 \times 8$

$\boxed{A\square = 56 \text{ sq m}}$

$$\begin{array}{r} 10 \\ 10 \\ 7 \\ + 7 \\ \hline \boxed{P = 34 \text{ m}} \end{array}$$

148 *Lesson 38*

c. Check your work.
- Find part J on page 156. That shows what you should have for each problem.
- Raise your hand if you got everything right.

EXERCISE 2 PROBLEM SOLVING
Comparison/Classification

a. Find part 2.
- For some of the problems, you'll make a number family with the name **difference.** For others, you'll make a family that does not have the name **difference.**
- Remember, if the problem compares two things or asks about the difference between two things, your family needs the name **difference.** Otherwise it doesn't.
b. Work problem A. Raise your hand when you're finished. (Observe students and give feedback.)
- (Write on the board:)

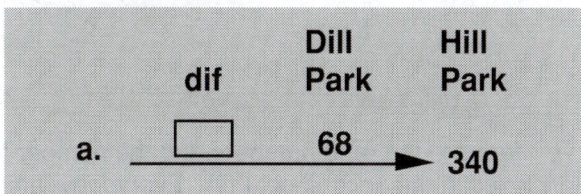

- Here's the family you should have. The problem asks about the difference. There were 272 fewer flowers in Dill Park.
c. Work problem B. Raise your hand when you're finished. (Observe students and give feedback.)
- (Write on the board:)

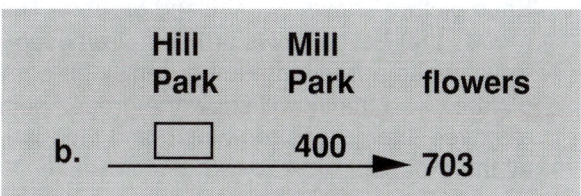

- Here's the family you should have. The problem does not tell about the difference or ask about the difference. There were 303 flowers in Hill Park.
d. Work problem C. Raise your hand when you're finished. (Observe students and give feedback.)
- (Write on the board:)

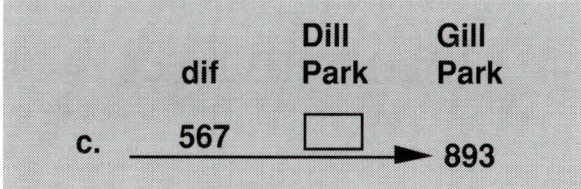

- Here's the family you should have. The problem tells about the difference. There were 326 trees in Dill Park.
e. Work problem D. Raise your hand when you're finished. (Observe students and give feedback.)
- (Write on the board:)

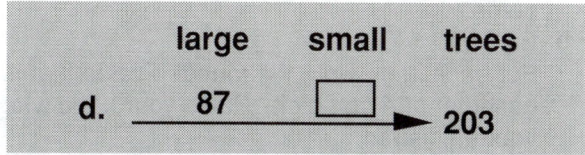

- Here's the family you should have. The problem does not tell about the difference or ask about the difference. There were 116 small trees.
- Raise your hand if you got all the problems right.

EXERCISE 3 LONG DIVISION
Estimated Quotient

a. Find part 3.
- For each problem, you'll say the problem for the tens. FIrst you'll round the number you divide by.
b. Problem A. What are you dividing by? (Signal.) *19.*
- How many tens does 19 round to? (Signal.) *2.*
- Say the estimation problem that divides by 2. (Signal.) *9 divided by 2.*
c. Problem B. What are you dividing by? (Signal.) *82.*
- How many tens does 82 round to? (Signal.) *8.*
- Say the estimation problem that divides by 8. (Signal.) *50 divided by 8.*
d. Your turn: Work the problems in part 3. Raise your hand when you're finished.
(Observe students and give feedback.)
Key:

$$a.\ 19\overline{)94}\ \boxed{4\tfrac{18}{19}}\quad -76\quad \overline{18}$$
$$b.\ 82\overline{)501}\ \boxed{6\tfrac{9}{82}}\quad -492\quad \overline{9}$$
$$c.\ 68\overline{)499}\ \boxed{7\tfrac{23}{68}}\quad -476\quad \overline{23}$$

e. Check your work.
- Find part K. That shows what you should have for the problems in part 3. Correct any mistakes.

EXERCISE 4 ESTIMATION
Tens, Hundreds and Thousands

a. Find part 4.
- I'll read what it says. Follow along: When you round the first digit of a numeral, the rounded value may be more than 9. When it is, you write the rounded value as 10. You must make sure that you show the correct number of zeros.

- The value shown is 970. If you round the number based on the first digit, you'll round 9 to 10. You replace the other digits with zeros. You write: 1000, not 100.
- Remember, if the 9 rounds up, write 10, but make sure you have the correct number of zeros.

b. Find part 5.
- For each item, write the rounded value that is based on the first digit. Raise your hand when you're finished.
  (Observe students and give feedback.)
- (Write on the board:)

a.	3000
b.	90,000
c.	10,000
d.	1000
e.	9000
f.	50

- Here's what you should have for each value. Values B, C and D had 9 as the first digit. C and D rounded to 10.
- Raise your hand if you got everything right.

EXERCISE 5 PROBLEM SOLVING
Ratio Tables

a. Find part 6.
  These are ratio problems that have three names. The names are not shown in bold type. The ratio numbers go in the first column of a table. Another number goes in the second column.
b. Work problem A. Make the table. Write answers to the questions. Raise your hand when you're finished. (Observe students and give feedback.)
- (Write on the board:)

a.	U.S.	2	
	foreign	7	105
	stamps	9	

- The names are U.S., foreign and stamps. The ratio numbers are 2, 7 and 9. The number of foreign stamps is 105. Everybody, how many U.S. stamps does Bernard have? (Signal.) *30.*
- Everybody, how many stamps does Bernard have in all? (Signal.) *135.*

c. Work problem B. Raise your hand when you're finished. (Observe students and give feedback.)
- (Write on the board:)

b.	hardback	5	
	paperback	4	
	books	9	720

- Here are the ratio numbers and the total number of books in the library. Everybody, how many hardback books are there? (Signal.) *400.*
- How many paperback books are there? (Signal.) *320.*

d. Work problem C. Raise your hand when you're finished. (Observe students and give feedback.)
- (Write on the board:)

c.	squirrels	6	
	foxes	1	95
	animals	7	

- Here are the ratio numbers and the number for foxes in the second column. Everybody, how many squirrels were counted? (Signal.) *570.*
- How many animals were counted? (Signal.) *665.*

EXERCISE 6 NUMBER FAMILIES
Fraction Pictures

a. Find part 7.
- Each picture shows equal-sized groups. The groups that have arrows are being removed. You'll make a fraction number family for each picture. The family will show the groups being removed, the groups remaining and the total for all the groups.
b. Item A. There are 5 groups of seeds. How many groups are being removed? (Signal.) *1.*
- Make the fraction number family showing the groups being removed as the first number. Raise your hand when you're finished.
  (Observe students and give feedback.)
- (Write on the board:)

removed	remaining	seeds
a. $\dfrac{1}{5}$	$\dfrac{4}{5}$ →	$\dfrac{5}{5}$

- Here's what you should have for item A. 1/5 of the seeds are being removed. 4/5 of the seeds remain. The big number is 5/5. That's all the seeds.
c. Make fraction number families for the rest of the items in part 7. Show the groups being removed as the first small number in your family. Raise your hand when you're finished.
  (Observe students and give feedback.)
d. (Write on board:)

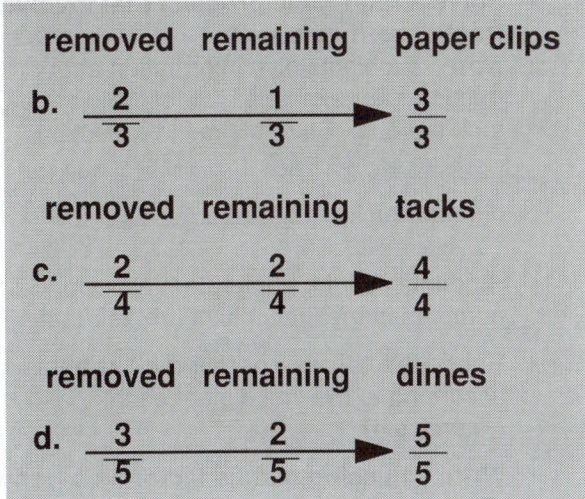

- Here's what you should have for items B, C and D.
e. Raise your hand if you got everything right.

EXERCISE 7 INDEPENDENT WORK

- (In addition to independent work in the textbook, assign **Bridge to Connecting Math Concepts** *Independent Worksheet* 19 as classwork or homework. Before beginning the next lesson, check the students' independent work.)

Lesson 40 – Test 4

Objectives

- Work a division problem that divides by a two-digit value and has a one-digit quotient. (Exercise 1)

- Perform on a mastery test of skills presented in lessons 31 through 39. (Exercise 2)

Note: Exercise 3 provides instructions for marking the test.

EXERCISE 1 LONG DIVISION
Estimated Quotient

a. Open your textbook to lesson 40 and find part 1.
- Sometimes the estimate for division problems gives you the wrong answer.
- All the estimate answers for problems in part 1 are wrong. They are either too big or too small.
- You know how to correct the wrong estimates. If you can't subtract, you make the number in the answer smaller. If you subtract and have a remainder that is too large, you make the number in the answer larger.
b. Work problem A. Raise your hand when you're finished. (Observe students and give feedback.)
- (Write on the board:)

$$23\overline{)187} \quad 8\tfrac{3}{23}$$
$$\underline{-184}$$
$$3$$

- Here's what you should have. The original estimate was 9. That's too large.
c. Work problem B. Raise your hand when you're finished. (Observe students and give feedback.)
- (Write on the board:)

$$67\overline{)340} \quad 5\tfrac{5}{67}$$
$$\underline{-335}$$
$$5$$

- Here's what you should have. The original estimate was 4. That's too small.
d. Work problem C. Raise your hand when you're finished. (Observe students and give feedback.)

- (Write on the board:)

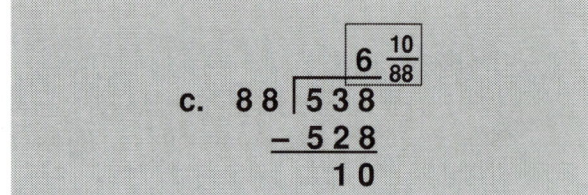

- Here's what you should have. The original estimate was 5. That's too small.

EXERCISE 2 TEST 4

Note: Students are not to use calculators.

a. This is a test. You should only have your textbook, a sharpened pencil and lined paper.
b. Find part 1 of test 4 in your textbook. √
c. Do the test on your own. Raise your hand when you've completed the test.
   (Observe students but do not give feedback.)

EXERCISE 3 MARKING THE TEST

a. (Collect the papers. Use the Answer Key to score the tests. Award scores as follows:)

Test 4 Percent Summary

SCORE	%	SCORE	%	SCORE	%
74	100	66	89	58	78
73	99	65	88	57	77
72	97	64	86	56	76
71	96	63	85	55	74
70	95	62	84	54	73
69	93	61	82	53	72
68	92	60	81	52	70
67	91	59	80		

b. (Complete the Test 4 Remedy Summary to determine whether remedies are needed.)
- (If more than 1/4 of the students did not pass a test part, present the remedy for that part before beginning lesson 41. Remedies appear at the end of the Test 4 Answer Key.)

EXERCISE 4 INDEPENDENT WORK

- (Assign **Bridge to Connecting Math Concepts** *Independent Worksheet* 20 as classwork or homework. Before beginning the next lesson, check the students' independent work.)

Lesson 41

Objectives

- Construct a fraction number family from a diagram. (Exercise 1)

- Find the area and the perimeter of a rectangle, triangle or parallelogram. (Exercise 2)

- **Solve a ratio-table problem that involves fractions.** (Exercise 3)
  Note: Problems give information about a fraction. For example: 2/5 of the workers took the bus. Students make a fraction

  number family: $\underset{bus}{\dfrac{2}{5}}$ $\underset{no\ bus}{\dfrac{3}{5}}$ ⟶ $\underset{workers}{\dfrac{5}{5}}$.

  Students use the numerators as ratio numbers in the first column of their table.

- Round two-, three-, or four-digit values according to the first digit of each. (Exercise 4)

- **Write a missing middle factor as a fraction.** (Exercise 5)
  Note: Problems are of the form: 3 ( ) = 11. Students first write the missing value as a fraction. Later, they write the mixed number it equals.

- **Use the angle symbol (∠) to write an equation that shows the degrees in an angle.** (Exercise 6)

EXERCISE 1 NUMBER FAMILIES
Fraction Pictures

a. Open your textbook to lesson 41 and find part 1.
- Some of the pictures show pies. Some show groups.
b. For each picture, you'll make a fraction number family. You'll show the fractions for the parts being **removed,** the parts **remaining** and the fraction for **all** the parts. That's the big number.
- Raise your hand when you've completed part 1. (Observe students and give feedback.)

c. (Write on the board:)

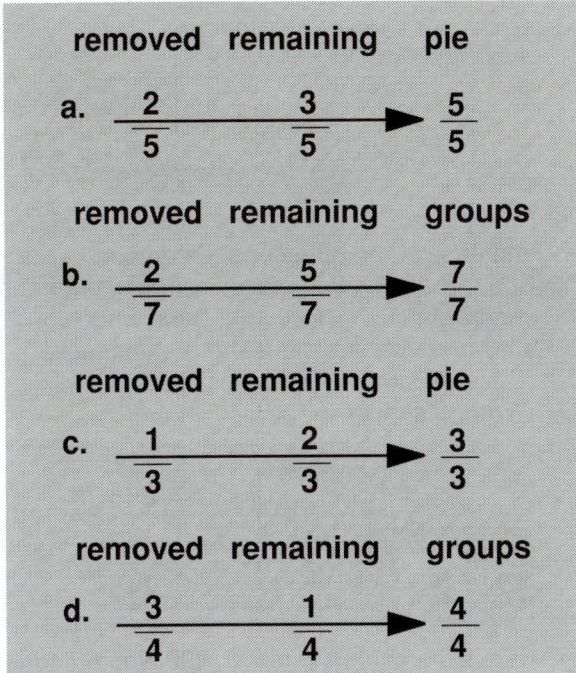

	removed	remaining	pie
a.	$\dfrac{2}{5}$	$\dfrac{3}{5}$ →	$\dfrac{5}{5}$

	removed	remaining	groups
b.	$\dfrac{2}{7}$	$\dfrac{5}{7}$ →	$\dfrac{7}{7}$

	removed	remaining	pie
c.	$\dfrac{1}{3}$	$\dfrac{2}{3}$ →	$\dfrac{3}{3}$

	removed	remaining	groups
d.	$\dfrac{3}{4}$	$\dfrac{1}{4}$ →	$\dfrac{4}{4}$

- Here's what you should have for each item.
d. Raise your hand if you got everything right.

EXERCISE 2 AREA AND PERIMETER
Parallelograms/Rectangles/Triangles

a. Find part 2.
- You'll find the area and the perimeter of each figure. Remember, for area, you use height. The height is marked with a dotted line. If there is no dotted line, one of the sides shows the height.
b. Work all the problems in part 2. Raise your hand when you're finished.
  (Observe students and give feedback.)

Key:

a. $A\triangle = \dfrac{b \times h}{2}$

$A\triangle = \dfrac{63 \times 20}{2} = \dfrac{1260}{2}$

$\boxed{A\triangle = 630 \text{ sq cm}}$

$\begin{array}{r} 25 \\ 52 \\ + 63 \\ \hline \end{array}$

$\boxed{P = 140 \text{ cm}}$

b. $A\square = b \times h$

$A\square = 15 \times 20$

$\boxed{A\square = 300 \text{ sq m}}$

$\begin{array}{r} 22 \\ 22 \\ 15 \\ + 15 \\ \hline \end{array}$

$\boxed{P = 74 \text{ m}}$

c. $A\triangle = \dfrac{b \times h}{2}$

$A\triangle = \dfrac{5 \times 12}{2} = \dfrac{60}{2}$

$\boxed{A\triangle = 30 \text{ sq yd}}$

$\begin{array}{r} 12 \\ 13 \\ + 5 \\ \hline \end{array}$

$\boxed{P = 30 \text{ yd}}$

c. Check your work.
- Find part J on page 163. That shows what you should have for each problem. Raise your hand if you got everything right.

EXERCISE 3 RATIOS AND PROPORTIONS
Tables

a. Find part 3.
- These are problems that require a fraction number family and a ratio table.
- Here's how to identify these problems: They give three names. They give a **fraction** and **a number that is not a fraction.** Remember, three names, a fraction and a number that is not a fraction.
b. Look at the sample problem: 2/5 of the workers in a company took a bus to work. The rest didn't. There was a total of 745 workers in the company. How many workers took a bus to work? How many workers didn't take a bus to work?
- The problem has three names. What are the names? (Call on individual students. Ideas: *bus, no bus, workers.*)
c. The problem gives a fraction. Everybody, what fraction? (Signal.) *2-fifths.*
- The problem gives a number that is not a fraction. What number? (Signal.) *745.*
- That's a number for the **second column** of your table.
- The problem gives a fraction, so the first thing you do is make a fraction number family with three names and three fractions.
- (Write on the board:)

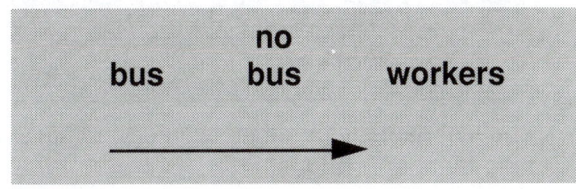

bus	no bus	workers
	→	

- Here are the names.
- What's the fraction for all the workers? (Signal.) *5-fifths.*
- What's the fraction for the workers that take the bus? (Signal.) *2-fifths.*
- What's the fraction for the workers that do not take the bus? (Signal.) *3-fifths.*

- (Write to show:)

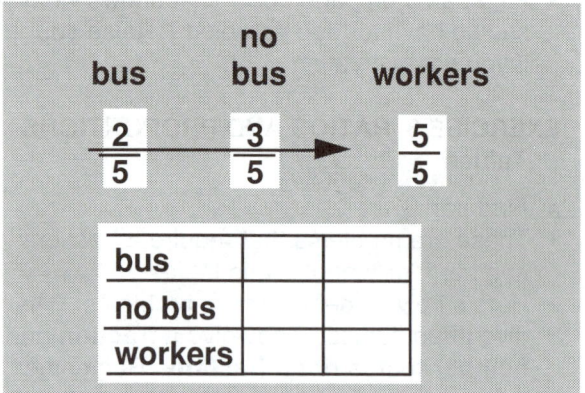

bus		
no bus		
workers		

d. Now we make the table with the same names.
- We use the numerators of the fractions as ratio numbers for the first column.
- I'll say the names and you tell me the ratio numbers.
- The first name is for workers who take the bus. What's the ratio number? (Signal.) *2.*
- (Write to show:)

bus	**2**	
no bus		
workers		

- The next name is for workers who do not take the bus. What's the ratio number? (Signal.) *3.*
- (Write to show:)

bus	**2**	
no bus	**3**	
workers		

- The last name is for all the workers. What's the ratio number? (Signal.) *5.*
- (Write to show:)

bus	**2**	
no bus	**3**	
workers	**5**	

- The problem gives a number for the second column. What number? (Signal.) *745.*
- What's the name for 745? (Signal.) *Workers.*

- (Write to show:)

bus	**2**	
no bus	**3**	
workers	**5**	**745**

e. Now you can figure out the missing numbers and answer the questions the problem asks. Copy what's on the board, including the fraction number family. Raise your hand when you've answered the questions the problem asks. (Observe students and give feedback.)
- (Write to show:)

bus	**2**	**298**
no bus	**3**	**447**
workers	**5**	**745**

- Here's the complete table. 298 workers took the bus. 447 workers did not take the bus. Remember, first make a fraction number family. Then put the numerators in the first column of the table. The numerators are ratio numbers.
f. Problem A. The underlined sentences tell about the fraction number family.
- Make the family with the names and all the fractions. Raise your hand when you've done that much.
(Observe students and give feedback.)
- (Write on the board:)

	new	used	cars
a.	$\dfrac{4}{7}$	$\dfrac{3}{7}$	$\dfrac{7}{7}$

- Here's what you should have. The names are **new, used** and total **cars.** 3/7 of the cars were used. So the fractions are 4/7, 3/7 and 7/7.
- Everybody, what are the ratio numbers? (Signal.) *4, 3 and 7.*
- Write the names and ratio numbers in a table. Put in the other number the problem tells about and figure out the number of used cars and total cars. Raise your hand when you're finished. (Observe students and give feedback.)

- (Write on the board:)

a.	new	4	64	1. 48 used cars
	used	3	48	2. 112 total cars
	cars	7	112	

- Check your work. The problem tells you that 64 cars were new. So there were 48 used cars and 112 total cars.
g. Problem B. The underlined sentences tell about the fraction number family. Make the family with the names and all the fractions. Raise your hand when you've done that much.
   (Observe students and give feedback.)
- (Write on the board:)

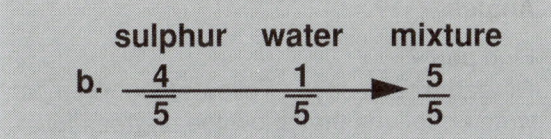

$$\text{sulphur} \quad \text{water} \quad \text{mixture}$$
$$\text{b.} \quad \frac{4}{5} \quad \quad \frac{1}{5} \quad \longrightarrow \quad \frac{5}{5}$$

- Here's what you should have. The fractions are 4/5, 1/5 and 5/5. Everybody, what are the ratio numbers? (Signal.) *4, 1 and 5.*
- Write the names and ratio numbers in a table. Put in the other number the problem tells about and figure out the weight of the sulphur and water. Raise your hand when you're finished.
   (Observe students and give feedback.)
- (Write on the board:)

b.	sulphur	4	60	1. 60 lb
	water	1	15	2. 15 lb
	mixture	5	75	

- Check your work. The problem tells you there were 75 pounds of mixture. So there were 60 pounds of sulphur and 15 pounds of water.
h. Problem C. Make the family with the names and all the fractions. Raise your hand when you've done that much.
   (Observe students and give feedback.)
- (Write on the board:)

$$\text{cornmeal} \quad \text{canned} \quad \text{corn}$$
$$\text{c.} \quad \frac{5}{8} \quad \quad \frac{3}{8} \quad \longrightarrow \quad \frac{8}{8}$$

- Here's what you should have. The names are **cornmeal, canned** and **corn** or **harvest.** The fractions are 5/8, 3/8 and 8/8. Everybody, what are the ratio numbers? (Signal.) *5, 3 and 8.*
- Write the names and ratio numbers in a table. Put in the other number the problem tells about and figure out the weight of the corn made into cornmeal and the weight of the total harvest. Raise your hand when you're finished.
   (Observe students and give feedback.)
- (Write on the board:)

c.	cornmeal	5	155	1. 155 lb
	canned	3	93	2. 248 lb
	corn	8	248	

- Check your work. The problem tells you 93 pounds of corn were canned. So 155 pounds was ground into cornmeal. The total harvest was 248 pounds.

EXERCISE 4 ESTIMATION
Tens, Hundreds and Thousands

a. Find part 4.
- For these problems, you'll round the values according to the first digit. Remember, if the first digit is 9, it may round to 10. When that happens, make sure you show the correct number of zeros.
b. Work problem A. Write the rounded values for each number and write the answer. Raise your hand when you're finished.
   (Observe students and give feedback.)
- (Write on the board:)

$$
\begin{array}{r}
\text{a.} \quad 500 \\
10,000 \\
+ \ \ 200 \\
\hline
\boxed{10,700}
\end{array}
$$

- Here's what you should have. 9983 rounds to 10 thousand. The estimation answer is 10,700. Raise your hand if you got it right.
c. Work the rest of the problems in part 4. Write the problem with the rounded numbers and write the answer. Raise your hand when you're finished. (Observe students and give feedback.)

d. (Write on the board:)

	b.	50	c.	2000
		100		1000
		+ 2000		+ 6000
		2150		**9000**

- Here's what you should have for B and C. Raise your hand if you got everything right.

EXERCISE 5 MULTIPLICATION
Missing Factor

a. Find part 5.
- I'll read what it says. Follow along: You know how to work multiplication problems that have a missing middle value. You work a division problem. The problem is: 4 times some value equals 20. That's 20 divided by 4.
- You can also write the missing value as a fraction. You just write the fraction for the division problem.
- The missing value for this equation is 20 divided by 4.
- You can see the fraction for 20 divided by 4. That's the missing value. 4 times 20/4 equals 20.
- Remember, say the division problem, then write the fraction for that problem.
b. Find part 6.
- For each problem, you'll say the division problem for the missing value. Then you'll say the fraction that shows the division.
c. Problem A: 3 times some value is 11. Say the division problem for the missing value. (Signal.) *11 divided by 3.*
- Say the fraction for 11 divided by 3. (Signal.) *11-thirds.*
- Problem B: 5 times some value is 6. Say the division problem for the missing value. (Signal.) *6 divided by 5.*
- Say the fraction for 6 divided by 5. (Signal.) *6-fifths.*
- Problem C: 9 times some value is 20. Say the division problem for the missing value. (Signal.) *20 divided by 9.*
- Say the fraction for 20 divided by 9. (Signal.) *20-ninths.*
- (Repeat step c until firm.)
d. Your turn: Copy the problems in part 6 and write the missing value as a fraction. Raise your hand when you're finished.
(Observe students and give feedback.)

e. (Write on the board:)

$$\text{a. } 3\left(\frac{11}{3}\right) = 11 \qquad \text{b. } 5\left(\frac{6}{5}\right) = 6$$

$$\text{c. } 9\left(\frac{20}{9}\right) = 20 \qquad \text{d. } 6\left(\frac{50}{6}\right) = 50$$

$$\text{e. } 2\left(\frac{19}{2}\right) = 19 \qquad \text{f. } 7\left(\frac{40}{7}\right) = 40$$

- Here's what you should have.
- Raise your hand if you got everything right.
f. Later you'll work the division problem and figure out the mixed number answer each fraction equals.

EXERCISE 6 GEOMETRY
Angles

a. Find part 7.
- I'll read what it says. Follow along: You can write equations that show the degrees for angles. You'll use the symbol shown to identify a lettered angle.
- You can see an equation that says: Angle J equals 61 degrees.
b. Look at equation A.
- Everybody, what does the equation say? (Signal.) *Angle F equals 108 degrees.*
- Read equation B. (Signal.) *Angle N equals 37 degrees.*
- From now on, you won't trace figures. You'll write equations to show the degrees for the lettered angles.
c. Find part 8.
- Your turn: Figure out the missing angles in the figure.
- Write the complete equations for angles K, P and T. (Observe students and give feedback.)
d. (Write on the board:)

$$\angle k = 100°$$
$$\angle p = 80°$$
$$\angle t = 100°$$

- Here's what you should have.

EXERCISE 7 INDEPENDENT WORK

- (In addition to independent work in the textbook, assign **Bridge to Connecting Math Concepts** *Independent Worksheet* 21 as classwork or homework. Before beginning the next lesson, check the students' independent work.)

Lesson 42

Objectives

- **Solve a set of problems that multiply more than two values, some of which are fractions.** (Exercise 1)

  Example: $2 \times \frac{1}{5} \times \frac{3}{8} = \boxed{}$.

- Write a missing middle factor as a fraction. (Exercise 2)

- **Read and write decimal values that involve tenths, hundredths or thousandths.** (Exercise 3)

- **Solve a set of ratio-table problems, some of which involve fractions.** (Exercise 4)

- **Write a fraction based on a diagram.** (Exercise 5)

  Example: **What's the fraction for**

  3 pieces? Students write the fraction 3/8.

EXERCISE 1 FRACTION MULTIPLICATION
Three Factors

a. Open your textbook to lesson 42 and find part 1.

b. I'll read what it says. Follow along: You've worked with fractions that have a denominator of 1. Any number over 1 equals the number. 5 over 1 equals 5. 243 over 1 equals 243.

- You can show that these fractions equal the whole number by using your calculator.

- To do that, you work the division problem for the fraction.

- Say the division problem for fraction A. (Signal.) *34 divided by 1.*

- Say the division problem for fraction B. (Signal.) *3 divided by 1.*

- Use your calculator. Work those problems. Write the answer. You don't have to write the whole equation. Raise your hand when you're finished.

- Everybody, what does 34 divided by 1 equal? (Signal.) *34.*

- What does 3 divided by 1 equal? (Signal.) *3.*

- Remember, the simple fraction for a whole number is the whole number over 1.

c. (Write on the board:)

$$\frac{2}{3} \times \frac{4}{5} \times \frac{7}{2} =$$

- For some problems, you have to multiply more than two values. You just multiply the first two values, then multiply the answer by the last value.

- (Point to the problem on the board.)

- Here's 2/3 times 4/5 times 7/2. On top we have 2 times 4. What's that? (Signal.) *8.*

- What's 8 times 7? (Signal.) *56.*

- (Write to show:)

$$\frac{2}{3} \times \frac{4}{5} \times \frac{7}{2} = \underline{56}$$

- On the bottom, we have 3 times 5. What's that? (Signal.) *15.*

- What's 15 times 2? (Signal.) *30.*

- (Write **30.**)

- So 2/3 times 4/5 times 7/2 equals 56/30.

d. Find part 2.

- For each problem, you have to multiply more than two values. Some of the values shown are whole numbers. If there's a fraction in the problem, you'll change any whole numbers in the problem into simple fractions. That's the whole number over 1.

- You don't have to rewrite your answer.

e. Work problem A. Raise your hand when you're finished. √

- On top, you multiplied 5 times 1 times 3. What's the answer? (Signal.) *15.*

- On the bottom, you multiplied 2 times 2 times 4. What's the answer? (Signal.) *16.*

- So the answer to the whole problem is 15/16.

f. Your turn: Work the rest of the problems in part 2. Raise your hand when you're finished. (Observe students and give feedback.)

g. Check your work.

- Problem B. You multiplied 2 times 6 times 7. What's the answer? (Signal.) *84.*

- If you wrote the whole number over 1, your answer is correct. 84 over 1 equals 84.

- Problem C. On top, you multiplied 2 times 3 times 5. What's the answer? (Signal.) *30.*

- On the bottom, you multiplied 1 times 2 times 8. What's the answer? (Signal.) *16.*

- So the answer to the whole problem is 30/16.

EXERCISE 2 MULTIPLICATION
Missing Factor

a. Find part 3.
• You're going to show the missing value in each problem as a fraction. Remember, you just say the division problem for the missing value. Then you write the fraction for that division problem.
b. Problem A: 8 times some value equals 648. Say the division problem for the missing value. (Signal.) *648 divided by 8.*
• Say the fraction for 648 divided by 8. (Signal.) *648-eighths.*
• Copy problem A and write the missing value as a fraction. Then use your calculator to figure out what that fraction equals. On the line below, rewrite the equation using the whole number. Raise your hand when you're finished. (Observe students and give feedback.)
• (Write on the board:)

a. $8\left(\frac{648}{8}\right) = 648$

$8 \ (\ 81 \ ) = 648$

• Here's what you should have. 648/8 is 81. So your bottom equation is: 8 times **81** equals 648.
c. Your turn: Copy problem B and write the missing fraction. Below, rewrite the equation using a whole number. Raise your hand when you're finished. (Observe students and give feedback.)
• (Write on the board:)

b. $7\left(\frac{434}{7}\right) = 434$

$7 \ (\ 62 \ ) = 434$

• Here's what you should have. 434/7 is 62. So your bottom equation is: 7 times **62** equals 434.
d. Your turn: Copy problem C and write both equations. Raise your hand when you're finished. (Observe students and give feedback.)
• (Write on the board:)

c. $3\left(\frac{1503}{3}\right) = 1503$

$3 \ (\ 501 \ ) = 1503$

• Here's what you should have. Raise your hand if you got everything right.
e. You're going to use your calculator to check your answers.
• (Point to problem A: 8(81) = 648.)
• Here's what you enter to check problem A: 8 times 81. Enter these values and see whether you end up with 648. If you do, the equation is right.
• Hold up your calculator when you have an answer. (Observe students and give feedback.)
• You end up with 648. So the answer's correct.
f. Your turn: Check problem B: 7 times 62 equals . . . See if you end up with 434. Hold up your calculator when you have an answer. (Observe students and give feedback.)
• You end up with 434. So the answer's correct.
g. Your turn: Check problem C: 3 times 501. Hold up your calculator when you have an answer. (Observe students and give feedback.)
• You end up with 1503. So the answer is correct.

EXERCISE 3 DECIMALS
Place Value

a. Find part 4.
• You're going to learn about decimal values. I'll read what it says. Follow along: All the values that are written before the decimal point are 1 or more.
• All values that come after the decimal point are less than 1.
• You can write a decimal point after any whole number. That's what your calculator does. You can see the whole numbers 203, 1 and 45.
• Listen: The columns in part 4 show how whole numbers and decimal values work. The columns for whole numbers are in purple. The columns for decimals are in green.
• Touch the numbers 1000 and 9999.
• Whole numbers between 1000 and 9999 have four digits. The first digit is the thousands digit.
• Touch the numbers 100 and 999.
• Whole numbers between 100 and 999 have three digits. The first digit is the hundreds digit.
• Touch the numbers 10 and 99.
• Whole numbers between 10 and 99 have two digits. The first digit is the tens digit.
• Touch the numbers 1 and 9.
• Whole numbers between 1 and 9 have one digit. That digit tells the number of ones.
• Remember, all the values that are written **before** the decimal point are 1 or more than 1. All the values that come **after** the decimal point are less than 1.

- If the decimal part ends one place after the decimal point, that part tells about tenths.
- Touch 1-tenth and 9-tenths. They end one place after the decimal point.
- If the decimal part ends two places after the decimal point, that part tells about hundredths.
- Touch 1-hundredth and 99-hundredths.
- If the decimal part ends three places after the decimal point, that part tells about thousandths.
- Touch 1-thousandth and 999-thousandths.

b. Find part 5.
- (Teacher reference:)

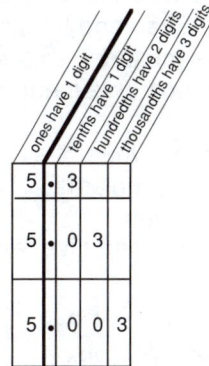

- Part 5 shows different numbers. All have a whole number and a decimal part.
- I'll read what it says. Follow along: Here are rules for reading decimal numbers: You read the whole number part. You say **and** for the decimal point. Then you read the decimal part.
- If the number ends **one** place after the decimal point, you say **tenths.**
- If the number ends **two** places after the decimal point, you say **hundredths.**
- If the number ends **three** places after the decimal point, you say **thousandths.**

c. Look at the first number. The number ends one place after the decimal point. So that number is 5 and 3-tenths. Everybody, read that number. (Signal.) *5 and 3-tenths.*

d. The next number ends two places after the decimal point. That's hundredths. Everybody, read that number. (Signal.) *5 and 3-hundredths.*
- (Repeat step d until firm.)

e. The last number ends 3 places after the decimal point. Everybody, read that number. (Signal.) *5 and 3-thousandths.*
- (Repeat step e until firm.)

f. Find part 6.
- Numeral A. How many places after the decimal point is the last digit? (Signal.) *Two.*
- So the decimal part is 35-hundredths. The numeral is 6 and 35-hundredths. Everybody, read numeral A. (Signal.) *6 and 35-hundredths.*

- Numeral B. Read it. (Signal.) *5 and 13-hundredths.*

g. Numeral C. How many places after the decimal point is the last digit? (Signal.) *One.*
- That's tenths, not hundredths. So numeral C is 6 and 4-tenths. Everybody, read numeral C. (Signal.) *6 and 4-tenths.*
- Read numeral D. (Signal.) *5 and 3-tenths.*

h. Numeral E is not tenths. Read numeral E. (Signal.) *7 and 41-hundredths.*

i. Read numeral F. (Signal.) *25 and 6-tenths.*

j. Numeral G. How many places after the decimal point is the last digit? (Signal.) *Three.*
- So numeral G is thousandths. Everybody, read numeral G. (Signal.) *3 and 189-thousandths.*
- Read numeral H. (Signal.) *18 and 502-thousandths.*
- (Repeat steps h-j until firm.)

k. Remember, a numeral that ends one place after the decimal point is tenths; two places after the decimal point is hundredths; three places after the decimal point is thousandths.

l. Write part O on your lined paper. √
- You're going to write decimal numbers. I'll say **and** for the decimal point. Each value is going to end in **4.** Remember, if it ends in 4-tenths, the 4 is one place after the decimal point. If it's 4-hundredths, the 4 is two places after the decimal point and you may need a zero.

m. Value A: 9 and 4-hundredths. Write it. √
- (Write on the board:)

a. 9.04

- Here it is. 9 and 4 hundredths.
- Value B: 9 and 4-tenths. Write it. √
- (Write on the board:)

b. 9.4

- Here it is.
- Value C: 9 and 4-thousandths. Write it. √
- (Write on the board:)

c. 9.004

- Here it is.
- Value D: 42 and 14-hundredths. Write it. √
- (Write on the board:)

d. 42.14

- Here it is.
- Value E: 1 and 84-thousandths. Write it. √

- (Write on the board:)

> **e.   1.084**

- Here it is.
n. Raise your hand if you got everything right.

EXERCISE 4 RATIOS AND PROPORTIONS
Tables

a. Find part 7.
- For these problems you'll make ratio tables. Some of these problems tell about a **fraction.** For those problems, **you make a fraction number family.** The numerators of the fractions are the ratio numbers. You don't have to make a number family for problems that do not tell about a fraction.
b. Your turn: Work problem A. Make sure you answer both questions. Raise your hand when you're finished.
  (Observe students and give feedback.)
- (Write on the board:)

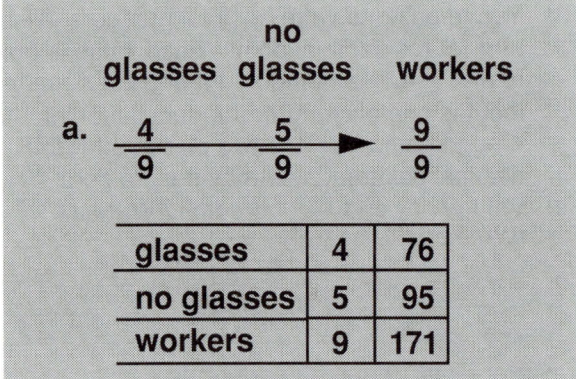

	no glasses	glasses	workers
a.	$\frac{4}{9}$	$\frac{5}{9}$ →	$\frac{9}{9}$

glasses	4	76
no glasses	5	95
workers	9	171

- Here's the number family and table you should have.
- Check your work. Problem A says that 4/9 of the workers in a factory wore glasses. The names are for **glasses, no glasses** and total **workers.** The ratio numbers are **4, 5** and **9.** The other number the problem gives is **76.** That's the number of workers who wore glasses. Everybody, how many workers did not wear glasses? (Signal.) *95.*
- How many workers were there in all? (Signal.) *171.*
- Raise your hand if you got everything right.
c. Your turn: Work problem B. Raise your hand when you're finished.
  (Observe students and give feedback.)

Key:

b. sand	7	266
cement	2	76
mixture	9	342

- Check your work. Problem B says that the ratio of sand to cement is 7 to 2. You did not make a fraction number family. The names are **sand, cement** and total **mixture.** The ratio numbers are **7, 2** and **9.** The other number the problem gives is **342.** That's the total number of pounds in the mixture. Everybody, how many pounds of sand are in the mixture? (Signal.) *266.*
- How many pounds of cement are in the mixture? (Signal.) *76.*
- Raise your hand if you got everything right.
d. Your turn: Work problem C. Raise your hand when you're finished.
  (Observe students and give feedback.)
Key:

	chicken pox	no pox	students
c.	$\frac{3}{8}$	$\frac{5}{8}$ →	$\frac{8}{8}$

chicken pox	3	75
no pox	5	125
students	8	200

- Check your work. Problem C says that 3/8 of the students caught chicken pox. You make a fraction number family. The names are for **chicken pox, no** chicken **pox** and total **students.** The fractions are **3/8, 5/8** and **8/8.** The ratio numbers are **3, 5** and **8.** The other number the problem gives is **125.** That's the number of students who did not catch chicken pox. Everybody, how many students caught chicken pox? (Signal.) *75.*
- How many students were there in all? (Signal.) *200.*
- Raise your hand if you got everything right.

EXERCISE 5 FRACTION ANALYSIS
From Pictures

a. Find part 8.
- The sample item shows a picture of a pie. There are no arrows to show the number of pieces that are being removed.

- If you know how many pieces are in the whole pie, you can make up fractions for the whole pie or for any number of pieces.
- Everybody, how many pieces are in the pie? (Signal.) *Five.*
- So what's the fraction for the whole pie? (Signal.) *5-fifths.*
- If 5/5 is the fraction for the whole pie, what's the fraction for three pieces of the pie? (Signal.) *3-fifths.*
- What's the fraction for one piece of the pie? (Signal.) *1-fifth.*
- What's the fraction for four pieces of the pie? (Signal.) *4-fifths.*

b. Item A shows a different pie. There are two questions. The first question asks about the fraction for the whole pie. The other question asks about the fraction for 2 pieces.
- Write answers to those questions. Raise your hand when you're finished.
- (Write on the board:)

$$\text{a. 1. } \frac{6}{6}$$
$$\text{2. } \frac{2}{6}$$

- Here's what you should have. The pie is divided into 6 parts. So the fraction for the whole pie is 6/6. The fraction for 2 pieces of the pie is 2/6.
- Raise your hand if you got it right.

c. Write answers to both questions for item B. Raise your hand when you're finished.
(Observe students and give feedback.)
- (Write on the board:)

$$\text{b. 1. } \frac{8}{8}$$
$$\text{2. } \frac{1}{8}$$

- Here's what you should have. The first question asks about the fraction for 8 pieces. That's the whole pie. The fraction is 8/8. The other question asks about the fraction for 1 piece. That's 1/8.

d. Items C and D show groups that are equal.
- Answer both questions for the items. Count carefully for item D. Raise your hand when you're finished.
(Observe students and give feedback.)

e. (Write on the board:)

$$\text{c. 1. } \frac{2}{3} \qquad \text{d. 1. } \frac{1}{7}$$
$$\text{2. } \frac{3}{3} \qquad \text{2. } \frac{6}{7}$$

- Here's what you should have for items C and D.
- For item C, the first question asks about the fraction for 2 groups. That's 2/3. The other question asks about the fraction for 3 groups. That's 3/3.
- For item D, the first question asks about the fraction for 1 group. That's 1/7. The other question asks about the fraction for 6 groups. That's 6/7.

f. Raise your hand if you got everything right.

EXERCISE 6 INDEPENDENT WORK

- (In addition to independent work in the textbook, assign **Bridge to Connecting Math Concepts** *Independent Worksheet* 22 as classwork or homework. Before beginning the next lesson, check the students' independent work.)

Lesson 43

EXERCISE 1 NUMBER FAMILIES
Improper Fractions

a. Open your textbook to lesson 43 and find part 1.
- These are fraction number families that do not have a fraction equal to 1 as the big number.
- You work these problems like any other number family problem. You figure out the missing number in the numerators. The denominators are the same for all the fractions.
b. Copy family A and figure out the missing number. Raise your hand when you're finished.
(Observe students and give feedback.)
- (Write on the board:)

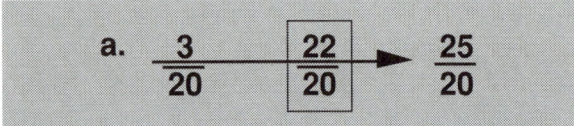

a. $\dfrac{3}{20}$ → $\boxed{\dfrac{22}{20}}$ → $\dfrac{25}{20}$

- Here's what you should have. The missing fraction is 22/20.
c. Your turn: Work the rest of the items in part 1. Copy each family. Write the missing fraction. Raise your hand when you're finished.
(Observe students and give feedback.)

d. (Write on board:)

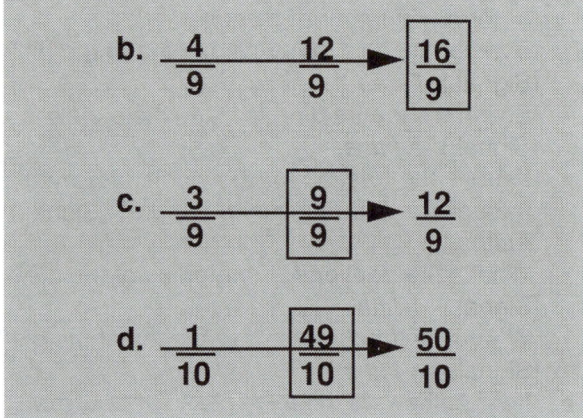

b. $\dfrac{4}{9}$ → $\dfrac{12}{9}$ → $\boxed{\dfrac{16}{9}}$

c. $\dfrac{3}{9}$ → $\boxed{\dfrac{9}{9}}$ → $\dfrac{12}{9}$

d. $\dfrac{1}{10}$ → $\boxed{\dfrac{49}{10}}$ → $\dfrac{50}{10}$

- Here's what you should have for each family.
e. Raise your hand if you got everything right.

EXERCISE 2 FRACTION MULTIPLICATION
Three Factors

a. Find part 2.
- For each problem, you have to multiply more than two values. Some of the values shown are whole numbers. If there's a fraction in the problem, change any whole number into a simple fraction and multiply. You don't have to rewrite your answer.
b. Work problem A. Raise your hand when you're finished.
- On top, you multiplied 3 times 5 times 2. What's the answer? (Signal.) 30.
- On the bottom, you multiplied 4 times 1 times 5. What's the answer? (Signal.) 20.
- So the answer to the whole problem is 30/20.
c. Your turn: Work the rest of the problems in part 2. Raise your hand when you're finished.
(Observe students and give feedback.)
d. Check your work.
- Problem B. On top, you multiplied 1 times 5 times 7. What's the answer? (Signal.) 35.
- On the bottom, you multiplied 3 times 6 times 2. What's the answer? (Signal.) 36.
- So the answer to the whole problem is 35/36.
- Problem C. You multiplied 2 times 5 times 7. What's the answer? (Signal.) 70.

EXERCISE 3 DECIMALS
Place Value

a. You're going to write decimal numbers on your lined paper. How many places do you show after the decimal point for **tenths?** (Signal.) One.
- How many places do you show for **hundredths?** (Signal.) Two.

- How many for **thousandths?** (Signal.) *Three.*
b. Write part O on your lined paper, then write A through F. √
c. Value A: 4 and 1-tenth. Write it. √
- Value B: 4 and 1-thousandth. Write it. √
- Value C: 6 and 2-hundredths. Write it. √
- Value D: 15-hundredths. Write it. √
- Value E: 4-tenths. Write it. √
- Value F: 15-thousandths. Write it. √
d. (Write on the board:)

a.	4.1
b.	4.001
c.	6.02
d.	.15
e.	.4
f.	.015

- Here's what you should have for each value. Raise your hand if you got all the items correct.
e. Your turn: Read value A. (Signal.)
  4 and 1-tenth.
- Value B. (Signal.) *4 and 1-thousandth.*
- Value C. (Signal.) *6 and 2-hundredths.*
- Value D. (Signal.) *15-hundredths.*
- Value E. (Signal.) *4-tenths.*
- Value F. (Signal.) *15-thousandths.*
- (Repeat step e until firm.)

EXERCISE 4   RATIOS AND PROPORTIONS
Tables

a. Find part 3.
- You'll make ratio tables for these problems. Some of these problems tell about a fraction. For those problems, you make a fraction number family. The top numbers of the fractions are the ratio numbers.
b. Your turn: Work problem A. Raise your hand when you're finished.
  (Observe students and give feedback.)
- (Write on the board:)

a.			
	water	2	24
	milk	5	60
	total	7	84

- Here's the table. You did not write a fraction number family.

- Check your work. Problem A says there are 2 cups of water for every 5 cups of milk. The ratio numbers are **2, 5** and **7.** The other number the problem gives is **84.** That's the total number of cups.
- Everybody, how many cups of water were used? (Signal.) *24.*
- How many cups of milk were used? (Signal.) *60.*
- Raise your hand if you got everything right.
c. Your turn: Work problem B. Raise your hand when you're finished.
  (Observe students and give feedback.)
  Key:

$$b. \quad \underset{ski\ suits}{\frac{5}{7}} \quad \underset{\substack{no \\ ski\ suits}}{\frac{2}{7}} \longrightarrow \underset{people}{\frac{7}{7}}$$

ski suits	5	95
no ski suits	2	38
people	7	133

- Check your work. Problem B says that 5/7 of the people wore ski suits. You made a fraction number family. The names are for people who wore **ski suits,** people who did **not** wear ski suits and total **people.** The fractions are 5/7, 2/7 and 7/7. The ratio numbers are **5, 2** and **7.** The other number the problem gives is **95.** That's the number of people who wore ski suits.
- Everybody, how many people were on vacation? (Signal.) *133.*
- How many people did not wear ski suits? (Signal.) *38.*
- Raise your hand if you got everything right.
d. Your turn: Work problem C. Raise your hand when you're finished.
  (Observe students and give feedback.)
  Key:

$$c. \quad \underset{late}{\frac{1}{5}} \quad \underset{\substack{on \\ time}}{\frac{4}{5}} \longrightarrow \underset{days}{\frac{5}{5}}$$

late	1	42
on time	4	168
days	5	210

- Check your work. Problem C says that on 1/5 of the school days, the bus arrived late. You made a fraction number family. The names are **late, on time** and total **days.** The fractions are 1/5, 4/5 and 5/5. The ratio numbers are **1, 4** and **5.** The other number the problem gives is **210.** That's the number of days in all.
- Everybody, on how many days did the bus arrive late? (Signal.) *42.*
- On how many days did the bus arrive on time? (Signal.) *168.*
- Raise your hand if you got everything right.

EXERCISE 5 FRACTION ANALYSIS
From Pictures

a. Find part 4.
- Each item asks two questions about a pie or equal-sized groups. Write answers to both questions for item A. Raise your hand when you're finished. √
- (Write on the board:)

$$\textbf{a.} \quad 1. \ \frac{2}{6}$$
$$2. \ \frac{5}{6}$$

- Here's what you should have. The first question asks about the fraction for 2 groups. That's 2/6. The other question asks about the fraction for 5 groups. That's 5/6.
- Raise your hand if you got it right.
b. Your turn: Write answers to the rest of the questions. Raise your hand when you're finished. (Observe students and give feedback.)
c. (Write on the board:)

$$\textbf{b.} \quad 1. \ \frac{6}{6}$$
$$2. \ \frac{1}{6}$$

$$\textbf{c.} \quad 1. \ \frac{4}{4} \qquad \textbf{d.} \quad 1. \ \frac{6}{8}$$
$$2. \ \frac{1}{4} \qquad \qquad 2. \ \frac{3}{8}$$

- Here's what you should have for each item.
d. Raise your hand if you got everything right.

EXERCISE 6 MULTIPLICATION
Missing Factor

a. Find part 5.
- You can read fractions as division problems even if the answer is less than 1.
b. Fraction A is 3/4. The division problem is 3 divided by 4. That's less than 1.
c. Fraction A: Read it as a division problem. (Signal.) *3 divided by 4.*
- Fraction B: Read it as a division problem. (Signal.) *1 divided by 7.*
- Fraction C: Read it as a division problem. (Signal.) *24 divided by 39.*
- (Repeat step c until firm.)
d. Find part 6.
- You're going to write the missing value in each problem as a fraction. Some of the fractions will be less than 1.
e. Problem A: 21 times some value equals 5. Say the division problem for the missing value. (Signal.) *5 divided by 21.*
- Say the fraction for 5 divided by 21. (Signal.) *5-twenty-firsts.*
f. Problem B: 18 times some value equals 9. Say the division problem for the missing value. (Signal.) *9 divided by 18.*
- Say the fraction for 9 divided by 18. (Signal.) *9-eighteenths.*
g. Your turn: Copy each problem and write the missing value in each problem as a fraction. Raise your hand when you're finished. (Observe students and give feedback.)
h. Check your work.
- I'll say each problem. You tell me the fraction for the missing value.
- Problem A: 21 times some fraction equals 5. What's the missing fraction? (Signal.) *5-twenty-firsts.*
- Problem B: 18 times some fraction equals 9. What's the fraction? (Signal.) *9-eighteenths.*
- Problem C: 7 times some fraction equals 11. What's the fraction? (Signal.) *11-sevenths.*
- Problem D: 19 times some fraction equals 4. What's the fraction? (Signal.) *4-nineteenths.*
- Problem E: 50 times some fraction equals 17. What's the fraction? (Signal.) *17-fiftieths.*
- Problem F: 39 times some fraction equals 75. What's the fraction? (Signal.) *75-thirty-ninths.*

EXERCISE 7 INDEPENDENT WORK

- (In addition to independent work in the textbook, assign **Bridge to Connecting Math Concepts** Independent Worksheet 23 as classwork or homework. Before beginning the next lesson, check the students' independent work.)

Lesson 44

EXERCISE 1 PROBLEM SOLVING
Fraction Number Families

a. Open your textbook to lesson 44 and find part 1.
- These are word problems that don't have pictures.
- For each problem, you'll make a fraction number family and then write the fraction that answers the question the problem asks.
b. Look at the sample problem. It says: A pie was divided into 6 equal parts. A person removed 4 parts. What is the fraction for the pie that remains?
- First, we make the fraction number family with three fractions. Then we write the fraction that answers the question.

- (Write on the board:)

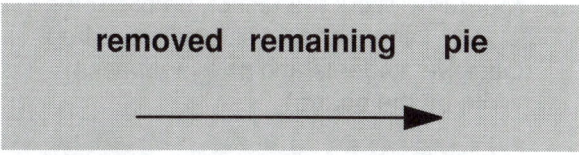

- Here are the names: **removed, remaining** and **pie.**
- The pie is divided into 6 equal parts. So what's the fraction for the big number? (Signal.) *6-sixths.*
- (Write: **6/6.**)
- What's the fraction for the parts that were removed? (Signal.) *4-sixths.*
- What's the fraction for the pie that remained? (Signal.) *2-sixths.*
- (Write to show:)

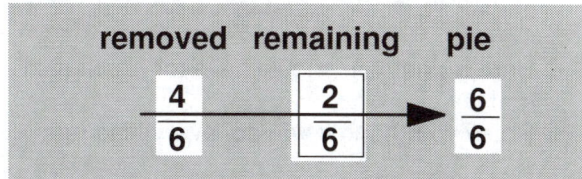

- The fraction that answers the question the problem asks is 2/6.
c. Problem A: Seeds were divided into 12 equal-sized groups. 7 of those groups were removed. What's the fraction for the groups that were not removed?
- Make the fraction number family. Write the answer to the question. Raise your hand when you're finished.
  (Observe students and give feedback.)
- (Write on the board:)

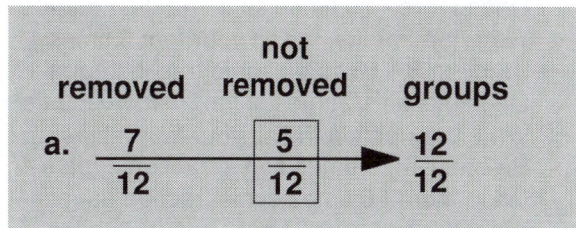

- Here's what you should have.
d. Work problem B. Raise your hand when you're finished. (Observe students and give feedback.)
- (Write on the board:)

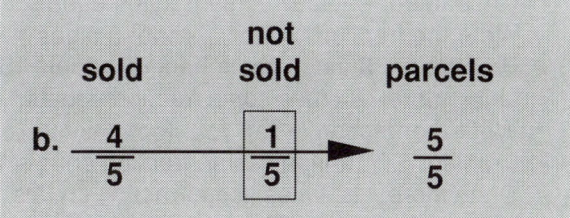

- Here's what you should have.
e. Your turn: Work the rest of the problems in part 1. Raise your hand when you're finished. (Observe students and give feedback.)
f. (Write on the board:)

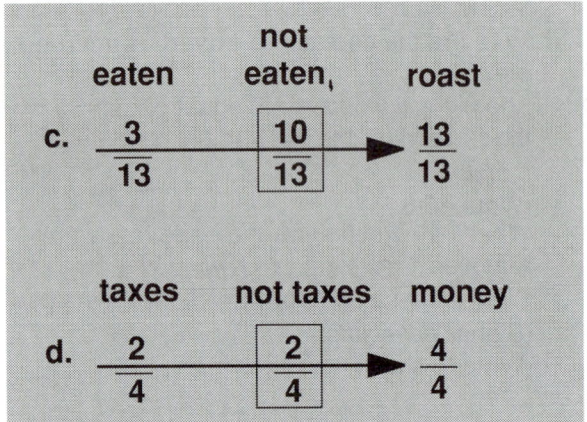

- Here's what you should have for problems C and D.
g. Raise your hand if you got everything right.

EXERCISE 2 MULTIPLICATION
Missing Factor

a. I'm going to say multiplication problems with a missing middle number. You'll write the missing value as a fraction.
- Here's a practice problem. Listen: 4 times some fraction equals 9. Once more: 4 times some fraction equals 9. Everybody, what's the fraction? (Signal.) *9-fourths.*
b. Write part O on your lined paper.
- Problem A: 18 times some fraction equals 7. Once more: 18 times some fraction equals 7. Write the fraction. (Tap your foot 3 times.)
- Problem B: 10 times some fraction equals 1. Once more: 10 times some fraction equals 1. Write the fraction. (Tap your foot 3 times.)
- Problem C: 6 times some fraction equals 40. Once more: 6 times some fraction equals 40. Write the fraction. (Tap your foot 3 times.)
- Problem D: 19 times some fraction equals 62. Once more: 19 times some fraction equals 62. Write the fraction. (Tap your foot 3 times.)
- Problem E: 4 times some fraction equals 11. Once more: 4 times some fraction equals 11. Write the fraction. (Tap your foot 3 times.)
- Problem F: 8 times some fraction equals 16. Once more: 8 times some fraction equals 16. Write the fraction. (Tap your foot 3 times.)
- Problem G: 12 times some fraction equals 15. Once more: 12 times some fraction equals 15. Write the fraction. (Tap your foot 3 times.)

- Problem H: 29 times some fraction equals 30. Once more: 29 times some fraction equals 30. Write the fraction. (Tap your foot 3 times.)
c. Check your work. I'll say each problem. You tell me the answer.
- Problem A: 18 times some fraction equals 7. What's the fraction? (Signal.) *7-eighteenths.*
- Problem B: 10 times some fraction equals 1. What's the fraction? (Signal.) *1-tenth.*
- Problem C: 6 times some fraction equals 40. What's the fraction? (Signal.) *40-sixths.*
- Problem D: 19 times some fraction equals 62. What's the fraction? (Signal.) *62-nineteenths.*
- Problem E: 4 times some fraction equals 11. What's the fraction? (Signal.) *11-fourths.*
- Problem F: 8 times some fraction equals 16. What's the fraction? (Signal.) *16-eighths.*
- Problem G: 12 times some fraction equals 15. What's the fraction? (Signal.) *15-twelfths.*
- Problem H: 29 times some fraction equals 30. What's the fraction? (Signal.) *30-twenty-ninths.*
- Raise your hand if you wrote all the correct fractions.

EXERCISE 3 DECIMALS
Place Value

a. You're going to write decimal numbers on your lined paper. How many places do you show after the decimal point for hundredths? (Signal.) *Two.*
- How many places do you show after the decimal point for tenths? (Signal.) *One.*
- How many for thousandths? (Signal.) *Three.*
b. Write part OO on your lined paper.
c. Value A: 7 and 8-tenths. Write it. √
- Value B: 3 and 2-thousandths. Write it. √
- Value C: 8 and 5-hundredths. √
- Value D: 22-hundredths. √
- Value E: 12 and 3-tenths. √
- Value F: 19-thousandths. √
d. (Write on the board:)

a.	7.8
b.	3.002
c.	8.05
d.	.22
e.	12.3
f.	.019

- Here's what you should have for each value. Raise your hand if you got all the items correct.
e. Your turn: Read each value. Value A. (Signal.) *7 and 8-tenths.*
- Value B. (Signal.) *3 and 2-thousandths.*

- Value C. (Signal.) *8 and 5-hundredths.*
- Value D. (Signal.) *22-hundredths.*
- Value E. (Signal.) *12 and 3-tenths.*
- Value F. (Signal.) *19-thousandths.*
- (Repeat step e until firm.)

EXERCISE 4 PRIME FACTORS

a. Find part 2.
- I'll read what it says. Follow along: You're going to work with prime numbers. Here's a rule about prime numbers: If you divide a **prime** number by anything other than 1 or the number itself, you won't get a whole-number answer.
- **5** is a prime number because it can't be evenly divided by anything but 1 or 5.
- **7** is a prime number because it can't be evenly divided by anything but 1 or 7.
- **16** is not a prime number because it can be divided by something other than 1 and 16. It can be divided by 2, or by 4, or by 8.
- You can see a list of some prime numbers: 2, 3, 5, 7, 11 and so forth.
- Prime numbers that are multiplied together are prime factors.
- You can show numbers that are not primes as prime factors.
- 45 is not a prime number. You can show it as prime factors.
- You can start with any two factors of 45: 45 equals 9 times 5. 9 is not a prime. So you rewrite 9 as prime factors: 3 times 3.
- You can see the prime factors of 45: 3 times 3 times 5.

b. Sample item A: 8 is not a prime number. You'll write it as prime factors. You could start with the factors 2 times 4.
   4 is not a prime number. So you rewrite 4 as prime factors. What factors are those? (Signal.) *2 and 2.*
- Write an equation to show the prime factors for 8. Remember, factors must equal 8 and every factor must be a prime number. Raise your hand when you're finished.
   (Observe students and give feedback.)
- (Write on the board:)

$$8 = 2 \times 2 \times 2$$

- Here's what you should have. 8 equals 2 times 2 times 2.
- Raise your hand if you got it right.

c. Sample item B: 48. That's 8 times 6. But neither 8 nor 6 are prime numbers. So you write the prime factors of 8 and the prime factors of 6. Remember the times signs.
- Write the prime factors of 48. Raise your hand when you're finished.
   (Observe students and give feedback.)
- (Write on the board:)

$$48 = 2 \times 2 \times 2 \times 2 \times 3$$

- Here's what you should have. 48 equals 2 times 2 times 2 times 2 times 3.
- You can have the factors in any order, but you need four **2s** and a **3.**
- Raise your hand if you got it right.

d. Find part 3.
- Some of the equations show prime factors that equal the number. But not all equations are right. Some of them show factors that are not primes. And some of them have factors that are wrong.

e. Problem A: 12 equals 2 times 2 times 3. Are all those factors primes? (Signal.) *Yes.*
- Does 2 times 2 times 3 equal 12? (Signal.) *Yes.*
- So Problem A is correct.

f. Figure out if B is correct. If it isn't, show the prime factors that equal 50. Raise your hand when you're finished.
   (Observe students and give feedback.)
- (Write on the board:)

b. $50 = 2 \times 5 \times 5$

- Here's what you should have.
- Item B showed 50 equal to 10 times 5. Does 50 equal 10 times 5? (Signal.) *Yes.*
- Are 10 and 5 prime numbers? (Signal.) *No.*
- You should have 50 equals 2 times 5 times 5.

g. Your turn: Figure out if the rest of the items are correct. If they aren't, write the correct equation. Raise your hand when you're finished.
   (Observe students and give feedback.)

h. (Write on the board:)

c. $40 = 2 \times 2 \times 2 \times 5$
e. $70 = 2 \times 5 \times 7$

- Here are the correct equations for items C and E. The factors for item C did not equal 40. The factors for item E were not prime.
- Raise your hand if you got it right.

EXERCISE 5   PROBLEM SOLVING
Fraction Comparison

a. Find part 4.
- These are sentences that compare two things by giving a fraction.
- The fraction tells whether the first thing named in the sentence is larger or smaller than the second thing named.
- The rules are in the box: If the fraction is **more than 1,** the first thing named is **larger.** If the fraction is **less than 1,** the first thing named is **smaller.**

b. Look at the first sample sentence: Jan is 5/8 as tall as Fran. Jan is named first.
- The fraction is **less than 1,** so **Jan is less** than Fran. Jan is shorter.
- Sample sentence 2: Jan is 8/7 the height of Ted. Who is named first? (Signal.) *Jan.*
- The fraction is **more than 1,** so **Jan** is taller than Ted.
- Sample sentence 3: Pile A is 7/5 the weight of pile B.
- Pile A is named first. The fraction is **more than 1.** So **pile A** is more than pile B.

c. Item A: Hilda's weight is 9/8 of Edna's weight.
- Is the fraction more than 1 or less than 1? (Signal.) *More than 1.*
- Which person weighs more? (Signal.) *Hilda.*
- Write **Hilda** for item A. √

d. Item B: Pile C is 4/5 the weight of pile M.
- Is the fraction more than 1 or less than 1? (Signal.) *Less than 1.*
- That tells whether pile C is bigger or smaller. Write the letter of the pile that is **bigger.** Raise your hand when you're finished.
- (Write on the board:)

> **b.   M**

- Here's what you should have. The fraction is less than 1. So pile C is less and pile M is more. You should have written M.

e. Item C: Pile J is 7/3 the weight of pile B.
- Is the fraction more than 1 or less than 1? (Signal.) *More than 1.*
- That tells you which pile is bigger. Write the letter of the bigger pile.
- (Write on the board:)

> **c.   J**

- Here's what you should have. J is bigger.

f. Your turn: Write the name of the larger thing named in the rest of the items. Raise your hand when you're finished.
(Observe students and give feedback.)

g. (Write on the board:)

> **d.   turtle**
> **e.   forest**

- Here's what you should have for items D and E.

EXERCISE 6   NUMBER FAMILIES
Improper Fractions

a. Find part 5.
- All of these number families have a fraction and the number 1.
- For some families, the number 1 is the big number. For others, it's a small number.
- You'll write the fraction that equals 1 with the correct denominator. Then you'll write the missing fraction.

b. Make a family with the three fractions for item A. Raise your hand when you're finished.
(Observe students and give feedback.)
- (Write on the board:)

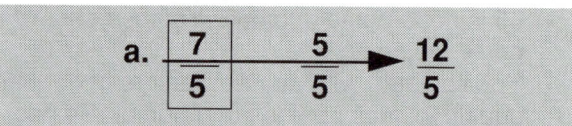

- Here's what you should have.

c. Your turn: Make families for the rest of the items in part 5. Raise your hand when you're finished. (Observe students and give feedback.)

d. (Write on the board:)

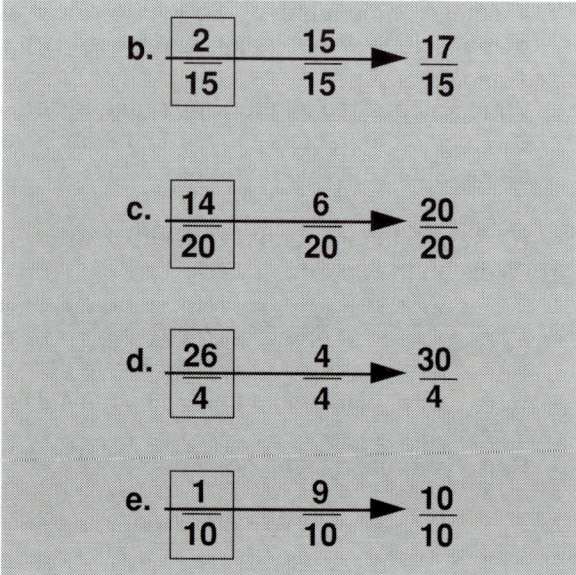

- Here's what you should have.
- e. Raise your hand if you got everything right.

EXERCISE 7 INDEPENDENT WORK

- (In addition to independent work in the textbook, assign **Bridge to Connecting Math Concepts** Independent Worksheet 24 as classwork or homework. Before beginning the next lesson, check the students' independent work.)

Lesson 45

Objectives

- **Write fractions with denominators of 10, 100 or 1000 from decimal numbers.** (Exercise 1)

- Use a fraction number family to work a problem that tells about numbers and asks about a fraction. (Exercise 2)

- Write the prime factors for a number that is not prime. (Exercise 3)

- Identify the larger entity named in sentences of the form: Hilda's weight is 9/8 of Edna's weight. (Exercise 4)

- Work a word problem by constructing a number-family table with column and row headings. (Exercise 5)

- Write a missing middle factor as a fraction. (Exercise 6)

EXERCISE 1 DECIMALS
As Fractions

a. Open your textbook to lesson 45 and find part 1.
- I'll read what it says. Follow along: You can write fractions for decimal numbers. You just read the decimal number and write the fraction for what you read.
- Here's 35-hundredths. You write the fraction: 35-hundredths.
- Here's 2-thousandths. You write the fraction: 2-thousandths.
- Here's 2-tenths. You write the fraction: 2-tenths.
- Remember, when you read the decimal value, you say the fraction you'll write.
b. Find part 2.
- The first column in the table is for decimal values. The second column is for fractions.
c. Item A. Everybody, read the decimal value. (Signal.) *5-tenths.*
- Copy the table headings and write both values for row A. Write the fraction 5/10. √
- (Write on the board:)

	Decimal	Fraction
a.	.5	$\dfrac{5}{10}$

- Here's what you should have.

d. Item B. Read the decimal number. (Signal.)
8-tenths.
• Write the decimal number and the fraction 8/10.
Then complete the rest of the table. Raise your
hand when you're finished.
(Observe students and give feedback.)
e. (Write to show:)

	Decimal	Fraction
a.	.5	$\frac{5}{10}$
b.	.8	$\frac{8}{10}$
c.	.08	$\frac{8}{100}$
d.	.003	$\frac{3}{1000}$
e.	.03	$\frac{3}{100}$

• Here's what you should have for each item.
f. Raise your hand if you got all the items correct.

EXERCISE 2 PROBLEM SOLVING
Fraction Number Families

a. Find part 3.
• These are problems that don't have pictures.
• For each problem, you'll make the fraction
number family and box the fraction that answers
the question.
b. Work problem A. Raise your hand when you're
finished. (Observe students and give feedback.)
• (Write on the board:)

	no	
onions	**onions**	**pizza**
a. $\frac{7}{9}$	$\boxed{\frac{2}{9}}$ ⟶	$\frac{9}{9}$

• Here's what you should have. 2/9 of the pizza
did not have onions.
c. Your turn: Work the rest of the problems in part
3. Raise your hand when you're finished.
(Observe students and give feedback.)

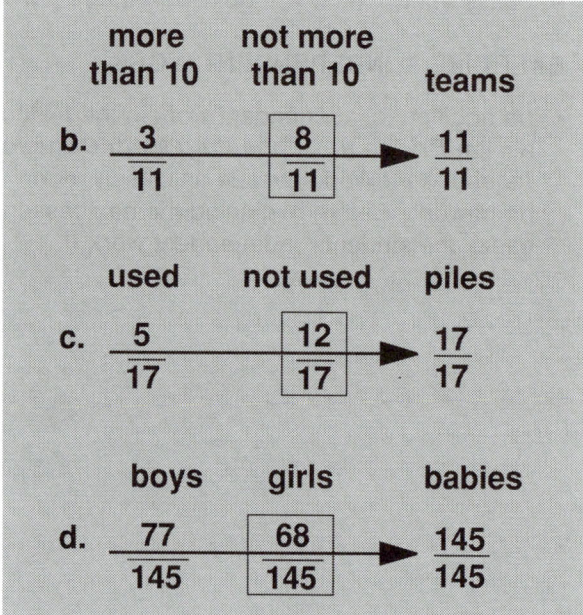

• Here's what you should have for each problem.
e. Raise your hand if you got everything right.

EXERCISE 3 PRIME FACTORS

a. Find part 4.
• These items show factors that are not primes.
You'll write equations that show only prime
factors. Remember, the factors must be primes
and the factors together must equal the first
number in the equation.
b. Write the prime factors for item A. Raise your
hand when you're finished.
• (Write on the board:)

> a.   56 = 2 x 2 x 2 x 7

• You rewrote 8 as 2 times 2 times 2.
c. Work the rest of the items in part 4. Raise your
hand when you're finished.
(Observe students and give feedback.)
d. (Write on the board:)

> b.   30 = 2 x 3 x 5
> c.   28 = 2 x 2 x 7
> d.   54 = 2 x 3 x 3 x 3

• Here's what you should have for B, C and D.
You can have the factors in any order.
e. Raise your hand if you got everything right.

EXERCISE 4 PROBLEM SOLVING
Fraction Comparison

a. Find part 5.
• Each sentence compares two things.
• If the fraction is **more than 1,** the first thing named in the sentence is **greater** than the second thing named in the sentence. If the fraction is **less than 1,** the first thing named is **less** than the other thing.
b. Item A: The cost of the couch was **3/7** as much as the cost of the bed.
• Is the fraction more than 1 or less than 1? (Signal.) *Less than 1.*
• Write the name that is more. √
• Everybody, which costs more, the couch or the bed? (Signal.) *The bed.*
c. Your turn: Write the names that are more for the rest of the items in part 5. Raise your hand when you're finished.
   (Observe students and give feedback.)
d. (Write on the board:)

> **a. bed**
> **b. dog**
> **c. Fran**
> **d. Jim**
> **e. Jay**

• Here's what you should have for each item.
e. Raise your hand if you got everything right.

EXERCISE 5 PROBLEM SOLVING
Number-Family Tables

a. Find part 6.
b. You'll make the table, put in the numbers the problem gives, figure out the missing numbers and write answers to the questions. When you write answers to the questions, make sure you write a number and a unit name. Raise your hand when you've completed the table and answered the questions.

Key:

	green	red	total
experienced	48	51	99
inexperienced	31	70	101
workers	79	121	200

Questions
a. the red crew
b. 79 workers
c. 99 workers
d. 200 workers
e. 31 workers

c. Find part J on page 179. That shows the table you should have. The black numbers are the ones the problem gives. You can see the answers to the questions. Check over your work.
d. Raise your hand if you got everything right.

EXERCISE 6 MULTIPLICATION
Missing Factor

a. Write part O on your paper.
• I'm going to say multiplication problems. For each problem, you'll write the answer. That's the missing fraction.
b. Problem A: 8 times some fraction equals 19. Once more: 8 times some fraction equals 19. Write the fraction. (Tap your foot 3 times.)
• Problem B: 45 times some fraction equals 7. Once more: 45 times some fraction equals 7. Write the fraction. (Tap your foot 3 times.)
• Problem C: 16 times some fraction equals 17. Once more: 16 times some fraction equals 17. Write the fraction. (Tap your foot 3 times.)
• Problem D: 12 times some fraction equals 90. Once more: 12 times some fraction equals 90. Write the fraction. (Tap your foot 3 times.)
• Problem E: 4 times some fraction equals 3. Once more: 4 times some fraction equals 3. Write the fraction. (Tap your foot 3 times.)
• Problem F: 75 times some fraction equals 1. Once more: 75 times some fraction equals 1. Write the fraction. (Tap your foot 3 times.)
• Problem G: 2 times some fraction equals 8. Once more: 2 times some fraction equals 8. Write the fraction. (Tap your foot 3 times.)
• Problem H: 33 times some fraction equals 2. Once more: 33 times some fraction equals 2. Write the fraction. (Tap your foot 3 times.)
c. Check your work. I'll say each problem. You tell me the fraction.
• Problem A: 8 times some fraction equals 19. What's the fraction? (Signal.) *19-eighths.*
• Problem B: 45 times some fraction equals 7. What's the fraction? (Signal.) *7-forty-fifths.*
• Problem C: 16 times some fraction equals 17. What's the fraction? (Signal.) *17-sixteenths.*
• Problem D: 12 times some fraction equals 90. What's the fraction? (Signal.) *90-twelfths.*
• Problem E: 4 times some fraction equals 3. What's the fraction? (Signal.) *3-fourths.*
• Problem F: 75 times some fraction equals 1. What's the fraction? (Signal.) *1-seventy-fifth.*
• Problem G: 2 times some fraction equals 8. What's the fraction? (Signal.) *8-halves.*
• Problem H: 33 times some fraction equals 2. What's the fraction? (Signal.) *2-thirty-thirds.*

EXERCISE 7 INDEPENDENT WORK

• (In addition to independent work in the textbook, assign **Bridge to Connecting Math Concepts** *Independent Worksheet* 25 as classwork or homework. Before beginning the next lesson, check the students' independent work.)

Lesson 46

Objectives

- **Write the X and Y values for points shown on the coordinate system.** (Exercise 1)

- **Write a number family with three names and a fraction for a sentence that compares.** (Exercise 2)

- **Use information shown in a number family table to work items that compare two quantities.** (Exercise 3)
  Note: Items are of the form: There are 14 more boats than trucks. Students refer to the table to find the number of boats, then complete the number family:

  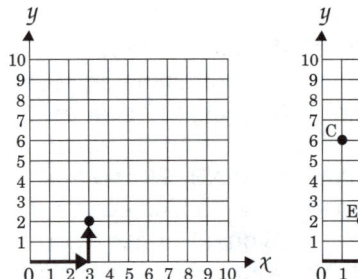

- Write the prime factors for a number that is not prime. (Exercise 4)

- Write a missing middle factor as a fraction. (Exercise 5)

- **Complete a table to show fractions and equivalent decimal values** (Exercise 6)

EXERCISE 1   COORDINATE SYSTEM
X and Y Values

a. Open your textbook to lesson 46 and find part 1.
- (Teacher reference:)

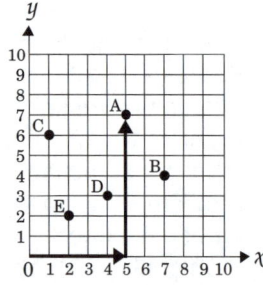

- I'll read what it says. Follow along: This kind of grid is a coordinate system. The coordinate system has an arrow for the X direction and an arrow for the Y direction.
- Point with your finger to show the X direction. (Students point to the right.)
- Point to show the Y direction. (Students point up.)
- You can tell about any point on the coordinate system by telling first about X and then about Y.

- Look at the coordinate system with **blue** arrows. The arrows show how to get to the point: X equals 3, Y equals 2. The point is 3 places to the **right of zero** and 2 places **up.**
- Touch the point. √
- Start at zero. Go 3 places in the X direction and then 2 places up. √
- The coordinate system with **green** arrows shows different points.
- Touch A on that coordinate system. √
- You can see the equations for point A: X equals 5, Y equals 7. That means: To get to point A, you go 5 places **from zero in the X direction,** then 7 places up for Y.

b. (Write on the board:)

> **B** $x =$, $y =$

c. Your turn: Write equations to show the X value and Y value for point B. Remember, write the number of places for X and the number of places for Y. Raise your hand when you've written what X equals and what Y equals. (Observe students and give feedback.)
- (Write to show:)

> **B** $x =$ **7**, $y =$ **4**

- Here's what you should have for point B. X equals 7. Y equals 4.
d. Write the X and Y values for C. Write what X equals and what Y equals. Raise your hand when you're finished. (Observe students and give feedback.)
- (Write on the board:)

> **C** $x =$ **1**, $y =$ **6**

- Here's what you should have. X equals 1. Y equals 6.
e. Write the X and Y values for D. Raise your hand when you're finished. √
- (Write on the board:)

> **D** $x =$ **4**, $y =$ **3**

- Here's what you should have. X equals 4. Y equals 3.
f. Write the X and Y values for E. Raise your hand when you're finished. √
- (Write on the board:)

> **E** $x =$ **2**, $y =$ **2**

- Here's what you should have. X equals 2. Y equals 2.
g. Raise your hand if you got everything right.

EXERCISE 2 PROBLEM SOLVING
Fraction Comparison

a. Find part 2.
- (Write on the board:)

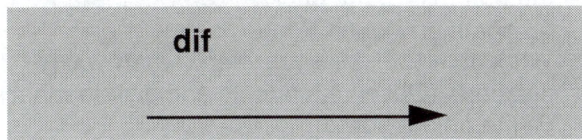

- For each sentence, you're going to make a number family that has three names and a fraction for one of the names.
- The name for the big number will be the name that is more.
- All the sentences compare two things. So the first name in your family is **difference.**
b. We'll do the sample sentence together.
- The weight of the boat was 5/4 the weight of the trailer.
- One of those names is more. Which is more? (Signal.) *The boat.*
- The boat is the big number. What fraction do I write for the boat? (Signal.) *5-fourths.*
- (Write to show:)

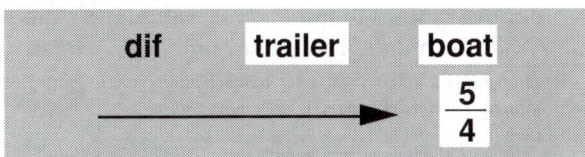

- Here's the family with the names where they belong and the fraction for the boat.
c. Sentence A: Car M went 4/5 the speed of car T.
- Make the number family with three names. Remember, the name for the big number is the name for the car that is **faster.**
- Raise your hand when you have the names and a fraction for car M.
  (Observe students and give feedback.)
- (Write on the board:)

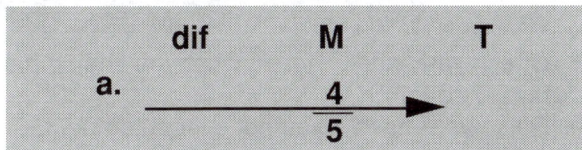

- Here's the family you should have. Raise your hand if you got it right.
d. Sentence B: Car T went 4/3 the speed of car Z.

- Make the family with three names and the fraction for car T. Raise your hand when you're finished. (Observe students and give feedback.)
- (Write on the board:)

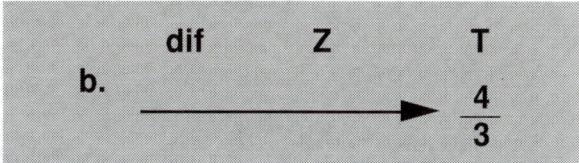

- Here's what you should have. **Car T** is the big number.
e. Your turn: Make families for the rest of the sentences in part 2. Remember, write three names and the fraction for the thing that is named first in the sentence. Raise your hand when you're finished.
  (Observe students and give feedback.)
f. (Write on the board:)

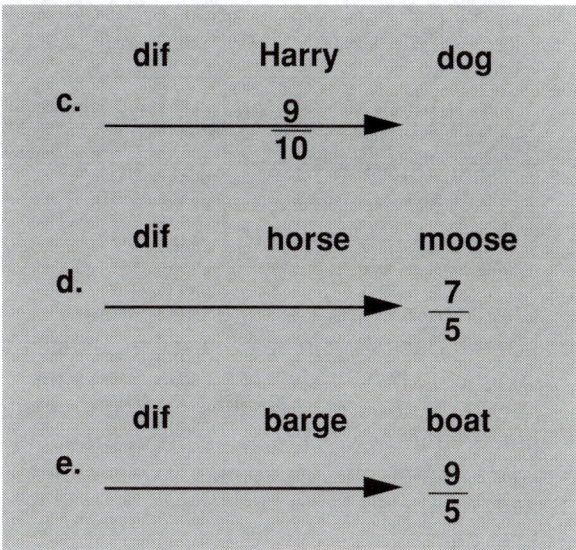

- Here's what you should have for each family.
g. Raise your hand if you got everything right.

EXERCISE 3 PROBLEM SOLVING
Tables with Comparison

a. Find part 3.
- The information for making this table gives you four numbers. Read the facts and make the table that has all the numbers the problem gives. Raise your hand when you've done that much. (Observe students and give feedback.)

b. (Write on the board:)

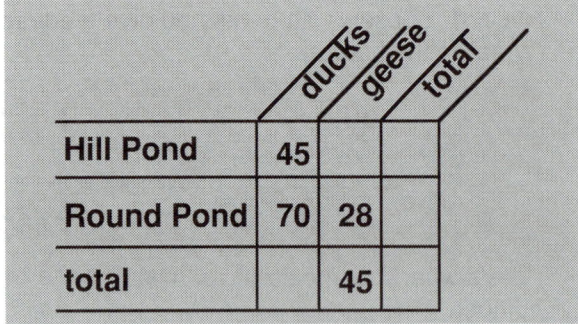

	ducks	geese	total
Hill Pond	45		
Round Pond	70	28	
total		45	

- Here's what you should have so far.
c. Figure out all the missing numbers. Raise your hand when you're finished.
  (Observe students and give feedback.)
d. (Write to show:)

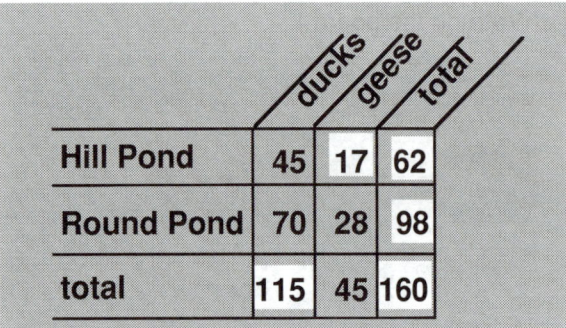

	ducks	geese	total
Hill Pond	45	17	62
Round Pond	70	28	98
total	115	45	160

- Check your work. Here are the numbers.
e. Listen: Each item in part 3 gives information for comparing two things. The table shows the number for **one** of those things. You'll make a number family with the name **difference** and figure out the missing number. These are tricky.
f. Item A: On Hill Pond there were 35 more bugs than geese.
- The table shows the number of geese on Hill Pond. Make the number family for geese, bugs and difference. Write the answer to item A. Raise your hand when you're finished.
  (Observe students and give feedback.)
- (Write on the board:)

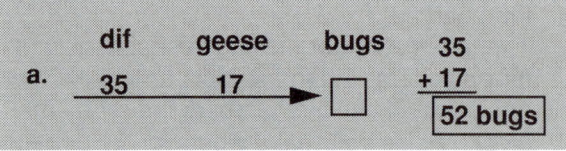

a.
dif   geese   bugs   35
35   17   →   ☐   + 17
                              52 bugs

- Here's what you should have.
- Everybody, how many bugs were on Hill Pond? (Signal.) *52.*
g. Work item B. Raise your hand when you're finished. (Observe students and give feedback.)

- (Write on the board:)

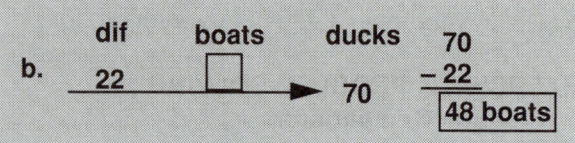

b.
dif   boats   ducks   70
22   ☐   →   70   − 22
                            48 boats

- Here's what you should have. The table gives the number for the ducks—70. So, there were 48 boats on Round Pond.
h. Work item C. Raise your hand when you're finished. (Observe students and give feedback.)
- (Write on the board:)

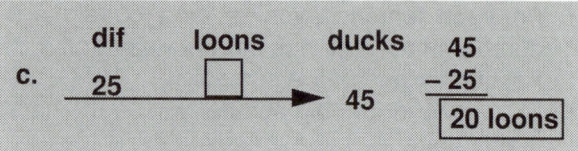

c.
dif   loons   ducks   45
25   ☐   →   45   − 25
                            20 loons

- Here's what you should have. There were 20 loons on Hill Pond.
i. Raise your hand if you got everything right.

EXERCISE 4 PRIME FACTORS

a. Find part 4.
- These items show factors that are not all primes. You'll write equations that show only prime factors. Remember, the factors must be primes and the factors together must equal the number.
b. Write the prime factors for number A. Raise your hand when you're finished.
- (Write on the board:)

> a. $18 = 2 \times 3 \times 3$

- You rewrote 6 as 2 times 3.
c. Work the rest of the items in part 4. Raise your hand when you're finished.
  (Observe students and give feedback.)
d. (Write on the board:)

> b. $24 = 2 \times 2 \times 2 \times 3$
> c. $63 = 3 \times 3 \times 7$
> d. $16 = 2 \times 2 \times 2 \times 2$

- Check your work. Here's what you should have for B through D.
e. Raise your hand if you got everything right.

EXERCISE 5 MULTIPLICATION
Missing Factor

a. I'm going to say multiplication problems. You'll write the missing value as a fraction. Write part O on your paper.

b. Problem A: 4 times some fraction equals 3. Once more: 4 times some fraction equals 3. Write the fraction. (Tap your foot 3 times.)
• Problem B: 8 times some fraction equals 40. Once more: 8 times some fraction equals 40. Write the fraction. (Tap your foot 3 times.)
• Problem C: 4 times some fraction equals 19. Once more: 4 times some fraction equals 19. Write the fraction. (Tap your foot 3 times.)
• Problem D: 51 times some fraction equals 8. Once more: 51 times some fraction equals 8. Write the fraction. (Tap your foot 3 times.)
• Problem E: 13 times some fraction equals 14. Once more: 13 times some fraction equals 14. Write the fraction. (Tap your foot 3 times.)
• Problem F: 17 times some fraction equals 1. Once more: 17 times some fraction equals 1. Write the fraction. (Tap your foot 3 times.)
• Problem G: 3 times some fraction equals 1. Once more: 3 times some fraction equals 1. Write the fraction. (Tap your foot 3 times.)
• Problem H: 2 times some fraction equals 10. Once more: 2 times some fraction equals 10. Write the fraction. (Tap your foot 3 times.)

c. Check your work. I'll say each problem. You tell me the fraction.
• Problem A: 4 times some fraction equals 3. What's the answer? (Signal.) *3-fourths.*
• Problem B: 8 times some fraction equals 40. What's the answer? (Signal.) *40-eighths.*
• Problem C: 4 times some fraction equals 19. What's the answer? (Signal.) *19-fourths.*
• Problem D: 51 times some fraction equals 8. What's the answer? (Signal.) *8-fifty-firsts.*
• Problem E: 13 times some fraction equals 14. What's the answer? (Signal.) *14-thirteenths.*
• Problem F: 17 times some fraction equals 1. What's the answer? (Signal.) *1-seventeenth.*
• Problem G: 3 times some fraction equals 1. What's the answer? (Signal.) *1-third.*
• Problem H: 2 times some fraction equals 10. What's the answer? (Signal.) *10-halves.*

EXERCISE 6 FRACTIONS AND DECIMALS
Conversion

a. Find part 5.
• The first column of the table is for decimal values. The second column is for fractions.

b. Row A. A decimal number is shown. What number? (Signal.) *73-hundredths.*
• Write the decimal value and the fraction for 73-hundredths. √
• (Write on the board:)

	Decimal	Fraction
a.	.73	$\frac{73}{100}$

• Here's what you should have.
c. Row B. A fraction is shown. What fraction? (Signal.) *19-thousandths.*
• Write the decimal number for 19-thousandths. Then complete the rest of the rows in part 5. Raise your hand when you're finished. (Observe students and give feedback.)
d. (Write to show:)

	Decimal	Fraction
a.	.73	$\frac{73}{100}$
b.	.019	$\frac{19}{1000}$
c.	.7	$\frac{7}{10}$
d.	.25	$\frac{25}{100}$
e.	.04	$\frac{4}{100}$
f.	.305	$\frac{305}{1000}$

• Here's what you should have for each item.
e. Raise your hand if you got all the items correct.

EXERCISE 7 INDEPENDENT WORK

• (In addition to independent work in the textbook, assign **Bridge to Connecting Math Concepts** *Independent Worksheet* 26 as classwork or homework. Before beginning the next lesson, check the students' independent work.)

Lesson 47

EXERCISE 1 FRACTIONS
Simplification

a. Open your textbook to lesson 47 and find part 1.
- (Teacher reference:)

$$\frac{28}{21} = \frac{2 \times 2 \times 7}{3 \times 7} = \frac{4}{3}$$

- I'll read what it says. Follow along: A fraction is **simplified** if it has the smallest numerator and the smallest denominator that is possible.
- Here are the steps: You start with a fraction that is not simplified.
- You show the **prime factors** for the numerator and the denominator.
- You cross out any **fractions that equal 1.**
- You can see the fraction that is not simplified. It's 28 over 21.
- Below, you can see the prime factors for the numerator and denominator.
- Below, the fraction that equals 1 is crossed out. 7/7 equals 1.
- Then you multiply the values that are **not crossed out.** The next equation shows the multiplication. On top, you have 2 times 2. That's 4. On the bottom, you have 3. So 28 over 21 equals 4/3.

- Remember, cross out the fractions that equal 1. Then multiply.
b. Find part 2.
- All these problems show a fraction that is not simplified and the prime factors for the numerator and denominator. You're going to cross out any fraction that equals 1 and then write the fraction you end up with.
c. Problem A. You're starting with 15/40. That fraction is not simplified. On top are the prime factors for 15—3 times 5. On the bottom are prime factors for 40—2 times 2 times 2 times 5. You're going to copy the problem. Cross out any fraction that equals 1. Then multiply the values that are not crossed out. Write what you end up with in the numerator and the denominator. Raise your hand when you're finished. (Observe students and give feedback.)
- (Write on the board:)

> **a.** $\dfrac{15}{40} = \dfrac{3 \times 5}{2 \times 2 \times 2 \times 5} = \dfrac{3}{8}$

- Here's what you should have. You crossed out 5/5 because it equals 1. On top, you have 3. On the bottom, you have 8. You ended up with 3/8. That's the **simplified fraction that equals 15/40.**
- Below, write the equation: 15/40 equals 3/8. Raise your hand when you're finished. √
- (Write to show:)

> **a.** $\dfrac{15}{40} = \dfrac{3 \times 5}{2 \times 2 \times 2 \times 5} = \dfrac{3}{8}$
>
> $\dfrac{15}{40} = \dfrac{3}{8}$

- Here's what you should have.
d. Problem B. You're starting with a fraction that is not simplified. What fraction is that? (Signal.) *10-twelfths.*
- You know it's not simplified because you can cross out a fraction that equals 1. Do it and write the simplified fraction that equals 10/12. Raise your hand when you've copied item B and written the simplified fraction.
  (Observe students and give feedback.)
- (Write on the board:)

> **b.** $\dfrac{10}{12} = \dfrac{2 \times 5}{2 \times 2 \times 3} = \dfrac{5}{6}$

- Here's what you should have. You crossed out 2/2. The simplified fraction that equals 10/12 is 5/6.
- Below, write the equation that shows the two equivalent fractions. Raise your hand when you're finished. √
- (Write to show:)

b. $\dfrac{10}{12} = \dfrac{\cancel{2} \times 5}{\cancel{2} \times 2 \times 3} = \dfrac{5}{6}$

$\dfrac{10}{12} = \dfrac{5}{6}$

- Here's what you should have. 10/12 equals 5/6.
e. Problem C. You're starting with a fraction that is not simplified. What fraction? (Signal.) *18-twelfths.*
- You know that 18/12 is not simplified because you can cross out a fraction that equals 1. In fact, you can cross out **two fractions that equal 1.** Do it. Then write what 18/12 equals. Below, write the equation that shows the two equivalent fractions. Raise your hand when you're finished. (Observe students and give feedback.)
- (Write on the board:)

c. $\dfrac{18}{12} = \dfrac{\cancel{2} \times 3 \times \cancel{3}}{\cancel{2} \times 2 \times \cancel{3}} = \dfrac{3}{2}$

$\dfrac{18}{12} = \dfrac{3}{2}$

- Here's what you should have. You crossed out 2/2 and 3/3. The simplified fraction that equals 18/12 is 3/2. Below, you should have the simple equation: 18/12 equals 3/2.
f. Raise your hand if you got everything right.

EXERCISE 2 PROBLEM SOLVING
Fraction Comparison

a. Find part 3.
- These are sentences that compare two things.
- You'll make a number family with three names. The name for the first small number is **difference.** You'll show the name that is **less** as the other small number and the name that is **more** as the big number.
b. Sentence A: The distance to the mountain was 5/8 the distance to the park.
- Make the number family with the thing that is farther away as the big number. Remember to show the fraction for mountain. Raise your hand when you're finished.

(Observe students and give feedback.)

- (Write on the board:)

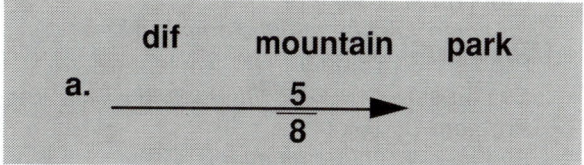

- Here's what you should have for sentence A.
c. Your turn: Make the family for sentence B. Show three names and one fraction. Raise your hand when you're finished.

(Observe students and give feedback.)
- (Write on the board:)

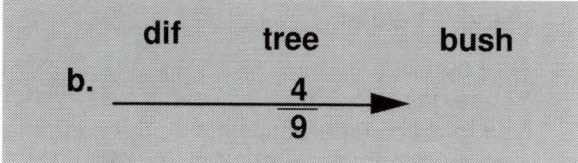

- Here's what you should have.
d. Your turn: Make families for the rest of the items in part 3. Raise your hand when you're finished. (Observe students and give feedback.)
e. (Write on the board:)

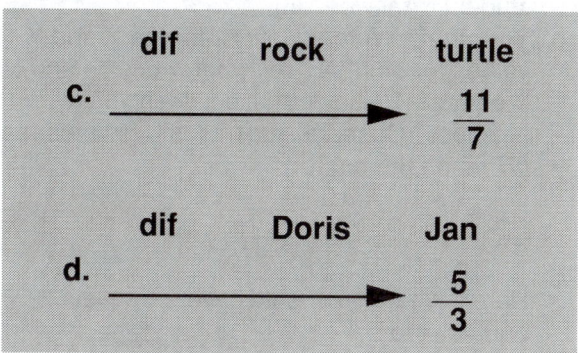

- Here's what you should have for items C and D.

EXERCISE 3 MULTIPLICATION
Missing Factor

a. Find part 4.
b. Write the complete equation for each item. Show the missing value as a fraction. Raise your hand when you're finished.

(Observe students and give feedback.)

Key:

a. $25 \left(\dfrac{17}{25} \right) = 17$ d. $2 \left(\dfrac{10}{2} \right) = 10$

b. $28 \left(\dfrac{13}{28} \right) = 13$ e. $5 \left(\dfrac{10}{5} \right) = 10$

c. $30 \left(\dfrac{42}{30} \right) = 42$

c. Check your work. Read each equation.
- Problem A. (Signal.)
  25 times 17-twenty-fifths equals 17.
- Problem B. (Signal.)
  28 times 13-twenty-eighths equals 13.
- Problem C. (Signal.)
  30 times 42-thirtieths equals 42.
- Problem D. (Signal.)
  2 times 10-halves equals 10.
- Problem E. (Signal.)
  5 times 10-fifths equals 10.
d. Raise your hand if you got everything right.

EXERCISE 4 COORDINATE SYSTEM
X and Y Values

a. Find part 5.
- You're going to write the X and Y values that tell how to reach different points on the coordinate system. Remember, the X axis goes to the right. The Y axis goes up. When you write the directions for reaching a letter, you start at zero. You write what X equals first, then what Y equals. One point has a Y value of zero. What's the letter for that point? (Call on a student. Answer: *A.*)
b. Your turn: Write equations for the X and Y values for point A. Tell what X equals and what Y equals. Raise your hand when you're finished. (Observe students and give feedback.)
- (Write on the board:)

> **A** $x = 2,$ $y = 0$

- Here's what you should have. X equals 2, Y equals zero.
c. Your turn: Write the X and Y values for the rest of the points. Raise your hand when you're finished. (Observe students and give feedback.)
d. (Write on the board:)

> **B** $x = 9,$ $y = 6$
> **C** $x = 4,$ $y = 11$
> **D** $x = 7,$ $y = 3$
> **E** $x = 3,$ $y = 9$

- Here's what you should have for points B through E.
e. Raise your hand if you got everything right.

EXERCISE 5 PROBLEM SOLVING
Tables with Comparison

a. Find part 6.
- The information for making this table tells you about four of the numbers. Read the facts and make the table with the numbers the problem gives. Raise your hand when you've done that much. (Observe students and give feedback.)
- (Write on the board:)

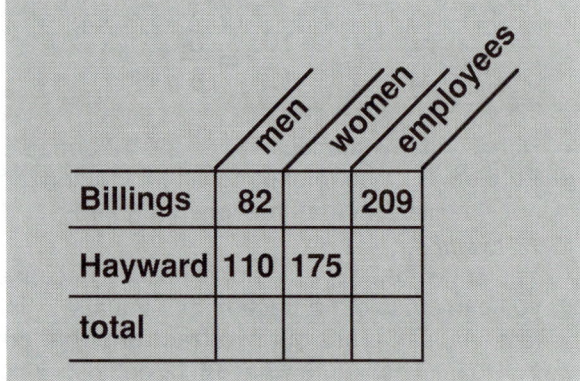

	men	women	employees
Billings	82		209
Hayward	110	175	
total			

- Here's what you should have so far.
b. Figure out all the missing numbers. Raise your hand when you're finished.
  (Observe students and give feedback.)
- (Write to show:)

	men	women	employees
Billings	82	127	209
Hayward	110	175	285
total	192	302	494

- Check your work. Here are the numbers.
c. Now work the items. These are tricky. For each item, you'll have to use **one** of the numbers in the table.
d. Item A: There are 65 fewer desks than total employees in the Billings Building.
- The table shows the number of employees in the Billings Building. Make the number family for desks and employees. Write the answer to item A. Raise your hand when you're finished. (Observe students and give feedback.)

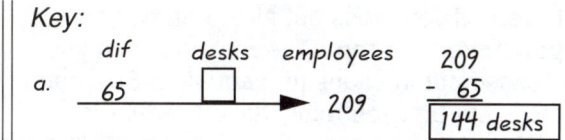

Key:

a.
```
     dif    desks   employees      209
     65    [  ] ──────► 209       - 65
                                  ┌────────┐
                                  │144 desks│
                                  └────────┘
```

- Everybody, how many desks are in the Billings Building? (Signal.) *144.*
- e. Your turn: Work the rest of the items in part 6. Raise your hand when you're finished. (Observe students and give feedback.)

Key:

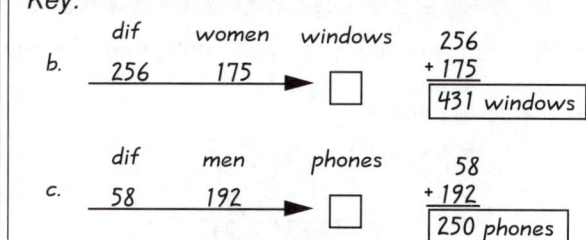

b.
```
     dif    women   windows       256
     256    175  ──────► [  ]     + 175
                                  ┌──────────┐
                                  │431 windows│
                                  └──────────┘
```

c.
```
     dif    men    phones         58
     58     192  ──────► [  ]     + 192
                                  ┌──────────┐
                                  │250 phones │
                                  └──────────┘
```

- f. Check your work.
- Item B: How many windows are in Hayward House? (Signal.) *431.*
- Item C: How many phones are in both buildings? (Signal.) *250.*

EXERCISE 6 FRACTIONS
As Decimals

- a. Find part 7.
- These are fractions. For each fraction, you'll write the decimal number. It's easy. The denominator of the fraction tells the number of places after the decimal point. If the denominator of the fraction is 100, how many places do you write after the decimal point? (Signal.) *Two.*
- The numerator of the fraction shows the digits you'll write for the decimal number.
- b. Fraction A: 305/100. What's the denominator? (Signal.) *100.*
- So how many places go after the decimal point? (Signal.) *Two.*
- The digits are 3, 0, 5. Write the digits 3, 0, 5 with a decimal point and two places after the decimal point. Raise your hand when you're finished.
- (Write on the board:)

> **a. 3.05**

- Here's the decimal value that equals 305/100. 3 and 5-hundredths.
- c. Fraction B: 246/10. How many places will be after the decimal point? (Signal.) *One.*

- Write the digits 2, 4, 6 so there's one place after the decimal point. Raise your hand when you're finished. (Observe students and give feedback.)
- (Write to show:)

> **a. 3.05**
> **b. 24.6**

- Here's what you should have: 24 and 6-tenths.
- Remember, the denominator of the fraction tells the number of places after the decimal point. The numerator shows the digits.
- d. Your turn: Write decimal values for the rest of the fractions in part 7. Raise your hand when you're finished. (Observe students and give feedback.)
- (Write to show:)

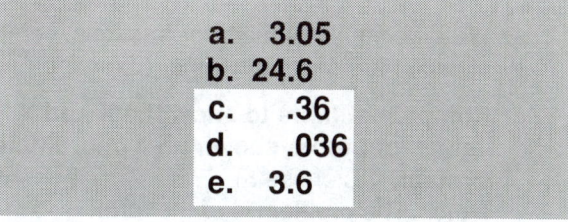

> **a. 3.05**
> **b. 24.6**
> **c.   .36**
> **d.   .036**
> **e. 3.6**

- Check your work. Here's what you should have for fractions C, D and E.
- Raise your hand if you got everything right.

EXERCISE 7 INDEPENDENT WORK

- (In addition to independent work in the textbook, assign **Bridge to Connecting Math Concepts** *Independent Worksheet* 27 as classwork or homework. Before beginning the next lesson, check the students' independent work.)

Lesson 48

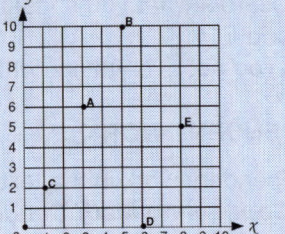
EXERCISE 1 FRACTIONS
Simplification: Numerator of 1

a. (Write on the board:)

$$\frac{6}{12} = \frac{\cancel{2} \times \cancel{3}}{\cancel{2} \times 2 \times \cancel{3}} =$$

- When you simplify fractions, you might cross out all the factors on top. Here's an example.
- You can see the factors 2 and 3 are crossed out on top.

- Listen: If you cross out all the numbers on top, **you have 1 on top.** The reason is that you've crossed out fractions that equal 1. So when you cross out 2/2, you really have 1 over 1 in its place. When you cross out 3/3, you really have 1 over 1 in its place.
- (Write to show:)

$$\frac{6}{12} = \frac{\overset{1}{\cancel{2}} \times \overset{1}{\cancel{3}}}{\cancel{2} \times 2 \times \cancel{3}} =$$

- You can see that when you multiply on top, you have 1 times 1. That's 1.
- (Write to show:)

$$\frac{6}{12} = \frac{\overset{1}{\cancel{2}} \times \overset{1}{\cancel{3}}}{\cancel{2} \times 2 \times \cancel{3}} = \frac{1}{2}$$

- Remember, if you cross out everything on top, you still have 1 on top. You **must write that 1 in the answer.**

b. Open your textbook to lesson 48 and find part 1.
- You're going to simplify these fractions, then write the equation below. For some problems, you'll cross out all the factors on top.
- Remember, you're replacing each factor with an invisible 1. You don't have to show the 1. But you **must have 1 in your answer**.

c. Problem A. You're starting with a fraction that is not simplified. What fraction? (Signal.) *10-thirtieths.*
- Cross out all the fractions that equal 1. Write the simplified fraction. Then write the simple equation below. Raise your hand when you've copied and worked item A.
  (Observe students and give feedback.)
- (Write on the board:)

$$\text{a.} \quad \frac{10}{30} = \frac{\cancel{2} \times \cancel{5}}{\cancel{2} \times 3 \times \cancel{5}} = \frac{1}{3}$$

$$\frac{10}{30} = \frac{1}{3}$$

- Here's what you should have. You crossed out 2/2 and 5/5. In the numerator, you have 1. In the denominator, you have 3. Your equation should be: 10/30 equals 1/3. Raise your hand if you got everything right.

d. Your turn: Copy and work problem B. Raise your hand when you're finished.
  (Observe students and give feedback.)

Key:

b. $\dfrac{4}{8} = \dfrac{\cancel{2} \times \cancel{2}}{\cancel{2} \times \cancel{2} \times 2} = \dfrac{1}{2}$

$\dfrac{4}{8} = \dfrac{1}{2}$

- Check your work.
- You should have crossed out the fractions 2/2 and 2/2. The simplified fraction is 1/2. The equation you wrote below is 4/8 equals 1/2.
- e. Your turn: Copy and work the rest of the problems in part 1. Raise your hand when you're finished.
  (Observe students and give feedback.)
  Key:

c. $\dfrac{12}{18} = \dfrac{\cancel{2} \times 2 \times \cancel{3}}{\cancel{2} \times 3 \times \cancel{3}} = \dfrac{2}{3}$

$\dfrac{12}{8} = \dfrac{2}{3}$

d. $\dfrac{28}{60} = \dfrac{\cancel{2} \times \cancel{2} \times 7}{\cancel{2} \times \cancel{2} \times 3 \times 5} = \dfrac{7}{15}$

$\dfrac{28}{60} = \dfrac{7}{15}$

- f. Check your work.
- Problem C. You should have crossed out 2/2 and 3/3. Everybody, what's the simplified fraction? (Signal.) *2-thirds.*
- The equation you wrote below is: 12/18 equals 2/3.
- Problem D. You should have crossed out 2/2 and 2/2. What's the simplified fraction? (Signal.) *7-fifteenths.*
- The equation you wrote below is: 28/60 equals 7/15.
- g. Raise your hand if you got everything right.

EXERCISE 2 PROBLEM SOLVING
Fraction Comparison

a. Find part 2.
- I'll read what it says. Follow along: You've worked with sentences that use a fraction to compare two things.
- Here's a rule about these comparisons: The thing that something is compared to **equals 1.**
- If something is compared to an elephant, the elephant is 1.
- If something is compared to a book, the book is 1.
- You can see a sentence: Doris weighed 3/4 as much as Edna.

- The sentence compares somebody to Edna. So Edna is 1.
- Next sentence: Doris ran 6/5 the distance that Tony ran.
- The sentence compares Doris to somebody else. Who is Doris compared to? (Signal.) *Tony.*
- So Tony is 1.
- Last sentence: The boat costs 3/5 as much as the car.
- The boat is compared to something. What's that? (Signal.) *The car.*
- The boat is compared to the car. So the car is 1 whole.
b. Find part 3.
- For item A, make a fraction number family with the three names and the fraction the sentence gives. Raise your hand when you've done that much. (Observe students and give feedback.)
- (Write on the board:)

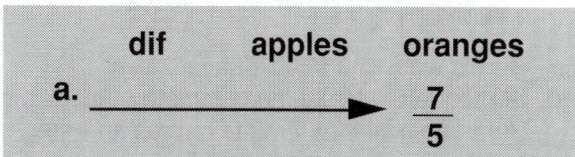

- Here's what you should have. There were 7/5 as many oranges as apples. **Oranges** is the big number. The fraction is 7/5.
- The sentence compares oranges to **apples,** so apples is 1 whole. You write 1 as a fraction. What fraction? (Signal.) *5-fifths.*
- Show the fraction for apples and figure out the fraction for the difference. Raise your hand when you're finished.
  (Observe students and give feedback.)
- (Write to show:)

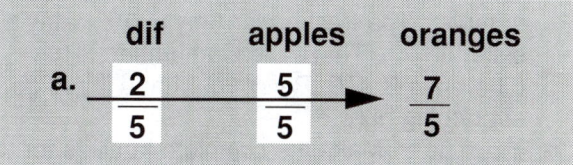

- Here's what you should have. Apples is 5/5. The difference is 2/5.
c. Sentence B: The weight of the apples was 4/3 the weight of the turnips.
- Make the fraction number family with three names and the fraction the sentence gives. Raise your hand when you've done that much. (Observe students and give feedback.)

- (Write on the board:)

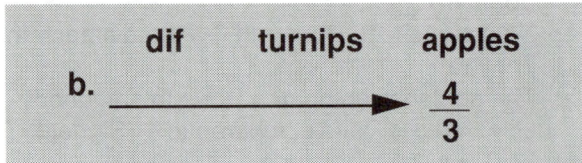

	dif	turnips	apples
b.	→		$\frac{4}{3}$

- Here's what you should have.
- The sentence compares the apples to the turnips, so the turnips equal 1.
- Write the fraction for turnips and the fraction for the difference. Raise your hand when you're finished. (Observe students and give feedback.)
- (Write to show:)

	dif	turnips	apples
b.	$\frac{1}{3}$	$\frac{3}{3}$ →	$\frac{4}{3}$

- Here's what you should have.
d. Sentence C. Make the complete number family. Remember, first put in the names and the fraction the problem gives. Then put in the fractions for the other name and the difference. Raise your hand when you're finished. (Observe students and give feedback.)
- (Write on the board:)

	dif	Amy	sister
c.	$\frac{4}{7}$	$\frac{3}{7}$ →	$\frac{7}{7}$

- Here's what you should have.
- Amy is 3/7 the age of her sister. **Amy** is a small number. The fraction for Amy is 3/7. Amy's sister is 7/7. That's the big number. The difference is 4/7. Raise your hand if you got everything right.
e. Your turn: Make the complete families for the rest of the items in part 3. Raise your hand when you're finished.
(Observe students and give feedback.)
- (Write on the board:)

	dif	F	C
d.	$\frac{1}{4}$	$\frac{4}{4}$ →	$\frac{5}{4}$

	dif	boat	pier
e.	$\frac{5}{9}$	$\frac{4}{9}$ →	$\frac{9}{9}$

- Here's what you should have for items D and E.

EXERCISE 3 MULTIPLICATION
Missing Factor

a. Find part 4.
b. Your turn: Write the complete equation for each item. Remember, write the missing value as a fraction. Raise your hand when you're finished. (Observe students and give feedback.)
Key:

a. $25\left(\frac{8}{25}\right) = 8$ d. $8\left(\frac{72}{8}\right) = 72$

b. $60\left(\frac{13}{60}\right) = 13$ e. $4\left(\frac{20}{4}\right) = 20$

c. $52\left(\frac{45}{52}\right) = 45$

c. Check your work. Read each equation.
- Problem A. (Signal.)
  25 times 8-twenty-fifths equals 8.
- Problem B. (Signal.)
  60 times 13-sixtieths equals 13.
- Problem C. (Signal.)
  52 times 45-fifty-seconds equals 45.
- Problem D. (Signal.)
  8 times 72-eighths equals 72.
- Problem E. (Signal.)
  4 times 20-fourths equals 20.
d. Raise your hand if you got everything right.

EXERCISE 4 FRACTIONS
As Decimals

a. Find part 5.
- This is a table that has fractions. You'll write the decimal value that equals each fraction. Remember, the denominator tells how many places come after the decimal point. The numerator shows the digits.
b. Copy the table. Write the decimal number that equals each fraction. Raise your hand when you're finished.
(Observe students and give feedback.)
c. (Write on the board:)

	Decimal
a.	1.03
b.	.019
c.	.7
d.	42.5
e.	.04
f.	3.05

d. Check your work. Here are the decimal values you should have. Raise your hand if you got everything right.

EXERCISE 5 COORDINATE SYSTEM
X and Y Values

a. Find part 6.
• You're going to complete the table that is shown next to the coordinate system. The first row is completed. The letter A refers to the point labeled A on your coordinate system.
• The table shows the X value for A. The X value is 3. That means you go 3 places from zero along the X axis.
• The Y value for point A is shown. That's the number of places you have to go up to reach point A. Y is 6.
b. The second row is for point B. The table shows the number for X. It does not show the number for Y.
• Touch point B on the coordinate system. √
• Figure out the number of places you go up for Y. Copy the table and complete the second row of the table.
  (Observe students and give feedback.)
• (Write on the board:)

	x	y
A	3	6
B	5	10
C		
D		
E		

• Here's what you should have for point B: X is 5; Y is 10.
c. Your turn. Complete the row for point C. Write the X value and the Y value. Raise your hand when you're finished.
  (Observe students and give feedback.)
• (Write to show:)

	x	y
A	3	6
B	5	10
C	1	2
D		
E		

• Here's what you should have for point C: X is 1; Y is 2.
d. Your turn. Write the X and Y values for the rest of the points shown on the coordinate system. Raise your hand when you're finished.
  (Observe students and give feedback.)
• (Write to show:)

	x	y
A	3	6
B	5	10
C	1	2
D	12	0
E	8	5

• Here's what you should have for points D and E. Raise your hand if you got everything right.

EXERCISE 6 PROBLEM SOLVING
Tables with Comparison

a. Find part 7.
• This is a table problem.
b. Read the facts. Make a table with the four numbers the problem gives. Raise your hand when you've done that much.
  (Observe students and give feedback.)
• (Write on the board:)

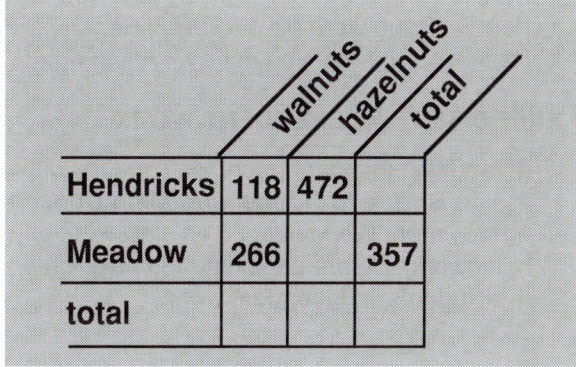

	walnuts	hazelnuts	total
Hendricks	118	472	
Meadow	266		357
total			

• Here's what you should have so far.
c. Figure out the missing numbers. Raise your hand when you're finished.
  (Observe students and give feedback.)

- (Write to show:)

	walnuts	hazelnuts	total
Hendricks	118	472	590
Meadow	266	91	357
total	384	563	947

- Here's what you should have.
d. Now answer the questions. You may have to make a difference number family. Raise your hand when you've answered the questions. (Observe students and give feedback.)

Key:

a.
$$\underset{307}{\text{dif}} \quad \underset{91}{\text{hazelnuts}} \quad \text{honeybees} \longrightarrow \square$$
$$\begin{array}{r} 307 \\ + 91 \\ \hline \boxed{398 \text{ honeybees}} \end{array}$$

b.
$$\underset{170}{\text{dif}} \quad \underset{\square}{\text{flowers}} \quad \text{nuts} \longrightarrow 590$$
$$\begin{array}{r} 590 \\ - 170 \\ \hline \boxed{420 \text{ flowers}} \end{array}$$

e. Find part J on page 191. That shows what you should have for each item.
- Item A: There were 398 honeybees in Meadow Park.
- Item B: There were 420 flowers in Hendricks Park.
f. Raise your hand if you got everything right.

EXERCISE 7 INDEPENDENT WORK

- (In addition to independent work in the textbook, assign **Bridge to Connecting Math Concepts** *Independent Worksheet* 28 as classwork or homework. Before beginning the next lesson, check the students' independent work.)

Lesson 49

Objectives

- **Make a number family that compares two values shown in a table.** (Exercise 1)
- Use prime factors to simplify a fraction. (Exercise 2)
- Write a fraction number family for a sentence that compares two things. (Exercise 3)
- **Write a fraction for a decimal value that is more than 1.** (Exercise 4)
- Complete a table to show the X and Y values of points shown on a coordinate system. (Exercise 5)
- **Solve a problem in which the missing middle factor is expressed as a letter.** (Exercise 6)
  Note: Problems are of the form 3f = 48. Students write: f = 48/3 = 16.

EXERCISE 1 PROBLEM SOLVING
Tables with Comparison

a. Open your textbook to lesson 49 and find part 1.
- Copy the table. √
- Statements A, B and C compare things named in the table. You're going to make the number family that shows each comparison.
b. Statement A: There were 15 more cars than trucks on J Street. What two things are compared in that statement? (Signal.) *Cars and trucks.*
- On which street were there 15 more cars than trucks? (Signal.) *J Street.*
- Find the two cells that the sentence tells about and put a little **X** in each cell. Raise your hand when you've done that much. √
- (Write on the board:)

	cars	trucks	vehicles
D St.	8	49	57
J St.	60 ˣ	45 ˣ	105
total	68	94	162

- Here's what you should have.
- Now make a number family for statement A: There were 15 more cars than trucks on J Street. 15 is the difference number. The other two numbers are shown in the table. Raise your hand when you've made your number family. (Observe students and give feedback.)
- (Write on the board:)

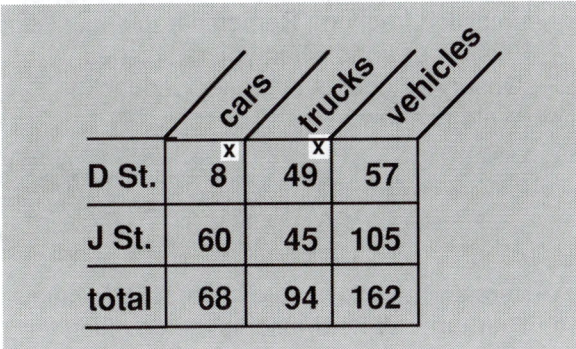

```
        dif     trucks    cars
  a.     15       45
       ──────────────────▶ 60
```

- Here's what you should have.
c. Statement B: There were 41 more trucks than cars on D Street.
- Find the two cells that the sentence tells about and put a little **X** in each cell. √
- (Change to show:)

	cars	trucks	vehicles
D St.	8 ˣ	49 ˣ	57
J St.	60	45	105
total	68	94	162

- Here's what you should have.
- Now make a number family for statement B: There were 41 more trucks than cars on D Street. Raise your hand when you're finished.
- (Write on the board:)

```
        dif      cars    trucks
  b.     41        8
       ──────────────────▶ 49
```

- Here's the family you should have.
d. Statement C: There were 48 fewer vehicles on D Street than on J Street.
- Put an **X** in the two cells that are compared in that statement. Raise your hand when you're finished. √

- (Change to show:)

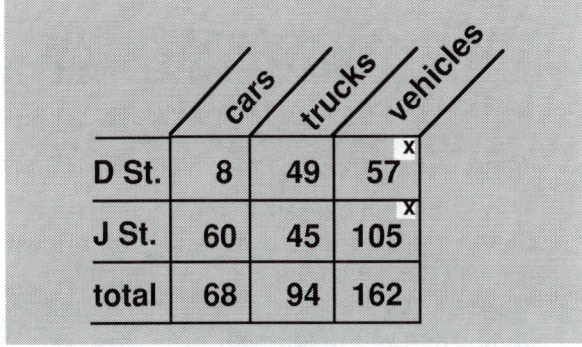

	cars	trucks	vehicles
D St.	8	49	57 ˣ
J St.	60	45	105 ˣ
total	68	94	162

- Here's what you should have.
- Now make a number family for statement C. Raise your hand when you're finished. (Observe students and give feedback.)
- (Write on the board:)

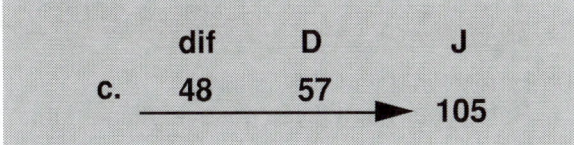

```
        dif       D        J
  c.     48       57
       ──────────────────▶ 105
```

- Here's what you should have.
e. Find part 2.
- For each statement, you'll put an **X** in each of the cells that is compared. Then you'll make a number family with the names for the two things compared in the statement.
f. Statement A: In August, there were 24 fewer red butterflies than butterflies that were not red.
- Copy the table. Make an **X** in the two cells that are being compared. Then make a number family with three names and three numbers. Raise your hand when you're finished. (Observe students and give feedback.)
- (Write on the board:)

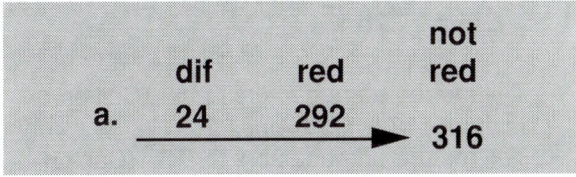

```
                          not
        dif      red      red
  a.     24       292
       ──────────────────▶ 316
```

- Here's the family you should have. The names are **red** and **not red.** The difference is 24.
g. Statement B: In September, there were 12 more red butterflies than butterflies that were not red.
- Make the **X**s. Make the number family. Raise your hand when you're finished. (Observe students and give feedback.)

- (Write on the board:)

	dif	not red	red
b.	12	498	510

(with arrow from 12 → 510)

- Here's the number family you should have.
h. Statement C: The total butterflies for August was 400 less than the total for September.
- Make the **X**s and the number family. Raise your hand when you're finished.
  (Observe students and give feedback.)
- (Write on the board:)

	dif	August	September
c.	400	608	1008

(with arrow from 400 → 1008)

- Here's what you should have.
i. Raise your hand if you got everything right.

EXERCISE 2 FRACTIONS
Simplification

a. Find part 3.
- None of the fractions in part 3 are simplified. You're going to simplify them. Here's how you'll do it. First you'll write an equal sign and show the prime factors for the numerator and the denominator. Then you'll cross out all the fractions that equal 1. Then you'll write the simplified fraction.
b. I'll work the sample fraction.
- (Write on the board:)

$$\frac{8}{18} = \underline{\qquad} = \underline{\quad}$$

- Tell me the prime factors for 8. (Call on a student.) *2 times 2 times 2.*
- Tell me the prime factors for 18. (Call on a student. Idea: *2 times 3 times 3.*)
- (Write to show:)

$$\frac{8}{18} = \frac{2 \times 2 \times 2}{2 \times 3 \times 3} = \underline{\quad}$$

c. Now we cross out any fractions that equal 1. Everyone, what's the fraction that equals 1? (Signal.) *2-halves.*
- (Cross out: **2/2**.)

- We have 4 in the numerator and 9 in the denominator.
- (Write to show:)

$$\frac{8}{18} = \frac{2 \times 2 \times 2}{2 \times 3 \times 3} = \frac{4}{9}$$

- 4/9 is the simplified fraction that equals 8/18.
d. Fraction A. Copy the fraction. Write the equal sign and the prime factors for 15 and 25. Raise your hand when you've done that much.
  (Observe students and give feedback.)
- (Write on the board:)

$$a. \quad \frac{15}{25} = \frac{3 \times 5}{5 \times 5}$$

- Here's what you should have so far. Now cross out any fractions that equal 1 and write the simplified fraction. Remember the equal sign. Raise your hand when you're finished.
  (Observe students and give feedback.)
- (Write to show:)

$$a. \quad \frac{15}{25} = \frac{3 \times \cancel{5}}{5 \times \cancel{5}} = \frac{3}{5}$$

- Here's what you should have. 15/25 equals 3/5.
e. Your turn: Write the complete equation for fraction B. Remember both equal signs. Raise your hand when you're finished.
  (Observe students and give feedback.)
- (Write on the board:)

$$b. \quad \frac{10}{40} = \frac{\cancel{2} \times \cancel{5}}{\cancel{2} \times 2 \times 2 \times \cancel{5}} = \frac{1}{4}$$

- Here's what you should have. The factors for 10 are 2 and 5. The factors for 40 are 2, 2, 2 and 5. You crossed out 2/2 and 5/5. The simplified fraction has 1 in the numerator and 4 in the denominator. 10/40 equals 1/4.
f. Work the rest of the problems in part 3. Show the answers as a fraction, even if it's more than 1. Remember both equal signs. Raise your hand when you're finished.
  (Observe students and give feedback.)

g. (Write on the board:)

$$c. \quad \frac{42}{35} = \frac{2 \times 3 \times \cancel{7}}{5 \times \cancel{7}} = \frac{6}{5}$$

$$d. \quad \frac{8}{16} = \frac{\cancel{2} \times \cancel{2} \times \cancel{2}}{\cancel{2} \times \cancel{2} \times \cancel{2} \times 2} = \frac{1}{2}$$

h. Here's what you should have.
• For problem C, you crossed out the fraction 7/7. The simplified fraction is 6/5. 42/35 equals 6/5.
• For problem D, you crossed out the fractions 2/2, 2/2 and 2/2. The simplified fraction is 1/2. 8/16 equals 1/2.
i. Raise your hand if you got everything right.

EXERCISE 3 PROBLEM SOLVING
Fraction Comparison

a. Find part 4.
• Each sentence compares something to something else.
• Remember, if something is compared to pile B, pile B equals 1. If something is compared to Edna, Edna is 1.
b. Make a complete number family for sentence A. Raise your hand when you're finished. (Observe students and give feedback.)
• (Write on the board:)

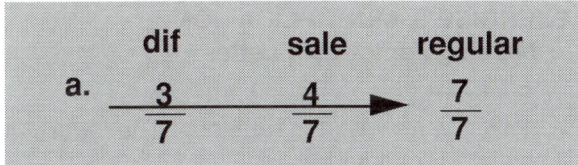

• Here's what you should have.
• The sentence gives the sale price of a couch as 4/7 of the regular price. The regular price of the couch is the big number. It's 1.
c. Make the complete number family for sentence B. Remember, first figure out which is the big number. Raise your hand when you're finished. (Observe students and give feedback.)
• (Write on the board:)

dif	potatoes	flour
b. $\dfrac{5}{3}$	$\dfrac{3}{3}$ ➡	$\dfrac{8}{3}$

• Here's what you should have. Flour is the big number. It's 8/3. Potatoes is 3/3. The difference is 5/3. Raise your hand if you got everything right.

d. Your turn: Make families for the rest of the items in part 4. Remember, first figure out which is the big number. Raise your hand when you're finished. (Observe students and give feedback.)
e. (Write on the board:)

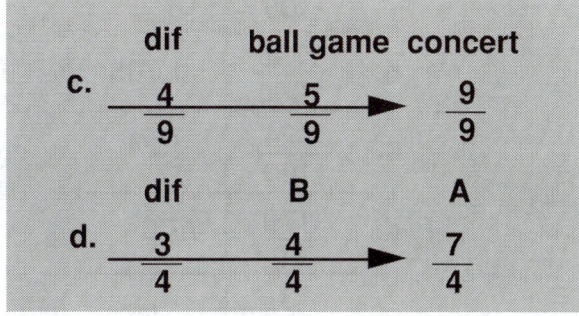

• Check your work. Here's what you should have for each family.
f. Raise your hand if you got everything right.

EXERCISE 4 DECIMALS
As Fractions

a. Find part 5.
• For this table, you'll write fractions for decimal values.
• Listen: If the decimal value is more than 1, write all the digits. Then write the correct denominator. If the decimal value shows tenths, the denominator is tenths.
b. Do row A. Show the fraction and the corresponding decimal value. Raise your hand when you're finished.
• (Write on the board:)

	Fraction	Decimal
a.	$\dfrac{304}{100}$	3.04

• Here's what you should have for row A.
c. Do the rest of the rows. Raise your hand when you're finished. (Observe students and give feedback.)

d. (Write to show:)

	Fraction	Decimal
a.	$\frac{304}{100}$	3.04
b.	$\frac{1705}{1000}$	1.705
c.	$\frac{124}{100}$	1.24
d.	$\frac{176}{10}$	17.6
e.	$\frac{3}{100}$	.03

- Here's what you should have for rows B through E. Raise your hand if you got everything right.

EXERCISE 5 COORDINATE SYSTEM
X and Y Values

a. Find part 6.
- The table shows the X values for some points on the coordinate system. No Y values are shown.
- You're going to complete the table and show the X and Y values for each point.
b. Complete the row for point A. Raise your hand when you're finished.
   (Observe students and give feedback.)
- (Write on the board:)

	x	y
A	6	4

- Here's what you should have: X is 6; Y is 4.
c. Complete the row for point B. Raise your hand when you're finished.
   (Observe students and give feedback.)
- (Write to show:)

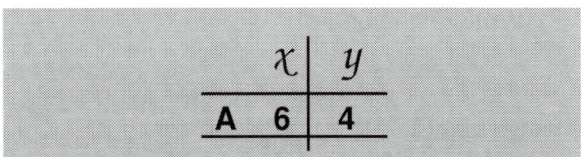

	x	y
A	6	4
B	8	6

- Here's what you should have: X is 8; Y is 6.
d. Row C doesn't show the value for either X or Y. Complete row C. Raise your hand when you're finished. (Observe students and give feedback.)

- (Write to show:)

	x	y
A	6	4
B	8	6
C	4	0

- Here's what you should have: X is 4; Y is zero.
e. Your turn. Complete the rest of the rows. Raise your hand when you're finished.
   (Observe students and give feedback.)
- (Write to show:)

	x	y
A	6	4
B	8	6
C	4	0
D	0	8
E	1	3

- Here's what you should have for points D and E.
- For D: X is zero; Y is 8.
- For E: X is 1; Y is 3.
f. Raise your hand if you got everything right.

EXERCISE 6 MULTIPLICATION
Missing Factor as a Letter

a. Find part 7.
- This is a new kind of problem. The letter F just stands for some fraction.
b. Look at the first sample problem. It says: 6 times some fraction equals 48. What does the problem say? (Signal.) *6 times some fraction equals 48.*
- The missing fraction is 48/6. You can see it written below. F equals 48/6. Then there's another equal sign and the number that 48/6 equals. 48 divided by 6 is 8. For sample problem 1, F equals 8.
c. Next sample problem. What does the problem say? (Signal.) *4 times some fraction equals 188.*
- What's the missing fraction? (Signal.) *188-fourths.*
- So F equals 188/4. That's 188 divided by 4.
- Work the problem and write the bottom equation to show the whole number or the mixed number that 188 divided by 4 equals. Raise your hand when you're finished.
   (Observe students and give feedback.)

- (Write on the board:)

$$f = \frac{188}{4} = 47$$

- Here's what you should have. 188 divided by 4 equals 47. So F equals 47.
d. Find problem A.
- Everybody, what does it say? (Signal.) *5 times some fraction equals 936.*
- Copy the problem. Below, write the equation to show the fraction that F equals and the whole number or the mixed number the fraction equals. Raise your hand when you're finished.
  (Observe students and give feedback.)
- (Write on the board:)

$$\textbf{a. } 5f = 936$$
$$f = \frac{936}{5} = 187\frac{1}{5}$$

- Here's what you should have for problem A. F equals 187 and 1/5.
e. Problem B. Everybody, what does it say? (Signal.) *4 times some fraction equals 338.*
- Copy the problem and write the complete equation to show what F equals. Raise your hand when you're finished.
  (Observe students and give feedback.)
- (Write on the board:)

$$\textbf{b. } 4f = 338$$
$$f = \frac{338}{4} = 84\frac{2}{4}$$

- Here's what you should have.
f. Your turn: Work the rest of the problems in part 7. Raise your hand when you're finished.
  (Observe students and give feedback.)
g. (Write on the board:)

$$\textbf{c. } 9f = 364$$
$$f = \frac{364}{9} = 40\frac{4}{9}$$

$$\textbf{d. } 8f = 450$$
$$f = \frac{450}{8} = 56\frac{2}{8}$$

- Here's what you should have for problems C and D.

EXERCISE 7 INDEPENDENT WORK

- (In addition to independent work in the textbook, assign **Bridge to Connecting Math Concepts** *Independent Worksheet* 29 as classwork or homework. Before beginning the next lesson, check the students' independent work.)

Lesson 50 – Test 5

Objectives

- Solve a problem in which the missing middle factor is expressed as a letter. (Exercise 1)

- Write a fraction for a decimal value that is more than 1. (Exercise 2)

- **Correct entries in a table to show the X and Y values of points shown on a coordinate system.** (Exercise 3)

- Perform on a mastery test of skills presented in lessons 41 through 49. (Exercise 4)

Note: Exercise 5 provides instructions for marking the test.

EXERCISE 1 MULTIPLICATION
Missing Factor as a Letter

a. Open your textbook to lesson 50 and find part 1.
- These are problems with a missing fraction. The F stands for the missing fraction.
b. Problem A. Everybody, what does it say? (Signal.) *5 times some fraction equals 384.*
- Copy the problem. Below, write the complete equation to show the fraction that F equals and the mixed number or the whole number for that fraction. Raise your hand when you're finished.
  (Observe students and give feedback.)
- (Write on the board:)

$$\textbf{a. } 5f = 384$$
$$f = \frac{384}{5} = 76\frac{4}{5}$$

- Here's what you should have. Raise your hand if you got everything right.
c. Work the rest of the problems in part 1. Raise your hand when you're finished.
  (Observe students and give feedback.)
- (Write on the board:)

> **b.** $3f = 384$
>
> $$f = \frac{384}{3} = 128$$
>
> **c.** $6f = 940$
>
> $$f = \frac{940}{6} = 156\frac{4}{6}$$
>
> **d.** $2f = 620$
>
> $$f = \frac{620}{2} = 310$$

- Here's what you should have for problems B, C and D. Raise your hand if you got everything right.

EXERCISE 2 DECIMALS
As Fractions

a. Find part 2.
- You're going to write fractions that equal decimal values. If the decimal shows tenths, the denominator of the fraction is 10. If the decimal shows hundredths, the denominator of the fraction is 100. The digits in the decimal number show the digits you'll write in the numerator of the fraction.
- For each item, you'll write an equation that shows the decimal value, an equal sign and the equivalent fraction.
b. Your turn: Write the equation for item A. Raise your hand when you're finished.
  (Observe students and give feedback.)
- (Write on the board:)

> **a.** $5.03 = \dfrac{503}{100}$

- Here's the equation: 5 and 3-hundredths equals 503/100. Raise your hand if you got it right.
c. Your turn: Write equations for the rest of the problems in part 2. Raise your hand when you're finished.
  (Observe students and give feedback.)

d. (Write on the board:)

> **b.** $42.8 = \dfrac{428}{10}$
>
> **c.** $.15 = \dfrac{15}{100}$
>
> **d.** $1.5 = \dfrac{15}{10}$

- Here's what you should have for items B, C and D.
- Raise your hand if you got everything right.

EXERCISE 3 COORDINATE SYSTEM
X and Y Values

a. Find part 3.
- A student completed the table. It is supposed to tell about points on the coordinate system. Unfortunately, the student made some mistakes.
b. Your turn: Check the X and Y values for each point. If either X or Y is wrong, write the corrected **row.** Show both the X and Y values for any row that is wrong. Don't write anything for rows that are correct. Raise your hand when you're finished.
  (Observe students and give feedback.)
- (Write on the board:)

x	y
A 2	6
C 0	3
D 3	9

- Here's what you should have. You should have found three mistakes.
- The student wrote the wrong X value for point A. The row for point A should show that X is 2 and Y is 6.
- The row for point C has the wrong Y value. The row for point C should show that X is zero and Y is 3.
- The row for point D is also wrong. Both the X and Y values are wrong. The row for point D should show that X is 3 and Y is 9.
- Raise your hand if you corrected all of the mistakes.

EXERCISE 4 TEST 5

> *Note:* **Students are not to use calculators for any part of the test.**

a. This is a test. You should only have your textbook, a sharpened pencil and lined paper on your desk.
b. Find part 1 of test 5 in your textbook. √
c. Do the test on your own. Raise your hand when you've completed the test.
   (Observe students but do not give feedback.)

EXERCISE 5 MARKING THE TEST

a. (Collect the students' papers. Use the Answer Key to score tests. Award scores for test 5 as follows:)

Test 5 Percent Summary					
SCORE	%	SCORE	%	SCORE	%
44	100	39	89	34	77
43	98	38	86	33	75
42	96	37	84	32	73
41	95	36	82	31	70
40	93	35	80		

b. (Complete the Test 5 Remedy Summary to determine whether remedies are needed. Reproducible Summary Sheets are at the back of the Teacher's Guide.)
• (If more than 1/4 of the students did not pass a test part, present the remedy for that part before beginning lesson 51. Remedies appear at the end of the Test 5 Answer Key.)

EXERCISE 5 INDEPENDENT WORK

• (Assign **Bridge to Connecting Math Concepts** *Independent Worksheet* 30 as classwork or homework. Before beginning the next lesson, check the students' independent work.)

Lesson 51

EXERCISE 1 DECIMALS
Mixed Numbers

a. Open your textbook to lesson 51 and find part 1.
- I'll read what it says. Follow along: You can write decimal numbers that are more than 1 as mixed numbers. You just read the decimal number. That tells you how to write the mixed number.

- You can see 3 and 14-hundredths written as a decimal number and as a mixed number.
- What's the next decimal number? (Signal.) *6 and 2-tenths.*
- You can see the mixed number 6 and 2/10 for that decimal number.

b. Find part 2.
- For each item, you'll show the complete equation with the decimal number and the mixed number it equals.

c. Write equation A. Raise your hand when you're finished. (Observe students and give feedback.)
- (Write on the board:)

> **a.** $1.07 = 1\frac{7}{100}$

- Here's what you should have: 1 and 7-hundredths equals 1 and 7/100.

d. Write equation B. Raise your hand when you're finished. (Observe students and give feedback.)
- (Write on the board:)

> **b.** $14.002 = 14\frac{2}{1000}$

- Here's what you should have: 14 and 2-thousandths equals 14 and 2/1000.

e. Your turn: Write equations for the rest of the items in part 2. Raise your hand when you're finished. (Observe students and give feedback.)
- (Write on the board:)

> **c.** $3.5 = 3\frac{5}{10}$
>
> **d.** $216.23 = 216\frac{23}{100}$

- Here's what you should have for items C and D.
- Raise your hand if you got everything right.

f. Remember, if you can read a decimal number that is more than 1, you know how to write it as a mixed number.

EXERCISE 2 DECIMALS
As Fractions

a. Find part 3.
- Each row of the table is supposed to show a decimal value and the **fraction** it equals.

b. Your turn: Copy the table and complete it. Raise your hand when you're finished. (Observe students and give feedback.)

Key:

	Decimal	Fraction
a.	8.5	$\frac{85}{10}$
b.	11.09	$\frac{1109}{100}$
c.	6.35	$\frac{635}{100}$
d.	.8	$\frac{8}{10}$
e.	25.1	$\frac{251}{10}$
f.	.047	$\frac{47}{1000}$

c. Check your work. Find part J on page 202. That shows what you should have for each row.
d. Raise your hand if you got everything right.

EXERCISE 3 PROBLEM SOLVING
Fraction Comparison/Classification

a. Find part 4.
• These are sentences that tell about number families. Some of the sentences tell about families that have the name **difference.** Some don't.
• Remember, if the sentence compares two things, the family has a difference number. If the sentence does not compare, there is no difference. You show the big number as 1 whole.
b. Sentence A: 3/5 of the red chairs were old. Raise your hand when you know whether the family has a difference number. √
• What's the answer? (Signal.) *No.*
• Sentence B: The red chairs were 3/4 the age of the blue chairs. Raise your hand when you know whether the family has a difference number. √
• What's the answer? (Signal.) *Yes.*
c. Go back to item A. Make the number family. Raise your hand when you're finished.
(Observe students and give feedback.)
• (Write on the board:)

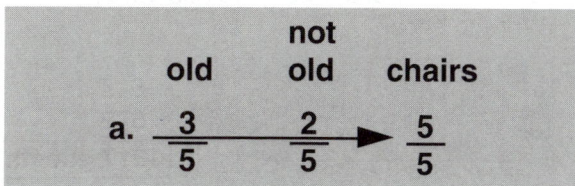

• Here's what you should have.
d. Your turn: Make families for all the sentences in part 4. Raise your hand when you're finished.
(Observe students and give feedback.)

e. (Write on the board:)

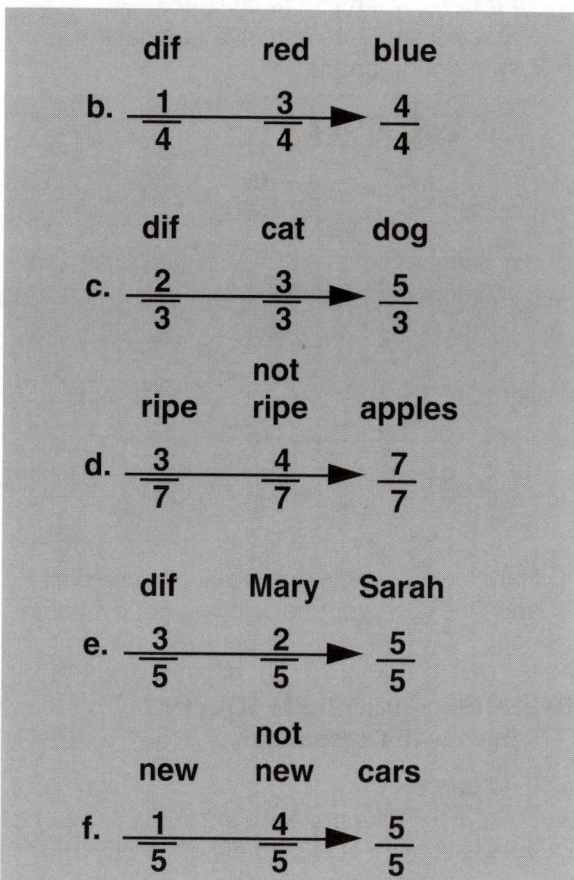

• Here's what you should have for each family.
f. Raise your hand if you got everything right.

EXERCISE 4 MULTIPLICATION
Missing Factor as a Letter

a. Find part 5.
• These are problems with a missing fraction. The F stands for the missing fraction.
b. Problem A says: 7 times some fraction equals 497.
• Copy the problem. Below, write the complete equation to show the fraction that F equals and the mixed number or the whole number for that fraction. Raise your hand when you're finished.
(Observe students and give feedback.)
• (Write on the board:)

> a. $7f = 497$
>
> $f = \frac{497}{7} = 71$

• Here's what you should have: F equals 71. Raise your hand if you got everything right.

c. Work the rest of the problems in part 5. Raise your hand when you're finished.
(Observe students and give feedback.)
- (Write on the board:)

> b. $3f = 762$
>
> $f = \dfrac{762}{3} = 254$
>
> c. $4f = 762$
>
> $f = \dfrac{762}{4} = 190\dfrac{2}{4}$
>
> d. $2f = 762$
>
> $f = \dfrac{762}{2} = 381$

- Here's what you should have for problems B, C and D. Raise your hand if you got everything right.

EXERCISE 5 PROBLEM SOLVING
Tables with Comparison

a. Find part 6.
- This table is supposed to show the number of guests staying at two hotels on Friday and Sunday. A lot of numbers are missing.
- The statements next to the table give you information about two of the missing numbers.
b. Statement 1: At hotel A, the number of guests on Friday was 211 less than the number of guests on Sunday.
- Copy the table. Put **X**s in the table to show the cells that are compared in statement 1. Write the family with names and one number. Raise your hand when you've done that much.
(Observe students and give feedback.)
Key:

	hotel A	hotel B	total
Friday	**x**	1243	
Sunday	672 **x**		
total			

- (Write on the board:)

- Here's the family you should have.
- Now look at the cells with **X**s. One cell has a number. Write that number in the family. Then figure out the missing number. Raise your hand when you're finished.
- (Write to show:)

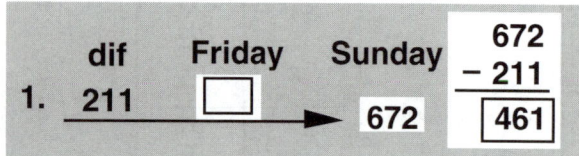

- Here's what you should have. The number of guests on Sunday is **672.** So the number on Friday is **461.** 461 goes in the cell for guests on Friday at hotel A. Put 461 in the right place in your table. √
c. Statement 2: On Sunday, 645 more guests were in hotel B than in hotel A.
- Put **X**s in the table to show the cells that are compared. Write the family with names and one number. Raise your hand when you've done that. (Observe students and give feedback.)
Key:

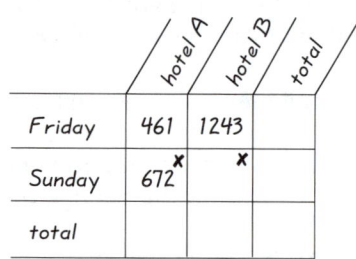

	hotel A	hotel B	total
Friday	461	1243	
Sunday	672 **x**	**x**	
total			

- (Write on the board:)

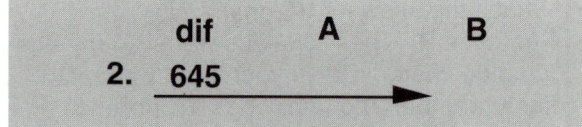

- Here's what you should have.
- Now look at the row for the guests on Sunday and write the number that's given for one of the hotels. Then figure out the missing number. Raise your hand when you're finished. (Observe students and give feedback.)
- (Write to show:)

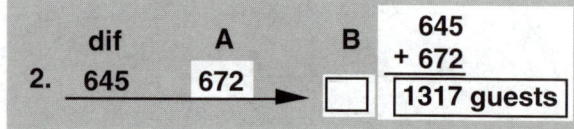

- Here's what you should have. The number of guests at hotel A is **672.** So the number at hotel B is **1317.** Write it in the table. Then figure out all the other numbers in the table. Raise your hand when you're finished.
  (Observe students and give feedback.)
d. Check your work.

Key:

	hotel A	hotel B	total
Friday	X 461	1243	1704
Sunday	X 672	X 1317	1989
total	1133	2560	3693

- Find part K. That shows what you should have. √
e. Make sure your table has all the right numbers.
- Write answers to all the questions in part 6. Raise your hand when you're finished.
  (Observe students and give feedback.)
f. Check your work.
- I'll read each question. You'll tell me the answer.
- A. How many guests stayed at hotel B on Sunday? (Signal.) *1317.*
- B. At hotel B, what was the total number of guests for both days? (Signal.) *2560.*
- C. On Sunday, were more guests staying at hotel A or hotel B? (Signal.) *Hotel B.*
- D. What was the total number of guests for both hotels on both days? (Signal.) *3693 guests.*
- E. On which day did fewer guests stay at hotel A? (Signal.) *Friday.*
g. Raise your hand if you got all those answers correct.

EXERCISE 6 FRACTIONS
Simplification

a. Find part 7.
- The fractions in part 7 are not simplified. You're going to simplify them.
- Remember how to do that. First, write an equal sign and show the prime factors for the numerator and the denominator. Then cross out all the fractions that equal 1. Then write an equal sign and the simplified fraction.
b. Fraction A. Copy the fraction, write the equal sign and the prime factors for 28 and 24. Raise your hand when you've done that much.
  (Observe students and give feedback.)

- (Write on the board:)

$$a. \ \frac{28}{24} = \frac{2 \times 2 \times 7}{2 \times 2 \times 2 \times 3}$$

- Here's what you should have so far. Now cross out any fractions that equal 1 and write the simplified fraction. Remember the equal sign. Raise your hand when you're finished.
  (Observe students and give feedback.)
- (Write to show:)

$$a. \ \frac{28}{24} = \frac{\cancel{2} \times \cancel{2} \times 7}{\cancel{2} \times \cancel{2} \times 2 \times 3} = \frac{7}{6}$$

- Here's what you should have: 28/24 equals 7/6.
c. Work the rest of the problems in part 7. Remember both equal signs. Raise your hand when you're finished.
  (Observe students and give feedback.)
d. (Write on the board:)

$$b. \ \frac{35}{10} = \frac{\cancel{5} \times 7}{2 \times \cancel{5}} = \frac{7}{2}$$

$$c. \ \frac{5}{50} = \frac{\cancel{5}}{2 \times \cancel{5} \times 5} = \frac{1}{10}$$

- Here's what you should have for items B and C.
- For item C, you crossed out everything on top. You should have written 1 in your answer: 5/50 equals 1/10.
e. Raise your hand if you got everything right.

EXERCISE 7 INDEPENDENT WORK

- (In addition to independent work in the textbook, assign **Bridge to Connecting Math Concepts** *Independent Worksheet* 31 as classwork or homework. Before beginning the next lesson, check the students' independent work.)

Lesson 52

EXERCISE 1 FRACTIONS
As Decimals and Mixed Numbers

a. Open your textbook to lesson 52 and find part 1.
- (Teacher reference:)

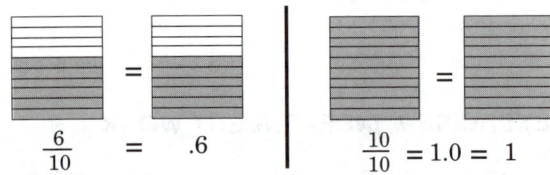

$$\frac{6}{10} = .6 \qquad \frac{10}{10} = 1.0 = 1$$

- I'll read what it says. Follow along: Decimal values work like fractions.
- You can see the equation 6/10 equals 6-tenths. The first picture shows the fraction 6-tenths. 10 parts with 6 of them shaded.
- Next to it is the decimal value 6/10. It's exactly the same amount as the fraction value. 10 parts with 6 of them shaded.
- Below, you can see the picture and the equation for 10-tenths. The fraction 10/10 equals the decimal value for 10-tenths. It's written as 1 with a decimal point and zero. And that equals 1.
- Remember, 10-tenths is 1 whole no matter how it is written.

- Below, you can see the picture and the equation for 14/10. The fraction 14/10 is 1 whole and 4 parts of the next unit. That's 1 and 4/10. You can see it written as a decimal and as a mixed number.
- Remember, if the fraction is more than 1, the decimal number is more than 1. And if the decimal number is more than 1, part of the decimal number is a whole number.

b. Find part 2.
- For each fraction, you'll write an equation. You'll copy the fraction and show the decimal number it equals. Then you'll show the mixed number it equals.

c. I'll do the sample item: 15/10.
- (Write on the board:)

$$\frac{15}{10} = 1.5 = 1\frac{5}{10}$$

- Here's the whole equation. 15/10 equals the decimal value 1 and 5-tenths. The mixed number is 1 and 5/10.

d. Item A: 99/10. Copy the fraction. Write the decimal number and the mixed number it equals. Raise your hand when you're finished.
- (Write on the board:)

$$\text{a.}\quad \frac{99}{10} = 9.9 = 9\frac{9}{10}$$

- Here's what you should have. 99/10 is 9 and 9-tenths. That equals 9 and 9/10.

e. Your turn: Write equations for the rest of the items in part 2. Raise your hand when you're finished. (Observe students and give feedback.)

f. (Write on the board:)

$$\text{b.}\quad \frac{541}{100} = 5.41 = 5\frac{41}{100}$$

$$\text{c.}\quad \frac{126}{10} = 12.6 = 12\frac{6}{10}$$

$$\text{d.}\quad \frac{409}{100} = 4.09 = 4\frac{9}{100}$$

- Check your work. Here's what you should have for items B, C and D.

EXERCISE 2 MULTIPLICATION
Missing Factor as a Letter

a. Find part 3.
- These are problems that have a missing fraction.

b. Copy each problem. Below, write the complete equation to show the fraction and the whole number or mixed number that F equals. Raise your hand when you're finished.
(Observe students and give feedback.)
Key:

a. $5f = 266$

$f = \dfrac{266}{5} = 53\dfrac{1}{5}$

b. $7f = 266$

$f = \dfrac{266}{7} = 38$

c. $2f = 49$

$f = \dfrac{49}{2} = 24\dfrac{1}{2}$

d. $3f = 996$

$f = \dfrac{996}{3} = 332$

c. Find part J on page 206. That shows what you should have for each problem.
d. Raise your hand if you got everything right.

EXERCISE 3 PROBLEM SOLVING
Fraction Comparison/Classification

a. Find part 4.
* You're going to make fraction number families for the sentences.
* Some of the sentences compare. You make a family with the name **difference** for those sentences. If a sentence doesn't compare, there is no difference.
b. Make the family for item A. Remember, make a **fraction** number family. Raise your hand when you're finished.
(Observe students and give feedback.)
* (Write on the board:)

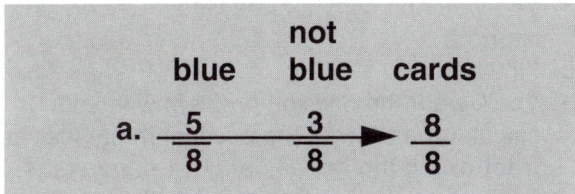

* Here's what you should have.   The sentence doesn't compare, so there is no difference. 5 out of every 8 cards are blue. The fraction for all of the cards is 8/8.
c. Make the family for item B. Raise your hand when you're finished.
(Observe students and give feedback.)
* (Write on the board:)

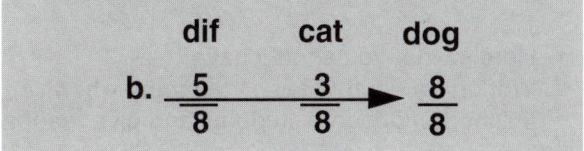

* Here's what you should have. The sentence compares, so there is a difference. The cat weighed 3/8 as much as the dog.
d. Your turn: Make families for the rest of the items in part 4. Raise your hand when you're finished. (Observe students and give feedback.)
e. (Write on the board:)

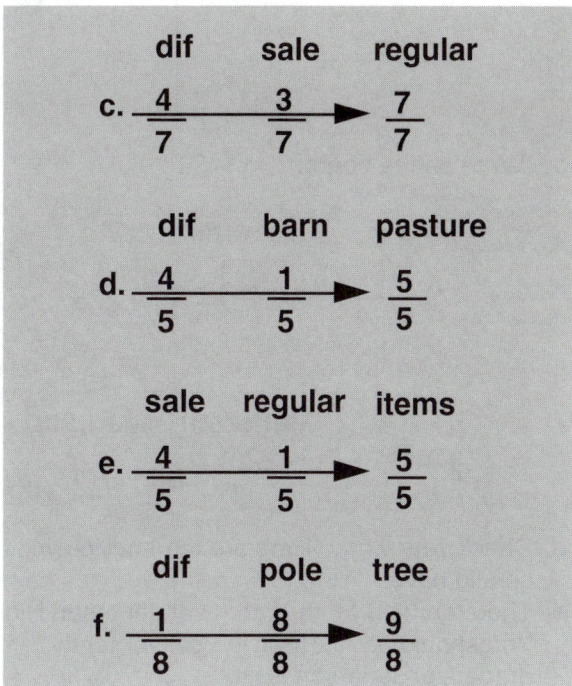

* Here's what you should have for the items.
* Sentences C, D and F compare, so those families have a difference.
f. Raise your hand if you got everything right.

EXERCISE 4 PROBLEM SOLVING
Tables with Comparison

a. Find part 5.
* The statements next to the table give you information about two of the missing numbers.
b. Copy the table. Put **X**s in the table to show the cells that are compared for statement 1. Make a number family. Figure out the missing number and copy that number in the table. Then do the same for statement 2. Raise your hand when your table has four numbers.
(Observe students and give feedback.)

Key:

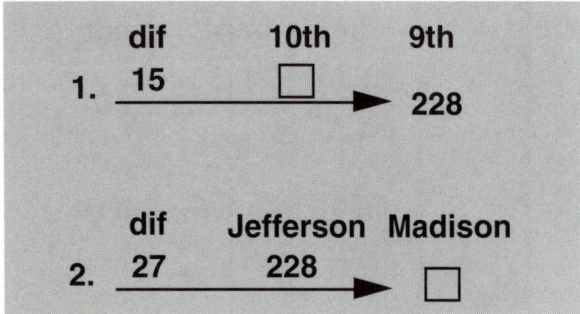

	9th graders	10th graders	total
Jefferson High	228 ✗	213 ✗	
Madison High	255 ✗	284	
total			

Key:

a. $\dfrac{10}{25} = \dfrac{2 \times \cancel{5}}{5 \times \cancel{5}} = \dfrac{2}{5}$ b. $\dfrac{4}{12} = \dfrac{\cancel{2} \times \cancel{2}}{\cancel{2} \times \cancel{2} \times 3} = \dfrac{1}{3}$

c. $\dfrac{12}{9} = \dfrac{2 \times 2 \times \cancel{3}}{3 \times \cancel{3}} = \dfrac{4}{3}$ d. $\dfrac{24}{30} = \dfrac{\cancel{2} \times 2 \times 2 \times \cancel{3}}{\cancel{2} \times \cancel{3} \times 5} = \dfrac{4}{5}$

c. (Write on the board:)

dif   10th   9th

1. 15   ▢ ➝ 228

dif   Jefferson   Madison

2. 27   228 ➝ ▢

d. Check your work. Here are the families you should have.
- There are 213 tenth-graders at Jefferson High. You should have 213 in the cell for tenth-graders at Jefferson High.
- There are 255 ninth-graders at Madison High. You should have 255 in the cell for ninth-graders at Madison High. Raise your hand if you got everything right.

e. Now figure out all the missing numbers in the table. Raise your hand when you're finished. (Observe students and give feedback.)

f. Check your work.
- Find part **16.** Make sure your table has all the right numbers. Later, you'll write the answers to all the questions in part 16. √

EXERCISE 5 FRACTIONS
Simplification

a. Find part 6.
- You're going to simplify these fractions by writing the prime factors.
- Remember, first write the prime factors. Then cross out any fractions that equal 1. Then multiply the factors that are left and write the simplified fraction.

b. Raise your hand when you've finished the problems in part 6. Remember the equal signs. (Observe students and give feedback.)

c. Check your work.
- Find part K. That shows what you should have for the problems in part 6.

d. Item A: 10/25 equals 2/5.
- Item B: 4/12 equals 1/3.
- Item C: 12/9 equals 4/3.
- Item D: 24/30 equals 4/5.

e. Correct any mistakes.

EXERCISE 6 DECIMALS
As Mixed Numbers

a. Find part 7.
- I'll read what it says. Follow along: When you work division problems on your calculator, the calculator shows mixed numbers as **decimal values.**
- When you divide 50 by 40 your calculator shows this answer: 1 and 25-hundredths. That's the same value as the mixed number 1 and 25/100.
- When you divide 20 by 8 your calculator shows this answer: 2 and 5-tenths. That's the same value as the mixed number 2 and 5/10.
- For some division problems, the calculator gives a very long answer. When you divide 10 by 3, you get this answer: 3 point 33 and then a bunch of other 3s. That's the decimal value for 3 and 1/3.
- You wouldn't copy that decimal number. You'd **round it.**

b. Find part 8.
- You'll work the division problem with your calculator and write the decimal remainder as a **fraction.** If the remainder is 14-hundredths, you'll show the fraction as 14/100.

c. Problem A: 5 divided by 2.
- Work the problem on your calculator. Write the decimal remainder as a fraction. Raise your hand when you're finished. (Observe students and give feedback.)
- (Write on the board:)

a. 2 $\dfrac{5}{10}$

- Here's what you should have.

d. Work problem B. Raise your hand when you're finished. (Observe students and give feedback.)

- (Write on the board:)

$$b.\ 2\ \frac{125}{1000}$$

- Here's what you should have.
e. Your turn: Work the rest of the items in part 8.
  Raise your hand when you're finished.
  (Observe students and give feedback.)
- (Write on the board:)

$$c.\ 3\ \frac{25}{100}$$

$$d.\ 3\ \frac{2}{10}$$

Here's what you should have for each item.
Raise your hand if you got everything right.

EXERCISE 7 INDEPENDENT WORK

- (In addition to independent work in the textbook, assign **Bridge to Connecting Math Concepts** *Independent Worksheet* 32 as classwork or homework. Before beginning the next lesson, check the students' independent work.)

Lesson 53

EXERCISE 1 COMPLEX FRACTIONS
Equal to 1

a. Open your textbook to lesson 53 and find part 1.
- I'll read what it says. Follow along: You've learned that fractions equal 1 if the top value is the same as the bottom value. That's true for complicated fractions that have a **fraction** over the same **fraction.**

- You can see 4/8 over 4/8.
- The fraction equals 1 because the value on top is the same as the value on the bottom.
- You can see some other fractions that equal 1. You can see some pretty fancy fractions.

b. (Write on the board:)

$$\frac{3}{3} = 1$$

- Here's 3/3. This fraction equals 1. The fraction has 3 in the numerator and 3 in the denominator.
- We'll rewrite each 3 as a fraction that equals 3. 6/2 equals 3. So we'll write 6/2 on top and 6/2 on the bottom.

c. (Change to show:)

$$\frac{\frac{6}{2}}{\frac{6}{2}} = 1$$

- Here's 6/2 over 6/2. We still have 3 over 3. So the fraction still equals 1 whole unit.

d. (Write on the board:)

$$\frac{\frac{10}{5}}{\frac{10}{5}} = 1$$

- Here's 10/5 over 10/5. It also equals 1.
- Listen: How many does 10/5 equal? (Signal.) *2.*
- So there are 2 on top and 2 on the bottom. The fraction 2/2 equals 1.
- Remember the rule: Any fraction that has the same value on top and on the bottom equals 1.

e. Find part 2.
- Each fraction is supposed to equal 1, but a value is missing in each fraction, either the top or the bottom value.

f. Touch problem A. √
- What's the top value? (Signal.) *5-eighths.*
- Say the whole fraction that equals 1. (Signal.) *5-eighths over 5-eighths.*
- Problem B. What's the bottom value? (Signal.) *7-thirds.*
- Say the whole fraction that equals 1. (Signal.) *7-thirds over 7-thirds.*

g. Your turn: Write the complete equation for each item. Raise your hand when you're finished. (Observe students and give feedback.)

h. Check your work.
- Read each equation.
- Equation A. (Signal.) *5-eighths over 5-eighths equals 1.*
- Equation B. (Signal.) *7-thirds over 7-thirds equals 1.*
- Equation C. (Signal.) *11-fifths over 11-fifths equals 1.*

EXERCISE 2 FRACTIONS
As Decimals and Mixed Numbers

a. Find part 3.
- For each item, you'll copy the fraction, then write the decimal number and the mixed number. Remember, if the fraction is more than 1, the decimal number is more than 1. It equals a mixed number.
b. Item A: 740/100. Copy the fraction. Write an equation to show the decimal number and the mixed number it equals. Raise your hand when you're finished. √
- (Write on the board:)

$$\text{a.} \quad \frac{740}{100} = 7.40 = 7\frac{40}{100}$$

- Here's what you should have.
c. Item B: 29/10. Raise your hand when you've written the complete equation. √
- (Write on the board:)

$$\text{b.} \quad \frac{29}{10} = 2.9 = 2\frac{9}{10}$$

- Here's what you should have.
d. Work item C. Raise your hand when you're finished. √
- (Write on the board:)

$$\text{c.} \quad \frac{163}{10} = 16.3 = 16\frac{3}{10}$$

- Here's what you should have.
e. Work item D. Raise your hand when you're finished. √
- (Write on the board:)

$$\text{d.} \quad \frac{422}{100} = 4.22 = 4\frac{22}{100}$$

- Here's what you should have.

EXERCISE 3 COMPLEX FRACTIONS
Equivalence

a. Find part 4.
- Each of these equations starts with a fraction that equals 1. You're going to create complicated fractions.
- Each fraction you create will have a numerator that equals the numerator shown in the first fraction and a denominator that equals the denominator of the first fraction.
b. Look at the first fraction in the sample problem.
- What's the numerator? (Signal.) *3.*
- What's the denominator? (Signal.) *3.*
- We're going to rewrite each 3 as a fraction that **equals 3.**
- The first fraction will have a denominator of 2. What fraction equals 3 and has a denominator of 2? (Signal.) *6-halves.*
- (Write on the board:)

$$\frac{3}{3} = \frac{\frac{6}{2}}{\frac{6}{2}}$$

- 6/2 over 6/2 equals the original fraction. It has 3 on the top and 3 on the bottom. So it equals 1.
c. For the next fraction, we'll rewrite 3 as a **different** fraction that equals 3.
- Everybody, what fraction equals 3 and has a denominator of 7? (Signal.) *21-sevenths.*
- (Write to show:)

$$\frac{3}{3} = \frac{\frac{6}{2}}{\frac{6}{2}} = \frac{\frac{21}{7}}{\frac{21}{7}}$$

- I've created two new fractions that equal 3/3.
d. Problem A. You will copy the equation and complete both fractions. For the first fraction, you'll rewrite 2 as a fraction that equals 2 and has a denominator of 5. For the other fraction, you'll rewrite 2 as a fraction that has a denominator of 8. Raise your hand when you've written the complete equation for problem A. (Observe students and give feedback.)
e. (Write on the board:)

$$\text{a.} \quad \frac{2}{2} = \frac{\frac{10}{5}}{\frac{10}{5}} = \frac{\frac{16}{8}}{\frac{16}{8}}$$

- Here's what you should have.

f. Work problem B. Rewrite each 7 as a fraction that equals 7 and has the denominator that is shown. Raise your hand when you're finished. (Observe students and give feedback.)
• (Write on the board:)

$$\text{b.} \quad \frac{7}{7} = \frac{\frac{21}{3}}{\frac{21}{3}} = \frac{\frac{63}{9}}{\frac{63}{9}}$$

• Here's what you should have.
g. Remember, a fraction equals 1 if the top value and the bottom value are the same.

EXERCISE 4 RATIOS AND PROPORTIONS
Tables

a. Find part 5.
• I'll read what it says. Follow along: Here's a difficult problem: The regular price of the bed is 8/5 the sale price. If the regular price of the bed is $560, what is the difference between the sale price and the regular price?
• The first sentence tells how to make a fraction number family. You can see that family.
• The big number is the regular price.
• The sentence compares the regular price to the sale price. So the sale price is 1 whole.
• The difference is 3/5.
• You know that the numerators are ratio numbers. You put them in a ratio table. Then you put in the number the problem gives for the regular price of the bed—$560.
• Now you can figure out the sale price of the bed and how much you'd save if you bought the bed on sale. That's the number for difference in the second column.
b. Find part 6.
• You're going to make a fraction number family and a ratio table for each problem.
c. Problem A: The regular price of a chair is 5/4 the sale price of the chair. If you bought the chair on sale, you'd save $12. What's the regular price of the chair? What's the sale price of the chair?
• The first sentence tells about the fraction number family. Remember, figure out which is the big number. You can write initials for sale price and regular price. Write **SP** for sale price and **RP** for regular price. Raise your hand when you've finished your number family. (Observe students and give feedback.)

• (Write on the board:)

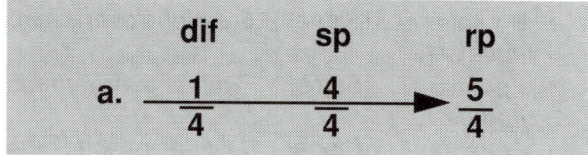

• Here's what you should have.
d. Now make a ratio table with the ratio numbers in the first column and the dollar amount the problem gives in the second column. The problem tells how much you'd save. That's the difference number. Make the ratio table with the names and four numbers. Raise your hand when you've done that much. (Observe students and give feedback.)
• (Write on the board:)

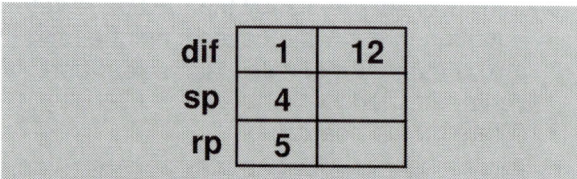

• Here's what you should have.
e. Now figure out the missing values in the second column and answer the questions. Raise your hand when you're finished. (Observe students and give feedback.)
• (Write to show:)

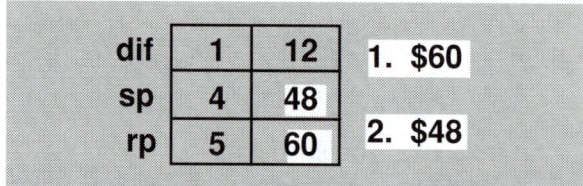

• Here's what you should have. The regular price is $60. The sale price is $48. Raise your hand if you got everything right.
f. Problem B: The elephant weighs 3/8 as much as the rock. The elephant weighs 6000 pounds. How much does the rock weigh? How much more does the rock weigh than the elephant weighs?
• The first sentence tells how to make a fraction number family. Make the family with initials. Write **E** for elephant and **R** for rock. Raise your hand when you've made the family. (Observe students and give feedback.)
• (Write on the board:)

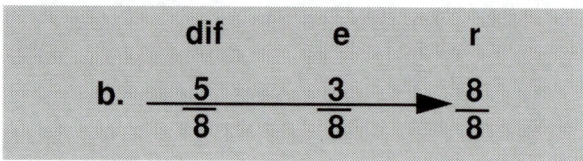

- Here's the family.
g. Now make a ratio table with ratio numbers in the first column and pounds in the second column. Raise your hand when you're finished with problem B.
(Observe students and give feedback.)
- (Write on the board:)

dif	5	10000
e	3	6000
r	8	16000

1. 16,000 lb

2. 10,000 lb

- Here's what you should have. The rock weighs 16,000 pounds. The rock weighs 10,000 pounds more than the elephant. Raise your hand if you got it right.
- That's a tough problem.
h. Problem C. Make the family. Write **TV** for the TV and write **SM** for the sewing machine. Raise your hand when you've written the fraction number family.
(Observe students and give feedback.)
- (Write on the board:)

$$c. \quad \overset{dif}{\underset{5}{\frac{2}{5}}} \quad \overset{TV}{\underset{5}{\frac{5}{5}}} \longrightarrow \overset{sm}{\underset{5}{\frac{7}{5}}}$$

- Check your work. Here's what you should have. The first sentence tells you that the sewing machine costs more than the TV. It's the big number. The TV is 1 whole—that's 5/5. The difference is 2/5.
i. Now make the ratio table and figure out the missing numbers. Show ratio numbers in the first column and dollars in the second. Raise your hand when you're finished.
(Observe students and give feedback.)
- (Write on the board:)

dif	2	112
TV	5	280
sm	7	392

1. $392

2. $112

- Here's the table you should have. The TV costs $280. So the sewing machine costs $392, and the difference is $112.

EXERCISE 5 DECIMALS
As Mixed Numbers

a. Find part 7.
- These are division problems. When you work them with your calculator, you get a decimal answer. You'll write the answer to each problem as a mixed number.
b. Work the problems in part 7. Show the answers as mixed numbers. Raise your hand when you're finished.
(Observe students and give feedback.)
- (Write on the board:)

$$a. \quad 6\frac{25}{100}$$
$$b. \quad 19\frac{5}{10}$$
$$c. \quad 4\frac{32}{100}$$
$$d. \quad 3\frac{125}{1000}$$

- Here's what you should have for each problem.
c. Raise your hand if you got everything right.

EXERCISE 6 FRACTIONS
Simplification

a. Find part 8.
b. You're going to simplify these fractions by writing the prime factors.
- Remember, first write the prime factors for the numerator and for the denominator. Then cross out any fractions that equal 1. Then multiply the factors that are left and write the simplified fraction. Raise your hand when you've finished the problems in part 8.
(Observe students and give feedback.)

Key:

$$a. \quad \frac{21}{28} = \frac{3 \times \cancel{7}}{2 \times 2 \times \cancel{7}} = \frac{3}{4} \qquad b. \quad \frac{8}{12} = \frac{\cancel{2} \times \cancel{2} \times 2}{\cancel{2} \times \cancel{2} \times 3} = \frac{2}{3}$$

$$c. \quad \frac{3}{18} = \frac{\cancel{3}}{2 \times \cancel{3} \times 3} = \frac{1}{6} \qquad d. \quad \frac{24}{30} = \frac{\cancel{2} \times 2 \times 2 \times \cancel{3}}{\cancel{2} \times \cancel{3} \times 5} = \frac{4}{5}$$

c. Check your work.
- Find part J on page 210. That shows what you should have for the problems in part 8.
d. Correct any mistakes.

EXERCISE 7 INDEPENDENT WORK

- (In addition to independent work in the textbook, assign **Bridge to Connecting Math Concepts** *Independent Worksheet* 33 as classwork or homework. Before beginning the next lesson, check the students' independent work.)

Lesson 54

EXERCISE 1 NUMBER RELATIONSHIPS
Fractions/Decimals/Mixed Numbers

a. Open your textbook to lesson 54 and find part 1.
- This is a new kind of table.
- The first column will show fractions with a denominator of 10, 100 or 1000.
- The middle column will show the same values written as decimal numbers.
- The last column will show the same values written as mixed numbers.
b. Copy the table and complete row A. Raise your hand when you've done that much.
  (Observe students and give feedback.)
- (Write on the board:)

	Fraction	Decimal	Mixed number
a.	$\frac{302}{100}$	3.02	$3\frac{2}{100}$

- Here's what you should have for row A.
c. Complete the rest of the table. Raise your hand when you're finished.
  (Observe students and give feedback.)

d. (Write to show:)

	Fraction	Decimal	Mixed number
a.	$\frac{302}{100}$	3.02	$3\frac{2}{100}$
b.	$\frac{58}{10}$	5.8	$5\frac{8}{10}$
c.	$\frac{1015}{1000}$	1.015	$1\frac{15}{1000}$
d.	$\frac{134}{10}$	13.4	$13\frac{4}{10}$

- Check your work. Here's what you should have.
e. For row B, you started with the mixed number. You wrote the decimal number. Then you wrote the fraction. Remember, the numerator of the fraction has the same digits as the decimal number. The denominator is 10.
- For row C, you started with the decimal number. You wrote the mixed number and the fraction. All show 1 and 15-thousandths.
- For row D, you started with the decimal number again. You wrote the mixed number and the fraction. All show 13 and 4-tenths.

EXERCISE 2 CIRCLES
Circumference and Diameter

a. Find part 2.
- (Teacher reference:)

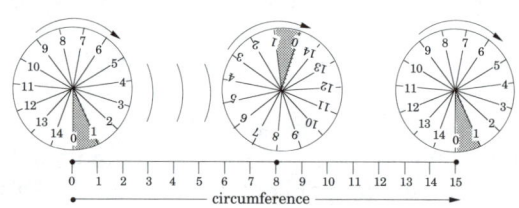

- You're going to learn about circles. I'll read what it says. Follow along: You've learned that the distance around any figure with straight sides is called the **perimeter.**
- A circle doesn't have any straight sides. So it doesn't have a perimeter. The distance around a circle is called the **circumference.**
- The first part of **circumference** has the same letters as the word **circle: C-I-R-C.**
- Everybody, what's the name for the distance around a circle? (Signal.) *Circumference.*
- The diagram shows the circumference as an arrow that goes all the way around the circle.

b. Listen: To figure out the distance around a circle, you **can't** add up the length of each side because there are no sides. But you can **measure** the circumference of a circle.
- One way is to roll the circle and see how long the path is when the circle **rotates one time.**
- The three pictures show a circle being rotated one time. You can see that the circle has spokes that are numbered from zero through 14.
- In the first picture, spoke zero is at the bottom. It's right above zero on the number line.
c. Touch the middle picture. √
- The middle picture shows where the circle is after it is turned more than half-way around. It has rolled 8 units on the number line. That's 8 spokes on the wheel.
- Touch the last picture. √
- The circle has zero on the bottom again. That means it has rotated all the way around. It has turned 14 units, plus 1 more to get back to zero. That's 15 units on the number line.
- Listen: The distance around the circle shown in the picture is 15 units. So the **circumference** of that circle is 15 units.
d. Here's the next fact about the circumference: The circumference of any circle is related to a line that goes through the center of that circle. That line is called the **diameter.**
- You can see the diameter. It goes through the widest part of the circle.
e. Everybody, what's the name of the line **through the center** of the circle? (Signal.) *Diameter.*
- What's the name for the distance **around** the circle? (Signal.) *Circumference.* (Repeat step e until firm.)
f. Find part 3.
- You can multiply the diameter of any circle by a particular number to get the circumference.
- The table shows the diameter and the circumference of different circles. You're going to figure out the fraction you multiply by.
g. Circle A. How many inches is the diameter? (Signal.) *7.*
- How many inches is the circumference? (Signal.) *22.*
- So you'll work the problem: **7 times some fraction equals 22.**
- Your turn: Write that equation with the **missing fraction.** Raise your hand when you've done that much. (Observe students and give feedback.)

- (Write on the board:)

$$a. \quad 7\left(\frac{22}{7}\right) = 22$$

- Here's what you should have. The fraction is 22/7. Raise your hand if you got it right.
- Use your calculator. Divide 22 by 7. Raise your hand when you have an answer displayed on your calculator. √
- You'll round the answer to **two decimal places.** That's 3 and 14-hundredths.
- Write the decimal answer. √
- (Write to show:)

$$a. \quad 7\left(\frac{22}{7}\right) = 22$$
$$\boxed{3.14}$$

h. Circle B. How many centimeters is the diameter? (Signal.) *21.*
- How many centimeters is the circumference? (Signal.) *66.*
- Write the multiplication equation that starts with 21. 21 times some fraction equals 66. Show the missing value as a fraction. Raise your hand when you've done that much. (Observe students and give feedback.)
- (Write on the board:)

$$b. \quad 21\left(\frac{66}{21}\right) = 66$$

- The equation you should have is 21 times 66/21 equals 66.
- Use your calculator and figure out the decimal value for the fraction. Round the number to two decimal places. Raise your hand when you've written the decimal number for circle B. √
- Everybody, what's the decimal number? (Signal.) *3 and 14-hundredths.*
- That's the same number you got for circle A.
i. Circle C. How many meters is the diameter? (Signal.) *5 and 9-hundredths.*
- How many meters is the circumference? (Signal.) *16.*
- Write the equation and show the fraction you multiply by. That equation starts with the number for the diameter. Raise your hand when you've done that much. √

- (Write on the board:)

$$c. \quad 5.09 \left(\frac{16}{5.09} \right) = 16$$

- Here's what you should have. When you divide 16 by 5 and 9/100, you must enter the decimal point—5, decimal point, zero, 9.
- Do the division and figure out the decimal number rounded to two decimal places. Raise your hand when you're finished.
  (Observe students and give feedback.)
- Everybody, what does that fraction equal? (Signal.) *3 and 14-hundredths.*
- That's the same number again.
j. For each circle, your fraction shows the **circumference divided by the diameter.** It's always the same number. Everybody, what's that number? (Signal.) *3 and 14-hundredths.*
k. Isn't it amazing that you'll always get that number when you divide the circumference by the diameter?
- Remember that number. You'll use it a lot.

EXERCISE 3 COMPLEX FRACTIONS
Equivalence

a. Find part 4.
- The fractions in these equations are supposed to equal 1, but either the top value or the bottom value is missing. Copy and complete each equation. Raise your hand when you're finished. (Observe students and give feedback.)
- (Write on the board:)

$$a. \ \frac{\frac{10}{5}}{\frac{10}{5}} = 1 \quad b. \ \frac{\frac{2}{3}}{\frac{2}{3}} = 1 \quad c. \ \frac{\frac{17}{4}}{\frac{17}{4}} = 1$$

Here's what you should have for each item.
b. Find part 5.
- These are equations that start with a fraction that equals 1. The denominators are shown for the fractions you'll create.
c. Problem A. For the first fraction, you'll rewrite 5 as a fraction that has a denominator of 3. For the next fraction, you'll rewrite 5 as a fraction that has a denominator of 9.
- Write the complete equation. Raise your hand when you're finished.
  (Observe students and give feedback.)

- (Write on the board:)

$$a. \ \frac{5}{5} = \frac{\frac{15}{3}}{\frac{15}{3}} = \frac{\frac{45}{9}}{\frac{45}{9}}$$

- Here's what you should have for A. 15/3 equals 5. 45/9 equals 5. Raise your hand if you got it right.
d. Write the complete equation for B. Raise your hand when you're finished.
  (Observe students and give feedback.)
- (Write on the board:)

$$b. \ \frac{6}{6} = \frac{\frac{30}{5}}{\frac{30}{5}} = \frac{\frac{54}{9}}{\frac{54}{9}}$$

- Here's what you should have for B.
e. Write the complete equation for C. Raise your hand when you're finished.
  (Observe students and give feedback.)
- (Write on the board:)

$$c. \ \frac{2}{2} = \frac{\frac{20}{10}}{\frac{20}{10}} = \frac{\frac{18}{9}}{\frac{18}{9}}$$

- Here's what you should have for each item.
f. For all the items, you started with fractions that equal 1. You created other fractions that equal 1. The top values and the bottom values are the same in all those fractions.

EXERCISE 4 RATIOS AND PROPORTIONS
Tables

a. Find part 6.
- For these problems, you'll make fraction number families and ratio tables.
b. Problem A: The area of the carpet was 3/2 the area of the living room. The installers had to cut 24 square meters from the carpet before it would fit in the living room. What's the area of the living room? What was the area of the carpet before the installers cut it?
- The first sentence tells about how to make a fraction number family. Make the family. You can use **C** for the area of the carpet and **LR** for the area of the living room. Raise your hand when you've done that much.
  (Observe students and give feedback.)

- (Write on the board:)

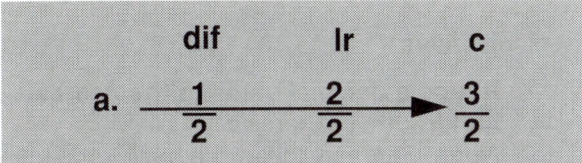

	dif	lr	c
a.	$\dfrac{1}{2}$	$\dfrac{2}{2}$ →	$\dfrac{3}{2}$

- Here's what you should have. The area of the carpet is the big number. The fraction for the area of the living room equals 1.
- Use the ratio numbers to make a table. Remember, the amount they cut from the carpet is the difference between the size of the carpet and the size of the living room. That number goes in the second column.
- Make the table. Figure out the missing numbers. Answer the questions with a number and a unit name. Raise your hand when you're finished. (Observe students and give feedback.)
- (Write on the board:)

dif	1	24	1. 48 sq m
lr	2	48	
c	3	72	2. 72 sq m

- Here's what you should have. The difference is 24 square meters. So the area of the living room is 48 square meters and the area of the carpet was 72 square meters before they cut it.
- How many square meters was it after they cut it? (Signal.) *48.*
- Make sure you have the numbers and the unit name square meters in your answers.
- c. Your turn: Work problem B. Make the family. You can use letters for names. Make the table. Figure out the missing numbers. Answer the questions with a number and a unit name. Raise your hand when you're finished. (Observe students and give feedback.)
- (Write on the board:)

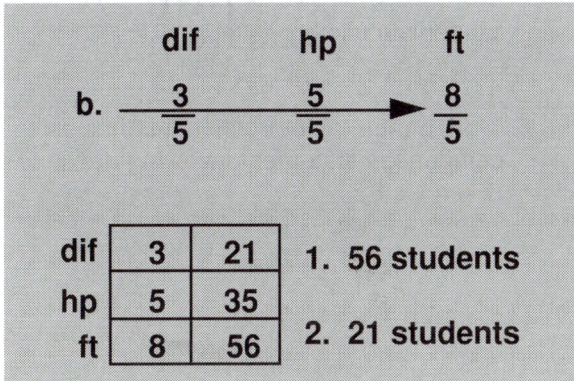

	dif	hp	ft
b.	$\dfrac{3}{5}$	$\dfrac{5}{5}$ →	$\dfrac{8}{5}$

dif	3	21	1. 56 students
hp	5	35	
ft	8	56	2. 21 students

- Here's what you should have. The field trip is the big number. Hockey is 5/5. The difference is 3/5. If there were 35 students at hockey practice, there were 56 on the field trip. There were 21 more students on the field trip than at hockey practice. The answers are 56 students and 21 students.
- d. Work problem C. Make the family. Make the table. Figure out the missing numbers. Answer the questions. Raise your hand when you're finished. (Observe students and give feedback.)
- (Write on the board:)

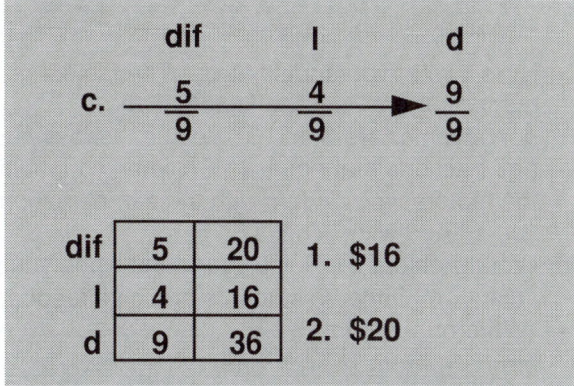

	dif	l	d
c.	$\dfrac{5}{9}$	$\dfrac{4}{9}$ →	$\dfrac{9}{9}$

dif	5	20	1. $16
l	4	16	
d	9	36	2. $20

- Here's what you should have. Dinner is the big number. Dinner is 9/9. The difference is 5/9. If dinner cost $36, lunch cost $16. The difference is $20. Dinner cost $20 more than lunch. Your answers should have dollar signs.
- e. Raise your hand if you got everything right.

EXERCISE 5 FRACTIONS
Simplification: Discrimination

a. Find part 7.
- I'll read what it says. Follow along: Some fractions cannot be simplified. A fraction cannot be simplified if it has no common factor in the numerator and denominator.
- You can see 14/15.
- The factors for 14 are 2 and 7.
- The factors for 15 are 3 and 5.
- No factor appears in **both** the numerator and denominator.
- The fraction cannot be simplified because you cannot cross out a fraction that equals 1.
- You can see another fraction that cannot be simplified: 21/10.
- The factors for the numerator are 3 and 7.
- The factors for the denominator are 2 and 5.
- No factor appears in **both** the numerator and denominator.
b. Find part 8.

- Some of the fractions cannot be simplified. For those fractions, you'll show the prime factors for the numerator and the denominator. But you won't cross out any fractions that equal 1. You'll show what you get when you multiply the factors. That should be the fraction you started with.

c. Work problem A. Raise your hand when you're finished. (Observe students and give feedback.)

- (Write on the board:)

$$a. \ \frac{8}{15} = \frac{2 \times 2 \times 2}{3 \times 5} = \frac{8}{15}$$

- Here's what you should have. The factors in the numerator are 2s. The factors in the denominator are 3 and 5. You can't cross out any fractions that equal 1. So when you multiply the factors, you end up with the same fraction you started with. That's 8/15.

d. Work problem B. Raise your hand when you're finished. (Observe students and give feedback.)

- (Write on the board:)

$$b. \ \frac{6}{7} = \frac{2 \times 3}{7} = \frac{6}{7}$$

- Here's what you should have. In the numerator, the factors are 2 and 3. In the denominator, the factor is 7. You can't cross out any fractions that equal 1. So when you multiply the factors, you end up with the same fraction you started with.

e. Your turn: Work the rest of the problems in part 8. Raise your hand when you're finished. (Observe students and give feedback.)

f. Check your work.

- Fraction C: 14/28. Can it be simplified? (Signal.) *Yes.*
- What does 14/28 equal? (Signal.) *1-half.*
- Fraction D: 24/36. Can it be simplified? (Signal.) *Yes.*
- What does 24/36 equal? (Signal.) *2-thirds.*

EXERCISE 6 INDEPENDENT WORK

- (In addition to independent work in the textbook, assign **Bridge to Connecting Math Concepts** *Independent Worksheet* 34 as classwork or homework. Before beginning the next lesson, check the students' independent work.)

Lesson 55

Objectives

- **Round a decimal value to the nearest hundredth.** (Exercise 1)
  Note: Students follow the basic procedure of looking at the digit after the hundredths digit. If that digit is 5 or more, students round the hundredths digit up.

- **Use the equation $\pi \times d = C$ to calculate the circumference of a circle.** (Exercise 2)
  Note: The symbol for pi (π) is introduced as part of this exercise.

- Use prime factors to simplify a fraction. (Exercise 3)

- **Add or subtract fractions with unlike denominators.** (Exercise 4)
  Note: Students write the numbers for counting by the smaller denominator. Then they see which is the first number that is reached when counting by the larger denominator.

Example:
$$\begin{aligned} \frac{7}{12} &= \frac{\square}{60} \\ + \frac{6}{15} &= + \frac{\square}{60} \end{aligned}$$

Students identify 60 as the first common number for 12 and for 15. By working the pair of equivalent-fraction problems, students figure out the missing numerators. Then they add and write the answer.

- **Multiply a fraction by a complex fraction that equals 1 to generate an equivalent fraction.** (Exercise 5)

Example:
$$\frac{2}{3} \left(\frac{\frac{4}{5}}{\frac{4}{5}} \right) =$$

Students multiply:
$$\frac{2}{3} \left(\frac{\frac{4}{5}}{\frac{4}{5}} \right) = \frac{\frac{8}{5}}{\frac{12}{5}}$$

- Work a ratio-table problem that has a comparison statement involving fractions. (Exercise 6)

EXERCISE 1 DECIMALS
Rounding

a. You're going to round decimal values to the nearest hundredth. That means the rounded number will have two digits after the decimal point.

b. Open your textbook to lesson 55 and find part 1.
- (Teacher reference:)

$$\downarrow \qquad\qquad \downarrow$$

2.38**5**1   2.38**4**1

2.38**5**1   2.38**4**1

2.39   2.38

c. The rules in part 1 tell how to round decimal values to hundredths. Follow along:
- Look at the **third** digit after the decimal point. If that digit is 5 or more, you do not copy the **hundredths** digit. You replace it with the digit that is 1 larger.
- In the first example, the digit after hundredths is 5. So the hundredths digit is changed from 8 to 9.
- If the third digit after the decimal point is less than 5, you just copy the hundredths digit.
- In the second example, the digit after hundredths is 4. So the hundredths digit is copied.
- Remember, if the digit after hundredths is 5 or more, change the hundredths digit. Otherwise, copy the hundredths digit.

d. Find part 2.

e. Sample 1. The second digit after the decimal point is 3. What's the third digit? (Signal.) *5.*
- Is that 5 or more? (Signal.) *Yes.*
- So I don't write 3. What do I write instead? (Signal.) *4.*
- (Write on the board:)

> ### 1.   7.14

- Here's that value rounded to the nearest hundredth. 7 and 135-thousandths rounds to 7 and 14-hundredths.
- Sample 2. The second digit after the decimal point is 6. What's the third digit? (Signal.) *4.*
- Is that 5 or more? (Signal.) *No.*
- So I just write 6 as the second digit.
- (Write on the board:)

> ### 2.   7.36

- Here's that value rounded to the nearest hundredth. 7 and 364-thousandths rounds to 7 and 36-hundredths.

f. Value A. What's the second digit after the decimal? (Signal.) *4.*
- Is the next digit 5 or more? (Signal.) *Yes.*
- So numeral A rounds to 3 and how many hundredths? (Signal.) *65.*

g. Value B. What's the second digit after the decimal? (Signal.) *5.*
- Is the next digit 5 or more? (Signal.) *No.*
- So numeral B rounds to 18 and how many hundredths? (Signal.) *5.*

h. Your turn: Write each value rounded to the nearest hundredth. Raise your hand when you're finished.
(Observe students and give feedback.)
Key:
a.	*3.65*	*b.*	*18.05*	*c.*	*10.27*
d.	*.63*	*e.*	*4.01*	*f.*	*4.02*

i. Check your work. Read each rounded value.
- Value A. (Signal.) *3 and 65-hundredths.*
- Value B. (Signal.) *18 and 5-hundredths.*
- Value C. (Signal.) *10 and 27-hundredths.*
- Value D. (Signal.) *63-hundredths.*
- Value E. (Signal.) *4 and 1-hundredth.*
- Value F. (Signal.) *4 and 2-hundredths.*

EXERCISE 2 CIRCLES
Circumference

a. Last time, you learned about circles.

b. What's the distance around the outside of a circle called? (Signal.) *Circumference.*
- What's the line through the center of the circle called? (Signal.) *Diameter.*
(Repeat step b until firm.)

c. You found that you can multiply the diameter of any circle by the same value to get the circumference.
- Raise your hand if you remember the value you multiply by. √
- Everybody, what value? (Signal.) *3 and 14-hundredths.*

d. Find part 3.
- (Teacher reference:)

$$\mathbf{d \times \pi = C}$$
$$\mathbf{\pi \times d = C}$$
$$\mathbf{3.14 \times d = C}$$

- I'll read what it says. Follow along: The diameter of any circle is related to the circumference of the circle. The circumference is always **3 and 14-hundredths times the diameter.**

- That's true of small circles and large circles.
- The number 3 and 14-hundredths has a special symbol.
- You can see that funny looking symbol. That's called **pi.**
- You can write an equation for working with circles as: Diameter times pi equals circumference.
- Or you can use the equation: Pi times diameter equals circumference.
- Both equations say the same thing: **3 and 14-hundredths times diameter equals circumference.**
- You'll use the equation that **starts with pi.**

e. Everybody, say that equation. (Signal.)
   Pi times diameter equals circumference.
- And what number does pi equal? (Signal.)
   3 and 14-hundredths.
   (Repeat step e until firm.)
f. Find part 4.
g. You'll work problem A. Copy the equation that's in the box. Below, put in the values you know for pi and the diameter. Figure out the circumference. You can use your calculator.
- When you figure out the circumference, write an equation that says, circumference equals so many inches. Raise your hand when you're finished. (Observe students and give feedback.)
- (Write on the board:)

> **a.** π x d = C
> 3.14 x 5 = C
> C = 15.7 in

- Here's what you should have. You multiplied: 3 and 14-hundredths times 5. Everybody, what's the circumference? (Signal.)
   15 and 7-tenths inches.
h. Problem B. Remember, use the equation that starts with the symbol pi. Put in the values you know for pi and the diameter. Figure out the circumference. Raise your hand when you've written an equation that shows how many meters C equals.
   (Observe students and give feedback.)
- (Write on the board:)

> **b.** π x d = C
> 3.14 x 13 = C
> C = 40.82 m

- Here's what you should have. You multiplied: 3 and 14-hundredths times 13. Everybody, what's the circumference? (Signal.)
   40 and 82-hundredths meters.

i. Your turn: Work the rest of the problems in part 4. Remember, start with the equation that has the symbol for pi. Raise your hand when you're finished. (Observe students and give feedback.)
   Key:

   c. π x d = C
      3.14 x 65 = C
      C = 204.1 ft

   d. π x d = C
      3.14 x 25 = C
      C = 78.5 in

j. Check your work.
- Problem C. What's the diameter? (Signal.)
   65 feet.
- What's the circumference? (Signal.)
   204 and 1-tenth feet.
- Problem D. What's the diameter? (Signal.)
   25 inches.
- What's the circumference? (Signal.)
   78 and 5-tenths inches.
k. Raise your hand if you got everything right.

EXERCISE 3 FRACTIONS
Simplification: Discrimination

a. Find part 5.
- Some of these fractions can be simplified. Some can't. Remember, if you can't cross out any fractions that equal 1, the fraction can't be simplified.
b. Your turn: Copy each fraction. Write the prime factors and cross out any fractions that equal 1. Then write the value you get when you multiply. Raise your hand when you're finished.
   (Observe students and give feedback.)
   Key:

   a. $\frac{40}{50} = \frac{2 \times 2 \times 2 \times 5}{2 \times 5 \times 5} = \frac{4}{5}$

   b. $\frac{6}{24} = \frac{2 \times 3}{2 \times 2 \times 2 \times 3} = \frac{1}{4}$

   c. $\frac{6}{25} = \frac{2 \times 3}{5 \times 5} = \frac{6}{25}$

   d. $\frac{1}{14} = \frac{1}{2 \times 7} = \frac{1}{14}$

c. Check your work.
- Fraction A: 40/50. Can it be simplified? (Signal.) *Yes.*
- You should have crossed out the fractions 2/2 and 5/5. The simplified fraction is 4/5.
- Fraction B: 6/24. Can it be simplified? (Signal.) *Yes.*
- You should have crossed out the fractions 2/2 and 3/3. The simplified value is 1/4.
- Fraction C. Can it be simplified? (Signal.) *No.*
- The factors for the numerator are 2 and 3. The factors for the denominator are 5 and 5. When you multiply, you get the fraction you started with: 6/25.
- Fraction D: 1/14. Can it be simplified? (Signal.) *No.*

- The factor for the numerator is 1. The factors for the denominator are 2 and 7. When you multiply, you get 1/14.
d. Raise your hand if you got everything right.

EXERCISE 4 FRACTION OPERATIONS
Unlike Denominators

a. Find part 6.
- (Teacher reference:)

$$\frac{7}{12} \qquad \frac{7}{12}\left(\frac{5}{5}\right) = \frac{35}{60}$$
$$+\frac{6}{15} \qquad +\frac{6}{15}\left(\frac{4}{4}\right) = +\frac{24}{60}$$
$$\blacksquare \qquad\qquad\qquad\qquad \frac{59}{60}$$

- The problem shown is 7/12 plus 6/15.
- I'll read what it says. Follow along: This problem can't be worked the way it is written because the denominators are not the same.
- To work the problem, **you find the lowest common denominator.** That's the first common number you reach when you count by the denominators.
- To find the lowest common denominator, start with the smaller denominator. That's **12.** Write the numbers for counting by 12.
- Then do the same thing for counting by 15. The first number that shows up for counting by 12 and counting by 15 is the lowest common denominator.
- You can use a calculator to find the first common number.
- Start with 12 plus plus. Then press equals. Each time you press equals, the calculator shows the next number you reach for counting by 12.
- You can see the numbers: 12, 24, 36, 48, 60, 72.
- Now you do the same thing with 15. Enter 15 plus plus. You stop as soon as you find a number that you reached counting by 12.
- You can see the numbers for 15: 15, 30, 45, 60.
- The first common number is 60. That's the common denominator. You write it for both fractions. Then you work the equivalent-fraction problems to figure out the numerators.
- 7/12 equals 35/60. 6/15 equals 24/60.
- Then you add and write the answer: 35/60 plus 24/60 equals 59/60.

- Remember the steps: Start with the smaller denominator. Write the numbers for counting by that denominator. Then do the same thing for the larger denominator. Stop when you find the first number common to both denominators. Write the common denominator for both fractions. Work the equivalent-fraction problems. Then write the answer to the problem.
b. Find part 7.
c. Problem A is 5/18 minus 1/14.
- The smallest denominator is 14. Use your calculator. Write down the first 10 numbers for counting by 14. Remember, enter 14 plus plus. Then press equals again and again.
- Write the numbers. Raise your hand when you're finished. √
- (Write on the board:)

14	28	42	56	70
84	98	112	126	140

- Here's what you should have.
- Now enter 18 plus plus, then press equals. Check each number you get. Stop when you get a number that you got counting by 14 and circle that number. Raise your hand when you've circled that number.
  (Observe students and give feedback.)
- Everybody, what's the first common number for 18 and 14? (Signal.) *126.*
- (Write on the board:)

$$\text{a.} \quad \frac{5}{18} \qquad = \frac{}{126}$$
$$\quad -\frac{1}{14} \qquad = -\frac{}{126}$$

- Here are the equivalent-fraction problems you work.
- Find the numerators. You can use your calculator. Raise your hand when you've done that much.
  (Observe students and give feedback.)
- (Write to show:)

$$\text{a.} \quad \frac{5}{18}\left(\frac{7}{7}\right) = \frac{35}{126}$$
$$\quad -\frac{1}{14}\left(\frac{9}{9}\right) = -\frac{9}{126}$$

- Here's what you should have. The fraction that equals 5/18 is 35/126. The fraction that equals 1/14 is 9/126.
- The problem you work now is 35/126 minus 9/126. Write the answer and box it. That's the answer to this problem. Raise your hand when you're finished.
  (Observe students and give feedback.)
- (Write to show:)

$$a. \quad \frac{5}{18}\left(\frac{7}{7}\right) = \frac{35}{126}$$

$$-\frac{1}{14}\left(\frac{9}{9}\right) = -\frac{9}{126}$$

$$\boxed{\frac{26}{126}}$$

- Here's what you should have: 26/126.
d. Copy problem B.
- The numbers for counting by 12 and by 15 are shown in part 6. Raise your hand when you know the common denominator. √
- Everybody, what's the common denominator? (Signal.) *60.*
- Write the two equivalent-fraction problems for the fractions and figure out the numerators. Raise your hand when you've written the fractions that equal 5/12 and 6/15.
  (Observe students and give feedback.)
- (Write on the board:)

$$b. \quad \frac{5}{12}\left(\frac{5}{5}\right) = \frac{25}{60}$$

$$+\frac{6}{15}\left(\frac{4}{4}\right) = +\frac{24}{60}$$

- Here's what you should have so far.
- Now write the answer and box it. Raise your hand when you're finished.
  (Observe students and give feedback.)
- Everybody, what does 25/60 plus 24/60 equal? (Signal.) *49-sixtieths.*

EXERCISE 5 COMPLEX FRACTIONS
Equivalence

a. (Write on the board:)

$$\frac{2}{3}\left(\cfrac{\frac{4}{5}}{\frac{4}{5}}\right) =$$

- Here's 2/3 multiplied by a fraction that equals 1. So you'll end up with a value that equals 2/3.
- Everybody, what's the fraction that equals 1 in this problem? (Signal.) *4-fifths over 4-fifths.*
- When you multiply, you'll get a fraction on the top and a fraction on the bottom.
b. On top we have 2 times 4/5. That's **8/5.**
- (Write to show:)

$$\frac{2}{3}\left(\cfrac{\frac{4}{5}}{\frac{4}{5}}\right) = \boxed{\frac{8}{5}}$$

- On the bottom, we have 3 times 4/5. That's **12/5.**
- (Write to show:)

$$\frac{2}{3}\left(\cfrac{\frac{4}{5}}{\frac{4}{5}}\right) = \cfrac{\frac{8}{5}}{\boxed{\frac{12}{5}}}$$

- All done. And remember, 2/3 equals 8/5 over 12/5, because we multiplied 2/3 by **1.**
c. Find part 8.
d. Problem A: 7/10 times 5/3 over 5/3. Copy the problem. Multiply and write the answer. Raise your hand when you're finished.
  (Observe students and give feedback.)
- (Write on the board:)

$$a. \quad \frac{7}{10}\left(\cfrac{\frac{5}{3}}{\frac{5}{3}}\right) = \cfrac{\frac{35}{3}}{\frac{50}{3}}$$

- Here's what you should have for problem A. We started with 7/10. So 35/3 over 50/3 equals 7/10.
e. Your turn: Work problem B. Raise your hand when you're finished.
  (Observe students and give feedback.)

- (Write on the board:)

$$b. \quad \frac{2}{5} \left(\frac{\frac{1}{9}}{\frac{1}{9}} \right) = \frac{\frac{2}{9}}{\frac{5}{9}}$$

- Here's what you should have. 2/5 times 1/9 over 1/9. Everybody, what's the answer? (Signal.) *2-ninths over 5-ninths*.
- What simple fraction does 2/9 over 5/9 equal? (Signal.) *2/5*.
- Raise your hand if you got everything right.

EXERCISE 6 RATIOS AND PROPORTIONS
Tables

a. Find part 9.
- These are problems that you can work if you make a fraction number family, then make a ratio table.
b. Problem A: The tank holds 7/5 as much as the barrel holds. The barrel holds 65 gallons. How much does the tank hold? How much more does the tank hold than the barrel?
- The first sentence tells about how to make a fraction number family. Make the family. You can write **T** for tank and **B** for barrel.   Raise your hand when you've done that much. (Observe students and give feedback.)
- (Write on the board:)

$$a. \quad \underset{\frac{2}{5}}{\overset{\textbf{dif}}{}} \quad \underset{\frac{5}{5}}{\overset{\textbf{b}}{}} \longrightarrow \underset{\frac{7}{5}}{\overset{\textbf{t}}{}}$$

- Here's what you should have. Use the ratio numbers and make a table. Figure out the missing numbers. Answer the questions with numbers and unit names. Raise your hand when you're finished. (Observe students and give feedback.)
- (Write on the board:)

dif	2	26	**1. 91 gal**
b	5	65	
t	7	91	**2. 26 gal**

- Here's what you should have. The problem tells you that the barrel holds 65 gallons. So the tank holds 91 gallons. The difference is 26 gallons.

That's how much more the tank holds than the barrel.

c. Your turn: Work problem B. You can use initials for the names. Remember to answer each question with a number and a unit name. Raise your hand when you're finished. (Observe students and give feedback.)
- (Write on the board:)

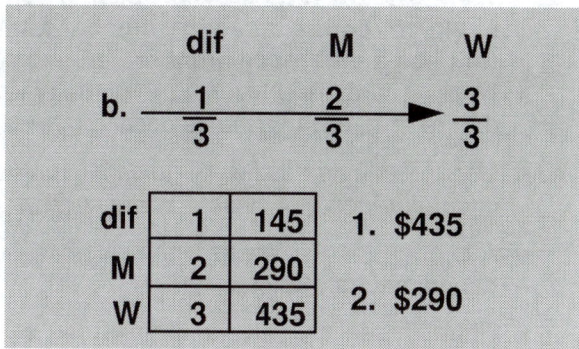

$$b. \quad \underset{\frac{1}{3}}{\overset{\textbf{dif}}{}} \quad \underset{\frac{2}{3}}{\overset{\textbf{M}}{}} \longrightarrow \underset{\frac{3}{3}}{\overset{\textbf{W}}{}}$$

dif	1	145	**1. $435**
M	2	290	
W	3	435	**2. $290**

- Check your work. Here's what you should have for problem B. The ratio numbers are 1, 2 and 3. Walter earned $145 more than Michael. So Michael earned $290. Walter earned $435.

EXERCISE 7 INDEPENDENT WORK

- (In addition to independent work in the textbook, assign *Bridge to Connecting Math Concepts Independent Worksheet* 35 as classwork or homework. Before beginning the next lesson, check the students' independent work.)

Lesson 56

EXERCISE 1 COMPLEX FRACTIONS
Equivalence

a. Open your textbook to lesson 56 and find part 1.
- In all these problems, the complicated fraction equals 1 because the top value is the same as the bottom value. So the answer you'll get when you multiply equals the value you start out with. The fractions will be equivalent.
b. Your turn: Copy each problem. Then multiply on the top and multiply on the bottom. Raise your hand when you've worked all the problems and shown the complete equations.
(Observe students and give feedback.)
Key:

a. $\dfrac{3}{7}\left(\dfrac{\frac{4}{5}}{\frac{4}{5}}\right) = \dfrac{\frac{12}{5}}{\frac{28}{5}}$ b. $\dfrac{9}{4}\left(\dfrac{\frac{1}{6}}{\frac{1}{6}}\right) = \dfrac{\frac{9}{6}}{\frac{4}{6}}$

c. $\dfrac{1}{8}\left(\dfrac{\frac{10}{7}}{\frac{10}{7}}\right) = \dfrac{\frac{10}{7}}{\frac{80}{7}}$ d. $\dfrac{5}{3}\left(\dfrac{\frac{2}{5}}{\frac{2}{5}}\right) = \dfrac{\frac{10}{5}}{\frac{6}{5}}$

c. Check your work.
- Find part J on page 224 of your textbook. That shows what you should have.
d. Problem A: 3/7 times 4/5 over 4/5. Everybody, what's the answer? (Signal.)
12-fifths over 28-fifths.
- What simple fraction does 12/5 over 28/5 equal? (Signal.) *3-sevenths.*

- Problem B: 9/4 times 1/6 over 1/6. Everybody, what's the answer? (Signal.)
9-sixths over 4-sixths.
- What simple fraction does 9/6 over 4/6 equal? (Signal.) *9-fourths.*
- Problem C: 1/8 times 10/7 over 10/7. Everybody, what's the answer? (Signal.)
10-sevenths over 80-sevenths.
- What simple fraction does 10/7 over 80/7 equal? (Signal.) *1-eighth.*
- Problem D: 5/3 times 2/5 over 2/5. Everybody, what's the answer? (Signal.)
10-fifths over 6-fifths.
- What simple fraction does 10/5 over 6/5 equal? (Signal.) *5-thirds.*
e. Raise your hand if you got everything right.

EXERCISE 2 RATIOS AND PROPORTIONS
Tables

a. Find part 2.
- For all these problems, you have to make a fraction number family, then a ratio table. Some problems have a difference number. Remember, if a sentence compares, there's a fraction for the difference. The thing you compare to is 1 whole.
b. Work problem A. You can write initials for the names in your number family. Raise your hand when you've finished the problem and answered the questions.
(Observe students and give feedback.)
- (Write on the board:)

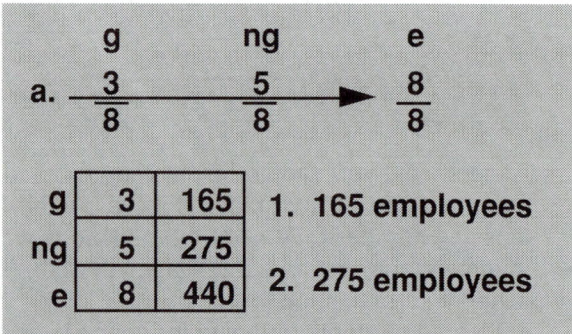

- Check your work. Here's what you should have.
- All the employees are 8/8. 3/8 of the workers wear glasses. 5/8 do not wear glasses. The ratio table shows the ratio numbers and the numbers for employees. There's a total of 440 employees. So 165 wear glasses and 275 do not wear glasses. Raise your hand if you got everything right.
c. Work problem B. You can use initials for the names. Raise your hand when you're finished.
(Observe students and give feedback.)

- (Write on the board:)

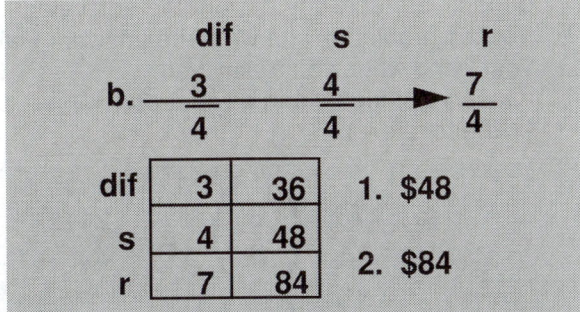

	dif	s	r
b.	$\frac{3}{4}$	$\frac{4}{4}$ →	$\frac{7}{4}$

dif	3	36	1. $48
s	4	48	
r	7	84	2. $84

- Check your work. Here's what you should have. The regular price is the big number. The sale price is 4/4. The difference is 3/4. The problem tells you that you save $36 if you buy the glasses on sale. That's the difference number. So the sale price is $48 and the regular price is $84. Raise your hand if you got everything right.
- You're working some difficult problems.
d. Your turn: Work the rest of the problems in part 2. Remember, if a problem compares, you need the name **difference.** Raise your hand when you're finished.
  (Observe students and give feedback.)
e. (Write on the board:)

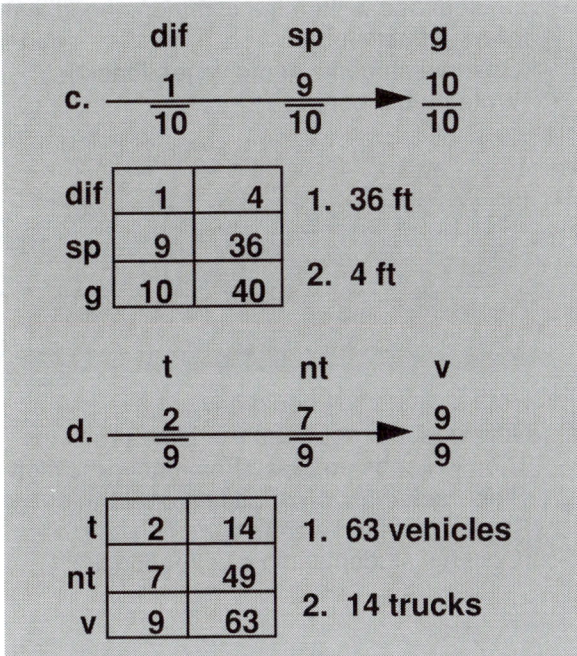

	dif	sp	g
c.	$\frac{1}{10}$	$\frac{9}{10}$ →	$\frac{10}{10}$

dif	1	4	1. 36 ft
sp	9	36	
g	10	40	2. 4 ft

	t	nt	v
d.	$\frac{2}{9}$	$\frac{7}{9}$ →	$\frac{9}{9}$

t	2	14	1. 63 vehicles
nt	7	49	
v	9	63	2. 14 trucks

f. Check your work. Here's what you should have for problems C and D.
- Problem C. The problem compares the swimming pool with the garden. The garden is the big number. The difference is 1/10. The problem tells you that the garden is 40 feet wide. So the swimming pool is 36 feet wide. The difference is 4 feet.

- Problem D. The problem does not compare. The names are **trucks, not trucks** and **vehicles.** The fractions are 2/9, 7/9 and 9/9. The problem tells you that 49 vehicles are **not** trucks. So, the total number of vehicles is 63 and 14 are trucks.

EXERCISE 3 CIRCLES
Circumference

a. (Write on the board:)

$$\pi$$

- Last time you learned a new symbol. Everybody, what's the name for this symbol? (Signal.) *Pi.*
- What number does pi equal? (Signal.) *3 and 14-hundredths.*
- Write the equation that starts with pi and tells how to find the circumference of any circle. Raise your hand when you're finished. √
- (Write on the board:)

$$\pi \times d = C$$

- Here's the equation: Pi times diameter equals circumference.
b. Find part 3.
- Each problem shows the diameter of a circle. You'll figure out the circumference. You'll use the equation: Pi times diameter equals circumference.
c. Work problem A. Remember to show the number and unit name for the circumference. Raise your hand when you're finished.
  (Observe students and give feedback.)
- (Write on the board:)

$$\begin{aligned} \textbf{a.} \quad & \pi \times d = C \\ & 3.14 \times 16 = C \\ & \boxed{C = 50.24 \text{ in}} \end{aligned}$$

- Here's what you should have. The diameter is 16 inches. The circumference is 50 and 24-hundredths inches.
d. Find the circumference for circle B. Raise your hand when you're finished.
  (Observe students and give feedback.)

- (Write on the board:)

$$b. \quad \pi \times d = C$$
$$3.14 \times 20 = C$$
$$\boxed{C = 62.8 \text{ m}}$$

- Here's what you should have. The diameter is 20 meters. The circumference is 62 and 8-tenths meters.
- e. Problem C: The diameter is 6 and 5-tenths feet. When you enter that number in your calculator, don't forget the decimal point.
- Work the problem. Raise your hand when you're finished.
  (Observe students and give feedback.)
- (Write on the board:)

$$c. \quad \pi \times d = C$$
$$3.14 \times 6.5 = C$$
$$\boxed{C = 20.41 \text{ ft}}$$

- Here's what you should have. The diameter is 6 and 5-tenths feet. The circumference is 20 and 41-hundredths feet.

EXERCISE 4 FRACTION OPERATIONS
Unlike Denominators

a. Find part 4.
- These are common-denominator problems. Remember the steps for working these problems. First, you write the numbers for counting by the smaller denominator. Then you find out which of those numbers is the first number you reach when you count by the larger denominator.
b. Problem A. The denominators are 12 and 9.
- Write down the first ten numbers for counting by 9. You don't need your calculator for that. Then use your calculator for counting by 12. Raise your hand when you've found the common denominator.
- Everybody, what's the common denominator? (Signal.) *36.*
- (Write on the board:)

$$a. \quad \frac{7}{12} = \frac{}{36}$$
$$- \frac{4}{9} = - \frac{}{36}$$

- Here's the problem with the common denominators. Copy it, work the equivalent-fraction problems, and write the answer. Raise your hand when you're finished.
  (Observe students and give feedback.)
- (Write to show:)

$$a. \quad \frac{7}{12} \left(\frac{3}{3} \right) = \frac{21}{36}$$
$$- \frac{4}{9} \left(\frac{4}{4} \right) = - \frac{16}{36}$$
$$\boxed{\frac{5}{36}}$$

- Here's what you should have: 7/12 minus 4/9 equals 5/36. Raise your hand if you got everything right.
c. Problem B. The denominators are 20 and 8.
- Write down the first ten numbers for counting by 8. Use your calculator to find the first number you reach when you count by 20.
- Raise your hand when you've copied the problem and written the common denominator for both fractions.
  (Observe students and give feedback.)
- (Write on the board:)

$$b. \quad \frac{1}{20} = \frac{}{40}$$
$$+ \frac{7}{8} = + \frac{}{40}$$

- Here's what you should have.
- Now work the equivalent-fraction problems. Then write the answer. Remember the sign that shows whether you're adding or subtracting. Raise your hand when you're finished.
  (Observe students and give feedback.)

- (Write to show:)

$$\text{b.} \quad \frac{1}{20} \left(\frac{2}{2}\right) = \frac{2}{40}$$

$$+ \frac{7}{8} \left(\frac{5}{5}\right) = + \frac{35}{40}$$

$$\frac{37}{40}$$

- Here's what you should have: 1/20 plus 7/8 equals 37/40. Raise your hand if you got everything right.

EXERCISE 5 DECIMALS
Rounding

a. Find part 5.
- You'll round each decimal value to the closest hundredth. Remember, you're rounding to hundredths, so the number you'll write will have **two** digits after the decimal point.
b. Numeral A. What's the hundredths digit? (Signal.) 7.
- What's the next digit? (Signal.) 6.
- Is that digit 5 or more? (Signal.) Yes.
- So do you round 7 to 8? (Signal.) Yes.
- Numeral A rounds to 5 and how many hundredths? (Signal.) 18.
c. Numeral B. What's the hundredths digit? (Signal.) 9.
- What's the next digit? (Signal.) 7.
- Is that digit 5 or more? (Signal.) Yes.
- So do you round up? (Signal.) Yes.
- This is tricky. You round 39-hundredths up. That's 40-hundredths. Numeral B rounds to 5 and 40-hundredths.
d. Numeral C. What's the hundredths digit? (Signal.) 7.
- What's the next digit? (Signal.) 4.
- Is that digit 5 or more? (Signal.) No.
- So numeral C rounds to how many hundredths? (Signal.) 27.
e. Your turn: Write each numeral rounded to the nearest hundredth. Raise your hand when you're finished. (Observe students and give feedback.)
  Key:
  a. 5.18   b. 5.40   c. .27
  d. 10.50   e. 5.15   f. 2.49

f. Check your work. Read each rounded numeral.
- Rounded numeral A. (Signal.) 5 and 18-hundredths.
- Rounded numeral B. (Signal.) 5 and 40-hundredths.
- Rounded numeral C. (Signal.) 27-hundredths.
- Rounded numeral D. (Signal.) 10 and 50-hundredths.
- Rounded numeral E. (Signal.) 5 and 15-hundredths.
- Rounded numeral F. (Signal.) 2 and 49-hundredths.
g. Raise your hand if you got everything right.

EXERCISE 6 INDEPENDENT WORK

- (In addition to independent work in the textbook, assign **Bridge to Connecting Math Concepts** *Independent Worksheet* 36 as classwork or homework. Before beginning the next lesson, check the students' independent work.)

Lesson 57

EXERCISE 1 RATIOS AND PROPORTIONS
Tables

a. Open your textbook to lesson 57 and find part 1.
- For all these problems, you have to make a fraction number family, then a ratio table. Some problems have a difference number. Remember, if a sentence compares, there's a fraction for the difference.
b. Work problem A. Raise your hand when you're finished. (Observe students and give feedback.)
- (Write on the board:)

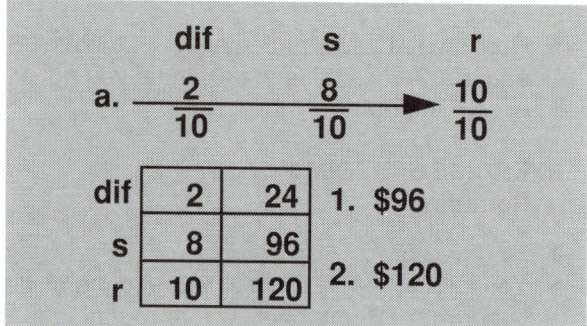

- Check your work. Here's what you should have. The regular price is the big number. The regular price is 10/10. The difference is 2/10. The problem tells you that you save $24 if you buy the jacket on sale. That's the difference number. So the sale price is $96 and the regular price is $120. Raise your hand if you got everything right.
c. Work problem B. Raise your hand when you're finished. (Observe students and give feedback.)
- (Write on the board:)

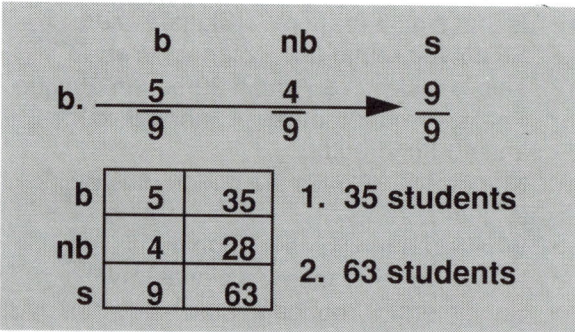

- Check your work. Here's what you should have.
- The problem does not compare. The names are **blue, not blue** and **students.** The problem tells you that 28 students are not wearing blue. So 35 students **are** wearing blue, and there is a total of 63 students. Raise your hand if you got everything right.
d. Work problem C. Raise your hand when you're finished. (Observe students and give feedback.)

- (Write on the board:)

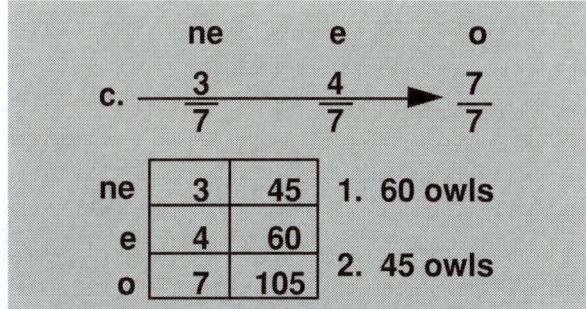

ne   e   o

c. $\dfrac{3}{7}$ $\dfrac{4}{7}$ ⟶ $\dfrac{7}{7}$

ne	3	45
e	4	60
o	7	105

1. 60 owls

2. 45 owls

- Check your work. Here's what you should have.
- The sentence does not compare. The names are **no eggs, eggs** and **owls.** The fractions are 3/7, 4/7 and 7/7. The problem tells you that there is a total of 105 owls. So 60 owls have produced eggs. 45 owls have not produced eggs. Raise your hand if you got everything right.

EXERCISE 2 INVERSE OPERATIONS
Single Equation

a. Find part 2.
- I'll read what it says. Follow along: If problems have the first number missing, you can figure out that number by working backwards and **undoing** the operation shown in the problem. You can see the rules for undoing.
- To undo **addition,** you **subtract** the same number.
- To undo **subtraction,** you **add** the same number.
- To undo **multiplication,** you **divide** by the same number.
- To undo **division,** you **multiply** by the same number.
b. Let's review. How would you undo subtracting 11? (Signal.) *Add 11.*
- How would you undo multiplying by 20? (Signal.) *Divide by 20.*
- How would you undo adding 56? (Signal.) *Subtract 56.*
- (Repeat step b until firm.)
c. Find part 3.
- We'll do the sample problems together.
- (Write on the board:)

1. ☐ + 11 = 34

2. ☐ x 4 = 40

- First you identify whether the operation is addition, subtraction, multiplication or division.

- (Point to: ☐ + 11 = 34.)
- What's the operation for this problem? (Signal.) *Addition.*
- The problem adds 11. How do you undo adding 11? (Signal.) *Subtract 11.*
- To undo, you start at the end of the problem, with 34. Everybody, what's 34 minus 11? (Signal.) *23.*
- That's the starting number.
- (Write to show:)

1. 23 + 11 = 34

2. ☐ x 4 = 40

- You can check your answer by seeing if the equation is true. 23 plus 11. Does it equal 34? (Signal.) *Yes.*
- So 23 is the right number.
d. Sample problem 2. First, you name the operation. What's the operation? (Signal.) *Multiplication.*
- What number does the problem multiply by? (Signal.) *4.*
- How do you undo multiplying by 4? (Signal.) *Divide by 4.*
- Start with the last number and say the division problem. (Signal.) *40 divided by 4.*
- What's the answer? (Signal.) *10.*
- (Write to show:)

1. 23 + 11 = 34

2. 10 x 4 = 40

- Remember, you start with the last number and work the problem for undoing the operation that is shown.
e. Problem A: Some number times 12 equals 96. What's the operation? (Signal.) *Multiplication.*
- What are you multiplying by? (Signal.) *12.*
- How do you undo multiplying by 12? (Signal.) *Divide by 12.*
- Say the problem you'll work. (Signal.) *96 divided by 12.*
- Figure out the answer and write the complete equation for problem A. Raise your hand when you're finished.
  (Observe students and give feedback.)
- (Write on the board:)

a. 8 x 12 = 96

- Here's what you should have. 96 divided by 12 is 8. The starting number is 8.
f. Problem B: Some number plus 36 equals 96.
- What's the operation? (Signal.) *Addition.*
- How much are you adding? (Signal.) *36.*
- How do you undo adding 36? (Signal.) *Subtract 36.*
- Say the problem you'll work. (Signal.) *96 minus 36.*
- Work it and write the complete equation. Raise your hand when you're finished. (Observe students and give feedback.)
- (Write on the board:)

> **b.** $\boxed{60} + 36 = 96$

- Here's the equation you should have: 60 plus 36 equals 96. The starting number is 60.
g. Work problem C. Raise your hand when you're finished. (Observe students and give feedback.)
- (Write on the board:)

> **c.** $\boxed{3} \times 46 = 138$

- Here's what you should have. 3 times 46 equals 138. You undid multiplying by 46. 138 divided by 46 is 3. 3 is the starting number.
h. Work problem D. Raise your hand when you're finished. (Observe students and give feedback.)
- (Write on the board:)

> **d.** $\boxed{338} - 138 = 200$

- Here's what you should have. 338 minus 138 equals 200. You undid subtracting 138. You worked the problem 200 plus 138. That equals 338. That's the starting number.

EXERCISE 3 FRACTION OPERATIONS
Unlike Denominators

a. Find part 4.
- These are common-denominator problems. The problems are written in a row. You'll write them as column problems. Then you'll find the common denominator and write the answer.
b. The first problem has denominators of 16 and 40.
- Use your calculator. Write the first ten numbers for 16. Find the first number that you reach counting by 40. Then work the equivalent-fraction problems and find the answer. Don't forget the minus sign. Raise your hand when you've completed the entire problem. (Observe students and give feedback.)

- (Write on the board:)

> **a.** $\dfrac{7}{16}\left(\dfrac{5}{5}\right) = \dfrac{35}{80}$
>
> $-\dfrac{3}{40}\left(\dfrac{2}{2}\right) = -\dfrac{6}{80}$
>
> $\boxed{\dfrac{29}{80}}$

- Here's what you should have. The common denominator is 80. 7/16 minus 3/40 equals 29/80. Make sure you have the minus sign and the work that is shown on the board.
c. Work problem B. Remember to show the complete work for the problem. Raise your hand when you're finished. (Observe students and give feedback.)
- (Write on the board:)

> **b.** $\dfrac{3}{14}\left(\dfrac{4}{4}\right) = \dfrac{12}{56}$
>
> $+\dfrac{7}{8}\left(\dfrac{7}{7}\right) = +\dfrac{49}{56}$
>
> $\boxed{\dfrac{61}{56}}$

- Here's what you should have. The common denominator is 56. 3/14 plus 7/8 equals 61/56. That's 1 and 5/56. Make sure you have a plus sign and the other work shown on the board.

EXERCISE 4 DECIMALS AND PERCENTS
Conversion

a. Find part 5.
- I'll read what it says. Follow along: You're going to work problems that refer to percent. Percents are related to hundredths decimal numbers. To change a hundredth decimal number into an equivalent percent, remove the decimal point and write the percent sign.
- You can see 56-hundredths.
- You can see the equivalent percent: 56 percent.
- You can see 1 and 78-hundredths.
- That's 178 percent.
- You can see 5 and 3-hundredths.
- That's 503 percent.

- You can see 4-hundredths.
- That's 4 percent.
- To go from percents to hundredths, you make sure you have two digits after the decimal point.
- 5 percent is 5-hundredths.
- 75 percent is 75-hundredths.
- 671 percent is 6 and 71-hundredths.
b. Find part 6.
- For each percent value, you'll write the equivalent hundredths decimal.
c. Item A: 37 percent. Write the decimal number. Raise your hand when you're finished.
- (Write on the board:)

> **a.** .37

- Here's the decimal number for 37 percent.
d. Your turn: Write decimal numbers for the rest of the percents in part 6. Remember, you need two places after the decimal point. Raise your hand when you're finished.
  (Observe students and give feedback.)
e. (Write on the board:)

> **b.** .08
> **c.** 1.56
> **d.** 2.09
> **e.** .99

- Check your work. Here's what you should have for each item. Raise your hand if you got everything right.
f. Find part 7.
- For each decimal number, you'll write the percent number and the percent sign.
g. Item A: 1 and 56-hundredths. That's 156 percent. Write the percent number. Raise your hand when you're finished. √
- (Write on the board:)

> **a.** 156%

- Here's what you should have.
h. Your turn: Write the percent numbers for the rest of the items in part 7. Raise your hand when you're finished.
  (Observe students and give feedback.)
i. (Write on the board:)

> **b.** 75%
> **c.** 1%
> **d.** 40%
> **e.** 306%

- Check your work. Here's what you should have.
- Item B: 75-hundredths. That's 75 percent.
- Item C: 1-hundredth. That's 1 percent.
- Item D: 40-hundredths. That's 40 percent.
- Item E: 3 and 6-hundredths. That's 306 percent.
j. Raise your hand if you got everything right.

EXERCISE 5 CIRCLES
Circumference

a. Find part 8.
- You'll figure out the circumference of each circle. You'll start with an equation.
b. Everybody, say that equation. (Signal.) *Pi times diameter equals circumference.*
- What's the number for pi? (Signal.) *3 and 14-hundredths.*
- (Repeat step b until firm.)
c. Find the circumference for circle A. Remember the unit name in the answer. Raise your hand when you're finished.
  (Observe students and give feedback.)
- (Write on the board:)

> **a.** $\pi \times d = C$
> $3.14 \times 7.8 = C$
> $\boxed{C = 24.492 \text{ in}}$

- Here's what you should have. Raise your hand if you got everything right.
d. Your turn. Work the rest of the problems in part 8. Raise your hand when you're finished.
  (Observe students and give feedback.)
e. (Write on the board:)

> **b.** $\pi \times d = C$
> $3.14 \times 50 = C$
> $\boxed{C = 157 \text{ yd}}$
>
> **c.** $\pi \times d = C$
> $3.14 \times 3.6 = C$
> $\boxed{C = 11.304 \text{ ft}}$

- Here's what you should have for circles B and C. Raise your hand if you got everything right.

EXERCISE 6 COMPLEX FRACTIONS
Equivalence

a. Find part 9.
- These are very difficult equivalent-fraction problems, but you can work them by writing the fraction that equals 1 as a fraction over the same fraction. The answer to each problem is the fraction that goes in the box.
b. Touch problem A. √
- Can you work the problem on the top? (Signal.) *No.*
- Can you work the problem on the bottom? (Signal.) *Yes.*
- I'll say the problem for the bottom numbers: 4 times **some fraction** equals 3. Figure out that fraction.
- Everybody, what's the fraction? (Signal.) *3-fourths.*
- So the fraction that equals 1 is 3/4 over 3/4. Copy the problem. Write that fraction. √
- (Write on the board:)

$$\text{a.} \quad \frac{8}{4} \left(\frac{\frac{3}{4}}{\frac{3}{4}} \right) = \frac{\square}{3}$$

- Here's what you should have so far.
- Now you can multiply and write the answer on top. Raise your hand when you're finished. (Observe students and give feedback.)
- (Write to show:)

$$\text{a.} \quad \frac{8}{4} \left(\frac{\frac{3}{4}}{\frac{3}{4}} \right) = \frac{\boxed{\frac{24}{4}}}{3}$$

- 8 times 3/4 equals 24/4. That's the answer. 24/4 equals **6.**
c. Problem B. Say the problem on the top. (Signal.) *9 times some fraction equals 3.*
- Your turn: Figure out the fraction.
- Everybody, what's the fraction? (Signal.) *3-ninths.*
- Copy the problem and write the fraction that equals 1. √
- (Write on the board:)

$$\text{b.} \quad \frac{9}{12} \left(\frac{\frac{3}{9}}{\frac{3}{9}} \right) = \frac{3}{\square}$$

- Here's what you should have so far.
- Multiply on the bottom and write the answer. Raise your hand when you're finished. (Observe students and give feedback.)
- (Write to show:)

$$\text{b.} \quad \frac{9}{12} \left(\frac{\frac{3}{9}}{\frac{3}{9}} \right) = \frac{3}{\boxed{\frac{36}{9}}}$$

- Here's what you should have . The answer is 36/9. That fraction equals **4.** Raise your hand if you got it right.
d. Your turn: Work problem C. Raise your hand when you're finished. (Observe students and give feedback.)
- (Write on the board:)

$$\text{c.} \quad \frac{2}{6} \left(\frac{\frac{9}{6}}{\frac{9}{6}} \right) = \frac{\boxed{\frac{18}{6}}}{9}$$

- The problem on the bottom is: 6 times some fraction equals 9. The fraction is 9/6. So the fraction that equals 1 is 9/6 over 9/6. You multiply 2 times 9/6 on top and you get 18/6. What whole number does that equal? (Signal.) *3.*
- Yes, 18/6 equals **3.**
e. Your turn: Work problem D. Raise your hand when you're finished. (Observe students and give feedback.)
- (Write on the board:)

$$\text{d.} \quad \frac{3}{12} \left(\frac{\frac{1}{3}}{\frac{1}{3}} \right) = \frac{1}{\boxed{\frac{12}{3}}}$$

- The problem on top is: 3 times some fraction equals 1. The fraction that equals 1 is 1/3 over 1/3. You multiply 12 times 1/3 on the bottom and you get 12/3. What whole number does that equal? (Signal.) *4.*

EXERCISE 7 INDEPENDENT WORK

- (In addition to independent work in the textbook, assign *Bridge to Connecting Math Concepts Independent Worksheet* 37 as classwork or homework. Before beginning the next lesson, check the students' independent work.)

Lesson 58

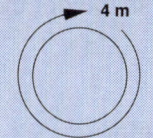 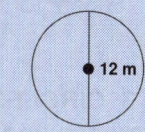
EXERCISE 1 COMPLEX FRACTIONS
Equivalence

a. Open your textbook to lesson 58 and find part 1.
- These are equivalent-fraction problems. Remember, you can work the problem on top if there are two numbers on top.

b. Touch problem A. √
- Can you start working the problem on top or on the bottom? (Signal.) *On the bottom.*
- Say the problem for the bottom numbers. (Signal.) *10 times some fraction equals 7.*
- Touch problem B. Can you start working the problem on top or on the bottom? (Signal.) *On top.*
- Say the problem for the top numbers. (Signal.) *11 times some fraction equals 2.*

- Touch problem C. Can you start working the problem on top or on the bottom? (Signal.) *On top.*
- Say the problem for the top numbers. (Signal.) *4 times some fraction equals 9.*

c. Go back to problem A and work it. Write the missing fraction that equals 1. Then multiply on top. Raise your hand when you're finished. (Observe students and give feedback.)
- (Write on the board:)

$$\text{a.} \quad \frac{3}{10} \left(\frac{\frac{7}{10}}{\frac{7}{10}} \right) = \frac{\boxed{\frac{21}{10}}}{7}$$

- Here's what you should have for problem A. The fraction that equals 1 is 7/10 over 7/10. When you multiplied on top, you got 21/10. That's the answer. 21/10 equals 2 and 1/10.

d. Your turn: Work problem B. Write the missing fraction of 1. Then multiply on the bottom to find the answer. Raise your hand when you're finished. (Observe students and give feedback.)
- (Write on the board:)

$$\text{b.} \quad \frac{11}{5} \left(\frac{\frac{2}{11}}{\frac{2}{11}} \right) = \frac{2}{\boxed{\frac{10}{11}}}$$

- Here's what you should have. The missing fraction of 1 is 2/11 over 2/11. When you multiplied on the bottom, you got 10/11. That's the answer.

e. Your turn: Work problem C. Write the missing fraction of 1. Then multiply on the bottom to find the answer. Raise your hand when you're finished. (Observe students and give feedback.)
- (Write on the board:)

$$\text{c.} \quad \frac{4}{3} \left(\frac{\frac{9}{4}}{\frac{9}{4}} \right) = \frac{9}{\boxed{\frac{27}{4}}}$$

- Here's what you should have. The missing fraction of 1 is 9/4 over 9/4. When you multiplied on the bottom, you got 27/4. That's the answer. It equals 6 and 3/4.

f. Your turn: Work problem D. Raise your hand when you're finished. (Observe students and give feedback.)

- (Write on the board:)

$$d. \quad \frac{8}{9} \left(\frac{\frac{1}{9}}{\frac{1}{9}} \right) = \frac{\boxed{\frac{8}{9}}}{1}$$

- Here's what you should have. The missing fraction of 1 is 1/9 over 1/9. When you multiplied on the top, you got 8/9. That's the answer.

EXERCISE 2 INVERSE OPERATIONS
Single Equation

a. Find part 2.
- These are problems that have the first number missing. You can work them by starting with the last number and undoing the operation that is shown.
b. If the problem shows adding 43, how do you undo it? (Signal.) *Subtract 43*.
- If the problem shows multiplying by 16, how do you undo it? (Signal.) *Divide by 16*.
- If the problem shows subtracting 77, how do you undo it? (Signal.) *Add 77*.
- (Repeat step b until firm.)
c. Problem A: What operation is shown in that problem? (Signal.) *Subtraction*.
- How do you undo subtracting 48? (Signal.) *Add 48*.
d. Problem B: What operation is shown in that problem? (Signal.) *Multiplication*.
- How do you undo multiplying by 56? (Signal.) *Divide by 56*.
e. Your turn: Work problem A. Raise your hand when you're finished.
  (Observe students and give feedback.)
- (Write on the board:)

a. $\boxed{61} - 48 = 13$

- Here's what you should have. 61 is the number you start with. 61 minus 48 equals 13.
f. Your turn: Work the rest of the problems in part 2. Raise your hand when you're finished.
  (Observe students and give feedback.)

g. (Write on the board:)

b. $\boxed{4} \times 56 = 224$

c. $\boxed{72} \div 3 = 24$

d. $\boxed{6} + 80 = 86$

e. $\boxed{5} \times 15 = 75$

- Here's what you should have for problems B, C, D and E. Raise your hand if you got everything right.

EXERCISE 3 CIRCLES
Circumference and Diameter

a. Find part 3.
b. (Write on the board:)

$$\pi \times d = C$$

- You've used this equation to work problems that ask about the circumference. You can use the same equation to figure out the diameter.
c. Problem A gives the circumference. What is it? (Signal.) *28 meters*.
- And what's pi? (Signal.) *3 and 14-hundredths*.
- (Write to show:)

a. $\pi \times d = C$

$3.14 \left(\quad \right) = 28$

- Here's the problem you work: 3 and 14-hundredths times some fraction equals 28. So the fraction is 28 over 3 and 14-hundredths.
- (Write to show:)

a. $\pi \times d = C$

$3.14 \left(\dfrac{28}{3.14} \right) = 28$

$\boxed{d = \qquad}$

- The diameter equals 28 over 3 and 14-hundredths.

- Your turn: Work problem A. Copy what's on the board. Use your calculator to figure out what 28 divided by 3 and 14-hundredths equals, and complete the equation for D. Round your answer to the nearest hundredth. Raise your hand when you're finished.
  (Observe students and give feedback.)
- (Write to show:)

a. $\pi \times d = C$

$$3.14 \left(\frac{28}{3.14} \right) = 28$$

$$\boxed{d = 8.92 \text{ m}}$$

- Here's what you should have. The circumference is 28. The diameter is 8 and 92-hundredths meters.
- d. For problem B, you'll figure out the **circumference.** The diameter is 12 meters. Start with the equation: Pi times diameter equals circumference. Raise your hand when you're finished.
  (Observe students and give feedback.)
- (Write on the board:)

b. $\pi \times d = C$

$$3.14 \times 12 = C$$

$$\boxed{C = 37.68 \text{ m}}$$

- Check your work. When you multiply pi by 12, you get 37 and 68-hundredths. So the circumference is 37 and 68-hundredths meters.
- e. Problem C asks about the diameter. Work the problem. Put in the value for pi and for C. Write the fraction for D. Round your answer to the nearest hundredth. Raise your hand when you're finished.
  (Observe students and give feedback.)
- (Write on the board:)

c. $\pi \times d = C$

$$3.14 \left(\frac{4}{3.14} \right) = 4$$

$$\boxed{d = 1.27 \text{ m}}$$

- Here's what you should have. The circumference is 4. The fraction you multiply by is 4 over 3 and 14-hundredths. Everybody, what's the diameter? (Signal.)
  1 and 27-hundredths meters.

f. Problem D asks about the circumference. Work the problem. Raise your hand when you're finished. (Observe students and give feedback.)
- (Write on the board:)

d. $\pi \times d = C$

$$3.14 \times 42 = C$$

$$\boxed{C = 131.88 \text{ m}}$$

- Here's what you should have. The diameter is 42. When you multiply by 3 and 14-hundredths, you get 131 and 88-hundredths. The circumference is 131 and 88-hundredths meters.

EXERCISE 4 FRACTION OPERATIONS
Unlike Denominators

a. Find part 4.
- These are common-denominator problems.
- The problems are written in a row. You'll write them as column problems. Then you'll find the common denominator.
b. The first problem has denominators of 10 and 15.
- Write the numbers for 10. Find the first number that you reach counting by 15. Then work the equivalent-fraction problems and find the answer. Don't forget the minus sign. Raise your hand when you've completed the entire problem.
  (Observe students and give feedback.)
- (Write on the board:)

a.
$$\frac{4}{10} \left(\frac{3}{3} \right) = \frac{12}{30}$$
$$-\frac{4}{15} \left(\frac{2}{2} \right) = -\frac{8}{30}$$
$$\boxed{\frac{4}{30}}$$

- Here's what you should have. The common denominator is 30. When you subtract 4/15 from 4/10, you end up with 4/30. Make sure you have the minus sign and the work that is shown on the board.
c. Work problem B. Remember to show the complete work for the problem. Raise your hand when you're finished.
  (Observe students and give feedback.)

• (Write on the board:)

$$\begin{aligned}
\text{b.} \quad & \frac{7}{8}\left(\frac{5}{5}\right) = \frac{35}{40} \\
+ & \frac{1}{10}\left(\frac{4}{4}\right) = +\frac{4}{40} \\
\hline
& \boxed{\frac{39}{40}}
\end{aligned}$$

• Here's what you should have. The common denominator is 40. When you add 7/8 and 1/10, you end up with 39/40. Make sure you have a plus sign and the other work shown on the board.

EXERCISE 5 DECIMALS AND PERCENTS
From Fractions

a. Find part 5.
• Each row of the table is supposed to show a fraction with a denominator of 100, a decimal value and a percent. Remember, 17-hundredths is 17 percent. That's 17 and a percent sign.
b. Copy and complete row A of the table. Raise your hand when you're finished.
(Observe students and give feedback.)
• (Write on the board:)

	Fraction	Decimal	%
a.	$\frac{19}{100}$	.19	19 %

• Here's what you should have. 19-hundredths is 19 percent.
c. Complete row B. Raise your hand when you're finished. (Observe students and give feedback.)
• (Write to show:)

	Fraction	Decimal	%
a.	$\frac{19}{100}$	.19	19 %
b.	$\frac{385}{100}$	3.85	385 %

• Here's what you should have. 385-hundredths is 3 and 85-hundredths, which is 385 percent.
d. Complete the rest of the table. Raise your hand when you're finished.
(Observe students and give feedback.)

e. (Write to show:)

	Fraction	Decimal	%
a.	$\frac{19}{100}$	.19	19 %
b.	$\frac{385}{100}$	3.85	385 %
c.	$\frac{100}{100}$	1.00	100 %
d.	$\frac{8}{100}$	.08	8 %
e.	$\frac{805}{100}$	8.05	805 %

• Here's what you should have for rows C, D and E.

EXERCISE 6 PROBLEM SOLVING
Inverse Operations: 2 Steps

a. Find part 6.
• You've solved problems that have more than one operation. You can write equations to work those problems. The equations show one operation at a time.
b. Look at the sample problem. It says: You start with 4 and multiply by 3. Then you add 5.
• You can see the two equations we need to figure out the number we end up with.
• The first equation tells about starting with 4 and multiplying by 3. The next equation adds 5. The box with the question mark shows where we'll write the answer to the whole problem.
• (Write on the board:)

$$4 \times 3 = \boxed{} \qquad ?$$
$$\boxed{} + 5 = \boxed{}$$

• I'll write the answer to the first equation as the beginning of the next equation. Everybody, what's 4 times 3? (Signal.) *12.*
• (Write to show:)

$$4 \times 3 = \boxed{12} \qquad ?$$
$$\boxed{12} + 5 = \boxed{}$$

• Now we can figure out the answer to the second equation. Everybody, what's 12 plus 5? (Signal.) *17.*

- (Write **17** in the remaining box.)
- That's the number you end up with if you start with 4 and multiply by 3, then add 5.
c. Problem A. You start with 26 and add 19. Then you divide by 5. The first sentence tells about the first equation.
- (Write on the board:)

a. $26 + 19 = \boxed{}$

- You'll figure out what 26 plus 19 equals. Then you'll use the same number and divide by 5.
- (Write to show:)

a. $26 + 19 = \boxed{}$
$\boxed{} \div 5 = \boxed{}$?

- Copy these equations. Figure out the answer to the first equation. Then put that number in the second equation and figure out the answer. Raise your hand when you're finished.
(Observe students and give feedback.)
- (Write to show:)

a. $26 + 19 = \boxed{45}$
$\boxed{45} \div 5 = \boxed{9}$?

- 26 plus 19 is 45. The second equation is 45 divided by 5. That's 9. And that's the answer.
d. Your turn: Write two equations for problem B. Don't complete the equations. Just show them with boxes. Raise your hand when you've done that much.
(Observe students and give feedback.)
- (Write on the board:)

b. $76 - 20 = \boxed{}$?
$\boxed{} \times 4 = \boxed{}$

- Here are the two equations. You start with 76 and subtract 20. Then you multiply that value by 4.
- Work the problems. Remember, you'll use the same number at the end of the first equation and at the beginning of the next equation. Raise your hand when you know the number you end up with. (Observe students and give feedback.)

- (Write to show:)

b. $76 - 20 = \boxed{56}$?
$\boxed{56} \times 4 = \boxed{224}$

- Here's what you should have. 76 minus 20 is 56. 56 times 4 is 224. Raise your hand if you got it right.
e. Your turn: Write the two equations for problem C. Then figure out the missing numbers. Raise your hand when you're finished.
(Observe students and give feedback.)
- (Write on the board:)

c. $344 \div 4 = \boxed{86}$?
$\boxed{86} \times 3 = \boxed{258}$

- Here are the two equations you should have. The number you end up with is 258. Raise your hand if you got everything right.

EXERCISE 7 INDEPENDENT WORK

- (In addition to independent work in the textbook, assign **Bridge to Connecting Math Concepts** Independent Worksheet 38 as classwork or homework. Before beginning the next lesson, check the students' independent work.)

Lesson 59

Objectives

- Use an inverse operation to undo a problem that has the first value missing. (Exercise 1)

- Complete a pair of equivalent fractions by generating a complex fraction that equals 1. (Exercise 2)

- **Write a statement of equality or inequality for two values.** (Exercise 3)

  Example: $\frac{7}{6}$ ■ **1**

  Students determine that 7/6 is more than 1. They write: $\frac{7}{6}$ **> 1** .

- Add or subtract fractions with unlike denominators. (Exercise 4)

- **Work a 3-step word problem that gives the starting number and specifies three operations.** (Exercise 5)

- **Complete a table to show fractions, decimals and corresponding percents.** (Exercise 6)

Note:

Fraction	□/100	Decimal	%
$\frac{2}{5}$			

The starting fraction does not have a denominator of 100. Students convert the fraction and then complete the row of the

table:

Fraction	□/100	Decimal	%
$\frac{2}{5}$	$\frac{40}{100}$	.40	40%

EXERCISE 1 INVERSE OPERATIONS
Single Equation

a. Open your textbook to lesson 59 and find part 1.
- These are problems that have the first number missing. You can work them by starting with the last number and undoing the operation that is shown.
b. If the problem shows multiplying by 24, how do you undo it? (Signal.) *Divide by 24.*
- If the problem shows adding 3, how do you undo it? (Signal.) *Subtract 3.*
- If the problem shows dividing by 20, how do you undo it? (Signal.) *Multiply by 20.*

- (Repeat step b until firm.)
c. Problem A: What operation is shown in that problem? (Signal.) *Division.*
- How do you undo dividing by 6? (Signal.) *Multiply by 6.*
d. Problem B: What operation is shown in that problem? (Signal.) *Subtraction.*
- How do you undo subtracting 203? (Signal.) *Add 203.*
e. Your turn: Work problem A. Use your calculator. Raise your hand when you're finished. (Observe students and give feedback.)
- (Write on the board:)

a. $\boxed{408} \div 6 = 68$

- Here's what you should have. The starting number is 408.
f. Your turn: Work the rest of the problems in part 1. Raise your hand when you're finished. (Observe students and give feedback.)
g. (Write on the board:)

b. $\boxed{343} - 203 = 140$

c. $\boxed{12} \times 56 = 672$

d. $\boxed{23} + 291 = 314$

- Here's what you should have for problems B, C and D. Raise your hand if you got everything right.
h. You can figure out whether the answers are correct by working the problem that is shown. For item A you'd work the problem: 408 divided by 6.
- Say the problem you'd work for item B. (Signal.) *343 minus 203.*
- Say the problem you'd work for item C. (Signal.) *12 times 56.*
- Use your calculator to work those problems. See if you end up with the number that's shown after the equal sign. Raise your hand when you're finished. (Observe students and give feedback.)

EXERCISE 2 COMPLEX FRACTIONS
Equivalence

a. Find part 2.
b. Work problem A. Raise your hand when you're finished. (Observe students and give feedback.)

228 Lesson 59

- (Write on the board:)

$$\text{a.} \quad \frac{3}{7} \left(\frac{\frac{6}{7}}{\frac{6}{7}} \right) = \frac{\boxed{\frac{18}{7}}}{6}$$

- Here's what you should have. The problem on the bottom is 7 times some fraction equals 6. That's 6/7. So what's the fraction that equals 1? (Signal.) *6-sevenths over 6-sevenths.*
- When you multiply on top, you get 18/7. Your turn: Write 18/7 as a mixed number and box it. Raise your hand when you're finished. √
- You should have written **2 and 4/7.**
- c. Your turn: Work problem B. Raise your hand when you're finished.
  (Observe students and give feedback.)
- (Write on the board:)

$$\text{b.} \quad \frac{8}{4} \left(\frac{\frac{10}{8}}{\frac{10}{8}} \right) = \frac{10}{\boxed{\frac{40}{8}}}$$

- Here's what you should have. The problem on top is 8 times some value equals 10. That's 10/8. So what's the fraction that equals 1? (Signal.) *10-eighths over 10-eighths.*
- When you multiply on the bottom, you get 40/8. Write what 40/8 equals. √
- You should have written **5.**
- d. Your turn: Work problem C. Write a fraction in the box, then write what that fraction equals. Raise your hand when you're finished.
  (Observe students and give feedback.)
- (Write on the board:)

$$\text{c.} \quad \frac{10}{5} \left(\frac{\frac{4}{5}}{\frac{4}{5}} \right) = \frac{\boxed{\frac{40}{5}}}{4} \quad \boxed{8}$$

- Here's what you should have. You end up with 40/5. That's 8.
- e. Raise your hand if you got all of them right.
- Good work. You're getting ready to work difficult ratio problems.

EXERCISE 3 INEQUALITY

- a. Find part 3.
- I'll read what it says. Follow along: An **equal sign** shows that the values on both sides of the sign are **equal.**
- An **inequality sign** is used to show that the sides are **not equal.**
- The side that is closer to the **bigger end** of the inequality sign is the side with the **bigger value.**
- The side closer to the **pointed end** of the sign is the side with the **smaller value.**
- You can see inequality signs. For the first sign, the left side is bigger. For the second sign, the right side is bigger.
- b. Find part 4.
- Each item shows two values. You'll copy the values and write the appropriate sign to show which side has the larger value or whether the sides are equal.
- Work problem A. Raise your hand when you're finished. (Observe students and give feedback.)
- (Write on the board:)

$$\text{a.} \quad 1 < \frac{7}{6}$$

- Here's what you should have: 1 is less than 7/6.
- c. Work problem B: Raise your hand when you're finished. (Observe students and give feedback.)
- (Write on the board:)

$$\text{b.} \quad \frac{10}{1} > 2$$

- Here's what you should have. 10 over 1 equals 10. 10 is more than 2.
- d. Touch item C. √
- The fraction is 10/5. What number does 10/5 equal? (Signal.) *2.*
- Item D has two fractions. You can see which fraction is more.
- Touch item E. √
- The fraction is 36/9. What number does 36/9 equal? (Signal.) *4.*
- Item F has two fractions — 13/14 and 14/13. Is 13/14 more than or less than 1? (Signal.) *Less than 1.*
- Is 14/13 more than or less than 1? (Signal.) *More than 1.*
- e. Your turn: Work problems C through F. For each item, copy the fractions and write the sign. Raise your hand when you're finished. (Observe students and give feedback.)

Key:

c. $\dfrac{10}{5} = 2$ d. $\dfrac{10}{5} < \dfrac{12}{5}$

e. $9 > \dfrac{36}{9}$ f. $\dfrac{13}{14} < \dfrac{14}{13}$

f. Find part J on page 236. That shows what you should have for each problem.

EXERCISE 4 FRACTION OPERATIONS
Unlike Denominators

a. Find part 5.
- These are common-denominator problems. The problems are written in a row. You'll write them as column problems. Then you'll find the common denominator.
b. The first problem has denominators of 12 and 16.
- Write the numbers for 12. Find the first number that you reach counting by 16. Then work the equivalent-fraction problems and find the answer. Don't forget the plus sign. Raise your hand when you've completed the entire problem.
  (Observe students and give feedback.)
- (Write on the board:)

a. $\dfrac{5}{12}\left(\dfrac{4}{4}\right) = \dfrac{20}{48}$

$+\dfrac{9}{16}\left(\dfrac{3}{3}\right) = +\dfrac{27}{48}$

$\boxed{\dfrac{47}{48}}$

- Here's what you should have. The common denominator is 48. When you add 5/12 and 9/16, you end up with 47/48. Make sure you have the plus sign and the work that is shown on the board.
c. Work problem B. Remember to show the complete work for the problems. Raise your hand when you're finished.
  (Observe students and give feedback.)

- (Write on the board:)

b. $\dfrac{11}{12}\left(\dfrac{3}{3}\right) = \dfrac{33}{36}$

$-\dfrac{8}{9}\left(\dfrac{4}{4}\right) = -\dfrac{32}{36}$

$\boxed{\dfrac{1}{36}}$

- Here's what you should have. The common denominator is 36. When you subtract 8/9 from 11/12, you end up with 1/36. Make sure you have a minus sign and the other work shown on the board.
d. Raise your hand if you got everything right.

EXERCISE 5 PROBLEM SOLVING
Inverse Operations: 3 Steps

a. Find part 6.
- These are problems that tell the number you start with and the steps you take in working the problem. Remember, you make a separate equation for each step.
b. Problem A: You start with 44 and divide by 4. Then you add 209. Then you divide by 10.
- For that problem, you write three equations. Write the first equation. Show a box for the missing value. Raise your hand when you've done that much.
  (Observe students and give feedback.)
- (Write on the board:)

a. $44 \div 4 = \square$

- Here's what you should have. You start with 44 and divide by 4.
- For the next equation, you start with a box, then add 209. For the last equation, you start with a box and divide by 10.
- (Write to show:)

a. $44 \div 4 = \square$
$\square + 209 = \square$
$\square \div 10 = \boxed{?}$

- Here are the equations.
- Copy the last two equations. Figure out the answer. Raise your hand when you're finished.
  (Observe students and give feedback.)

- (Write to show:)

$$
\begin{aligned}
\text{a. } 44 \div 4 &= \boxed{11} \\
\boxed{11} + 209 &= \boxed{220} \qquad \text{?} \\
\boxed{220} \div 10 &= \boxed{22}
\end{aligned}
$$

- Here's what you should have. You end up with 22. Raise your hand if you got everything right.
- c. Your turn: Work problem B. Write three equations. Raise your hand when you're finished with the problem.
  (Observe students and give feedback.)
- (Write on the board:)

$$
\begin{aligned}
\text{b. } 800 \div 4 &= \boxed{200} \\
\boxed{200} - 200 &= \boxed{0} \qquad \text{?} \\
\boxed{0} \times 6 &= \boxed{0}
\end{aligned}
$$

- Here's what you should have. You end up with zero. Raise your hand if you got everything right.
- d. Your turn: Work problem C. Write three equations. Raise your hand when you're finished with the problem.
  (Observe students and give feedback.)
- (Write on the board:)

$$
\begin{aligned}
\text{c. } 65 \times 7 &= \boxed{455} \\
\boxed{455} + 50 &= \boxed{505} \qquad \text{?} \\
\boxed{505} \div 5 &= \boxed{101}
\end{aligned}
$$

- Here's what you should have. You end up with 101.

EXERCISE 6 DECIMALS AND PERCENTS
From Fractions

a. Find part 7.
- Each row shows a fraction that does not have a denominator of 100.
- You're going to write the equivalent fraction that has a denominator of 100. Then you'll write the decimal number and the percent.
b. Item A. The fraction is 2/5. 2/5 equals a fraction that has a denominator of 100. Write the equation and figure out the fraction. Raise your hand when you're finished.
  (Observe students and give feedback.)
- (Write on the board:)

$$
\text{a. } \frac{2}{5} \left(\frac{20}{20} \right) = \frac{40}{100}
$$

- Here's the equation you used. The fraction that equals 1 is 20/20. So the fraction that equals 2/5 is 40/100.
- Copy the table and complete row A. Raise your hand when you're finished.
  (Observe students and give feedback.)
- (Write on the board:)

	$\frac{\square}{100}$	Decimal	%
a. $\dfrac{2}{5}$	$\dfrac{40}{100}$	.40	40%

- Here's what you should have. 2/5 equals 40/100. That's the decimal value 40-hundredths and that's 40 percent.
c. Your turn: Complete the rest of the items in part 7. Raise your hand when you're finished.
  (Observe students and give feedback.)
d. (Write to show:)

	$\frac{\square}{100}$	Decimal	%
a. $\dfrac{2}{5}$	$\dfrac{40}{100}$	.40	40%
b. $\dfrac{5}{4}$	$\dfrac{125}{100}$	1.25	125%
c. $\dfrac{1}{2}$	$\dfrac{50}{100}$	.50	50%
d. $\dfrac{11}{10}$	$\dfrac{110}{100}$	1.10	110%

- Check your work. Here's what you should have for rows B through D.
e. Raise your hand if you got everything right.

EXERCISE 7 INDEPENDENT WORK

- (In addition to independent work in the textbook, assign *Bridge to Connecting Math Concepts Independent Worksheet* 39 as classwork or homework. Before beginning the next lesson, check the students' independent work.)

Lesson 60 – Test 6

EXERCISE 1 FRACTIONS
Simplification: Mixed Numbers

a. Open your textbook to lesson 60 and find part 1.
- I'll read what it says. Follow along: You've worked with fractions that can be simplified.
- Some mixed numbers have fractions that can be simplified.
- You can see 5 and 2/4.
- To simplify this mixed number, we keep the whole number and simplify the fraction.
- 2/4 equals 2 over 2 times 2. A fraction of 2/2 is crossed out.
- So the simplified mixed number for 5 and 2/4 is 5 and 1/2.
b. Find part 2.
- These are mixed numbers. Some of them have a **fraction** that can be simplified.
c. The fraction for mixed number A can be simplified. Simplify the fraction and rewrite the mixed number with a simplified fraction. It will be 13 and the simplified fraction. Raise your hand when you're finished.
  (Observe students and give feedback.)
- (Write on the board:)

a. $\frac{8}{16} = \frac{\cancel{2} \times \cancel{2} \times \cancel{2}}{\cancel{2} \times \cancel{2} \times \cancel{2} \times 2} = \frac{1}{2}$
$\boxed{13\frac{1}{2}}$

- Here's what you should have. The simplified fraction is 1/2. The mixed number with the simplified fraction is 13 and 1/2.
d. Problem B. If the fraction can't be simplified, write the mixed number with the fraction shown. Raise your hand when you're finished.
  (Observe students and give feedback.)
- (Write on the board:)

b. $\frac{11}{35} = \frac{11}{5 \times 7} = \frac{11}{35}$
$\boxed{4\frac{11}{35}}$

- Here's what you should have: 4 and 11/35.
e. Your turn: Work the rest of the problems in part 2. If the fraction can be simplified, do the simplification and write the mixed number with the simplified fraction. If the fraction cannot be simplified, copy the mixed number the way it is written. Raise your hand when you're finished.
  (Observe students and give feedback.)
f. (Write on the board:)

c. $\frac{18}{36} = \frac{\cancel{2} \times \cancel{3} \times \cancel{3}}{\cancel{2} \times 2 \times \cancel{3} \times \cancel{3}} = \frac{1}{2}$
$\boxed{3\frac{1}{2}}$

d. $\frac{9}{28} = \frac{3 \times 3}{2 \times 2 \times 7} = \frac{9}{28}$
$\boxed{4\frac{9}{28}}$

e. $\frac{14}{35} = \frac{2 \times \cancel{7}}{5 \times \cancel{7}} = \frac{2}{5}$
$\boxed{7\frac{2}{5}}$

f. $\frac{72}{81} = \frac{2 \times 2 \times 2 \times \cancel{3} \times \cancel{3}}{3 \times 3 \times \cancel{3} \times \cancel{3}} = \frac{8}{9}$
$\boxed{14\frac{8}{9}}$

- Here's what you should have. Mixed numbers C, E and F can be simplified.
- Mixed number C: 3 and 18/36 simplifies to 3 and 1/2.

- Mixed number D: 4 and 9/28 cannot be simplified.
- Mixed number E: 7 and 14/35 simplifies to 7 and 2/5.
- Mixed number F: 14 and 72/81 simplifies to 14 and 8/9.

EXERCISE 2 CIRCLES
Circumference and Diameter

a. Find part 3.
- For some of these problems, you'll figure out the diameter. For others, you'll figure out the circumference.
- You'll use your calculator. If your answer has more than two decimal places, you'll round it two places.
- Remember, start with the equation that has the symbol for pi.
b. Everybody, say that equation. (Signal.)
   Pi times diameter equals circumference.
- And what does pi equal? (Signal.)
   3 and 14-hundredths.
- (Repeat step b until firm.)
c. Work problem A. Remember, show the answer as an equation that tells what **d** equals. Round the answer to hundredths. Raise your hand when you're finished.
   (Observe students and give feedback.)
- (Write on the board:)

> **a.** $\pi \times d = C$
> $$3.14\left(\frac{8}{3.14}\right) = 8$$
> $d = 2.55\ ft$

- Here's what you should have. The problem asks about the diameter. 3 and 14-hundredths times some fraction equals 8. Everybody, what's the fraction? (Signal.) *8 over 3 and 14-hundredths.*
- You should have written: D equals 2 and 55-hundredths feet. Raise your hand if you got everything right.
d. Work the rest of the problems in part 3. Raise your hand when you're finished.
   (Observe students and give feedback.)

b. $\pi \times d = C$
   $3.14 \times 200 = C$
   $\boxed{C = 628\ cm}$

c. $\pi \times d = C$
   $3.14\left(\frac{17}{3.14}\right) = 17$
   $\boxed{d = 5.41\ mi}$

d. $\pi \times d = C$
   $3.14 \times 24 = C$
   $\boxed{C = 75.36\ ft}$

e. Check your work.
- Find part J on page 238. That shows what you should have for each problem.
- Raise your hand if you got everything right.

EXERCISE 3 DECIMALS AND PERCENTS
From Fractions

a. Find part 4.
- Each row shows a fraction that does not have a denominator of 100. You're going to write the equivalent fraction that has a denominator of 100. Then you'll write the decimal number and the percent.
b. Item A. The fraction is 9/20. 9/20 equals a fraction that has a denominator of 100.
- Write the equation and figure out the fraction. Raise your hand when you're finished.
   (Observe students and give feedback.)
- (Write on the board:)

> **a.** $\dfrac{9}{20}\left(\dfrac{5}{5}\right) = \dfrac{45}{100}$

- Here's the equation you used. The fraction that equals 1 is 5/5. So the fraction that equals 9/20 is 45/100.
- Copy and complete row A. Raise your hand when you're finished. √
- (Write on the board:)

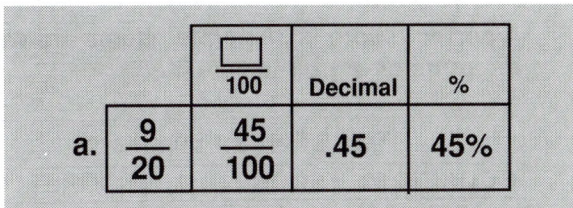

	$\dfrac{\square}{100}$	Decimal	%
a. $\dfrac{9}{20}$	$\dfrac{45}{100}$	.45	45%

- Here's what you should have. 9/20 equals 45/100. That's the decimal value 45-hundredths and that's 45 percent.
c. Your turn: Complete the rest of the items in part 4. Raise your hand when you're finished.
   (Observe students and give feedback.)

d. (Write to show:)

	$\boxed{}$ / 100	Decimal	%
a.	$\dfrac{9}{20}$ $\dfrac{45}{100}$	.45	45%
b.	$\dfrac{7}{5}$ $\dfrac{140}{100}$	1.40	140%
c.	$\dfrac{3}{2}$ $\dfrac{150}{100}$	1.50	150%
d.	$\dfrac{1}{4}$ $\dfrac{25}{100}$	.25	25%

- Check your work. Here's what you should have for items B through D.
- e. Raise your hand if you got everything right.

EXERCISE 4 TEST 6

Note: Students are to use calculators only for parts 7 and 8 of the test.

a. This is a test. You should only have your textbook, a sharpened pencil and lined paper on your desk.
b. Find part 1 of test 6 in your textbook. √
c. Do the test on your own. Raise your hand when you've completed part 6.
(Observe students but do not give feedback.)
- (After students complete part 6, permit them to use a calculator to complete parts 7 and 8. Observe students but do not give feedback.)

EXERCISE 5 MARKING THE TEST

a. (Collect the students' papers. Use the Answer Key to score the tests. Award scores for test 6 as follows:)

A perfect score is 30 points. Items are worth the number of points specified.

Test 6 Percent Summary					
SCORE	%	SCORE	%	SCORE	%
55	100	49	89	43	78
54	98	48	87	42	76
53	96	47	85	41	75
52	95	46	84	40	73
51	93	45	82	39	71
50	91	44	80		

b. (Complete the Test 6 Remedy Summary to determine whether remedies are needed. Reproducible Summary Sheets are at the back of the Teacher's Guide.)
- (If more than 1/4 of the students did not pass a test part, present the remedy for that part before beginning lesson 61. Remedies appear at the end of the Test 6 Answer Key.)

EXERCISE 6 INDEPENDENT WORK

- (Assign **Bridge to Connecting Math Concepts** *Independent Worksheet* 40 as classwork or homework. Before beginning the next lesson, check the students' independent work.)

Test Preparation Reminder

- Begin these lessons 7–10 school days before standardized testing begins. Present one lesson a day.

- If standardized testing begins more than ten days from now, present lesson 71 next.

- For test-preparation lessons that do not require a full period, continue with the regular **Bridge to Connecting Math Concepts** lesson presentation in the time that remains.

- Lesson worksheets begin at the end of the Test Preparation lessons.

Lesson 61

Objectives

- Write a statement of equality or equality for two values. (Exercise 1)

- Simplify mixed numbers. (Exercise 2)

- **Compare two fractions with unlike denominators.** (Exercise 3)

- Complete a pair of equivalent fractions by generating a complex fraction that equals 1. (Exercise 4)

- **Use inverse operations to solve a pair of equations and figure out the starting number.** (Exercise 5)

- Use the equation $\pi \times d = C$ to figure out the circumference or diameter of a circle. (Exercise 6)

- **Write an equation that shows a fraction, an equivalent fraction with a denominator of 100 and an equivalent percent value.** (Exercise 7)

EXERCISE 1 INEQUALITY

a. Open your textbook to lesson 61 and find part 1.
- Each item shows two values. You'll copy the values and write the appropriate sign to show which side has the larger value or whether the sides are equal. Remember, if one of the sides is larger, make the sign so the big end is closer to that side.

b. Work all the items in part 1. Raise your hand when you're finished.
(Observe students and give feedback.)
Key:

a. $\dfrac{13}{13} = \dfrac{7}{7}$ b. $\dfrac{5}{8} < \dfrac{8}{5}$

c. $1 = \dfrac{60}{60}$ d. $\dfrac{65}{64} > \dfrac{64}{65}$

c. Find part J on page 243. That shows what you should have for each item.

EXERCISE 2 FRACTIONS
Simplification: Mixed Numbers

a. Find part 2.
- These are mixed numbers. Some of them have fractions that can be simplified.

b. If the fraction can be simplified, do the simplification and write the mixed number with that fraction. If the fraction cannot be simplified, copy the mixed number the way it is written. Raise your hand when you're finished.
(Observe students and give feedback.)

c. (Write on the board:)

a. $\dfrac{55}{100} = \dfrac{5 \times 11}{2 \times 2 \times 5 \times 5} = \dfrac{11}{20}$

$\boxed{3\,\dfrac{11}{20}}$

b. $\dfrac{8}{10} = \dfrac{2 \times 2 \times 2}{2 \times 5} = \dfrac{4}{5}$

$\boxed{2\,\dfrac{4}{5}}$

c. $\boxed{7\,\dfrac{8}{19}}$

d. $\dfrac{21}{49} = \dfrac{3 \times 7}{7 \times 7} = \dfrac{3}{7}$

$\boxed{4\,\dfrac{3}{7}}$

- Here's what you should have. Raise your hand if you got everything right.

EXERCISE 3 FRACTIONS
Comparison: Unlike Denominators

a. Find part 3.
- (Teacher reference:)

$$\frac{3}{4}\left(\frac{2}{2}\right) = \frac{6}{8}$$
$$\frac{5}{8} \qquad = \frac{5}{8}$$
$$\boxed{\frac{3}{4} > \frac{5}{8}}$$

- I'll read what it says. Follow along: You can compare some pairs of fractions by examining them. Other pairs are more difficult. You can compare those fractions by finding the lowest common denominator. When the denominators are the same, the fraction with the larger numerator is the larger fraction.
- You can see a problem: Which is more, 3/4 or 5/8?
- We write the fractions in a column and figure out the lowest common denominator. That's 8.

- When both fractions have a denominator of 8, you can see that the larger fraction is 6/8.
- 3/4 equals 6/8. The problem asks: Which is more, 3/4 or 5/8? You can see the inequality statement showing that 3/4 is more than 5/8.
b. Find part 4.
- For both these problems, you'll find the lowest common denominator and then write the answer to the question as an inequality statement.
c. Problem A: Which is more, 2/3 or 3/5?
- Your turn: Write the two fractions in a column; find the lowest common denominator and complete both equations. Raise your hand when you've done that much.
(Observe students and give feedback.)
- (Write on the board:)

$$a. \quad \frac{2}{3}\left(\frac{5}{5}\right) = \frac{10}{15}$$

$$\frac{3}{5}\left(\frac{3}{3}\right) = \frac{9}{15}$$

- Here's what you should have so far. The lowest common denominator is 15.
- The problem asks: Which is more, 2/3 or 3/5? Write a statement with an inequality sign. Show 2/3 and 3/5 in that order. √
- (Write to show:)

$$a. \quad \frac{2}{3}\left(\frac{5}{5}\right) = \frac{10}{15}$$

$$\frac{3}{5}\left(\frac{3}{3}\right) = \frac{9}{15}$$

$$\boxed{\frac{2}{3} > \frac{3}{5}}$$

- Here's the answer. 2/3 is more than 3/5. That's because 2/3 is 10/15 but 3/5 is only 9/15.
d. Work problem B. Remember to answer the question with an inequality statement that starts with 9/12. Raise your hand when you're finished. (Observe students and give feedback.)

- (Write on the board:)

$$b. \quad \frac{9}{12} = \frac{9}{12}$$

$$\frac{5}{6}\left(\frac{2}{2}\right) = \frac{10}{12}$$

$$\boxed{\frac{9}{12} < \frac{5}{6}}$$

- Here's what you should have for problem B. The lowest common denominator is 12. 5/6 is more because it equals 10/12.
e. Remember, to compare fractions that do not have the same denominator, you can just find the lowest common denominator.

EXERCISE 4 COMPLEX FRACTIONS
Equivalence

a. Find part 5.
b. Work problem A. Show the answer as a fraction. Then write it as a whole number or a mixed number. Raise your hand when you're finished. (Observe students and give feedback.)
- (Write on the board:)

$$a. \quad \frac{7}{4}\left(\frac{\frac{9}{7}}{\frac{9}{7}}\right) = \frac{9}{\boxed{\frac{36}{7}}} \quad \boxed{5\frac{1}{7}}$$

- Here's what you should have.
c. Your turn: Work the rest of the problems in part 5. Remember, show the answer as a fraction. Then rewrite it as a whole number or a mixed number. Raise your hand when you're finished. (Observe students and give feedback.)

d. (Write on the board:)

$$\text{b.} \quad \frac{8}{3} \left(\frac{\frac{2}{3}}{\frac{2}{3}} \right) = \frac{\boxed{\frac{16}{3}}}{2} \quad \boxed{5\tfrac{1}{3}}$$

$$\text{c.} \quad \frac{5}{9} \left(\frac{\frac{5}{9}}{\frac{5}{9}} \right) = \frac{\boxed{\frac{25}{9}}}{5} \quad \boxed{2\tfrac{7}{9}}$$

$$\text{d.} \quad \frac{6}{10} \left(\frac{\frac{1}{6}}{\frac{1}{6}} \right) = \frac{1}{\boxed{\frac{10}{6}}} \quad \boxed{1\tfrac{4}{6}}$$

- Check your work. Here's what you should have.
e. Raise your hand if you got all of them right.

EXERCISE 5 INVERSE OPERATIONS
2-Step Problems

a. Find part 6.
- (Teacher reference:)

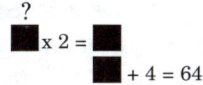

- I'll read what it says. Follow along: This problem shows a mystery number. The question mark above the **first box** shows that the first number is the mystery number.
- Look at the problem. It has two equations. The first equation tells you that you start with the mystery number and multiply by 2.
- Then you add 4 to that answer. You end up with 64.
- To work problems of this type, you start at the **end** of the second equation and work **backward.** You **undo** each step that was done.
- Everybody, what's the last number in the second equation? (Signal.) *64.*
- What did we do in the second equation to get 64? (Signal.) *Added 4.*
- We added 4. How do you undo adding 4? (Signal.) *Subtract 4.*
- So you work the problem 64 minus 4.
- The answer is 60. You can see the bottom equation completed, and the 60 is also written at the end of the **first** equation.
- Now we'll work backward from 60 to figure out the mystery number.
- We multiplied by 2 in the first equation. How do we undo multiplying by 2? (Signal.) *Divide by 2.*

- When we work the problem 60 divided by 2, we'll know the mystery number. Raise your hand when you've figured out the answer. √
- Everybody, what's the mystery number? (Signal.) *30.*
- Remember, start at the end. Work backward and undo each step.
d. Find part 7.
- Each problem in part 7 has two equations.
e. Problem A. Everybody, what's the last number in the second equation? (Signal.) *48.*
- What did you do in the second equation to get to 48? (Signal.) *Multiplied by 3.*
- How do you undo multiplying by 3? (Signal.) *Divide by 3.*
- Copy the equations. Figure out the answer to 48 divided by 3 and write the answer in **both equations.**
- Then work the first equation. Raise your hand when you're finished.
(Observe students and give feedback.)
- (Write on the board:)

$$\text{a.} \quad \overset{?}{\boxed{4}} + 12 = \boxed{16}$$
$$\boxed{16} \times 3 = 48$$

- Everybody, what's the mystery number? (Signal.) *4.*
f. Problem B. What's the last number in the second equation? (Signal.) *400.*
- What did you do to get to 400? (Signal.) *Subtracted 58.*
- How do you undo subtracting 58? (Signal.) *Add 58.*
- Copy the equations and write the answer to 400 plus 58 in both equations. Then undo the first equation and figure out the mystery number. Raise your hand when you're finished. (Observe students and give feedback.)
- (Write on the board:)

$$\text{b.} \quad \overset{?}{\boxed{248}} + 210 = \boxed{458}$$
$$\boxed{458} - 58 = 400$$

- You undo subtracting 58 by adding 58. The answer is 458. When you work the first equation, you undo adding 210. That's 458 **minus 210.** The answer is 248. That's the mystery number. Raise your hand if you got it right.

g. Your turn: Copy problem C. Undo what you did to reach the last number. Remember to write the answer in both equations. Then undo the first equation and figure out the mystery number. Raise your hand when you're finished.
(Observe students and give feedback.)
- (Write on the board:)

> ?
> c. $\boxed{12}$ x 5 = $\boxed{60}$
>
> $\boxed{60}$ + 16 = 76

- You undo adding 16 by subtracting 16. 76 minus 16 is 60. You undo multiplying by 5 by dividing by 5. 60 divided by 5 is 12. That's the mystery number. Raise your hand if you got it right.

EXERCISE 6 CIRCLES
Circumference and Diameter

a. Find part 8.
- For some of these problems, you'll figure out the diameter. For others, you'll figure out the circumference. You'll use your calculator. If your decimal answer is more than two places, you'll round it to two decimal places.
Remember, start with the equation that has the symbol for pi.
b. Everybody, say that equation. (Signal.)
Pi times diameter equals circumference.
- And what does pi equal? (Signal.)
3 and 14-hundredths.
(Repeat step b until firm.)
c. Work problem A. Remember, show the answer as an equation that tells what D equals. Raise your hand when you're finished.
(Observe students and give feedback.)
- (Write on the board:)

> a. $\pi \times d = C$
>
> $3.14 \left(\dfrac{30}{3.14} \right) = 30$
>
> $d = 9.55$ ft

- Here's what you should have. The problem asks about the diameter. 3 and 14-hundredths times some fraction equals 30. Everybody, what's the fraction? (Signal.) *30 over 3 and 14-hundredths.*
- You should have written: D equals 9 and 55-hundredths feet. Raise your hand if you got everything right.

d. Work the rest of the problems in part 8. Raise your hand when you're finished.
(Observe students and give feedback.)
Key:

> b. $\pi \times d = C$
> $3.14 \left(\dfrac{9.2}{3.14} \right) = 9.2$
> $\boxed{d = 2.93 \text{ in}}$

> c. $\pi \times d = C$
> $3.14 \times 70 = C$
> $\boxed{C = 219.8 \text{ cm}}$

> d. $\pi \times d = C$
> $3.14 \left(\dfrac{70}{3.14} \right) = 70$
> $\boxed{d = 22.29 \text{ yd}}$

e. Check your work. Find part K. That shows what you should have for each problem.
- Raise your hand if you got everything right.

EXERCISE 7 PERCENTS
From Fractions

a. Find part 9.
- For each fraction, you'll write the equivalent fraction that has a denominator of 100. Then you'll write the percent value that equals the fraction. The sample item shows the type of equation you'll write.
- The sample fraction is 3/5. Say the hundredths fraction. (Signal.) *60-hundredths.*
- Say the percent. (Signal.) *60 percent.*
b. Work the items in part 9. Raise your hand when you're finished.
(Observe students and give feedback.)
c. Check your work.
- Fraction A is 10/5. Say the hundredths fraction. (Signal.) *200-hundredths.*
- Say the percent. (Signal.) *200 percent.*
- Fraction B is 1/2. Say the hundredths fraction. (Signal.) *50-hundredths.*
- Say the percent. (Signal.) *50 percent.*
- Fraction C is 3/4. Say the hundredths fraction. (Signal.) *75-hundredths.*
- Say the percent. (Signal.) *75 percent.*
- Fraction D is 12/10. Say the hundredths fraction. (Signal.) *120-hundredths.*
- Say the percent. (Signal.) *120 percent.*
- Fraction E is 4/25. Say the hundredths fraction. (Signal.) *16-hundredths.*
- Say the percent. (Signal.) *16 percent.*
d. Raise your hand if you got everything right.

EXERCISE 8 INDEPENDENT WORK

- (In addition to independent work in the textbook, assign **Bridge to Connecting Math Concepts** *Independent Worksheet* 41 as classwork or homework. Before beginning the next lesson, check the students' independent work.)

Lesson 62

Objectives

- **Discriminate between problems that require a ratio table and those that require only a ratio equation.** (Exercise 1)

- Compare two fractions with unlike denominators. (Exercise 2)

- **Rewrite a fraction as a mixed number.** (Exercise 3)

  Note: Problems are of the form: $= \dfrac{18}{\boxed{\dfrac{340}{9}}}$

  Rewriting the boxed fraction is a component of complex ratio problems.

- **Work a ratio-equation problem that involves complex fractions.** (Exercise 4)
  For example:

$$\dfrac{\text{pies}}{\text{muffins}} \quad \dfrac{5}{8} \left(\dfrac{\frac{11}{8}}{\frac{11}{8}} \right) = \dfrac{\frac{55}{8}}{11}$$

- Use inverse operations to solve a pair of equations and figure out the starting number. (Exercise 5)

- **Use the equation π x d = C to figure out the circumference or radius of a circle.** (Exercise 6)

EXERCISE 1 RATIOS AND PROPORTIONS
Table Discrimination

a. Open your textbook to lesson 62 and find part 1.
- I'll read what it says. Follow along: Some ratio problems require a table. Those are problems that have **three names.** Problems that have only two names do not require a table.
- You can see a problem that requires a table: The ratio of <u>girls</u> to <u>children</u> is 1 to 3. If there are 60 children, how many <u>boys</u> are there?
- The three names are underlined. Notice that one of them is in the question.
- The names are **girls, boys** and **children.** The problem requires a table because it has three names.
- You can see a problem that does not require a table: The ratio of <u>girls</u> to <u>children</u> is 2 to 5. If there are 20 girls, how many children are there?
- The names are **girls** and **children.**

- Remember, if there are **three names**, you need a table. If there are only two names, you don't need a table. Also remember that one of the names may be in the question.
b. Find part 2.
c. Read problem A. Write the words **table** or **no table** to tell how you'll work the problem. Raise your hand when you've done that much. (Observe students and give feedback.)
- Problem A says: There are 3 sunny days for every 4 days that are not sunny. If there are 84 total days, how many are sunny days?
- The problem names three things: **sunny days, days that are not sunny,** and **total days.** You need a table.
d. Read problem B. Write **table** or **no table.** Raise your hand when you've done that much. (Observe students and give feedback.)
- Problem B says: The machine makes 11 garments every 3 hours. How many hours will it take for the machine to make 99 garments?
- There are only two names: **garments** and **hours.** You don't need a table.
e. Read problem C. Write **table** or **no table.** Raise your hand when you've done that much. (Observe students and give feedback.)
- Problem C says: In a pond, 3 of every 4 fish are bass. If there are 90 fish that are not bass, how many fish are in the pond?
- The problem names three things: **bass, not bass** and **fish.** You need a table.
f. Remember, if the problem names three things, you need a table.
g. Your turn: For each problem in part 2, make the ratio equation or the table. Don't **work** the problem, but show the names in a ratio equation or a table. Also, show the numbers the problem gives. Raise your hand when you've done that much. (Observe students and give feedback.)
h. (Write on the board:)

a.
sunny	3	
not sunny	4	
days		84

b. $\dfrac{\text{garments}}{\text{hr}} \quad \dfrac{11}{3} \left(\quad \right) = \dfrac{99}{\boxed{}}$

c.
bass	3	
not bass		90
fish	4	

- Here's what you should have for each problem. Raise your hand if you got everything right.
i. You'll complete the problems as part of your independent work.

EXERCISE 2 FRACTIONS
Comparison: Unlike Denominators

a. Find part 3.
- To work these problems, you find the lowest common denominator, then answer the question with a statement that shows the two fractions the problem names.
b. Work problem A. Raise your hand when you're finished. (Observe students and give feedback.)
- (Write on the board:)

$$\text{a. } \frac{6}{5} \left(\frac{7}{7}\right) = \frac{42}{35}$$
$$\frac{8}{7} \left(\frac{5}{5}\right) = \frac{40}{35}$$
$$\boxed{\frac{6}{5} > \frac{8}{7}}$$

- Here's what you should have. You compared 6/5 and 8/7. The common denominator is 35. 6/5 is larger because it equals 42/35. That's more than 40/35.
c. Work problem B. Raise your hand when you're finished. (Observe students and give feedback.)
- (Write on the board:)

$$\text{b. } \frac{3}{4} \left(\frac{3}{3}\right) = \frac{9}{12}$$
$$\frac{5}{6} \left(\frac{2}{2}\right) = \frac{10}{12}$$
$$\boxed{\frac{3}{4} < \frac{5}{6}}$$

- Here's what you should have. You compared 3/4 and 5/6. The common denominator is 12. 5/6 is larger because it equals 10/12. That's more than 9/12.
d. Raise your hand if you got both of these problems right.

EXERCISE 3 RATIOS AND PROPORTIONS
Fractional Answers

a. Find part 4.
- The boxed fraction is the answer to a ratio problem. You can't see the work for the ratio problem, but the boxed fraction is the answer to the problem. You're going to write the answer as a mixed number or a whole number.
b. Look at item A. Is the boxed fraction on top or on the bottom? (Signal.) *On the bottom.*
- Read the **boxed fraction** as a division problem. (Signal.) *340 divided by 9.*
- Write the division problem and figure out the whole number and the fraction remainder. Raise your hand when you're finished. (Observe students and give feedback.)
- (Write on the board:)

$$\text{a. } \quad 9 \overline{)340} \quad \boxed{37\tfrac{7}{9}}$$

- Here's what you should have. The fraction 340/9 equals 37 and 7/9. That's the answer to the problem.
c. Problem B. Read the boxed fraction as a division problem. (Signal.) *91 divided by 7.*
- Write the fraction as a division problem and figure out the answer. Raise your hand when you're finished.
(Observe students and give feedback.)
- (Write on the board:)

$$\text{b. } \quad 7 \overline{)91} \quad \boxed{13}$$

- Here's what you should have. The fraction 91/7 equals 13.
d. Your turn: Work problem C. Figure out the whole number and the fraction remainder. Raise your hand when you're finished. (Observe students and give feedback.)
- Check your work. Read the division problem you wrote. (Signal.) *99 divided by 5.*
- What's the answer? (Signal.) *19 and 4-fifths.*
e. Work problem D. Raise your hand when you're finished. (Observe students and give feedback.)
- Check your work. Read the division problem you wrote. (Signal.) *295 divided by 8.*
- What's the answer? (Signal.) *36 and 7-eighths.*

EXERCISE 4 RATIOS AND PROPORTIONS
Fractional Answers

a. Find part 5.
- I'll read what it says. Follow along: Some ratio problems are difficult because you have to figure out a complicated fraction that equals 1.
- You can see a problem: There are 5 pies for every 8 muffins. If there are 11 muffins, how many pies are there?
- The first sentence tells the names and the numbers: 5 pies for every 8 muffins. We write 5 for pies and 8 for muffins.
- The rest of the problem tells us that we have 11 muffins. We have to figure out how many pies we have. We write the parentheses and equal sign and 11 for muffins.
- We have two numbers on the bottom, so we can work that problem: 8 times some fraction equals 11. That's 11/8.
- So the fraction that equals 1 is 11/8 over 11/8. You can see that fraction in the equation.
- Now we work the problem on top. The answer is **55/8** pies. That's 6 and 7/8 pies — almost 7 pies.

b. Find part 6.
c. Problem A: The ratio of perch to trout is 5 to 7. If there are 9 perch, how many trout are there?
- Write the names and the rest of the problem. Figure out the answer. Write it as a fraction and box it. Raise your hand when you're finished.
(Observe students and give feedback.)
- (Write on the board:)

$$\text{a.}\quad \frac{\text{perch}}{\text{trout}}\quad \frac{5}{7}\left(\frac{\frac{9}{5}}{\frac{9}{5}}\right)=\frac{9}{\boxed{\frac{63}{5}}}$$

- Here's what you should have. The fraction that equals 1 is 9/5 over 9/5. When you multiply on the bottom, you get 63/5. That's the answer. There are 63/5 trout. That's 12 and 3/5 trout. I would hate to be that 3/5 trout!

d. Problem B: There are 5 potatoes needed for every 2 dinners. If there are 7 dinners, how many potatoes are needed?
- Write the names and the rest of the problem. Figure out the answer and box it. Raise your hand when you're finished.
(Observe students and give feedback.)

- (Write on the board:)

$$\text{b.}\quad \frac{\text{potatoes}}{\text{dinners}}\quad \frac{5}{2}\left(\frac{\frac{7}{2}}{\frac{7}{2}}\right)=\frac{\boxed{\frac{35}{2}}}{7}$$

- Here's what you should have. The fraction that equals 1 is 7/2 over 7/2. When you multiply on top, you get 35/2. That's the answer. There are 35/2 potatoes. That's 17 and 1/2 potatoes.

e. Your turn: Work the rest of the problems in part 6. Box the answers. Raise your hand when you're finished.
(Observe students and give feedback.)

f. (Write on the board:)

$$\text{c.}\quad \frac{\text{servings}}{\text{pounds}}\quad \frac{8}{6}\left(\frac{\frac{10}{8}}{\frac{10}{8}}\right)=\frac{10}{\boxed{\frac{60}{8}}}$$

$$\text{d.}\quad \frac{\text{planks}}{\text{cabinets}}\quad \frac{4}{3}\left(\frac{\frac{7}{3}}{\frac{7}{3}}\right)=\frac{\boxed{\frac{28}{3}}}{7}$$

g. Check your work. Here's what you should have for problems C and D.
- The answer to problem C is 60/8. There were 60/8 servings. That's 7 and 4/8 servings.
- The answer to problem D is 28/3. There are 28/3 planks. That's 9 and 1/3 planks.

h. Raise your hand if you got all the problems right.

EXERCISE 5 INVERSE OPERATIONS
2-Step Problems

a. Find part 7.
- (Teacher reference:)

$$\text{a.}\quad \overset{?}{\blacksquare}-100=\blacksquare$$
$$\blacksquare \times 4 = 340$$

- These are problems that have a mystery number. You figure out the answer by starting at the end of the second equation and undoing each step.
- How would you undo adding 300? (Signal.) *Subtract 300.*
- How would you undo multiplying by 13? (Signal.) *Divide by 13.*
- How would you undo subtracting 88? (Signal.) *Add 88.*

b. Look at problem A. The last number in the second equation is 340.

- What did you do to get to 340? (Signal.) *Multiplied by 4.*
- How do you undo multiplying by 4? (Signal.) *Divide by 4.*
- Copy problem A and figure out the mystery number. Remember, start with 340. Work backward until you've found the number. Raise your hand when you're finished.
  (Observe students and give feedback.)
- (Write on the board:)

$$\begin{array}{c} ? \\ \text{a. } \boxed{185} - 100 = \boxed{85} \\ \boxed{85} \times 4 = 340 \end{array}$$

- Here's what you should have. You undid multiplying by 4. 340 divided by 4 is 85. Then you undid subtracting 100. You added 100. The mystery number is 185. Raise your hand if you got everything right.
- c. Your turn: Work problem B. Raise your hand when you've found the mystery number.
  (Observe students and give feedback.)
- (Write on the board:)

$$\begin{array}{c} ? \\ \text{b. } \boxed{139} - 97 = \boxed{42} \\ \boxed{42} - 11 = 31 \end{array}$$

- You undid subtracting 11 by adding 11. 31 plus 11 is 42. Then you undid subtracting 97 by adding 97. Everybody, what's 42 plus 97? (Signal.) *139.*
- That's the mystery number.

EXERCISE 6 CIRCLES
Circumference and Radius

a. Find part 8.
- I'll read what it says. Follow along: You've worked circumference problems that tell about the diameter or ask about the diameter.
- Some problems tell about the **radius** or ask about the **radius.** The radius is one-half the diameter.
- You can see the diameter divided into 2-halves. Each half is a radius.
- To calculate the **radius,** you first find the **diameter.** Then you divide by 2.

- If the problem **tells about the radius,** you multiply the radius by 2 to get the diameter. Then you work the problem.
- Remember, the radius is half the diameter.
b. Find part 9.
- Each problem tells about the radius or asks about the radius.
c. Problem A. The circumference of a circle is 62 and 8-tenths centimeters. What is the radius of the circle?
- Use the equation for the circumference of a circle. Find the **diameter,** then divide by 2 to find the radius. Raise your hand when you've answered the question problem A asks.
  (Observe students and give feedback.)
- (Write on the board:)

$$\begin{array}{c} \text{a. } \pi \times d = C \\ 3.14 \left(\dfrac{62.8}{3.14} \right) = 62.8 \\ d = 20 \text{ cm} \\ \boxed{r = 10 \text{ cm}} \end{array}$$

- Here's what you should have. Raise your hand if you got everything right.
d. Problem B. The radius of the circle is 75 inches. What is the circumference?
- The problem tells about the radius. It's 75. What do you do to find the diameter? (Signal.) *Multiply by 2.*
- Find the diameter. Then find the circumference. Raise your hand when you're finished.
  (Observe students and give feedback.)
- (Write on the board:)

$$\begin{array}{c} \text{b. } \pi \times d = C \\ 3.14 \times 150 = C \\ \boxed{C = 471 \text{ in}} \end{array}$$

- Here's what you should have. The diameter is 150 inches. The circumference is 471 inches. Raise your hand if you got everything right.
e. Your turn: Work the rest of the problems in part 9. Raise your hand when you're finished.
  (Observe students and give feedback.)

- (Write on the board:)

c. $\pi \times d = C$

$3.14 \times 50 = C$

$\boxed{C = 157 \text{ ft}}$

d. $\pi \times d = C$

$3.14 \left(\dfrac{942}{3.14} \right) = 942$

$d = 300 \text{ m}$

$\boxed{r = 150 \text{ m}}$

- Here's what you should have for problems C and D. Remember, the **radius** is 1-half the diameter.

EXERCISE 7 INDEPENDENT WORK

a. Do the independent work for lesson 62.
- Finish the problems in part 2. Then do parts 10 through 13.
- (Assign *Bridge to Connecting Math Concepts* *Independent Worksheet* 42 as classwork or homework. Before beginning the next lesson, check the students' independent work.)

Lesson 63

Objectives

- Work a ratio-equation problem that involves complex fractions. (Exercise 1)

- **Compare a decimal value and a fraction to determine which is greater.** (Exercise 2) *Example:* Which is more, .4 or 3/8? Students write:

$$\frac{4}{10} \left(\frac{4}{4} \right) = \frac{16}{40}$$

$$\frac{3}{8} \left(\frac{5}{5} \right) = \frac{15}{40}$$

$$\boxed{.4 > \frac{3}{8}}$$

- Use inverse operations to solve a pair of equations and figure out the starting number. (Exercise 3)

- **Work a mixed set of equivalent-fraction problems, some of which require multiplying by a complex fraction equal to 1.** (Exercise 4)

- Discriminate between problems that require a ratio table and those that require only a ratio equation. (Exercise 5)

- Use the equation $\pi \times d = C$ to figure out the circumference or radius of a circle. (Exercise 6)

EXERCISE 1 RATIOS AND PROPORTIONS
Whole-Number or Mixed-Number Answers

a. Open your textbook to lesson 63 and find part 1.
- These are word problems. To solve them, you'll write a complicated fraction that equals 1. You'll write the answer to the question the problem asks as a fraction. Then you'll work the division problem to figure out the whole number or mixed number that the answer equals.
b. Problem A: A mixture uses 6 parts of silver for every 11 parts of gold. There are 52 parts of silver in the mixture. How many parts of gold are there? Work the problem and box the fraction that answers the question. Then write that fraction as a whole number or mixed number and a unit name. Raise your hand when you know how many parts of gold there are. (Observe students and give feedback.)

- (Write on the board:)

$$\text{a.} \quad \frac{\text{silver}}{\text{gold}} \quad \frac{6}{11} \left(\frac{\frac{52}{6}}{\frac{52}{6}} \right) = \frac{52}{\frac{572}{6}} \quad 95\frac{2}{6} \text{ parts}$$

- Here's what you should have. The fraction that equals 1 is 52/6 over 52/6. The answer is 572/6. That's 95 and 2/6 parts of gold.
- c. Your turn: Work problem B. Raise your hand when you're finished.
  (Observe students and give feedback.)
- (Write on the board:)

$$\text{b.} \quad \frac{\text{birds}}{\text{oz}} \quad \frac{5}{2} \left(\frac{\frac{88}{5}}{\frac{88}{5}} \right) = \frac{88}{\frac{176}{5}} \quad 35\frac{1}{5} \text{ oz}$$

- Here's what you should have. There were 5 birds for every 2 ounces of birdseed. If there was a total of 88 birds, how much birdseed was there? The answer is 176/5. That's 35 and 1/5 ounces of birdseed. Raise your hand if you got it right.

EXERCISE 2 FRACTIONS AND DECIMALS
Comparison

a. Find part 2.
- These problems require you to compare fractions with decimal values. To work the problem, first **write the decimal value as a fraction.** Then find the lowest common denominator for that fraction and the other fraction.
b. Problem A asks: Which is more, 3/8 or 4-tenths?
- Write the two values in a column but show 4-tenths as a fraction. Then find the lowest common denominator and write the statement that tells which value is greater. Raise your hand when you're finished.
  (Observe students and give feedback.)
- (Write on the board:)

$$\text{a.} \quad \frac{3}{8} \left(\frac{5}{5} \right) = \frac{15}{40}$$
$$\frac{4}{10} \left(\frac{4}{4} \right) = \frac{16}{40}$$
$$\boxed{\frac{3}{8} < .4}$$

- Here's what you should have. The lowest common denominator is 40. 3/8 is 15/40; 4/10 is 16/40. Your statement should show that 3/8 is less than 4-tenths.
c. For problem B, you'll compare 12-hundredths and 1/6. The simplest way to find the common denominator is to see if 6 goes into 100, if it goes into 200, 300, and so forth.
- Don't write out all the numbers for counting by 6. Just do the division for the different hundreds numbers.
- Work the rest of the problems in part 2. Remember, show the decimal value as a fraction. Raise your hand when you're finished.
  (Observe students and give feedback.)
d. (Write on the board:)

$$\text{b.} \quad \frac{12}{100} \left(\frac{3}{3} \right) = \frac{36}{300}$$
$$\frac{1}{6} \left(\frac{50}{50} \right) = \frac{50}{300}$$
$$\boxed{.12 < \frac{1}{6}}$$

$$\text{c.} \quad \frac{6}{10} \left(\frac{7}{7} \right) = \frac{42}{70}$$
$$\frac{4}{7} \left(\frac{10}{10} \right) = \frac{40}{70}$$
$$\boxed{.6 > \frac{4}{7}}$$

- Here's what you should have for problems B and C.

EXERCISE 3 INVERSE OPERATIONS
2-Step Problems

a. Find part 3.
b. Copy problem A. Work backwards. Figure out the mystery number. Raise your hand when you're finished.
  (Observe students and give feedback.)
- (Write on the board:)

$$\text{a.} \quad \boxed{125}^{?} - 100 = \boxed{25}$$
$$\boxed{25} \times 3 = 75$$

- Here's what you should have. The mystery number is 125.
c. Work problem B. Raise your hand when you're finished. (Observe students and give feedback.)
- (Write on the board:)

$$\overset{?}{\boxed{215}} \times 2 = \boxed{430}$$
$$\boxed{430} - 420 = 10$$

- Here's what you should have. The mystery number is 215.
d. Raise your hand if you got everything right.

EXERCISE 4   COMPLEX FRACTIONS
Discrimination

a. Find part 4.
   I'll read what it says. Follow along: You can figure out the answer to all ratio problems by showing the fraction that equals 1 as a fraction over a fraction. But for some problems, it's faster if you don't.
- You look at the problem you start with.
- If you know that the answer to that problem is a whole number, you just write a simple fraction.
- You can see an equivalent-fraction problem: 5/9 equals 35 over box.
- 5 times some value equals 35. You know the missing value is 7. So the fraction that equals 1 is 7/7.
- For the next equation, the problem on top is: 5 times some value equals 425.
- You know the missing value is a whole number, but you don't know what number it is. So you could write the missing value as the fraction 425/5. Or you could figure out what 425 divided by 5 equals and write that number.
- For the last equation, you work the problem: 5 times some value equals 3. You know the missing value is **not** a whole number. So you write the fraction 3/5. The fraction that equals 1 is 3/5 over 3/5.
b. Find part 5.
- For some of these problems, you'll write a simple fraction that equals 1. For others, you'll write a fraction over a fraction.
c. Problem A. In the denominator, you have 4 times some value equals 94. Do you write the missing value as a fraction or a whole number? (Signal.) *A fraction.*

- Problem B. In the numerator, you have 3 times some value equals 18. Do you write the missing value as a fraction or a whole number? (Signal.) *A whole number.*
- Problem C. In the numerator, you have 7 times some value equals 91. Do you write the missing value as a fraction or a whole number? (Signal.) *A fraction.*
d. Remember, you'll always get the correct answer by showing the fraction that equals 1 as a fraction over the same fraction. But it's faster to use a simple fraction if you can.
e. Your turn: Work problem A. Show the answer as a whole number or a mixed number. Raise your hand when you're finished.
   (Observe students and give feedback.)
- (Write on the board:)

$$\text{a.} \quad \frac{10}{4} \left(\frac{\frac{94}{4}}{\frac{94}{4}} \right) = \frac{\boxed{\frac{940}{4}}}{94} \quad \boxed{235}$$

- Here's what you should have. The answer is 940/4. That's 235.
f. Work the rest of the problems in part 5. Raise your hand when you're finished.
   (Observe students and give feedback.)
   Key:

$$\text{b.} \quad \frac{3}{7} \left(\frac{6}{6} \right) = \frac{18}{\boxed{42}} \qquad \text{d.} \quad \frac{9}{8} \left(\frac{5}{5} \right) = \frac{45}{\boxed{40}}$$

$$\text{c.} \quad \frac{7}{3} \left(\frac{\frac{91}{7}}{\frac{91}{7}} \right) = \frac{91}{\boxed{\frac{273}{7}}} \quad \boxed{39} \qquad \text{e.} \quad \frac{9}{6} \left(\frac{\frac{92}{6}}{\frac{92}{6}} \right) = \frac{\boxed{\frac{828}{6}}}{92} \quad \boxed{138}$$

$$\left[\text{or} \quad \frac{7}{3} \left(\frac{13}{13} \right) = \frac{91}{\boxed{39}} \right] \qquad \text{f.} \quad \frac{12}{5} \left(\frac{10}{10} \right) = \frac{\boxed{120}}{50}$$

g. Find part J on page 253. That shows what you should have.
h. Problem B. The answer is 42.
- Problem C. The answer is 39.
- Problem D. The answer is 40.
- Problem E. The answer is 138.
- Problem F. The answer is 120.
i. Raise your hand if you got them all right.

EXERCISE 5 RATIOS AND PROPORTIONS
  Table Discrimination

a. Find part 6.
• You'll figure out whether you need a table to work each problem. Remember, if the problem gives only two names, you don't need a table. If the problem gives three names, you need a table.
b. Read problem A. Write **table** or **no table.** Don't work the problem. Raise your hand when you're finished. √
• Problem A says: There are 3 sunny days for every 5 days that are not sunny. If there are 66 sunny days, how many days are not sunny? What did you write? (Signal.) *No table.*
• There are only two names: **Sunny** and **not sunny.** So you don't need a table.
c. Write **table** or **no table** for the rest of the problems in part 6. Raise your hand when you're finished.
  (Observe students and give feedback.)
  Key:
    b. no table
    c. table

d. Check your work.
• Problem B says: The ratio of girls to boys is 6 to 5. If there are 30 boys, how many girls are there? What did you write? (Signal.) *No table.*
• There are only two names. You don't need a table.
• Problem C: At a school, there are students and employees. The ratio of employees to total **people** is 4 to 9. If there are 603 total people in the school, how many are employees? How many are students? What did you write? (Signal.) *Table.*
• The problem names **three** things, so you need a table.
e. Your turn: For each problem in part 6, make the ratio equation or the table. Don't work the problem, but show the names in a ratio equation or a table. Also, show the numbers the problem gives. Raise your hand when you've done that much. (Observe students and give feedback.)

f. (Write on the board:)

a. $\dfrac{\text{sunny}}{\text{not sunny}} \quad \dfrac{3}{5} \left(\quad\right) = \dfrac{66}{\square}$

b. $\dfrac{\text{girls}}{\text{boys}} \quad \dfrac{6}{5} \left(\quad\right) = \dfrac{\square}{30}$

c.

students		
employees	4	
people	9	603

• Here's what you should have for each problem. Raise your hand if you got everything right.
g. You'll complete the problems as part of your independent work.

EXERCISE 6 CIRCLES
  Circumference and Radius

a. Find part 7.
• One of these problems asks about the radius. Some tell about the radius. Remember, the radius is one-half the diameter. So if the problem tells about the radius, you multiply by 2 to get the diameter.
• If the problem gives the circumference and asks about the radius, you find the diameter. Then you divide by 2 to get the radius.
• You always start with the equation for the circumference of the circle.
b. Use your calculator to work problem A. Raise your hand when you're finished.
  (Observe students and give feedback.)
• (Write on the board:)

a. $\pi \times d = C$

$3.14 \left(\dfrac{125.6}{3.14}\right) = 125.6$

$d = 40 \text{ in}$

$\boxed{r = 20 \text{ in}}$

• Here's what you should have. The problem gave information about the circumference so you figured out the diameter. Then you divided by 2 for the radius. For a circle with a circumference of 125 and 6-tenths inches, the radius is 20 inches.

c. Work the rest of the problems in part 7. Raise your hand when you're finished.
(Observe students and give feedback.)
- (Write on the board:)

b. $\pi \times d = C$

$3.14 \times 942 = C$

$\boxed{C = 2957.88 \text{ cm}}$

c. $\pi \times d = C$

$3.14 \times 1256 = C$

$\boxed{C = 3943.84 \text{ cm}}$

- Here's what you should have for problems B and C. Raise your hand if you got everything right.

EXERCISE 7 INDEPENDENT WORK

a. Do the independent work for lesson 63.
- Finish the problems in part 6. Then do parts 8 through 14.

- (Assign **Bridge to Connecting Math Concepts** *Independent Worksheet* 43 as classwork or homework. Before beginning the next lesson, check the students' independent work.)

Lesson 64

Objectives

- Compare a decimal value and a fraction to determine which is greater. (Exercise 1)

- Use inverse operations to solve a pair of equations and figure out the starting number. (Exercise 2)

- **Use the equation $\pi \times d = C$ to figure out the circumference, radius or diameter of a circle.** (Exercise 3)

- **Work a 2-step word problem that asks about the starting number and specifies two operations.** (Exercise 4)
  Note: Problems are of the form: You start with a mystery number and subtract 18. Then you multiply by 5. You end up with 45. What's the starting number?

- Work a ratio-equation problem that involves complex fractions. (Exercise 5)

- Discriminate between problems that require a ratio table and those that require only a ratio equation. (Exercise 6)

EXERCISE 1 FRACTIONS AND DECIMALS
Comparison

a. Open your textbook to lesson 64 and find part 1.
- For these problems you'll compare fractions with decimal values. Remember, write the decimal value as a fraction, then find the lowest common denominator. Then write a statement that compares the original values. Raise your hand when you're finished.
(Observe students and give feedback.)
Key:

a. $\dfrac{2}{8}\left(\dfrac{25}{25}\right) = \dfrac{50}{200}$

$\dfrac{25}{100}\left(\dfrac{2}{2}\right) = \dfrac{50}{200}$

$\boxed{\dfrac{2}{8} = .25}$

b. $\dfrac{5}{9}\left(\dfrac{10}{10}\right) = \dfrac{50}{90}$

$\dfrac{5}{10}\left(\dfrac{9}{9}\right) = \dfrac{45}{90}$

$\boxed{\dfrac{5}{9} > .5}$

c. $\dfrac{3}{4}\left(\dfrac{25}{25}\right) = \dfrac{75}{100}$

$\dfrac{74}{100} = \dfrac{74}{100}$

$\boxed{\dfrac{3}{4} > .74}$

b. Find part J on page 257. That shows what you should have.

EXERCISE 2 INVERSE OPERATIONS
2-Step Problems

a. Find part 2.
b. These are mystery-number problems. Copy them and figure out each mystery number. Raise your hand when you're finished.
   (Observe students and give feedback.)
c. (Write on the board:)

> ?
> a. $\boxed{82}$ x 2 = $\boxed{164}$
> $\boxed{164}$ – 64 = 100
>
> ?
> b. $\boxed{505}$ – 501 = $\boxed{4}$
> $\boxed{4}$ x 3 = 12
>
> ?
> c. $\boxed{122}$ + 28 = $\boxed{150}$
> $\boxed{150}$ – 100 = 50

• Check your work. Here's what you should have for each problem. Raise your hand if you got everything right.

EXERCISE 3 CIRCLES
Circumference/Diameter/Radius

a. Find part 3.
• Some of these problems tell about the diameter or ask about the diameter. Others tell or ask about the radius. Remember, the radius is one-half the diameter.
b. Work problem A. Raise your hand when you're finished. (Observe students and give feedback.)
• (Write on the board:)

> a. π x d = C
> $3.14 \left(\frac{628}{3.14} \right) = 628$
> $\boxed{d = 200 \text{ cm}}$

• Here's what you should have. The problem asks about the diameter.
c. Work problem B. Raise your hand when you're finished. (Observe students and give feedback.)

• (Write on the board:)

> b. π x d = C
> 3.14 x 11.6 = C
> $\boxed{C = 36.42 \text{ cm}}$

• Here's what you should have. The problem gives information about the radius. So the diameter is 2 times 5 and 8-tenths. That's 11 and 6-tenths centimeters. The circumference is 36 and 42-hundredths centimeters.
d. Your turn: Work the rest of the problems in part 3. Raise your hand when you're finished.
   (Observe students and give feedback.)
e. (Write on the board:)

> c. π x d = C
> 3.14 x 120 = C
> $\boxed{C = 376.8 \text{ yd}}$
>
> d. π x d = C
> 3.14 x 250 = C
> $\boxed{C = 785 \text{ m}}$
>
> e. π x d = C
> $3.14 \left(\frac{94.2}{3.14} \right) = 94.2$
> d = 30 in
> $\boxed{r = 15 \text{ in}}$

• Here's what you should have for problems C, D and E.
• The circumference of circle C is 376 and 8-tenths yards.
• The circumference of circle D is 785 meters.
• The radius of circle E is 15 inches.
f. Raise your hand if you got everything right.

EXERCISE 4 PROBLEM SOLVING
Inverse Operations: 2-Steps

a. Find part 4.
• These are word problems. You'll work them by writing two equations, then undoing each step to figure out the mystery number. For these problems, the first sentence tells how to write the first equation. The first number in that equation is the mystery number.

b. **Problem A:** You start with the mystery number and subtract 18.
- Write the equation for the first sentence. Show a question mark and box for the number you start with. Write a box for the number you end up with. Raise your hand when you've done that much. (Observe students and give feedback.)
- (Write on the board:)

$$\text{a. } \boxed{}^{\,?} - 18 = \boxed{}$$

- Here's the first equation. It shows that you start with the mystery number and subtract 18.
- The next sentences tell about the second equation. The problem says that you multiply by 5. You end up with 45.
- The starting number for that equation is a box.
- (Write to show:)

$$\text{a. } \boxed{}^{\,?} - 18 = \boxed{}$$
$$\boxed{} \times 5 = 45$$

- Here's the second equation. It shows multiplying by 5 and ending up with 45.
- Copy the equation, but don't work the problem. √
c. **Problem B.** The first sentence tells about the first equation: You start with the mystery number and multiply by 9.
- Write the first equation. Raise your hand when you've done that much.
  (Observe students and give feedback.)
- (Write on the board:)

$$\text{b. } \boxed{}^{\,?} \times 9 = \boxed{}$$

- Here's what you should have.
- The rest of the problem says: Then you add 15. You end up with 69. Write the second equation. Remember, it begins with a box. Raise your hand when you're finished.
  (Observe students and give feedback.)
- (Write to show:)

$$\text{b. } \boxed{}^{\,?} \times 9 = \boxed{}$$
$$\boxed{} + 15 = 69$$

- Here's what you should have.
d. **Your turn:** Read problems C and D. Write two equations for each problem. Raise your hand when you've written both equations for C and D. (Observe students and give feedback.)
e. (Write on the board:)

$$\text{c. } \boxed{}^{\,?} - 20 = \boxed{}$$
$$\boxed{} \times 4 = 84$$

$$\text{d. } \boxed{}^{\,?} \times 3 = \boxed{}$$
$$\boxed{} \times 5 = 105$$

- Here's what you should have for problems C and D.
f. Now figure out the starting number for each problem. Raise your hand when you're finished. (Observe students and give feedback.)
g. (Write to show:)

$$\text{a. } \boxed{27}^{\,?} - 18 = \boxed{9}$$
$$\boxed{9} \times 5 = 45$$

$$\text{b. } \boxed{6}^{\,?} \times 9 = \boxed{54}$$
$$\boxed{54} + 15 = 69$$

$$\text{c. } \boxed{41}^{\,?} - 20 = \boxed{21}$$
$$\boxed{21} \times 4 = 84$$

$$\text{d. } \boxed{7}^{\,?} \times 3 = \boxed{21}$$
$$\boxed{21} \times 5 = 105$$

- Here's what you should have.

EXERCISE 5 RATIOS AND PROPORTIONS
Mixed-Number Answers

a. Find part 5.
- You're going to work these ratio problems and write each answer as a mixed number and a unit name.

b. Problem A: If 6 bottles fill 1 container, how many containers do 43 bottles fill?
• Your turn: Work problem A. Remember to box your answer. Raise your hand when you have a **fraction** for the answer.
  (Observe students and give feedback.)
• (Write on the board:)

$$\text{a. } \frac{\text{bottles}}{\text{container}} \quad \frac{6}{1} \left(\frac{\frac{43}{6}}{\frac{43}{6}} \right) = \frac{43}{\boxed{\frac{43}{6}}}$$

• Check your work. Here's what you should have. When you multiply, you get 43/6.
• You're going to figure out the mixed number and the unit for the answer. Do it. Raise your hand when you're finished.
  (Observe students and give feedback.)
• You divided 43 by 6. The answer is 7 and 1/6. What's the unit name? (Signal.) *Containers.*
• That means you'd fill 7 whole containers and 1/6 of the next container. Raise your hand if you wrote **7 and 1/6 containers** as the answer.
c. Your turn: Work problem B. Write the answer as a mixed number and a unit name. Raise your hand when you're finished.
  (Observe students and give feedback.)
• (Write on the board:)

$$\text{b. } \frac{\text{mi}}{\text{hr}} \quad \frac{24}{7} \left(\frac{\frac{6}{7}}{\frac{6}{7}} \right) = \frac{\boxed{\frac{144}{7}}}{6}$$

• Check your work. Here's the ratio problem. When you multiply, you get the fraction 144/7. You wrote that fraction as a division problem. The answer is 20 and 4/7. What's the unit name? (Signal.) *Miles.*
• That means Jimmy would have walked **20 miles and 4/7** of the next mile.
d. Your turn: Work problem C. Write the answer as a mixed number and a unit name. Raise your hand when you're finished.
  (Observe students and give feedback.)
• (Write on the board:)

$$\text{c. } \frac{\text{containers}}{\text{oz}} \quad \frac{5}{18} \left(\frac{\frac{8}{5}}{\frac{8}{5}} \right) = \frac{8}{\boxed{\frac{144}{5}}}$$

• Check your work. Here's the ratio problem. When you multiply, you get the fraction 144/5. You wrote that fraction as a division problem. The answer is 28 and 4/5. What's the unit name? (Signal.) *Ounces.*
• That means 8 containers would hold **28 and 4/5 ounces.**
e. Your turn: Work problem D. Write the answer as a mixed number and a unit name. Raise your hand when you're finished.
  (Observe students and give feedback.)
• (Write on the board:)

$$\text{d. } \frac{\text{hr}}{\text{rooms}} \quad \frac{8}{6} \left(\frac{\frac{14}{8}}{\frac{14}{8}} \right) = \frac{14}{\boxed{\frac{84}{8}}}$$

• Check your work. Here's the ratio problem. When you multiply, you get the fraction 84/8. You wrote that fraction as a division problem. The answer is 10 and 4/8. What's the unit name? (Signal.) *Rooms.*
• That means in 14 hours, **10 and 4/8 rooms** would be painted.

EXERCISE 6 RATIOS AND PROPORTIONS
Table Discrimination

a. Find part 6.
• For some of these problems, you need a ratio table. For others you don't. Remember, if the problem gives three names, you need a table. If the problem gives only two names, you don't need a table.
b. Make the ratio equation or the table for each problem. Don't work the problems, just show the names and the numbers each problem gives. Raise your hand when you've done that much. (Observe students and give feedback.)

Key:

a. $\dfrac{\text{blue}}{\text{red}} \dfrac{4}{5} \left(\right) = \dfrac{\boxed{}}{2000}$

b. $\dfrac{\text{secretaries}}{\text{departments}} \dfrac{3}{4} \left(\right) = \dfrac{\boxed{}}{48}$

c.
red	3	
yellow	7	
buttons		3000

d.
cashiers	3	
not cashiers		
employees	11	121

c. Find part K. That shows what you should have for each problem. Raise your hand if you got everything right.

EXERCISE 7 INDEPENDENT WORK

a. Do the independent work for lesson 64.
- Finish the problems in part 6. Then do parts 7 through 13.
- (Assign *Bridge to Connecting Math Concepts* *Independent Worksheet* 44 as classwork or homework. Before beginning the next lesson, check the students' independent work.)

Lesson 65

Objectives

- **Add a whole number and fractions with unlike denominators.** (Exercise 1)

- **Determine whether a missing factor is more than 1 or less than 1, then figure out the fraction it equals.** (Exercise 2)

- **Compare a factor and the product in a multiplication equation.** (Exercise 3)
  Note: Problems are of the form:
  $A \times 1.09 = B$. Students apply the rule that if A is multiplied by more than 1, the product is more than A. For this example:
  $A \times 1.09 = B$, B is the larger value.
  Conversely, if A is multiplied by less than 1, the product is less than A. For this example:
  $A \times 1/3 = B$, B is the smaller value.

- **Given the area of a rectangle and the length for one pair of sides, figure out the length of the other pair of sides.** (Exercise 4)

- Work a ratio-equation problem that involves complex fractions. (Exercise 5)

- Work a 2-step word problem that asks about the starting number and specifies two operations. (Exercise 6)

EXERCISE 1 FRACTION OPERATIONS
Whole Numbers and Fractions

a. Open your textbook to lesson 65 and find part 1.
- These are common denominator problems that have a whole number and fractions.
b. Problem A: 1/2 plus 3 plus 3/4.
- Write the three values in a column. Find the lowest common denominator. Raise your hand when you've written the common denominator for all three values.
  (Observe students and give feedback.)
- (Write on the board:)

$$
\begin{array}{rcl}
\textbf{a.} \quad \dfrac{1}{2} & = & \overline{4} \\[2mm]
3 & = & \overline{4} \\[2mm]
+\,\dfrac{3}{4} & = & +\,\overline{4} \\
\hline
\end{array}
$$

- Here's what you should have so far. The lowest common denominator is 4.
- Now work the equivalent-fraction problems. Write the answer to the whole problem as a mixed number. Raise your hand when you're finished. (Observe students and give feedback.)
- (Write to show:)

a. $\dfrac{1}{2}\left(\dfrac{2}{2}\right) = \dfrac{2}{4}$

$3 = \dfrac{12}{4}$

$+\dfrac{3}{4} = +\dfrac{3}{4}$

$\dfrac{17}{4} = \boxed{4\,\tfrac{1}{4}}$

- Here's what you should have. The answer is 4 and 1/4. Raise your hand if you got everything right.
c. Work problem B. Raise your hand when you're finished. (Observe students and give feedback.)
- (Write on the board:)

b. $2 = \dfrac{20}{10}$

$\dfrac{3}{10} = \dfrac{3}{10}$

$+\dfrac{1}{5}\left(\dfrac{2}{2}\right) = +\dfrac{2}{10}$

$\dfrac{25}{10} = \boxed{2\,\tfrac{5}{10}}$

- Here's what you should have. The answer is 2 and 5/10.
d. Work problem C. Raise your hand when you're finished. (Observe students and give feedback.)

- (Write on the board:)

c. $\dfrac{1}{4}\left(\dfrac{3}{3}\right) = \dfrac{3}{12}$

$\dfrac{1}{6}\left(\dfrac{2}{2}\right) = \dfrac{2}{12}$

$+\ 1 = +\dfrac{12}{12}$

$\dfrac{17}{12} = \boxed{1\,\tfrac{5}{12}}$

- Here's what you should have. The answer is 1 and 5/12.

EXERCISE 2 MULTIPLICATION
Missing Factor Compared to 1

a. Find part 2.
- I'll read what it says. Follow along: When you multiply by a value, you may end up with more than you started with. You may end up with the same value you started with. Or you may end up with less than you started with.
- Here are the rules:
- If you end up with more than you started with, the value you multiply by is more than 1.
- If you end up with the same value you started with, the value you multiply by is 1.
- If you end up with less than you started with, the value you multiply by is less than 1.
b. Listen: If you end up with less than you start with, what do you know about the value you multiply by? (Signal.) *It's less than 1.*
- If you end up with more than you start with, what do you know about the value you multiply by? (Signal.) *It's more than 1.*
c. Find part 3.
- The value you multiply by in each problem is not shown. You're going to indicate whether it is more than 1 or less than 1.
d. Look at the sample problem. You start out with 90 and multiply by M. You end up with 91.
- Did you end up with more or less than you started with? (Signal.) *More.*
- So what do you know about M? (Signal.) *It's more than 1.*
- (Write on the board:)

$$M > 1$$

- Here's the statement that shows M is more than 1.
e. Problem A. What number are you starting with? (Signal.) *70.*
- What number are you ending with? (Signal.) *17.*
- Is that more or less than you start with? (Signal.) *Less.*
- So what do you know about M? (Signal.) *It's less than 1.*
- Yes, M is less than 1.
- Write the inequality statement for M. √
- (Write on the board:)

> **a.   M < 1**

- Here's what you should have.
f. Write the inequality statement for problem B. Remember, if you start with more than you end with, you multiply by more than 1. Raise your hand when you're finished.
- (Write on the board:)

> **b.   M < 1**

- Here's what you should have. You end up with less than you start with, so M is less than 1.
g. Your turn: Write the statement with M and 1 for the rest of the problems in part 3. Raise your hand when you're finished.
(Observe students and give feedback.)
h. (Write on the board:)

> **c.   M > 1**
> **d.   M < 1**

- Here's what you should have for problems C and D. Raise your hand if you got everything right.
i. For each item, write the equation that shows what M equals. Write the missing value as a fraction. Then check it to see whether it is more than 1 or less than 1. Raise your hand when you're finished.
(Observe students and give feedback.)
- (Write to show:)

> **a. M = $\frac{17}{70}$ b. M = $\frac{1}{2}$**
>
> **c. M = $\frac{123}{5}$ d. M = $\frac{11}{56}$**

- Here's what you should have for each problem.
- For problem A: M is 17/70. That's less than 1.

- For problem B: M is 1/2. That's less than 1.
- For problem C: M is 123/5. That's more than 1.
- For problem D: M is 11/56. That's less than 1.
j. Remember, you don't have to work the problem to figure out whether the number you multiply by is more than 1 or less than 1.

EXERCISE 3   MULTIPLICATION
Comparison: Factor and Product

a. Find part 4.
- These problems are different. Each problem tells what you multiply R by to get M.
- Some problems show the value you multiply by as a fraction. Others show it as a decimal value.
- For each problem, you'll tell which value is larger, R or M. **Remember, if you multiply R by more than 1, M is larger. If you multiply R by less than 1, M is smaller.**
b. Problem A: R is multiplied by 1 and 3-hundredths to get M. Is R multiplied by more than 1 or less that 1? (Signal.) *More than 1.*
- Write R or M to show which is larger. Raise your hand when you're finished.
- Everybody, which is larger, R or M? (Signal.) *M.*
c. Work problem B. Raise your hand when you're finished. (Observe students and give feedback.)
- R is multiplied by 30/29. That's more than 1. So which is larger, R or M? (Signal.) *M.*
d. Work the rest of the problems in part 4. Raise your hand when you're finished.
(Observe students and give feedback.)
e. Check your work.
- Problem C: R is multiplied by 2-tenths. Which is larger, R or M? (Signal.) *R.*
- Problem D: R is multiplied by 3 over 1. So, which is larger, R or M? (Signal.) *M.*
- Problem E: R is multiplied by 1/3. Which is larger, R or M? (Signal.) *R.*
- Problem F: R is multiplied by 17/12. So, which is larger, R or M? (Signal.) *M.*
f. Raise your hand if you got everything right.

EXERCISE 4   AREA
Missing Length

a. Find part 5.
- This is a new kind of problem. The area of the rectangles are shown, but the length of a side is missing.
b. The sample problem shows the area and the height.
- One way to work the problem is to rewrite the equation for the area so it ends with a number for the area.

- (Write on the board:)

$$b \times h = A$$

- That's base times height equals area.
- Now you can put in the numbers the problem gives and figure out the missing number.
- (Write to show:)

$$b \times h = A$$
$$b \times 16 = 192$$

- You can figure out the missing number by undoing. You multiplied by 16 to get to 192. So you work the division problem 192 divided by 16. The answer is 12. So the length of the base is 12 feet—not **square** feet.
- (Write to show:)

$$b \times h = A$$
$$b \times 16 = 192$$
$$\boxed{b = 12 \text{ ft}}$$

c. Your turn: Work problem A. Start with the equation for the area of a rectangle, but rewrite it so you end with the area. Then put in the numbers the problem gives and figure out the length of the base. Raise your hand when you're finished.
   (Observe students and give feedback.)
- (Write on the board:)

$$\text{a. } b \times h = A$$
$$b \times 21 = 210$$
$$\boxed{b = 10 \text{ in}}$$

- Here's what you should have. The base is 10 inches long. Raise your hand if you got everything right.
d. Work problem B. It gives the length of the base. You have to figure out the height. Raise your hand when you're finished.
   (Observe students and give feedback.)
- (Write on the board:)

$$\text{b. } 33 \times h = 396$$
$$33 \left(\frac{396}{33} \right) = 396$$
$$\boxed{h = 12 \text{ m}}$$

- Here's what you should have. The missing value is 12 meters.
e. Remember how to work these problems.

EXERCISE 5 RATIOS AND PROPORTIONS
Whole-Number or Mixed-Number Answers

a. Find part 6.
- These are ratio problems. You're going to write each answer as a mixed number or whole number with a unit name.
b. Problem A asks: How many bags contain 41 pounds?
- Your answer will have a mixed number and the word **bags.** Your turn: Work problem A. Raise your hand when you're finished.
   (Observe students and give feedback.)
- (Write on the board:)

$$\text{a. } \frac{\text{bags}}{\text{lb}} \quad \frac{8}{5} \left(\frac{\frac{41}{5}}{\frac{41}{5}} \right) = \frac{\boxed{\frac{328}{5}}}{41}$$

- Check your work. Here's the ratio problem and the answer. You should have written **65 and 3/5 bags.**
c. Problem B asks: How long will it take to make 21 tractors?
- Your answer will have a number and unit name for the time it will take. Work problem B. Raise your hand when you're finished.
   (Observe students and give feedback.)
- (Write on the board:)

$$\text{b. } \frac{\text{tractors}}{\text{days}} \quad \frac{7}{2} \left(\frac{3}{3} \right) = \frac{21}{\boxed{6}}$$

- Check your work. Here's the ratio problem. The problem asks, how long will it take to make 21 tractors? Everybody, what's the answer? (Signal.) *6 days.*
- Yes, 6 **days.**
d. Problem C asks: How far does the train go in 2 hours?
- Your answer will have a mixed number and unit name for how far it will go. Work problem C. Raise your hand when you're finished.
   (Observe students and give feedback.)

- (Write on the board:)

$$\text{c.} \quad \frac{\text{mi}}{\text{hr}} \quad \frac{128}{3} \left(\frac{\frac{2}{3}}{\frac{2}{3}} \right) = \frac{\frac{256}{3}}{2}$$

- Check your work. Here's the ratio problem. The problem asks, how far does the train go in 2 hours? Everybody, what's the answer? (Signal.) *85 and 1-third miles.*
- The train travels **85 and 1/3 miles** in 2 hours.

EXERCISE 6 PROBLEM SOLVING
Inverse Operations: 2 Steps

a. Find part 7.
- These are word problems that tell about mystery numbers. For each problem, you'll write two equations.
b. Read problem A. Write both equations. Remember, the mystery number is a box with a question mark. Raise your hand when you've written the equations with boxes. Don't work the problem.
(Observe students and give feedback.)
- (Write on the board:)

$$\text{a.} \quad \overset{?}{\boxed{}} + 52 = \boxed{}$$
$$\boxed{} \times 2 = 678$$

- Here are the equations you should have.
- Now figure out the mystery number. You can use your calculator. Raise your hand when you're finished.
(Observe students and give feedback.)
- (Write to show:)

$$\text{a.} \quad \overset{?}{\boxed{287}} + 52 = \boxed{339}$$
$$\boxed{339} \times 2 = 678$$

- Check your work. Here's what you should have. The mystery number is 287.
c. Your turn: Write equations for problem B and figure out the mystery number. Raise your hand when you're finished.
(Observe students and give feedback.)

- (Write on the board:)

$$\text{b.} \quad \overset{?}{\boxed{49}} \div 7 = \boxed{7}$$
$$\boxed{7} + 77 = 84$$

- Here's what you should have.
d. Your turn: Work problems C and D. Raise your hand when you're finished.
(Observe students and give feedback.)
Key:

$$\text{c.} \quad \overset{?}{\boxed{37}} \times 16 = \boxed{592}$$
$$\boxed{592} - 506 = 86$$

$$\text{d.} \quad \overset{?}{\boxed{1}} \times 30 = \boxed{30}$$
$$\boxed{30} + 280 = 310$$

e. Find part J on page 261. That shows what you should have for problems C and D.
- Problem C. What's the starting number? (Signal.) *37.*
- Problem D. What's the starting number? (Signal.) *1.*
f. Raise your hand if you got everything right.

EXERCISE 7 INDEPENDENT WORK

- (In addition to independent work in the textbook, assign **Bridge to Connecting Math Concepts** *Independent Worksheet* 45 as classwork or homework. Before beginning the next lesson, check the students' independent work.)

Lesson 66

Objectives

- Determine whether a missing factor is more than 1 or less than 1, then figure out the fraction it equals. (Exercise 1)

- Add a whole number and fractions with unlike denominators. (Exercise 2)

- Compare a factor and the product in a multiplication equation. (Exercise 3)

- **Solve a set of multiplication/division word problems by making ratio equations.** (Exercise 4)

- Work a 2-step word problem that asks about the starting number and specifies two operations. (Exercise 5)

- Given the area of a rectangle and the length for one pair of sides, figure out the length of the other pair of sides. (Exercise 6)

EXERCISE 1 MULTIPLICATION
Missing Factor Compared to 1

a. Open your textbook to lesson 66 and find part 1.
- Each problem shows the starting number and the ending number. You have to indicate whether the number you multiply by is more than 1 or less than 1.
- If you end up with more than you start out with, what do you know about the number you multiply by? (Signal.) *It's more than 1.*
- If you end up with less than you start out with, what do you know about the number you multiply by? (Signal.) *It's less than 1.*
b. Problem A. Write the inequality statement about M. Raise your hand when you're finished. √
- (Write on the board:)

> **a. M < 1**

- Here's what you should have. 11 is less than 35, so M is less than 1.
c. Figure out what the missing number **is** and check whether it is less than 1. Show the missing number as a fraction. Raise your hand when you're finished.
(Observe students and give feedback.)

- (Write to show:)

> **a. M < 1**
>
> $M = \dfrac{11}{35}$

- Here's what you should have. The missing number is 11/35. That's less than 1.
d. Work problem B. Write a statement that tells whether M is more or less than 1. Then write what M equals. Raise your hand when you're finished. (Observe students and give feedback.)
- (Write on the board:)

> **b. M > 1**
>
> $M = \dfrac{6}{5}$

- Here's what you should have. M is more than 1. M equals 6/5.
e. Work problem C. Write the inequality statement about M. Then write what M equals. Raise your hand when you're finished.
(Observe students and give feedback.)
- (Write on the board:)

> **c. M < 1**
>
> $M = \dfrac{23}{24}$

- Here's what you should have. M is less than 1. M equals 23/24.
f. Raise your hand if you got everything right.

EXERCISE 2 FRACTION OPERATIONS
Whole Numbers and Fractions

a. Find part 2.
- These are common denominator problems that have a whole number as one of the values.
b. Work problem A. Show the answer as a mixed number. Raise your hand when you're finished.
(Observe students and give feedback.)

- (Write on the board:)

$$\text{a.} \quad \frac{1}{4}\left(\frac{5}{5}\right) = \frac{5}{20}$$

$$\frac{1}{5}\left(\frac{4}{4}\right) = \frac{4}{20}$$

$$+\,1 \qquad\quad = +\frac{20}{20}$$

$$\frac{29}{20} = \boxed{1\frac{9}{20}}$$

- Here's what you should have. 1/4 plus 1/5 plus 1 equals 1 and 9/20.
- c. Work problem B. Raise your hand when you're finished. (Observe students and give feedback.)
- (Write on the board:)

$$\text{b.} \quad \frac{3}{8} \qquad\quad = \frac{3}{8}$$

$$\frac{3}{2}\left(\frac{4}{4}\right) = \frac{12}{8}$$

$$+\,3 \qquad\quad = +\frac{24}{8}$$

$$\frac{39}{8} = \boxed{4\frac{7}{8}}$$

- Here's what you should have. 3/8 plus 3/2 plus 3 equals 4 and 7/8.

EXERCISE 3 MULTIPLICATION
Comparison: Factor and Product

a. Find part 3.
- Each problem tells what N is multiplied by to get T. Remember, if N is multiplied by more than 1, T is larger. If N is multiplied by less than 1, N is larger.
b. Write N or T for each problem to show which is larger. Raise your hand when you're finished. (Observe students and give feedback.)
c. Check your work.
- Problem A: N is multiplied by 9 and 9-tenths. Which is larger? (Signal.) *T.*
- Problem B: N is multiplied by 99-hundredths. Which is larger? (Signal.) *N.*

- Yes, N is larger than T.
- Problem C: N is multiplied by 7/8. Which is larger? (Signal.) *N.*
- Problem D: N is multiplied by 1/2. Which is larger? (Signal.) *N.*
- Problem E: N is multiplied by 3/2. Which is larger? (Signal.) *T.*
- Problem F: N is multiplied by 5/4. Which is larger? (Signal.) *T.*
- Problem G: N is multiplied by 4/5. Which is larger? (Signal.) *N.*
d. Raise your hand if you got everything right.

EXERCISE 4 PROBLEM SOLVING
Multiplication/Division as Proportion

a. Find part 4.
- I'll read what it says. Follow along: Some word problems use the word **each.** The word **each** means **1.** 45 miles **each** hour means 45 miles in **1** hour. If they make 156 cars **each** day, they make 156 cars in **1** day.
- You have worked problems that tell about **each** or **1** as multiplication problems or division problems. You can also work them as ratio problems. The advantage of working them as ratio problems is that you will always work the problem the right way, even though the problem may be difficult.
- I'll read the problem: There were 8 ounces in each can. If there were 648 ounces, how many cans were there?
- This problem could be hard to work if you don't use a ratio equation. When you work it as a ratio equation, it's easy.
- You can see the equation. The names are **ounces** and **cans.**
- The first fraction shows 8 ounces in 1 can. The other fraction shows the number of ounces and a box for the number of cans.
- Remember, if the problem tells about **each,** you can work it as a multiplication problem or a division problem, but the problem may be easier to work as a ratio problem.
b. Find part 5.
- These problems have the word **each.** Some **ask** about **each.** That's asking about **1.** Some **tell** about **each.** That's telling about **1.** You'll write ratio equations for all these problems.
c. Work problem A. Make a ratio equation. Show the answer as a whole number or a mixed number and a unit name. Raise your hand when you're finished.
(Observe students and give feedback.)

- (Write on the board:)

$$\textbf{a.} \quad \frac{\text{tubs}}{\text{gal}} \quad \frac{1}{280} \left(\frac{6}{6} \right) = \frac{6}{\boxed{1680}}$$

- Check your work. Problem A asks: How many gallons do 6 tubs hold? What's the answer? (Signal.) *1680 gallons.*
d. Work problem B. Raise your hand when you're finished. (Observe students and give feedback.)
- (Write on the board:)

$$\textbf{b.} \quad \frac{\text{spots}}{\text{ladybugs}} \quad \frac{6}{1} \left(\frac{64}{64} \right) = \frac{384}{\boxed{64}}$$

- Check your work. Problem B asks: How many ladybugs were there? What's the answer? (Signal.) *64 ladybugs.*
e. Work problem C. Raise your hand when you're finished. (Observe students and give feedback.)
- (Write on the board:)

$$\textbf{c.} \quad \frac{\text{cookies}}{\text{boxes}} \quad \frac{144}{6} \left(\frac{\frac{1}{6}}{\frac{1}{6}} \right) = \frac{\boxed{\frac{144}{6}}}{1}$$

- Check your work. Problem C asks: How many cookies were in each box? What's the answer? (Signal.) *24 cookies.*
f. Work problem D. Raise your hand when you're finished. (Observe students and give feedback.)
- (Write on the board:)

$$\textbf{d.} \quad \frac{\text{bottles}}{\text{pt}} \quad \frac{1}{5} \left(\frac{\frac{72}{5}}{\frac{72}{5}} \right) = \frac{\boxed{\frac{72}{5}}}{72}$$

- Check your work. Problem D asks: How many bottles are filled? What's the answer? (Signal.) *14 and 2-fifths bottles.*

EXERCISE 5 PROBLEM SOLVING
Inverse Operations: 2 Steps

a. You've learned rules for undoing. Listen: How would you undo multiplying by 4? (Signal.) *Divide by 4.*
- And how would you undo dividing by 4? (Signal.) *Multiply by 4.*
b. Find part 6.

- These are undoing problems that have a step that divides.
c. Problem A says: You start with a number and multiply by 7. Then you divide by 5. You end up with 14. What's the mystery number?
- Write two equations with boxes. Raise your hand when you've done that much. √
- What's the last number? (Signal.) *14.*
- To get to 14, you divide by 5. How do you undo dividing by 5? (Signal.) *Multiply by 5.*
- Say the problem for that part. (Signal.) *14 times 5.*
- Work it. Write the answer, and figure out the mystery number for problem A. Raise your hand when you're finished.
  (Observe students and give feedback.)
- (Write on the board:)

$$\textbf{a.} \quad \boxed{}^{?} \times 7 = \boxed{70}$$
$$\boxed{70} \div 5 = 14$$

- Check your work. 14 times 5 is 70. 70 divided by 7 is the mystery number. Everybody, what's the mystery number? (Signal.) *10.*
- (Write: **10.**)
d. Your turn: Work problem B. Raise your hand when you're finished.
  (Observe students and give feedback.)
- (Write on the board:)

$$\textbf{b.} \quad \boxed{}^{?} \div 3 = \boxed{576}$$
$$\boxed{576} \div 8 = 72$$

- Check your work. You undid dividing by 8 by multiplying: 72 times 8. The answer is 576. You undid dividing by 3 by multiplying: 576 times 3. Everybody, what's the mystery number? (Signal.) *1728.*
- (Write: **1728.**)
e. Your turn: Work problem C. Raise your hand when you're finished.
  (Observe students and give feedback.)
- (Write on the board:)

$$\textbf{c.} \quad \boxed{75}^{?} + 100 = \boxed{175}$$
$$\boxed{175} \div 5 = 35$$

- Here's what you should have for problem C. Raise your hand if you got everything right.

EXERCISE 6 AREA
Missing Length

a. Find part 7.
- These are area problems that have either the base or the height missing. Remember, rewrite the equation so it says: Base times height equals area. Then put in the numbers you know and figure out the missing number.
b. Work problem A. Raise your hand when you're finished. (Observe students and give feedback.)
- (Write on the board:)

> a. $b \times h = A$
>
> $12 \times \left(\dfrac{84}{12}\right) = 84$
>
> $h = 7 \text{ m}$

- Here's what you should have. The height is 84/12. That's **7 meters.**
c. Work problem B. Raise your hand when you're finished. (Observe students and give feedback.)
- (Write on the board:)

> b. $b \times h = A$
>
> $b \times 20 = 480$
>
> $b = 24 \text{ ft}$

- Here's what you should have. You figured out the base. That's **24 feet.**
d. Work problem C. Raise your hand when you're finished. (Observe students and give feedback.)
- (Write on the board:)

> c. $b \times h = A$
>
> $19 \times \left(\dfrac{266}{19}\right) = 266$
>
> $h = 14 \text{ in}$

- Here's what you should have. You figured out the height. That's **14 inches.**
e. Raise your hand if you got everything right.

EXERCISE 7 INDEPENDENT WORK

- (In addition to independent work in the textbook, assign **Bridge to Connecting Math Concepts** Independent Worksheet 46 as classwork or homework. Before beginning the next lesson, check the students' independent work.)

Lesson 67

Objectives

- **Use inverse operations to solve multistep word problems.** (Exercise 1)
- Solve a set of multiplication/division word problems by making ratio equations. (Exercise 2)
- **Determine whether a missing factor in a decimal multiplication problem is more than 1 or less than 1, then figure out the missing value.** (Exercise 3)

$$90\,(\blacksquare) = 91.8$$

Note: Problems are of the form $\boxed{.98 \text{ or } 1.02}$ Students first choose the answer that must be correct (1.02). They then work the division problem: $91.8 \div 90$ on their calculator to check that the value is correct.

- **Work a mixed set of problems that involve multiplication, addition or subtraction of a whole number and fractions.** (Exercise 4)
- Use comparison facts to figure out numbers for a number-family table. (Exercise 5)

EXERCISE 1 PROBLEM SOLVING
Inverse Operations: Multistep

a. Open your textbook to lesson 67 and find part 1.
- You've worked mystery number problems that have two equations. You can work much harder problems that have many equations. You just have to remember that every equation **has a box as the first number.**
b. Problem A: You start with some number. Then you multiply by 5. Then you subtract 16. Then you add 11. Then you divide by 3. You end up with 30. What's the mystery number?
- Every sentence that begins with the word **then** tells about an equation that has a box as the first number. For this problem, you'll need four equations. See if you can write all the equations for this problem. Raise your hand when you've done that much. √
- (Write on the board:)

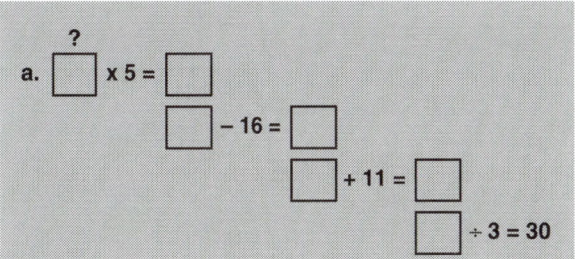

- Here's what you should have. Make sure your equations are correct. Then just work backward until you get to the mystery number. Raise your hand when you're finished.
  (Observe students and give feedback.)
- (Write to show:)

```
        ?
a.  [19] x 5 = [95]
          [95] – 16 = [79]
                [79] + 11 = [90]
                      [90] ÷ 3 = 30
```

- Here's what you should have. The mystery number is 19. Raise your hand if you got it right.
c. Your turn: Write equations for problem B. Raise your hand when you've done that much. √
- (Write on the board:)

```
        ?
b.  [ ] ÷ 3 = [ ]
          [ ] – 20 = [ ]
                [ ] x 2 = 8
```

- Here's what you should have. This problem has three equations.
- Your turn: Figure out the mystery number. Raise your hand when you're finished.
  (Observe students and give feedback.)
- (Write to show:)

```
        ?
b.  [72] ÷ 3 = [24]
          [24] – 20 = [4]
                [4] x 2 = 8
```

- Here's what you should have. The mystery number is 72. Raise your hand if you got everything right.
d. Work problem C. Raise your hand when you're finished. (Observe students and give feedback.)

- (Write on the board:)

```
        ?
c.  [0] + 56 = [56]
          [56] ÷ 7 = [8]
                [8] x 20 = 160
```

- Here's what you should have. The mystery number is zero.

EXERCISE 2 PROBLEM SOLVING
Multiplication/Division as Proportion

a. Find part 2.
- I'll read what it says. Follow along: You've worked with word problems that use the word **each** to mean **1**. **Each** box is **1** box. Sometimes the word **per** is used just like **each.**
- I'll read the problem: A machine makes 36 baseball cards in 3 seconds. How many **cards per second** does the machine make?
- The question asks, how many cards does the machine make in 1 second? You can see the ratio. **Cards** and **seconds** are the names. The first fraction is 36-thirds. The last fraction has **1** for seconds.
b. Find part 3.
- You'll work problems that have the word **per.** Some ask about **per.** That's asking about **1.** Some tell about **per.** That's telling about **1.**
c. Work problem A. Make a ratio equation. Show the answer as a number and a unit name. Raise your hand when you're finished.
  (Observe students and give feedback.)
- (Write on the board:)

$$ \text{a.}\quad \frac{\text{mi}}{\text{hr}}\ \frac{64}{1}\left(\frac{8}{8}\right) = \frac{\boxed{512}}{8} $$

- Check your work. Here's the equation you should have. Problem A asks: How far does the train travel in 8 hours? What's the answer? (Signal.) *512 miles.*
d. Work problem B. Raise your hand when you're finished. (Observe students and give feedback.)
- (Write on the board:)

$$ \text{b.}\quad \frac{\textbf{students}}{\textbf{classes}}\ \frac{120}{5}\left(\frac{\frac{1}{5}}{\frac{1}{5}}\right) = \frac{\boxed{\frac{120}{5}}}{1} $$

- Check your work. Here's the equation you should have. Problem B asks: How many students per class are there? What's the answer? (Signal.) *24 students.*
e. Work problem C. Raise your hand when you're finished. (Observe students and give feedback.)
- (Write on the board:)

$$\text{c. } \frac{\text{seeds}}{\text{holes}} \quad \frac{4}{1} \left(\frac{\frac{136}{4}}{\frac{136}{4}} \right) = \frac{136}{\boxed{\frac{136}{4}}}$$

- Check your work. Here's the equation you should have. Problem C asks: If he plants 136 seeds, how many holes does he need? What's the answer? (Signal.) *34 holes.*

EXERCISE 3 DECIMAL MULTIPLICATION
Missing Factor

a. Find part 4.
- (Teacher reference:)

a. $43\,(\blacksquare) = 38.27$ b. $56.1\,(\blacksquare) = 50.49$

     $\boxed{.89 \text{ or } 1.12}$ $\boxed{.9 \text{ or } 1.1}$

- Each problem shows two answers. One of the answers is more than 1 and one is less than 1. You'll figure out whether you end up with more or less than you start with. You'll indicate the correct answer. Then you'll work the problem on your calculator to check the answer.
b. Problem A: 43 times some value equals 38 and 27-hundredths. The choices for the missing value are 89-hundredths and 1 and 12-hundredths.
- One of those values is less than 1 and one is more than 1. Figure out which answer must be correct and write it. Then use your calculator and work the original division problem. Raise your hand when you've written the answer and checked it.
 (Observe students and give feedback.)
- Check your work. You worked the problem 38 and 27-hundredths divided by 43. The missing value is **89-hundredths.** Raise your hand if you got it right.
c. Work problem B. Write the correct answer. Then work the original division problem on your calculator. Raise your hand when you're finished. (Observe students and give feedback.)
- Problem B: 56 and 1-tenth times some value is 50 and 49-hundredths. The choices are 9-tenths and 1 and 1-tenth. Which is the correct choice? (Signal.) *9-tenths.*

- You worked the problem 50 and 49-hundredths divided by 56 and 1-tenth. The answer is **9-tenths.**
d. Your turn: Work the rest of the problems in part 4. Raise your hand when you're finished. (Observe students and give feedback.)
e. (Write on the board:)

     c. 1.02   d. 2.2

- Here's what you should have for problems C and D. Raise your hand if you got everything right.

EXERCISE 4 FRACTION OPERATIONS
Whole Numbers and Fractions

a. Find part 5.
- Some of these problems multiply. Some add or subtract.
- For the multiplication problems, you just write the whole number as a simple fraction over 1. For the other problems, you find a common denominator.
b. Work the problems in part 5. Show the answer as a whole number or a mixed number. Raise your hand when you're finished.
 (Observe students and give feedback.)
Key:

$$a. \ \frac{3}{1} \times \frac{4}{6} \times \frac{7}{8} = \frac{84}{48} = \boxed{1\frac{36}{48}}$$

$$b. \ \begin{array}{l} 3 = \dfrac{72}{24} \\[6pt] \dfrac{1}{6}\left(\dfrac{4}{4}\right) = \dfrac{4}{24} \\[6pt] +\dfrac{7}{8}\left(\dfrac{3}{3}\right) = +\dfrac{21}{24} \\ \hline \dfrac{97}{24} = \boxed{4\frac{1}{24}} \end{array}$$

$$c. \ \begin{array}{l} \dfrac{15}{2} = \dfrac{15}{2} \\[6pt] -3 = -\dfrac{6}{2} \\ \hline \dfrac{9}{2} = \boxed{4\frac{1}{2}} \end{array}$$

$$d. \ \frac{2}{5} \times \frac{10}{1} \times \frac{1}{2} = \frac{20}{10} = \boxed{2}$$

c. Find part J on page 268. That shows what you should have for each problem.

EXERCISE 5 PROBLEM SOLVING
Tables with Comparison

a. Find part 6.
- This is a table problem. The first three facts tell about numbers for the table. Fact 4 gives you information that helps you get one more fact. But you'll have to make a number family for fact 4.
b. Make a table. Put in the four numbers for the facts. Raise your hand when you've done that much. (Observe students and give feedback.)
- (Write on the board:)

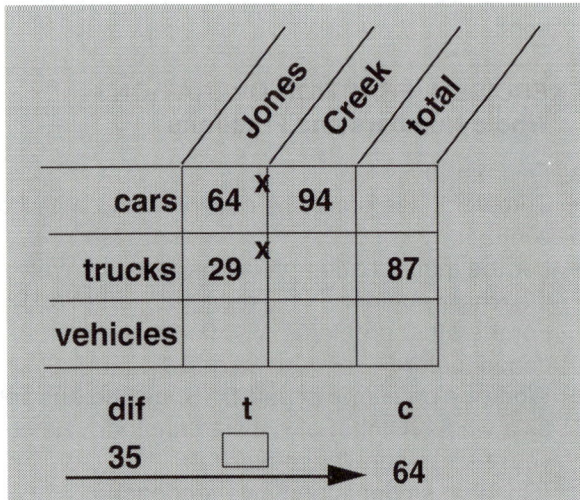

- Here's what you should have so far. For fact 4, you made a number family showing the number of cars and trucks on Jones Road. There were 64 cars. There were 35 fewer trucks than cars. So there were 29 trucks. Raise your hand if you got everything right for the facts.
c. Figure out the missing numbers and answer the questions. Raise your hand when you're finished. (Observe students and give feedback.)
- (Write to show:)

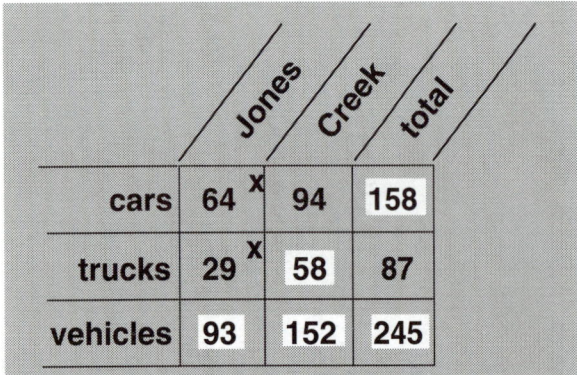

d. I'll read each question. Tell me the answer.
- Item A: On which road were there more trucks? (Signal.) *Creek Road.*

- Item B: What was the total number of cars? (Signal.) *158 cars.*
- Item C: What was the total number of vehicles on Creek Road? (Signal.) *152 vehicles.*

EXERCISE 6 INDEPENDENT WORK

- (In addition to independent work in the textbook, assign **Bridge to Connecting Math Concepts** *Independent Worksheet* 47 as classwork or homework. Before beginning the next lesson, check the students' independent work.)

Lesson 68

EXERCISE 1 AREA
Discrimination: Missing Length

a. Open your textbook to lesson 68 and find part 1.
- Some of these problems ask about the area. Some of them tell about the area and ask about either the base or the height.
- Remember, the units for area are square units. Units for a side are not square units but regular length units.
b. Work all the problems in part 1. Use the equation base times height equals area for all the problems. Raise your hand when you're finished. (Observe students and give feedback.)

Key:

a. $b \times h = A$
    $12 \times 11 = A$
    $\boxed{A = 132 \text{ sq ft}}$

b. $b \times h = A$
    $15 \left(\frac{105}{15}\right) = 105$
    $\boxed{h = 7 \text{ in}}$

c. $b \times h = A$
    $b \times 10 = 62$
    $\boxed{b = 6\frac{2}{10} \text{ in}}$

d. $b \times h = A$
    $17 \times 20 = A$
    $\boxed{A = 340 \text{ sq m}}$

c. Find part J on page 272. That shows what you should have for each item. Raise your hand if you got everything right.

EXERCISE 2 DECIMAL MULTIPLICATION
Missing Factor

a. Find part 2.
- Each problem shows two choices for the missing number.
- Figure out which choice is correct. Write the missing number. Then work the problem on your calculator and see if the answer you chose is correct.
b. Work the problems in part 2. Raise your hand when you're finished.
   (Observe students and give feedback.)
c. (Write on the board:)

a.	1.3	d.	1.02
b.	.9	e.	.98
c.	1.02	f.	.98

- Here's what you should have for each item. Raise your hand if you got everything right.

EXERCISE 3 PROBLEM SOLVING
Multiplication/Division as Proportion

a. Find part 3.
- The problems in part 3 refer to **each** or **per**. You'll work each problem as a ratio problem. Remember to show the answer as a number and a unit name.
b. Work the problems in part 3. Raise your hand when you're finished.
   (Observe students and give feedback.)
Key:

a. $\dfrac{tablets}{boxes} \dfrac{8}{1} \left(\dfrac{\frac{512}{8}}{\frac{512}{8}}\right) = \dfrac{512}{\frac{512}{8}} \boxed{64 \text{ boxes}}$

b. $\dfrac{tubs}{gal} \dfrac{1}{61} \left(\dfrac{\frac{549}{61}}{\frac{549}{61}}\right) = \dfrac{\frac{549}{61}}{549} \boxed{9 \text{ tubs}}$

c. $\dfrac{engines}{cylinders} \dfrac{1}{6} \left(\dfrac{\frac{234}{6}}{\frac{234}{6}}\right) = \dfrac{\frac{234}{6}}{234} \boxed{39 \text{ engines}}$

d. $\dfrac{sec}{buttons} \dfrac{5}{6} \left(\dfrac{\frac{1}{6}}{\frac{1}{6}}\right) = \dfrac{\frac{5}{6}}{1} \boxed{\frac{5}{6} \text{ sec}}$

e. $\dfrac{mi}{hr} \dfrac{580}{7} \left(\dfrac{\frac{1}{7}}{\frac{1}{7}}\right) = \dfrac{\frac{580}{7}}{1} \boxed{82\frac{6}{7} \text{ mi}}$

c. Find part K. That shows what you should have for each problem.

EXERCISE 4 RATIOS AND PROPORTIONS
Decimals/Percents

a. Find part 4.
- I'll read what it says. Follow along: You've worked ratio problems that involve fractions. You make a fraction number family.
- You can do the same thing for problems that use decimal values or percents. You make a fraction number family.
- I'll read the sentence that gives a decimal value: 3-tenths of the children had new shoes.
- You can see the fraction number family with 3-tenths written as a fraction.
- I'll read the sentence that gives a percent value: The rainfall in August was 40 percent of the rainfall in July.
- You can see the fraction number family with 40 percent written as a fraction.
- Remember, rewrite the decimal value or the percent value as a fraction.
b. Find part 5.
- For each problem, you'll make a fraction number family and a ratio table.
c. Problem A: 45 percent of the windows in an apartment building were dirty. There were 300 windows in the building. How many were dirty? How many were clean?
- Make the fraction number family with 45 percent as a fraction. Raise your hand when you've done that much.
  (Observe students and give feedback.)
- (Write on the board:)

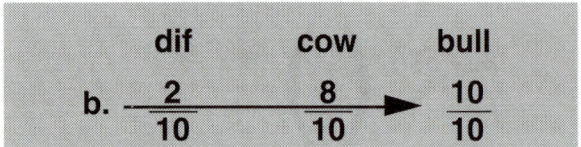

- Here's what you should have. The fraction for all the windows is 100/100.
- You won't work the rest of the problem now, but make sure your number family is correct.
d. Read problem B. Make the fraction number family. Raise your hand when you've done that much. (Observe students and give feedback.)
- (Write on the board:)

- Here's what you should have. The fraction for the bull is 10/10. The fraction for the cow is 8/10. The difference is 2/10.
e. Read problem C. Make the fraction number family. Raise your hand when you've done that much. (Observe students and give feedback.)
- (Write on the board:)

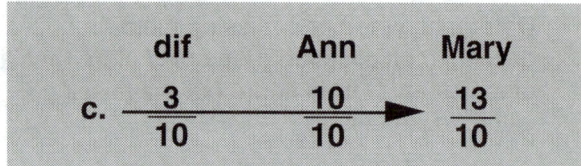

- Here's what you should have. Mary's the big number. Her fraction is 13/10. Ann is 10/10.
f. Read problem D. Make the fraction number family. Raise your hand when you've done that much. (Observe students and give feedback.)
- (Write on the board:)

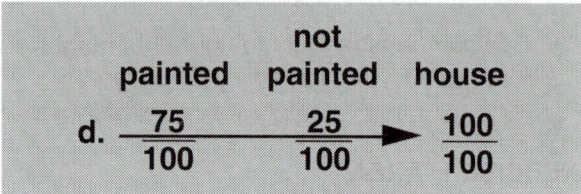

- Here's what you should have. The fraction for the whole house is 100/100. The painted part is 75/100. The unpainted part is 25/100.
g. As part of your independent work, you'll work the ratio-table problems and answer the questions.

EXERCISE 5 PROBLEM SOLVING
Inverse Operations: Multistep

a. Find part 6.
- These are mystery-number problems that involve more than two equations. You'll write one equation for each step. If the problem tells you that you add 17, you'll write an equation for that step. If the problem tells you that you divide by 3, you'll write an equation for that step. Each equation except the last one has a box at the beginning and another box at the end.
b. Your turn: Write all the equations with boxes for problem A. Raise your hand when you've done that much.
(Observe students and give feedback.)

- (Write on the board:)

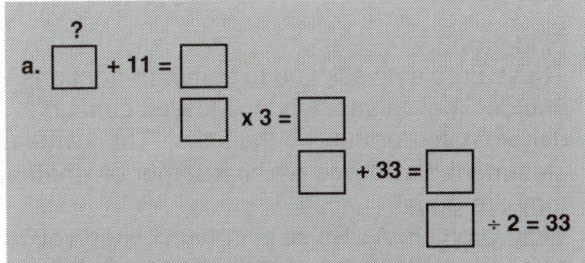

- Here's what you should have. Raise your hand if you got it right.
- c. Write equations with boxes for problem B. Raise your hand when you've done that much. (Observe students and give feedback.)
- (Write on the board:)

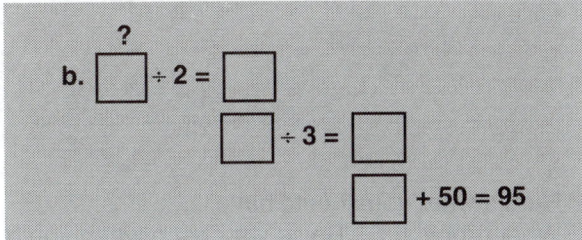

- Here's what you should have. Raise your hand if you got it right.
- d. As part of your independent work, you'll figure out the mystery number for each problem.

EXERCISE 6 PROBLEM SOLVING
Tables with Comparison

a. Find part 7.
- The four facts give information for numbers in your table. For one of the facts, you'll have to make a number family and use one of the numbers you put in the table. That fact compares two cells of the table.
b. Put in the numbers for the four facts. Raise your hand when you've done that much. (Observe students and give feedback.)

- (Write on the board:)

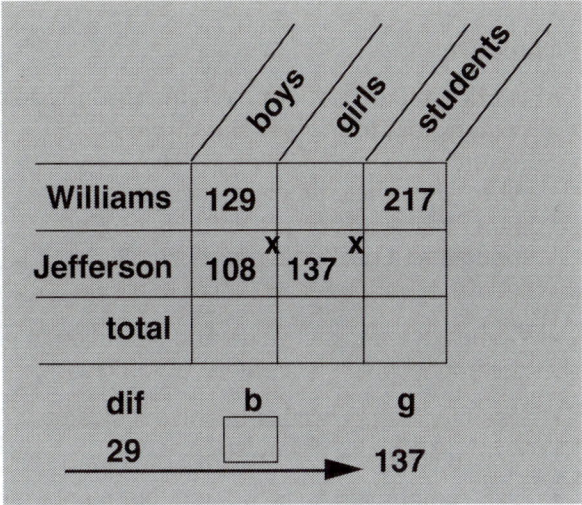

- Here's what you should have so far. Fact 4 tells you that at Jefferson, there are 29 fewer boys than girls. There are 108 boys at Jefferson.
c. Make sure your numbers are correct. Then complete the table and answer the questions. For two of them, you'll need to make a number family. Raise your hand when you're finished. (Observe students and give feedback.)
d. Check your work.
- Item A asks: How many teachers are there at Williams School? What's the answer? (Signal.) *29 teachers.*
- Item B asks: How many students are in Davis School? What's the answer? (Signal.) *401 students.*
- Item C asks: How many students are in Jefferson school? What's the answer? (Signal.) *245 students.*

EXERCISE 7 INDEPENDENT WORK

a. Do the independent work for lesson 68.
- Finish the problems for part 5. Complete the equations for part 6. Then do parts 8 and 9.
- (Assign **Bridge to Connecting Math Concepts** *Independent Worksheet* 48 as classwork or homework. Before beginning the next lesson, check the students' independent work.)

Lesson 69

Objectives

- **Work a mixed set of problems that involve rectangles, triangles, and circles.** (Exercise 1)
  Note: For circle problems, students find either the radius or circumference. For triangle and rectangle problems, students find the area, perimeter, or length of a side.

- Compare a decimal value and a fraction to determine which is greater. (Exercise 2)

- Use inverse operations to solve multistep word problems. (Exercise 3)

- Use comparison facts to figure out numbers for a number-family table. (Exercise 4)

- Work a mixed set of problems that involve multiplication, addition or subtraction of a whole number and fractions. (Exercise 5)

- Work a ratio-table problem that involves a decimal or a percent value. (Exercise 6)

EXERCISE 1 AREA/PERIMETER/ CIRCUMFERENCE
Discrimination Set

a. Open your textbook to lesson 69 and find part 1.
- These problems involve circles, triangles and rectangles.
b. Work the problems. Raise your hand when you're finished.
  (Observe students and give feedback.)
  Key:

  a. $\pi \times d = C$
  $3.14\left(\frac{300}{3.14}\right) = 300$
  $d = 95.54$
  $\boxed{r = 47.77 \ cm}$

  b. $\pi \times d = C$
  $3.14 \times 15 = C$
  $\boxed{C = 47.1 \ mi}$

  c. $A\triangle = \frac{b \times h}{2}$
  $A\triangle = \frac{25 \times 12}{2} = \frac{300}{2}$
  $\boxed{A\triangle = 150 \ sq \ in}$

  d. $b \times h = A$
  $20 \times 38 = A$
  $\boxed{A = 760 \ sq \ m}$
  $\begin{array}{r} 38 \\ 38 \\ 20 \\ +20 \\ \hline \end{array}$
  $\boxed{P = 116 m}$

  e. $b \times h = A$
  $11 \times 40 = A$
  $\boxed{A = 440 \ sq \ ft}$

  f. $b \times h = A$
  $b \times 13 = 182$
  $\boxed{b = 14 \ in}$

  $\begin{array}{r} 15 \\ 20 \\ +25 \\ \hline \end{array}$
  $\boxed{P = 60 \ in}$

c. Find part J on page 277. That shows what you should have for each problem.

EXERCISE 2 FRACTIONS AND DECIMALS
Comparison

a. Find part 2.
- These problems ask you to compare fractions and decimal values. Find the lowest common denominator for the two fractions. Then write a statement that shows which is larger or whether they are equal.
b. Work problem A. Raise your hand when you're finished. (Observe students and give feedback.)
- (Write on the board:)

  a. $\frac{8}{10} \left(\frac{2}{2}\right) = \frac{16}{20}$

  $\frac{17}{20} = \frac{17}{20}$

  $\boxed{.8 < \frac{17}{20}}$

- Here's what you should have.
c. Work problem B. Raise your hand when you're finished. (Observe students and give feedback.)
- (Write on the board:)

  b. $\frac{9}{25} \left(\frac{4}{4}\right) = \frac{36}{100}$

  $\frac{34}{100} = \frac{34}{100}$

  $\boxed{\frac{9}{25} > .34}$

- Here's what you should have.
d. Your turn: Work the rest of the problems in part 2. Raise your hand when you're finished. (Observe students and give feedback.)

Key:

c. $\frac{7}{20}\left(\frac{5}{5}\right)=\frac{35}{100}$ d. $\frac{3}{2}\left(\frac{5}{5}\right)=\frac{15}{10}$

$\frac{41}{100}=\frac{41}{100}$ $\frac{15}{10}=\frac{15}{10}$

$\boxed{\frac{7}{20}<.41}$ $\boxed{\frac{3}{2}=1.5}$

e. $\frac{3}{5}\left(\frac{20}{20}\right)=\frac{60}{100}$

$\frac{65}{100}=\frac{65}{100}$

$\boxed{\frac{3}{5}<.65}$

e. (Write on the board:)

> c. $\frac{7}{20}<.41$
>
> d. $\frac{3}{2}=1.5$
>
> e. $\frac{3}{5}<.65$

- Here are the statements you should have for problems C, D and E. Raise your hand if you got everything right.

EXERCISE 3 PROBLEM SOLVING
Inverse Operations: Multistep

a. Find part 3.
- These are mystery-number problems.
b. Write equations with boxes. Figure out the mystery number. Raise your hand when you've worked all the problems in part 3.
  (Observe students and give feedback.)
c. (Write on the board:)

> ?
> a. $\boxed{2222} - 2 = \boxed{2220}$
> $\boxed{2220} \div 20 = \boxed{111}$
> $\boxed{111} - 11 = 100$
>
> ?
> b. $\boxed{48} \div 4 = \boxed{12}$
> $\boxed{12} - 10 = \boxed{2}$
> $\boxed{2} \div 2 = 1$
>
> ?
> c. $\boxed{14} + 56 = \boxed{70}$
> $\boxed{70} \div 70 = \boxed{1}$
> $\boxed{1} - 1 = 0$

- Here's what you should have. Raise your hand if you got everything right.

EXERCISE 4 PROBLEM SOLVING
Tables with Comparison

a. Find part 4.
- This is a table problem.
b. Read the facts. Figure out the four numbers they tell about. You'll need number families to figure out two of the numbers. Raise your hand when you have a table with four numbers.
  (Observe students and give feedback.)
- (Write on the board:)

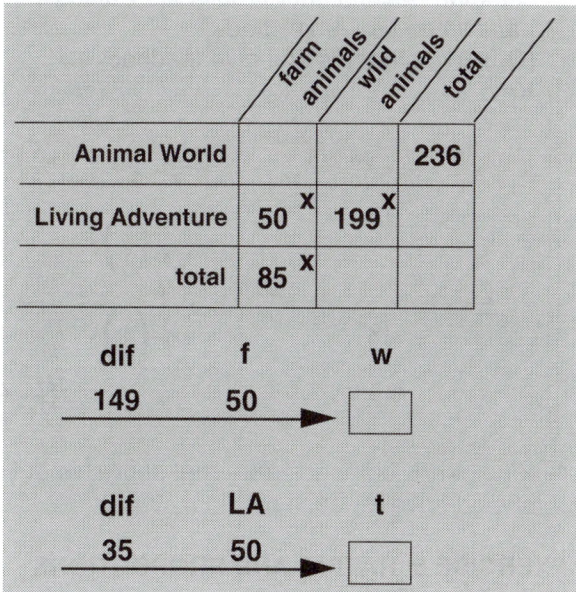

	farm animals	wild animals	total
Animal World			236
Living Adventure	50 ˣ	199 ˣ	
total	85 ˣ		

dif 149   f 50   w → ☐

dif 35   LA 50   t → ☐

- Here's the table you should have.
c. Now figure out the rest of the numbers and answer the questions. Raise your hand when you're finished.
  (Observe students and give feedback.)
- (Write to show:)

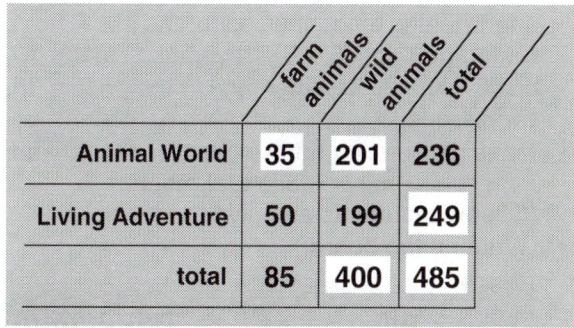

	farm animals	wild animals	total
Animal World	35	201	236
Living Adventure	50	199	249
total	85	400	485

- Here are the numbers for the table.
d. Item A: Everybody, in which park are there more farm animals? (Signal.) *Living Adventure.*
- Item B: What's the total number of animals at Living Adventure? (Signal.) *249.*

- Item C: The number 400 tells about the total for what kind of animals? (Signal.) *Wild animals.*
- Item D: What's the total number of farm animals? (Signal.) *85.*
- Item E: How many more animals are in Living Adventure park than Animal World? (Signal.) *13.*

EXERCISE 5 FRACTION OPERATIONS
Whole Numbers and Fractions

a. Find part 5.
b. Work the problems. In multiplication problems, show the whole number as a simple fraction. Work addition or subtraction problems as common-denominator problems. Raise your hand when you're finished.
   (Observe students and give feedback.)

Key:

a. $\frac{1}{1} \times \frac{3}{1} \times \frac{12}{10} = \frac{36}{10} = \boxed{3\frac{6}{10}}$

b. $\frac{2}{7} \times \frac{5}{3} \times \frac{2}{1} = \boxed{\frac{20}{21}}$

c. $\frac{15}{3} = \frac{15}{3}$
$-2 = -\frac{6}{3}$
$\frac{9}{3} = \boxed{3}$

d. $\frac{1}{8}\left(\frac{5}{5}\right) = \frac{5}{40}$
$4 = \frac{160}{40}$
$+\frac{3}{5}\left(\frac{8}{8}\right) = +\frac{24}{40}$
$\frac{189}{40} = \boxed{4\frac{29}{40}}$

c. Find part K on page 278. That shows what you should have for each problem.

EXERCISE 6 RATIOS AND PROPORTIONS
Decimals/Percents

a. Find part 6.
- These are ratio-table problems that refer to percents or decimal values. To work the problems, you just change the percent or decimal value into a fraction.
b. Problem A: The cost of the refrigerator was 1 and 5-tenths times the cost of the TV. The refrigerator cost $705. How much did the TV cost? What's the difference in the price of the two products?
- Make the fraction number family. Raise your hand when you've done that much.
   (Observe students and give feedback.)
- (Write on the board:)

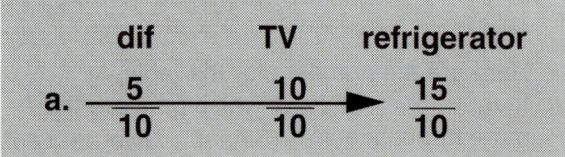

	dif	TV	refrigerator
a.	$\frac{5}{10}$	$\frac{10}{10}$ ⟶	$\frac{15}{10}$

- Here's what you should have. The refrigerator is the big number — 15/10. The TV is 10/10. The difference is 5/10.
c. Your turn: Write number families for the rest of the items in part 6. Raise your hand when you've done that much.
   (Observe students and give feedback.)
d. (Write on the board:)

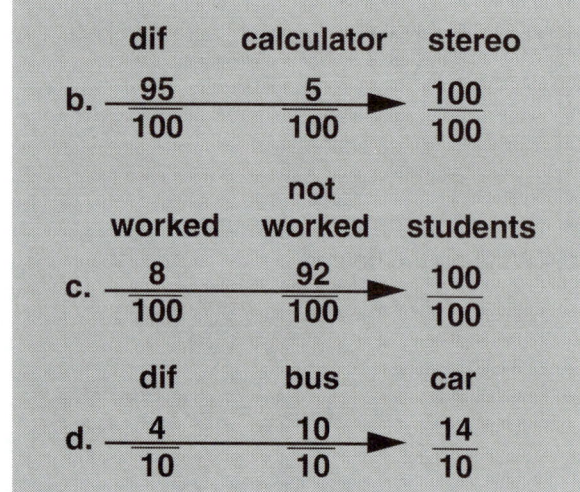

	dif	calculator	stereo
b.	$\frac{95}{100}$	$\frac{5}{100}$ ⟶	$\frac{100}{100}$

	worked	not worked	students
c.	$\frac{8}{100}$	$\frac{92}{100}$ ⟶	$\frac{100}{100}$

	dif	bus	car
d.	$\frac{4}{10}$	$\frac{10}{10}$ ⟶	$\frac{14}{10}$

- Here are the families you should have for problems B, C and D. Raise your hand if you got everything right.
e. As part of your independent work, you'll make the ratio tables and answer the questions.

EXERCISE 7 INDEPENDENT WORK

a. Do the independent work for lesson 69.
- Complete the problems for part 6. Then work parts 7 and 8.
- (Assign *Bridge to Connecting Math Concepts Independent Worksheet* 49 as classwork or homework. Before beginning the next lesson, check the students' independent work.)

Lesson 70 – Test 7

EXERCISE 1 TEST 7

Note: Students are not to use a calculator for any part of the test except part 5.

a. Open your textbook to test 7 and find part 1. √
b. This is a test. You should only have your textbook, a sharpened pencil and lined paper on your desk.
c. Do the test on your own. Raise your hand when you've completed part 4.
   (Observe students but do not give feedback.)
- (After students complete part 4, permit them to use a calculator to complete part 5. Observe students but do not give feedback.)

EXERCISE 2 MARKING THE TEST

a. (Collect the students' papers. Use the Answer Key to score the tests. Award scores for test 7 as follows:)

Test 7 Percent Summary					
SCORE	%	SCORE	%	SCORE	%
47	100	42	89	37	79
46	98	41	87	36	77
45	96	40	85	35	74
44	94	39	83	34	72
43	91	38	81	33	70

b. (Complete the Test 7 Remedy Summary to determine whether remedies are needed. Reproducible Summary Sheets are at the back of the Teacher's Guide.)
- (If more than 1/4 of the students did not pass a test part, present the remedy for that part before presenting the end-of-program test. Remedies appear at the end of the Test 7 Answer Key.)

Final Test

EXERCISE 3 INDEPENDENT WORK

- (Assign **Bridge to Connecting Math Concepts** *Independent Worksheet* 50 as classwork or homework. Before beginning the next lesson, check the students' independent work.)

CUMULATIVE TEST REMINDER

Present the Final Cumulative Test that appears in Appendix A of the Teacher's Guide. Also, provide remedies for students who do not pass the test.

Test Preparation for
CMC Bridge

- Start these lessons at least 7–10 days before standardized testing is to begin. Present one lesson a day.
- For test-prep lessons that do not require a full period, continue with the regular CMC lesson presentation in the time that remains.

Test Preparation

LESSON 1

> **Materials Note:**
>
> Each student will need:
> - lined paper
> - test booklet and Multiple-Choice Response Sheet

EXERCISE 1 DISTRIBUTION
Multiplying Numbers and Unknowns

a. (Write on the board):

> $$5 \times A =$$

- For some problems, you're going to multiply numbers by letters. The letters stand for numbers, but you don't know what the numbers are, so you can't figure out a number answer.
b. Everybody, read this problem. (Signal.) *5 times A.*
- Listen: 5 times A equals 5A.
- What does 5 times A equal? (Signal.) *5A.*
c. What does 5 times C equal? (Signal.) *5C.*
- What does 11 times M equal? (Signal.) *11M.*
- What does one-half times R equal? (Signal.) *One-half R.*
- What does 37 times T equal? (Signal.) *37T.*
d. I'll tell you the answer to a tougher problem: T times 37. That equals 37T.
e. Your turn: What does T times 37 equal? (Signal.) *37T.*
- What does M times one-half equal? (Signal.) *One-half M.*
- What does R times 5 equal? (Signal.) *5R.*
- What does 18 times P equal? (Signal.) *18P.*
- What does S times 400 equal? (Signal.) *400S.*
- What does 5 times C equal? (Signal.) *5C.*
f. (Repeat step e until firm.)
g. (Write on the board:)

> 1. $\dfrac{4}{5} \times Q =$ 2. $Z \times 59 =$
>
> 3. $T \times \dfrac{2}{7} =$ 4. $29 \times R =$

- Copy each of the problems and write what it equals.
- (Observe students and give feedback.)

- (Write to show:)

> 1. $\dfrac{4}{5} \times Q = \dfrac{4}{5} Q$ 2. $Z \times 59 = 59Z$
>
> 3. $T \times \dfrac{2}{7} = \dfrac{2}{7} T$ 4. $29 \times R = 29R$

h. Check your work. Here's what you should have. Read each problem and your answer.
- Item 1. (Signal.) *4-fifths times Q equals 4-fifths Q.*
- Item 2. (Signal.) *Z times 59 equals 59Z.*
- Item 3. (Signal.) *T times 2-sevenths equals 2-sevenths T.*
- Item 4. (Signal.) *29 times R equals 29R.*

EXERCISE 2
Quadrilaterals

a. (Direct students to find lesson 1 in their **test booklets** and write their names and indicate lesson 1 on their answer sheets.)
- Find the figures above item 1 in your **test booklet.** √
- These are the names of 4-sided figures. You know some of them. We can describe them by telling about their angles and sides.
b. Touch the first figure. √
- What's the name of that figure? (Signal.) *A rectangle.*
- A **rectangle** has 4 sides. Each pair of sides can be different lengths. All angles are 90 degrees.
c. Touch the next figure. √
- A **square** is a rectangle with all sides the same length.
- What's the name for a rectangle with all sides the same length? (Signal.) *A square.*
d. Touch the next figure. √
- A **parallelogram** is like a rectangle. One pair of lines goes in one direction. The top line and the bottom line are parallel because they go in the same direction. The other pair of parallel lines goes in another direction. Both those lines go up and to the right, but the angles of this parallelogram are not 90 degrees.
- Touch the sharp angles on the parallelogram. √
- Touch the wide angles on the parallelogram. √
e. Listen: A square is a parallelogram. A rectangle is a parallelogram.
- Are squares parallelograms? (Signal.) *Yes.*
- Are rectangles parallelograms? (Signal.) *Yes.*

- Touch the parallelogram again. √
- Is that parallelogram a rectangle? (Signal.) *No.*
  That's right, the parallelogram is not a rectangle because its angles are not 90-degree angles.
- Is that parallelogram a square? (Signal.) *No.*
  That's right, the parallelogram is not a square because its angles are not 90-degree angles, **and** the length of the sides are not all the same.
f. Touch the last figure. √
- That's a **rhombus.** A rhombus is a parallelogram that has all sides the same length.
- Does a square have sides that are all the same length? (Signal.) *Yes.*
- Is a square a parallelogram? (Signal.) *Yes.*
- So, is a square a rhombus? (Signal.) *Yes.*
g. Touch the first figure in the box again. √
- Is that rectangle a parallelogram? (Signal.) *Yes.*
  Yes, both pairs of sides of a rectangle are parallel.
- Is it a rhombus? (Signal.) *No.*
  The lengths of the sides of the rectangle are not the same, so it's not a rhombus.
h. Touch the figure labeled **square.** √
- Is that square a parallelogram? (Signal.) *Yes.*
- Is it a rhombus? (Signal.) *Yes.*
  A square has two pairs of parallel lines that are all the same length, so it's a rhombus.
i. Touch the figure labeled parallelogram. √
- Is it a parallelogram? (Signal.) *Yes.*
- Is it a rhombus? (Signal.) *No.*
  The lengths of all the sides of that parallelogram aren't equal, so it isn't a rhombus.
- Is it a rectangle? (Signal.) *No.*
j. Touch the last figure. √
- Is it a rectangle? (Signal.) *No.*
- Is it a square? (Signal.) *No.*
- Is it a parallelogram? (Signal.) *Yes.*
- Is it a rhombus? (Signal.) *Yes.*
  Both pairs of sides are parallel and the sides are of equal length, so it's a rhombus. The angles are not 90 degrees, so it isn't a rectangle or a square.

k. Find item 1. √
- You're going to mark the best answer for items 1 through 7.
l. Item 1. The choices are: A, a square; B, a parallelogram; C, a rhombus; D, a rectangle; E, all of the above.
- Is the figure for item 1 a square? (Signal.) *Yes.*
- Is the figure for item 1 a parallelogram? (Signal.) *Yes.*
- Is the figure for item 1 a rhombus? (Signal.) *Yes.*
- Is the figure for item 1 a rectangle? (Signal.) *Yes.*
  So, the answer is E, all of the above.
- Mark the answer for item 1 on your answer sheet. √
m. Item 2. The choices are: F, a parallelogram; G, a rhombus; H, a rectangle; I, a parallelogram and a rhombus; J, a parallelogram and a rectangle.
- Is the figure for item 2 a parallelogram? (Signal.) *Yes.*
- Is the figure for item 2 a rhombus? (Signal.) *Yes.*
- Is the figure for item 2 a rectangle? (Signal.) *No.*
  So the answer for item 2 is choice I, a parallelogram and a rhombus.
n. Mark the answer for item 2, then work items 3 through 7.
- Raise your hand when you're finished.
- (Observe students and give feedback.)
o. Check your work.
- Item 3. What letter did you mark? (Signal.) *A.*
  Yes, the figure for 3 is a parallelogram.
- Item 4. What letter did you mark? (Signal.) *J.*
  The figure for 4 is not a parallelogram, a rhombus, or a rectangle, so the answer is J, none of the above.
- Item 5. What letter did you mark? (Signal.) *B.*
  The figure for 5 is a parallelogram and a rectangle.
- Item 6. What letter did you mark? (Signal.) *G.*
  The figure for 6 is a parallelogram.
- Item 7. What letter did you mark: (Signal.) *C.*
  The figure for 7 is a rhombus.

EXERCISE 3
Fractions to Mixed Numbers

a. Find item 9. √
- Remember how to change these fractions into mixed numbers: You just work the fraction as a division problem.

b. Item 9. Say the division problem. (Signal.) *17 divided by 5.*
- Figure out the answer and mark the correct choice for item 9. Don't mark an answer for item 8.
- Put your pencils down when you've marked the answer for item 9. √
- Check your work.
- What's the mixed number for 17-fifths? (Signal.) *3 and 2-fifths.*
- What letter did you mark? (Signal.) *B.*
- Answer A, 2 and 7 fifths, is also equal to 17 fifths. But it's not the right answer because the fraction is more than 1.

c. Item 10. Say the division problem. (Signal.) *15 divided by 2.*
- Figure out the answer and mark it. √
- Item 10. What's the mixed number for 15-halves? (Signal.) *7 and one-half.*
- What letter did you mark? (Signal.) *I.*

d. Item 11. Say the division problem. (Signal.) *9 divided by 5.*
- Figure out the answer and mark it. √
- Item 11. What's the mixed number for 9-fifths? (Signal.) *One and 4-fifths.*
- What letter did you mark? (Signal.) *C.*

e. Mark answers for items 12 through 15.
- Put your pencils down when you've marked answers through item 15. √

f. Check your work.

g. Item 12: What's the mixed number for 39-fourths? (Signal.) *9 and 3-fourths.*
- What letter did you mark? (Signal.) *F.*

h. Item 13: What's the mixed number for 77-ninths? (Signal.) *8 and 5-ninths.*
- What letter did you mark? (Signal.) *A.*

i. Item 14: What's the mixed number for 26-eighths? (Signal.) *3 and 2-eighths.*
- What letter did you mark? (Signal.) *I.*

j. Item 15: What's the mixed number for 40-sevenths? (Signal.) *5 and 5-sevenths.*
- What letter did you mark? (Signal.) *C.*

EXERCISE 4
Dimensions and Views

a. Touch the object and the drawings before item 17 in your **test booklet.** √
- (Teacher reference:)

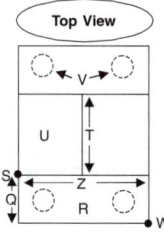
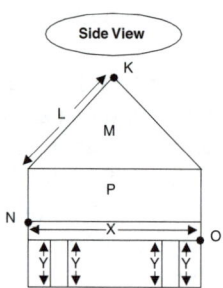

- This object has a roof made of two rectangles. One of them is labeled number 2. Touch it. √
- The line where the two sides of the roof meet is labeled number one. Touch it. √
- The walls are rectangles that are the same size. The front looks like the back. One wall is labeled number 9. Touch it. √
- The house has 2 porches, one on each side of the house. One porch roof is labeled number 7. Touch it. √
- The front edge of the porch roof is labeled number 6. Touch it. √
- The wall of the house next to the porch is labeled number 5. Touch it. √
- The triangular part of the roof is labeled number 4. Touch it. √

b. Touch the word **top,** above the house. √
- If you follow the arrow down from the word **top,** you would be looking at it from the **top view.** Touch the picture of the **top view.** √

c. Touch the word **side.** √
• If you follow the arrow from the word **side,** you'd be looking at it from the **side view.** Touch the picture of the **side view.** √
d. Go back to the picture of the house. Touch the roofline labeled number 1. √
e. Now touch the line labeled **T** in the drawing of the top view. √
• Line T shows the roofline. Line T shows the same part of the house that line 1 shows.
f. Touch the side view. √
g. Find the line or the point that shows the roofline in the side view. √
• Everybody, which letter in the side view shows the roofline? (Signal.) *K.*
Yes, point K shows the roofline. It is represented by a point.
• Go back to the house.
h. Touch the part of the roof labeled number 2. √
i. Look at item 17. I'll read it: Which letter in the top view represents the roof?
• Raise your hand when you know which letter in the top view shows the part of the house labeled number 2. √
• Everybody, which letter shows the roof of the house? (Signal.) *U.*
Mark the correct choice for item 17. Then work item 18.
• Raise your hand when you've marked answers for items 17 and 18. Don't mark an answer for item 16.
• (Observe students and give feedback.)
j. Check your work.
• Item 17. Which letter in the top view represents the roof?
• What letter did you mark? (Signal.) *D.*
Yes, choice D was letter U.
• Item 18. What letter did you mark? (Signal.) *F.*
So, which letter in the side view represents the roof? (Signal.) *L.*
Yes, line L of the side view represents the roof. Choice F was letter L.
k. Work items 19 through 22.
• Put your pencil down when you've marked answers through item 22.
• (Observe students and give feedback.)
l. Check your work.
• Item 19. What letter did your mark? (Signal.) *B.*
Yes, part M represents the part of the object labeled number 4.

• Item 20. What letter did you mark? (Signal.) *H.*
Yes, line Z represents the part of the object labeled number 4.
• Item 21. What letter did you mark? (Signal.) *B.*
Yes, point W represents the corner of the porch roof labeled number 8.
• Item 22. What letter did you mark? (Signal.) *F.*
Yes, point N represents the line of the object labeled number 6.
m. Items 23 through 26 tell about a letter on the top view or the side view, and ask about a part on the picture that has numbers. Work items 23 and 24.
• Raise your hand when you've marked answers through item 24.
• (Observe students and give feedback.)
n. Check your work.
• Item 23. What letter did you mark? (Signal.) *C.*
Yes, letter R represents the part of the object labeled number 7.
• Item 24. What letter did you mark? (Signal.) *J.*
Yes, letter K represents the part of the object labeled number 1. One was not one of the choices, so you should have marked J, none of the above.
o. I'll read item 25: The line for letter X in the side view represents one of the numbered parts of the object. What is the best description of that part?
• Touch the line labeled X in the side view. √
• Could that line represent a pillar? (Signal.) *No.*
• Could that line represent the roof of the house? (Signal.) *No.*
• Could that line represent the porch? (Signal.) *Yes.*
p. (Repeat step o until firm.)
q. Mark the answer for item 25. Then work item 26.
• Raise your hand when you've marked answers through item 26.
• (Observe students and give feedback.)
r. Check your work.
• Item 25. What letter did you mark? (Signal.) *C.*
• Item 26. What letter did you mark? (Signal.) *G.*
Yes, the best description for the dotted lines for V is the pillars holding up the porch roof.

EXERCISE 5
Exponents

a. (Write on the board:)

1.	$= 6^3$
2.	$= 1^4$
3. $f \times f \times f \times f \times f$	$=$
4. 5×5	$=$
5.	$= M^5$
6. $9 \times 9 \times 9$	$=$

- Some of these problems show a base number and an exponent. Other problems show the multiplication.
- The base number that's shown for item 1 is **6.** The exponent is **3.** I'll read the base number and exponent for item 1: 6 to the third.
- Your turn: Read the base number and the exponent for item 1. (Signal.) *6 to the third.*
b. Item 2. What is the exponent? (Signal.) *4.*
- What is the base number? (Signal.) *One.*
- Read the base number and the exponent. (Signal.) *One to the fourth.*
c. (Repeat step b until firm.)
d. Item 5. What is the base number? (Signal.) *M.*
- What is the exponent? (Signal.) *5.*
- Read the base number and the exponent. (Signal.) *M to the fifth.*
e. You'll write complete equations for each item. You'll copy what's shown and write what goes on the other side of the equals sign.
f. Look at item 1. Read the base and exponent again. (Signal.) *6 to the third.*
- That means you'll multiply 6 by itself 3 times.
- How many times will you show 6? (Signal.) *3.*
- The complete equation is: 6 times 6 times 6 **equals** 6 to the third.
- What's the complete equation for item 1? (Signal.) *6 times 6 times 6 equals 6 to the third.*
g. Item 2. One to the fourth is shown. How many times will you show one? (Signal.) *4.*
- Say the complete equation for item 2. Get ready. (Signal.) *One times one times one times one equals one to the fourth.*

h. Item 3 shows: f times f times f times f times f.
- F is the base number. How many times is f shown? (Signal.) *5.*
- So, what's the base number and the exponent in the answer? (Signal.) *F to the fifth.*
- Say the complete equation for item 3. (Signal.) *F times f times f times f times f equals f to the fifth.*
i. (Repeat step h until firm.)
j. Item 4 shows: 5 times 5.
- What's the base number? (Signal.) *5.*
- What's the exponent? (Signal.) *2.*
- Say the base number and the exponent. (Signal.) *5 to the second.*
- Say the complete equation for item 4. (Signal.) *5 times 5 equals 5 to the second.*
k. (Repeat step j until firm.)
l. Item 5 shows: M to the fifth.
- How many times will you show M? (Signal.) *5.* Say the complete equation for item 5. (Signal.) *M times M times M times M times M equals M to the fifth.*
m. (Repeat step l until firm.)
n. Item 6 shows: 9 times 9 times 9.
- What's the base number? (Signal.) *9.*
- What's the exponent? (Signal.) *3.*
- Say the base number and the exponent. (Signal.) *9 to the third.*
- Say the complete equation for item 6. (Signal.) *9 times 9 times 9 equals 9 to the third.*
o. (Repeat step n until firm.)
p. Your turn: Write the complete equation for item 1.
- Raise your hand when you're finished.
- (Observe students and give feedback.)
- (Write to show:)

1. $6 \times 6 \times 6$	$= 6^3$

- Check your work. Here's what you should have.
- Read the equation for item 1. Get ready. (Signal.) *6 times 6 times 6 equals 6 to the third.*
q. Work items 2 through 6.
- Raise your hand when you're finished.
- (Observe students and give feedback.)
- (Write to show:)

2. $1 \times 1 \times 1 \times 1$	$= 1^4$
3. $f \times f \times f \times f \times f$	$= f^5$
4. 5×5	$= 5^2$
5. $M \times M \times M \times M \times M$	$= M^5$
6. $9 \times 9 \times 9$	$= 9^3$

- Check your work. Here's what you should have.

EXERCISE 6
Classification of Angles

a. You're going to learn the names of angles. The name of a 90-degree angle is a **right angle.** What's the name of a 90-degree angle? (Signal.) *A right angle.*
• Yes, a right angle.

b. Angles that are **less** than 90 degrees are called **acute angles.** What are they? (Signal.) *Acute angles.*

c. Angles that are **more** than 90 degrees are called **obtuse angles.** What are they? (Signal.) *Obtuse angles.*

d. What's a 90-degree angle called? (Signal.) *A right angle.*
• What's an angle **less** than 90 degrees called? (Signal.) *Acute angle.*
• What's an angle **more** than 90 degrees called? (Signal.) *Obtuse angle.*

e. (Repeat step d until firm.)

f. (Write on the board:)

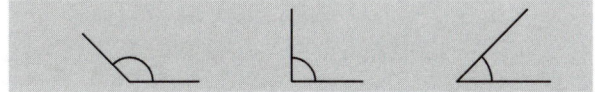

g. (Touch the first angle.) Is this angle equal to, more than, or less than 90 degrees? (Signal.) *More than 90 degrees.*
• So, it's an **obtuse** angle. What kind of angle is it? (Signal.) *Obtuse angle.*

h. (Touch the second angle.) Is this angle equal to, more than, or less than 90 degrees? (Signal.) *Equal to 90 degrees.*
• So, is this angle a right angle, obtuse angle, or acute angle? (Signal.) *Right angle.*

i. (Touch the last angle.) Is this angle equal to, more than, or less than 90 degrees? (Signal.) *Less than 90 degrees.*
• So, is this angle a right angle, obtuse angle, or acute angle? (Signal.) *Acute angle.*

j. This time, I'll touch an angle. You'll tell me if it's a right angle, an obtuse angle, or an acute angle.
• (Touch the right angle.) Is this a right angle, obtuse angle, or acute angle? (Signal.) *Right angle.*

• (Touch the acute angle.) Is this a right angle, obtuse angle, or acute angle? (Signal.) *Acute angle.*
• (Touch the obtuse angle.) Right, obtuse, or acute angle? (Signal.) *Obtuse angle.*
• (Touch the acute angle.) Right, obtuse, or acute angle? (Signal.) *Acute angle.*

k. (Write to show:)

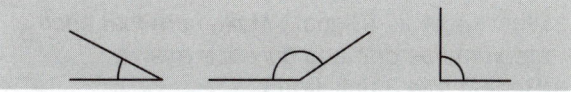

l. I'll touch the angle. You tell me: right angle, obtuse angle, or acute angle.
• (Touch the obtuse angle.) What kind of angle is this? (Signal.) *Obtuse angle.*
• (Touch the right angle.) What kind of angle is this? (Signal.) *Right angle.*
• (Touch the obtuse angle.) What kind of angle is this? (Signal.) *Obtuse angle.*
• (Touch the acute angle.) What kind of angle is this? (Signal.) *Acute angle.*

m. (Repeat step l until firm.)

EXERCISE 7
Test-Taking Rules
What to Do When You're Running Out of Time

a. Let me tell you about the tests you're going to take about two weeks from now. They are hard. You may not know how to work some of the problems, and you probably won't have enough time to work all of the other problems. But it's really important to do as well as you can on the tests. There are three rules that you have to remember when you are taking the tests. These rules will help you do as well as you can.

b. Rule 1: Work the problems that you can work.
• What's rule 1? (Signal.) *Work the problems that you can work.*

c. (Repeat rule 1 until firm.)

d. That means if you know how to work a problem, write it down on a scratch sheet and work it. Don't guess or try to work hard problems in your head. Write them down and work them.
• What do you write on your scratch sheet if the problem is **254 minus 125?** (Signal.) *254 minus 125.*

e. (Repeat step d until firm.)

f. Rule 2: If you don't know how to work a problem, skip it and come back to it.
- What's rule 2? (Signal.) *If you don't know how to work a problem, skip it and come back to it.*

g. (Repeat rule 2 until firm.)

h. Yes, you skip problems you don't know how to work and come back to them if you have time.

i. Rule 3: Make sure that each problem has **one** and **only** one answer.
- What's rule 3? (Signal.) *Make sure that each problem has one and only one answer.*

j. (Repeat rule 3 until firm.)

k. Here's what rule 3 means: When there is only one minute left to finish the test, you need to mark one answer for each of the items that don't have answers.
- Look at your answer sheet. Does item 8 have an answer? (Signal.) *No.*
- So if there was one minute left, you'd mark an answer for item 8.
- Does item 15 have an answer? (Signal.) *Yes.*
  So, would you mark an answer for item 15? (Signal.) *No.*
- Does item 16 have an answer? (Signal.) *No.*
  So, would you mark an answer for item 16 if there was only one minute left? (Signal.) *Yes.*
- Look on your answer sheet and find the **next** item that needs an answer marked. √
  Everybody, what's the next item that needs an answer marked? (Signal.) *27.*
- Look on your answer sheet and find the **last** item that needs an answer marked. √
  Everybody, what's the last item? (Signal.) *56.*

l. (Repeat step k until firm.)

Test Preparation Lesson 2

Materials Note:

Each student will need:
- lined paper
- test booklet and Multiple-Choice Response Sheet

EXERCISE 1
Distribution

a. (Write on the board:)

$$384 = 300 + 80 + 4$$

- I rewrote the whole number **3 hundred 84** as the three parts that are added together to get 384: 3 hundred plus 80 plus 4.
- Say the addition for 384. (Signal.) *3 hundred plus 80 plus 4.*
b. Your turn: Say the addition for the number **59.** (Signal.) *50 plus 9.*
- Say the addition for the number **2 hundred 73.** (Signal.) *2 hundred plus 70 plus 3.*
- Say the addition for the number **one hundred 80.** (Signal.) *One hundred plus 80.*
- Say the addition for the number **3 thousand 4 hundred 50.** (Signal.) *3 thousand plus 4 hundred plus 50.*
- Say the addition for the number **45.** (Signal.) *40 plus 5.*
- (Write to show:)

$$45$$
$$40 + 5$$

- This shows 45 rewritten as 40 plus 5.
c. (Write on the board:)

45×3	$=$
$(40 + 5) \times 3$	$=$
$(40 \times 3) + (5 \times 3)$	$=$

- This shows how you multiply 45 times 3 when you write 45 as 40 plus 5: 45 times 3 equals 40 plus 5 **times** 3. That equals 40 times 3 **plus** 5 times 3.
- Your turn: Say the new multiplication for 45 times 3. (Signal.) *40 times 3 plus 5 times 3.*
- The parentheses around the two parts show that you get an answer for each part and then add the two answers.
- Copy the problem and work it. Write what 40 times 3 equals. Write what 5 times 3 equals. Then add the answers.

- (Observe students and give feedback.)
d. (Write to show:)

45×3	$=$
$(40 + 5) \times 3$	$=$
$(40 \times 3) + (5 \times 3)$	$=$
$120 + 15$	$= 135$

- Check your work.
- You added 120 and 15. What is the answer? (Signal.) *135.*
- That's the same answer you get if you multiply 45 times 3.
e. New problem.
- (Write on the board:)

$$174 \times 3$$

- Say the addition for **one hundred 74.** (Signal.) *One hundred plus 70 plus 4.*
- (Write to show:)

$$174 \qquad \times 3 =$$
$$(100 + 70 + 4) \times 3 =$$

- One hundred 74 times 3 equals 100 plus 70 plus 4 **times** 3.
- So you'll write below: one hundred times 3, plus 70 times 3, plus 4 times 3.
- Tell me what you'll write below. (Signal.) *One hundred times 3 plus 70 times 3 plus 4 times 3.*
- Copy what's on the board and write the new problem and the answer. Remember, make parentheses around each part. Then get the answer for each part. Then add up the answers.
- (Write to show:)

174	$\times$	3	$=$
$(100 + 70 + 4)$	$\times$	3	$=$
$(100 \times 3) +$	$(70 \times 3) +$	(4×3)	$=$
$300 +$	$210 +$	12	$= 522$

- Here's what you should have.
- What does one hundred 74 times 3 equal? (Signal.) *522.*

f. New problem.
- (Write on the board:)

3	×	27	=

- Say the addition for 27. Get ready. (Signal.) *20 plus 7.*
- (Write to show:)

3	×	27	=
3	×	(20 + 7)	=

- I'll say the new problem for 3 times 27: 3 times 20, plus 3 times 7.
- Your turn: Say the new problem. (Signal.) *3 times 20, plus 3 times 7.*
- Copy what's on the board. Below, write the new problem and figure out the answer. Remember the parentheses.
- Raise your hand when you're finished.
- (Observe students and give feedback.)
- (Write to show:)

3	×	27	=
3	×	(20 + 7)	=
(3 × 20)	+	(3 × 7)	=
60	+	21	= 81

- Check your work. Here's what you should have.
- What does 3 times 27 equal? (Signal.) *81.*

g. (Write on the board:)

1.	6 × 730	=
2.	378 × 2	=
3.	8 × 1603	=
4.	417 × 9	=

h. Read item 1. Get ready. (Signal.) *6 times 7 hundred 30.*
- Say the addition for 730. (Signal.) *700 plus 30.*
- Say the new problem for 6 times 700 **plus** 30. (Signal.) *6 times 700 plus 6 times 30.*

i. (Repeat step h until firm.)

j. Read item 2. Get ready. (Signal.) *3 hundred 78 times 2.*
- Say the addition for 378. (Signal.) *300 plus 70 plus 8.*
- Say the new problem for 300 plus 70 plus 8 times 2. (Signal.) *300 times 2 plus 70 times 2 plus 8 times 2.*

k. (Repeat item j until firm.)

l. Read item 3. Get ready. (Signal.) *8 times one thousand 6 hundred 3.*
- Say the addition for 1603. (Signal.) *One thousand plus 6 hundred plus 3.*
- Say the new problem for 8 times 1000 plus 600 plus 3. (Signal.) *8 times 1000 plus 8 times 600 plus 8 times 3.*

m. (Repeat step l until firm.)

n. Read item 4. Get ready. (Signal.) *4 hundred 17 times 9.*
- Say the addition for 417. (Signal.) *4 hundred plus 10 plus 7.*
- Say the new problem for 400 plus 10 plus 7 times 9. (Signal.) *400 times 9 plus 10 times 9 plus 7 times 9 equals.*

o. (Repeat step n until firm.)

p. Work these problems. First, copy the problem. Below, write the equation with the addition for the bigger number. Write the new problem. Then write what each part equals and write the answer.
- Raise your hand when you're finished.
- (Observe students and give feedback.)

q. (Write on the board:)

1.	6	×	730			=	
	6	×	(700 + 30)			=	
	(6 × 700)	+	(6 × 30)			=	
	4200	+	180			=	4380
2.	378	×	2			=	
	300 + 70 + 8) ×		2			=	
	(300 × 2)	+	(70 × 2)	+	(8 × 2)	=	
	600	+	140	+	16	=	756
3.	8	×	1603			=	
	8	×	(1000 + 600 + 3)			=	
	8 × 1000)	+	(8 × 600)	+	(8 × 3)	=	
	8000	+	4800	+	24	=	12,824
4.	417	×	9			=	
	(400 + 10 + 7) ×		9			=	
	(400 × 9)	+	(10 × 9)	+	(7 × 9)	=	
	3600	+	90	+	63	=	3753

r. Check your work. Here's what you should have.
- Item 1. What does 6 times 7 hundred 30 equal? (Signal.) *4 thousand 3 hundred 80.*
- Item 2. What does 3 hundred 78 times 2 equal? (Signal.) *756.*
- Item 3. What does 8 times one thousand 6 hundred 3 equal? (Signal.) *12 thousand 8 hundred 24.*
- Item 4. What does 4 hundred 17 times 9 equal? (Signal.) *3 thousand 7 hundred 53.*

EXERCISE 2
Quadrilaterals

a. (Direct students to find lesson 2 in their **test booklets** and write their names and indicate lesson 2 on their answer sheets.)
• Find the figures above item 1 in your **test booklet.** √
• (Teacher reference:)

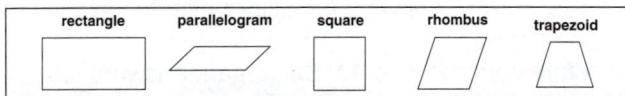

• You can see the names of the 4-sided figures. You know some of them. We can describe them by telling about their angles and sides.
b. Touch the figure labeled **rectangle.** √
• A **rectangle** has 4 sides. Each pair of sides can be different lengths. All angles are 90 degrees.
c. Touch the figure labeled **parallelogram.** √
• One pair of lines goes in one direction. Those lines are the top line and the bottom line. The other pair of parallel lines goes in another direction.
d. Touch the figure labeled **square.** √
• A **square** is like a rectangle but all sides are the same length.
e. Touch the figure labeled **rhombus.** √
• A **rhombus** is a parallelogram that has all sides the same length.
f. Touch the figure labeled **trapezoid.** √
• What's the name for that figure? (Signal.) *Trapezoid.*
• A **trapezoid** has one pair of lines that goes in the same direction. Those are the top line and the bottom line of this trapezoid. The other two lines do not go in the same direction.

g. Touch the figure for item 1. √
• First you'll tell me the name for figures 1 through 8. Then you will mark the answers.
• Figure 1. What kind of figure is it? (Signal.) *A rhombus.*
Yes, it's a rhombus. Rhombuses are also parallelograms.
• Figure 2. What kind of figure is it? (Signal.) *A trapezoid.*
Yes, a trapezoid has one pair of sides that are parallel.
• Figure 3. What kind of figure is it? (Signal.) *A rectangle.*
Yes, a rectangle. Rectangles are also parallelograms.
• Figure 4. What kind of figure is it? (Signal.) *A parallelogram.*
• Figure 5. What kind of figure is it? (Signal.) *A square.*
Yes, a square. Squares are also rectangles, parallelograms, **and** rhombuses.
• Figure 6. What kind of figure is it? (Signal.) *A trapezoid.*
• Figure 7. What kind of figure is it? (Signal.) *A rhombus.*
Yes, a rhombus. Rhombuses are also parallelograms.
• Figure 8. What kind of figure is it? (Signal.) *A rectangle.*
Yes, a rectangle. Rectangles are also another kind of figure.
• Everybody, what other kind of figure is a rectangle? (Signal.) *A parallelogram.*
h. (Repeat step g until firm.)
i. Go back to item 1 and mark the choice that best describes each figure on your answer sheet.
• Raise your hand when you've marked answers for items 1 through 8.
• (Observe students and give feedback.)
j. Check your work.
k. Item 1. What letter did you mark? (Signal.) *D.*
• What kind of figure is it: (Signal.) *A rhombus.*
Yes, figure 1 is a rhombus and a parallelogram.
l. Item 2. What letter did you mark? (Signal.) *H.*
• What kind of figure is it? (Signal.) *A trapezoid.*
Yes, the figure in 2 is a trapezoid.

m. Item 3. What letter did you mark? (Signal.) *C.*
- What kind of figure is it? (Signal.) *A rectangle.*
  Yes, the figure in 3 is a rectangle.
n. Item 4. What letter did you mark? (Signal.) *F.*
- What kind of figure is it? (Signal.) *A parallelogram.*
  Yes, the figure in 4 is a parallelogram.
o. Item 5. What letter did you mark? (Signal.) *B.*
- What kind of figure is it? (Signal.) *A square.*
  Yes, the figure in 5 is a square, a parallelogram, and a rectangle.
p. Item 6. What letter did you mark? (Signal.) *J.*
- What kind of figure is it? (Signal.) *A trapezoid.*
  Yes, the figure in 16 is not a rhombus and a trapezoid. It's also not a parallelogram or a square, so the answer is J, none of the above.
q. Item 7. What letter did you mark? (Signal.) *C.*
- What kind of figure is it? (Signal.) *A rhombus.*
  Yes, the figure in 7 is a rhombus and a parallelogram.
r. Item 8. What letter did you mark? (Signal.) *F.*
- What kind of figure is it? (Signal.) *A rectangle.*
  Yes, the figure in 8 is a parallelogram and a rectangle.

EXERCISE 3
Exponents

a. Find item 11 in your **test booklet.** √
- (Teacher reference:)

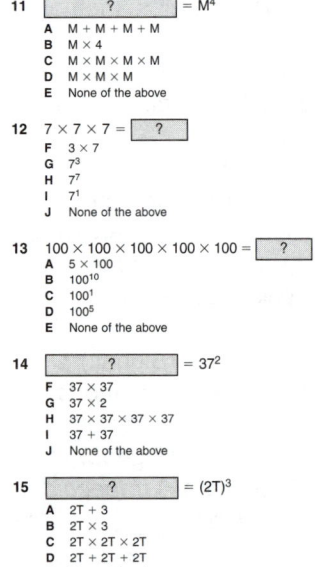

- For items 11 through 15, you're going to figure out the complete equation with the multiplication and the base number and the exponent it equals.
b. Item 11. What's the base number? (Signal.) *M.*
- What's the exponent? (Signal.) *4.*
- Read the base number and the exponent. (Signal.)
  M to the fourth.

- How many times will M be shown? (Signal.) *4.*
- Say the multiplication for M to the fourth. (Signal.)
  M times M times M times M.
c. Item 12. 7 times 7 times 7.
- What's the base number? (Signal.) *7.*
- What's the exponent? (Signal.) *3.*
- Say the base number and the exponent for item 12. (Signal.) *7 to the third.*
  Yes, item 12 shows the multiplication for 7 to the third power.
d. Work items 11 and 12. Don't mark answers for items 9 and 10.
- Raise your hand when you've marked the answers through item 12.
- (Observe students and give feedback.)
e. Check your work.
f. Item 11 shows **M to the fourth.**
- What letter did you mark? (Signal.) *C.*
  Yes, M times M times M times M equals M to the fourth power.
g. Item 12 shows **7 times 7 times 7.**
- What letter did you mark? (Signal.) *G.*
  Yes, 7 times 7 times 7 equals 7 to the third power.
h. Item 13. What's the base number? (Signal.) *100.*
- What's the exponent? (Signal.) *5.*
- Say the base number and the exponent for item 13. (Signal.) *100 to the fifth.*
  Yes, item 13 shows the multiplication for 100 to the fifth power.
i. Item 14 shows **37 to the second.**
- What's the base number? (Signal.) *37.*
- What's the exponent? (Signal.) *2.*
- Say the multiplication for 37 to the second power. Get ready. (Signal.) *37 times 37.*
j. Item 15 shows **2T to the third.**
- What's the base number? (Signal.) *2T.*
- What's the exponent? (Signal.) *3.*
  Say the multiplication for 2T to the third power. Get ready. (Signal.) *2T times 2T times 2T.*
k. Work items 13 through 15.
- Raise your hand when you've marked answers through item 15.
- (Observe students and give feedback.)
l. Check your work.
m. Item 13 shows **100 times 100 times 100 times 100 times 100.**
- What letter did you mark? (Signal.) *D.*
  Yes, it equals 100 to the fifth power.
n. Item 14 shows **37 to the second.**
- What letter did you mark? (Signal.) *F.*
  Yes, 37 times 37 **equals** 37 to the second power.
o. Item 15 shows **2T to the third power.**
- What letter did you mark? (Signal.) *C.*
  Yes, 2T times 2T times 2T **equals** 2T to the third power.

EXERCISE 4
Average

a. Find the sample problem above item 17. √
- You're going to find the average number. To find the average number, you add up all the amounts and you divide by the number of the amounts. Once more: Add up all the amounts, divide by the number of amounts. If there are seven amounts, you divide by 7. If there are two amounts, you divide by 2.
- If there are 11 amounts, what do you divide by? (Signal.) *11.*
- If there are 3 amounts, what do you divide by? (Signal.) *3.*

b. I'll read the sample problem.
- There are 5 amounts: $5, $6, $2, $12, $10. Those amounts show how much Jimmy earned on five different days.
- To find the average, you add up the 5 amounts. That's 35 dollars.
- Then you divide. What number do you divide 35 by? (Signal.) *5.*
- What's the answer? (Signal.) *7.*
  So, the average amount Jimmy earned on the five days was $7.
- Remember, add up the amounts. Then divide by the number of amounts.

c. Item 17. The high temperature on four days was 11 degrees, 15 degrees, 10 degrees, and 12 degrees. What was the average high temperature for the four-day period?
- Tell me what you'll do first. (Call on a student. Ideas: *Add; add 11 plus 15 plus 10 plus 12.*) Yes, you add the 4 temperatures.
- Then what will you do? (Signal.) *Divide by 4.*
- Figure out the total for the amounts. Then divide by the number of amounts. Mark the right answer. √
- Check your work. You added up the amounts. That's 48. You divided by 4.
- What's the average temperature for the four-day period? (Signal.) *12 degrees.*
- What letter did you mark? (Signal.) *C.*

d. Work problems 18 and 19.
- Pencils down when you've marked the answers through item 19.
- (Observe students and give feedback.)

e. Check your work.

f. Item 18. The table shows how far a runner ran on Monday through Friday.
- What's the average distance the runner ran each day? (Signal.) *8 miles.*
- What letter did you mark? (Signal.) *G.*
- Item 19. The table shows the number of people in 8 cars that were on the highway. What's the average number of people in a car? (Signal.) *3.*
- What letter did you mark? (Signal.) *C.*

EXERCISE 5
Shapes—3 Dimensions

a. Find the shapes in the box above item 21. √
- (Teacher reference:)

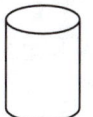 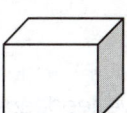  

- These are objects that are not flat. They have special names.
- Touch the first object. It's called a cylinder. What's it called? (Signal.) *A cylinder.*
- The next object is called a cube. It's like a square, but it has 6 sides. Think of an ice cube. What's the name of that shape? (Signal.) *A cube.*
- The next object is called a sphere. It's like a circle, but a circle is flat. A sphere is round on all sides, like a ball. What's the name of that shape? (Signal.) *A sphere.*
- The last object is called a cone. It's shaped like an ice cream cone. What's the name of that object? (Signal.) *A cone.*

b. Touch the first object in the box again. √
- What's the name? (Signal.) *A cylinder.*
- What's the name of the next object? (Signal.) *A cube.*
- What's the name of the next object? (Signal.) *A sphere.*
- What's the name of the last object? (Signal.) *A cone.*

c. Find the shapes below the box. The shapes have numbers under them.
- Items 21 through 25 ask about the objects with numbers under them.

d. Item 21 asks: Which number shows the small sphere?
- Find the small sphere and mark the right choice. Don't mark an answer for item 20. √
- What letter did you mark? (Signal.) *B.*
- Yes, object 2 is the small sphere. That's choice B.

e. Item 22 asks: Which number shows the large cylinder?
• Mark the right choice. √
• What letter did you mark? (Signal.) *H.*
• Yes, object 6 is the large cylinder. You should have marked choice H.
f. Work items 23 through 25.
• Raise your hand when you're finished.
• (Observe students and give feedback.)
g. Check your work.
h. Item 23 asks: Which number shows the large cube?
• What letter did you mark? (Signal.) *A.*
Yes, object 1 is the large cube.
i. Item 24 asks: Which number shows the large cone?
• What letter did you mark? (Signal.) *I.*
Yes, object 9 is a large cone.
j. Item 25 asks: Which number shows the large sphere?
• What letter did you mark? (Signal.) *B.*
Yes, object 12 is the large sphere.

EXERCISE 6
Mixed Numbers to Improper Fractions

a. Find item 27. √
• These are mixed numbers. You'll change them into improper fractions.
b. Item 27. 8 and 2-fifths.
• What's the denominator? (Signal.) *5.*
• Say the multiplication problem to figure out how many fifths are in **8 wholes.** Get ready. (Signal.) *5 times 8.*
• What's the answer? (Signal.) *40.*
• You have 40-fifths plus 2-fifths.
• What's the improper fraction? (Signal.) *42-fifths.*

c. Item 28. What's the denominator? (Signal.) *9.*
• Say the multiplication problem to figure out how many ninths are in 3 wholes. Get ready. (Signal.) *9 times 3.*
• What's the answer? (Signal.) *27.*
• You have 27-ninths plus 2-ninths.
• So what's the improper fraction? (Signal.) *29-ninths.*
d. Item 29. What's the denominator? (Signal.) *3.*
• Say the multiplication problem to figure out how many thirds are in 2 wholes. Get ready. (Signal.) *3 times 2.*
• What's the answer? (Signal.) *6.*
• You have 6-thirds plus one-third.
• So what's the improper fraction? (Signal.) *7-thirds.*
e. Item 30. What's the denominator? (Signal.) *4.*
• Say the multiplication problem to figure out how many fourths are in 10 wholes. (Signal.) *4 times 10.*
• What's the answer? (Signal.) *40.*
• You have 40-fourths plus 3-fourths.
• So what's the improper fraction? Get ready. (Signal.) *43-fourths.*
f. Your turn: mark the answer to items 27 through 30. Don't mark an answer for item 26.
• (Observe students and give feedback.)
g. Check your work.
h. Item 27. What letter did you mark? (Signal.) *E.*
Yes, 8 and 2-fifths equals 42-fifths. 42-fifths was not one of the choices, so the answer is E, none of these.
i. Item 28. What letter did you mark? (Signal.) *G.*
Yes, 3 and 2-ninths equals 29-ninths.
j. Item 29. What letter did you mark? (Signal.) *C.*
Yes, 2 and one-third equals 7-thirds.
k. Item 30. What letter did you mark? (Signal.) *F.*
Yes, 10 and 3-fourths equals 43-fourths.

EXERCISE 7
Classification of Angles

a. (Write on the board:)

> **right angle**
> **acute angle**
> **obtuse angle**

- You've learned the names for different angles.
b. Which angle is **less** than 90 degrees? (Signal.) *Acute angle.*
- Which angle is **more** than 90 degrees? (Signal.) *Obtuse angle.*
- Which angle is **exactly** 90 degrees? (Signal.) *Right angle.*
c. (Repeat step b until firm.)
d. Touch item 31. √
- (Teacher reference:)

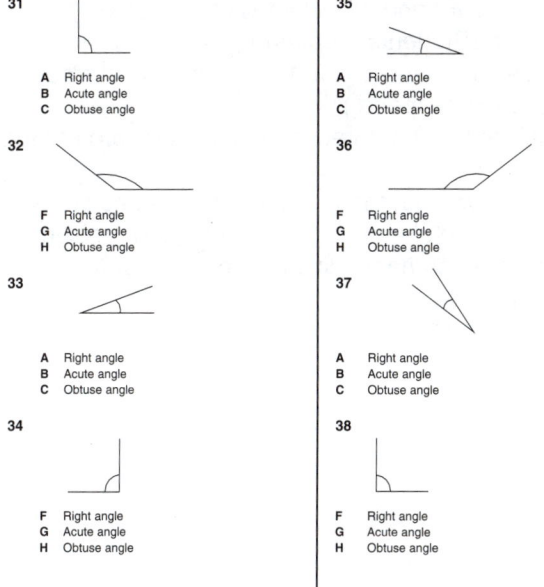

31

A Right angle
B Acute angle
C Obtuse angle

32

F Right angle
G Acute angle
H Obtuse angle

33

A Right angle
B Acute angle
C Obtuse angle

34

F Right angle
G Acute angle
H Obtuse angle

35

A Right angle
B Acute angle
C Obtuse angle

36

F Right angle
G Acute angle
H Obtuse angle

37

A Right angle
B Acute angle
C Obtuse angle

38

F Right angle
G Acute angle
H Obtuse angle

- For each angle, you'll mark the name **right** angle, **acute** angle, or **obtuse** angle. Raise your hand when you've marked answers for items 31 through 38.
- (Observe students and give feedback.)
e. Check your work. For each item, you'll tell me what letter you marked. Then you'll tell me whether the angle is right, acute, or obtuse.
f. Item 31. What letter did you mark? (Signal.) *A.*
- What kind of angle is angle 31? (Signal.) *Right.*

g. Item 32. What letter did you mark? (Signal.) *H.*
- What kind of angle is angle 32? (Signal.) *Obtuse.*
h. Item 33. What letter did you mark? (Signal.) *B.*
- What kind of angle is angle 33? (Signal.) *Acute.*
i. Item 34. What letter did you mark? (Signal.) *F.*
- What kind of angle is angle 34? (Signal.) *Right.*
j. Item 35. What letter did you mark? (Signal.) *B.*
- What kind of angle? (Signal.) *Acute.*
k. Item 36. What letter did you mark? (Signal.) *H.*
- What kind of angle? (Signal.) *Obtuse.*
l. Item 37. What letter did you mark? (Signal.) *B.*
- What kind of angle? (Signal.) *Acute.*
m. Item 38. What letter did you mark? (Signal.) *F.*
- What kind of angle? (Signal.) *Right.*

EXERCISE 8
Test-Taking Rules

What to Do When You're Running Out of Time

a. There are 3 rules that help you do well when you're taking a test.
b. Rule 1: Work the problems that you can work.
- What's rule 1? (Signal.) *Work the problems that you can work.*
c. (Repeat rule 1 until firm.)
d. That means if you know how to work a problem, write it down on a scratch sheet and work it.
- What do you write on your scratch sheet if the problem is **45 times 17**? (Signal.) *45 times 17.*
e. (Repeat step d until firm.)
f. Rule 2: If you don't know how to work a problem, skip it and come back to it.
- What's rule 2? (Signal.) *If you don't know how to work a problem, skip it and come back to it.*
g. (Repeat rule 2 until firm.)
h. Rule 3: Make sure that each problem has one and only one answer.
- What's rule 3? (Signal.) *Make sure that each problem has one and only one answer.*
i. (Repeat rule 3 until firm.)
j. Who remembers **when** you should start marking the items that don't have answers? (Call on a student. Idea: *When there's one minute left.*)
- Everybody, when do you start marking items that don't have answers? (Signal.) *When there's one minute left.*

k. Look at your answer sheet and find the first item you'd mark if there was only one minute left.
• Everybody, what's the number of the first item you'd mark? (Signal.) *9.*
• Find the next item you'd mark. What's the number of the next item? (Signal.) *10.*
• What's the number of the next item you'd mark? (Signal.) *16.*
• What's the number of the next item you'd mark? (Signal.) *20.*
• What's the number of the next item you'd mark? (Signal.) *26.*
• What's the number of the next item you'd mark? (Signal.) *39.*
• What's the number of the **last** item you'd mark? (Signal.) *56.*
l. (Repeat step k until firm.)
m. We're going to pretend you're taking a test, and there's not much time left. You haven't answered items 9, 10, 16, 20, 26, and 39 through 56. So, when I tell you there's only one minute left, you'll mark one and only one answer for each of those numbers. There really aren't any problems; you're just guessing. But if this were a real test, you wouldn't have time to work the problems, so you would have to guess.
n. Get ready to mark the answers. After you're finished, we'll find out how many more points you scored by following the rules for doing well on tests.
o. Here we go. There is only one minute left. Finish up your paper.
p. (Reinforce students who quickly fill in one and only one answer for items 9, 10, 16, 20, 26, and 39 through 56 on the answer sheet.)
• (Prompt students who don't fill answers in quickly to start marking answers more quickly.)
• (Alert students who have more than one answer filled in on any item to make sure that there's only one answer per item.)
q. (After one minute, say:) Stop working.
• Item 9. What letter did you mark? (Signal.) (Students respond.)
The make-believe answer is **B.** Make an X by item 9 if you didn't mark B. Make a C by item 9 if you marked B.
• Raise your hand if you got it right. √
• Item 10. The answer is **F.** Raise your hand if you got it right. √

• Make a C by item 10 if you marked F. If you didn't mark F, make an X.
r. For the rest of the items you just marked, I'll tell you the make-believe answer.
• Raise your hand if you marked that answer. Remember to make a C or an X for each item.
• Item 16. The answer is **I.** (Acknowledge students who raise their hands.)
• Item 20. The answer is **H.** (Acknowledge students who raise their hands.)
• Item 26. The answer is **I.** (Acknowledge students who raise their hands.)
• Item 39. The answer is **E.** (Acknowledge students who raise their hands.)
• (Repeat for items: 40, H; 41, B; 42, J; 43, D; 44, G; 45, A; 46, H; 47, C; 48, I; 49, A; 50, G; 51, B; 52, F; 53, A; 54, J; 55, D; 56, H.)
s. Count the Cs you wrote next to the numbers.
• Raise your hand if you got eight or more of them correct. (Students respond.)
• Raise your hand if you got six or seven of them correct. (Students respond.)
• Raise your hand if you got five of them correct. (Students respond.)
t. (Repeat for four, three, two, one, and zero of them correct.)
• Most of you got three or four of the items correct. You wouldn't have gotten any of these problems correct if you hadn't filled in answers for those items.

Test Preparation Lesson 3

EXERCISE 1
Multiplying Numbers and Unknowns

a. (Write on the board:)

1. $8 \times V =$	2. $T \times 36 =$
3. $\dfrac{8}{3} \times L =$	4. $Q \times \dfrac{4}{7} =$

• Item 1: 8 times V. The answer is 8V.
b. Your turn: Read each problem and say the answer.
• Item 1. (Signal.) *8 times V equals 8V.*
• Item 2. (Signal.) *T times 36 equals 36T.*
• Item 3. (Signal.) *8-thirds times L equals 8-thirds L.*
• Item 4. (Signal.) *Q times 4-sevenths equals 4-sevenths Q.*
c. Copy each of the problems and write what it equals.
• (Observe students and give feedback.)
• (Write to show:)

1. $8 \times V = \boxed{8V}$	2. $T \times 36 = \boxed{36T}$
3. $\dfrac{8}{3} \times L = \boxed{\dfrac{8}{3}L}$	4. $Q \times \dfrac{4}{7} = \boxed{\dfrac{4}{7}Q}$

• Check your work. Here's what you should have.

EXERCISE 2 Dimensions and Views

a. (Direct students to find lesson 3 in their **test booklets** and write their names and indicate lesson 3 on their answer sheets.)
• Find the object and the drawings in your **test booklet** and touch the object. √

• (Teacher reference:)

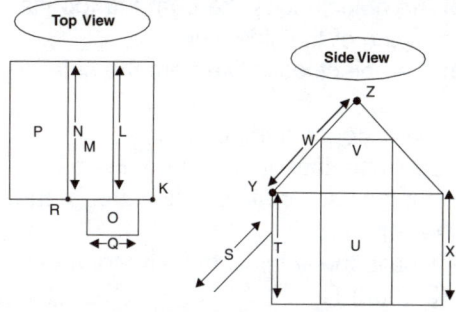

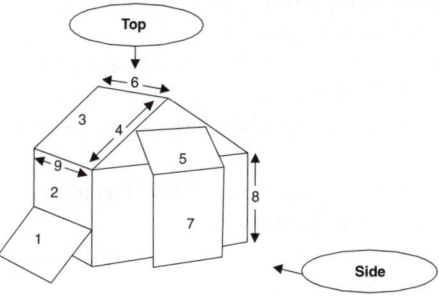

• This object looks like a barn. The roof has 2 rectangular parts. The roofline, where the parts of the roof come together, is labeled number 6. Touch it. √
• The rectangular part of the roof you can see is labeled number 3. Touch it. √
• The edge of the roof is labeled number 4. Touch it. √
• The walls are rectangles. The front wall is labeled number 2. Touch it. √
• The line where the front wall and the roof meet is labeled number 9. Touch it. √
• A rectangular wall is leaning diagonally on the front wall. It is labeled number 1. Touch it. √
• A slanted wall is leaning against the side of the barn, resting on another wall. Those parts are labeled numbers 5 and 7. Touch them. √
• The line where two walls meet is labeled number 8. Touch it. √
b. Touch the word **top,** above the object. √
  If you follow the arrow pointing from the word **top,** you'd be looking at the object from which view? (Signal.) *The top view.*
• Touch the word **side.** √
  If you follow the arrow from the word **side,** you'd be looking at the object from which view? (Signal.) *The side view.*
c. (Repeat step b until firm.)

d. Touch the drawing of the top view of the object. √
That's what the object looks like from the top view.
- Touch the drawing of the side view. √
That's what the object looks like from the side view.
- Back to the barn. Touch number 1. √
Number 1 labels the lean-to in front of the barn.
e. Raise your hand when you know the letter of the lean-to in the **top** view. √
- Everybody, what's the letter of the lean-to in the **top** view? (Signal.) *P.*
Yes, P represents number 1.
f. Raise your hand when you know the letter of the lean-to in the **side** view. √
- Everybody, what's the letter of the lean-to in the **side** view? (Signal.) *S.*
g. Touch the part of the barn labeled number 6. √
Number 6 is the roofline.
- Raise your hand when you know the letter of the roofline in the **side** view. √
- Everybody, what's the letter of the roofline in the side view? (Signal.) *Z.*
Yes, point Z represents number 6.
h. Raise your hand when you know the letter of the roofline in the **top** view. √
- Everybody, what's the letter of the roofline in the **top** view? (Signal.) *L.*
Yes, line L represents number 6.
i. Touch number 7. Raise your hand when you know the letter for number 7 in the **side** view. √
- Everybody, what's the letter for 7 in the side view? (Signal.) *U.*
j. Raise your hand when you know the letter for 7 in the **top** view. √
- Everybody, what's the letter for number 7 in the top view? (Signal.) *Q.*
Yes, line Q represents number 7.
k. Touch number 8. √
- Raise your hand when you know the letter for number 8 in the **top** view. √
- Everybody, what's the letter for number 8 in the top view? (Signal.) *K.*
Yes, point K represents number 8.
l. Raise your hand when you know the letter for number 8 in the **side** view. √
- Everybody, what's the letter? (Signal.) *X.*
Yes, line X represents number 8.
m. (Repeat steps d through l until firm.)

n. Your turn: Work items 1 through 10.
- Pencils down when you're finished.
- (Observe students and give feedback.)
o. Check your work.
p. Item 1: Which letter in the top view represents the roof?
- What letter did you mark? (Signal.) *A.*
Yes, choice A was part M.
q. Item 2: Which letter in the side view represents the roofline?
- What letter did you mark? (Signal.) *I.*
Yes, choice I was point Z.
r. Item 3: Which letter in the side view represents the part of the object labeled number 2?
- What letter did you mark? (Signal.) *C.*
Yes, choice C was line T.
s. Item 4: Which letter in the top view represents the line of the object labeled number 8?
- What letter did you mark? (Signal.) *I.*
Yes, choice I was point K.
t. Item 5: Which letter in the side view represents the lean-to?
- What letter did you mark? (Signal.) *A.*
Yes, choice A was line S.
u. Item 6: Which letter in the top view represents the part of the object labeled number 7?
- What letter did you mark? (Signal.) *G.*
Yes, choice G was line Q.
v. Item 7: The line for the letter X in the side view represents one of the numbered parts of the object.
- Everybody, what letter did you mark? (Signal.) *E.*
Yes, letter X represents the line of the object numbered 8. 8 was not one of the choices, so you should have marked E, none of the above.
w. Item 8: The point for letter Y in the side view represents one of the numbered parts of the object.
- Everybody, what letter did you mark? (Signal.) *H.*
Yes, point Y represents the line of the object numbered 9.
x. Item 9: The line for letter W in the side view represents one of the numbered parts of the object. What is the best description of that part?
- What letter did you mark? (Signal.) *B.*
Yes, line W, in the side view could represent the roof of the barn or the edge of the roof. The roof of the barn was the only choice listed, so that's the best description.

y. Item 10: The line for letter L, in the top view, represents one of the parts of the object. What is the best description of that part?
- What letter did you mark? (Signal.) *F.*
  Yes, the best description for the line for letter L is the roofline.

EXERCISE 3
Quadrilaterals

a. Find the names above item 12. √
- You've learned the names of different 4-sided figures.
- I'll read each name. Follow along: rectangle, parallelogram, trapezoid, square, rhombus.
- (Teacher reference:)

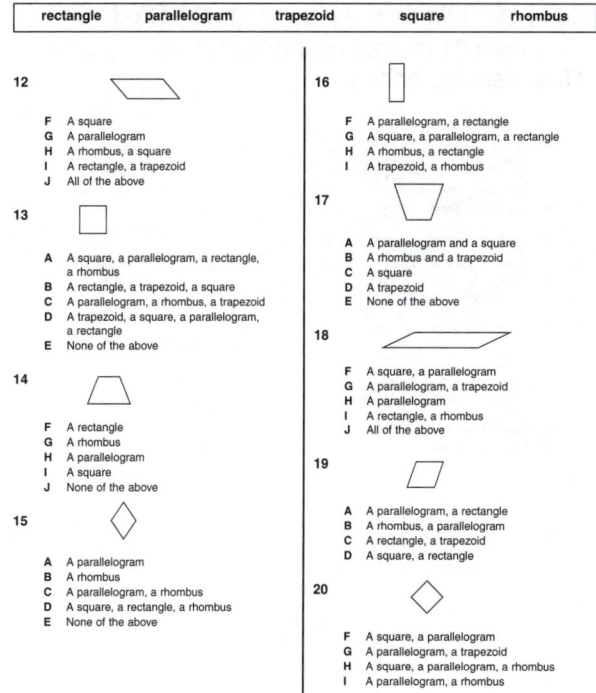

rectangle	parallelogram	trapezoid	square	rhombus

12

F A square
G A parallelogram
H A rhombus, a square
I A rectangle, a trapezoid
J All of the above

13

A A square, a parallelogram, a rectangle, a rhombus
B A rectangle, a trapezoid, a square
C A parallelogram, a rhombus, a trapezoid
D A trapezoid, a square, a parallelogram, a rectangle
E None of the above

14

F A rectangle
G A rhombus
H A parallelogram
I A square
J None of the above

15

A A parallelogram
B A rhombus
C A parallelogram, a rhombus
D A square, a rectangle, a rhombus
E None of the above

16

F A parallelogram, a rectangle
G A square, a parallelogram, a rectangle
H A rhombus, a rectangle
I A trapezoid, a rhombus

17

A A parallelogram and a square
B A rhombus and a trapezoid
C A square
D A trapezoid
E None of the above

18

F A square, a parallelogram
G A parallelogram, a trapezoid
H A parallelogram
I A rectangle, a rhombus
J All of the above

19

A A parallelogram, a rectangle
B A rhombus, a parallelogram
C A rectangle, a trapezoid
D A square, a rectangle

20

F A square, a parallelogram
G A parallelogram, a trapezoid
H A square, a parallelogram, a rhombus
I A parallelogram, a rhombus

b. One of the figures for items 12 through 20 is a rectangle.
- Raise your hand when you know the number for that figure. √
- Everybody, what's the number for the rectangle? (Signal.) *16.*
- Two of the figures are squares. What are the numbers of the squares? (Signal.) *13 and 20.*
- Two of the other figures are rhombuses. What are the numbers of the rhombuses? (Signal.) *15 and 19.*
- Two of the other figures are parallelograms. What are the numbers of the parallelograms? (Signal.) *12 and 18.*
- Two of the figures are trapezoids. What are the numbers of the trapezoids? (Signal.) *14 and 17.*
c. (Repeat step b until firm.)

d. You're going to mark the best choice for each figure.
- Raise your hand when you've marked answers for items 12 through 20. Don't mark an answer for item 11.
- (Observe students and give feedback.)
e. Check your work.
f. Item 12. What letter did you mark? (Signal.) *G.*
- What kind of figure is it? (Signal.) *A parallelogram.*
  Yes, the figure in 12 is a parallelogram.
g. Item 13. What letter did you mark? (Signal.) *A.*
- What kind of figure is it? (Signal.) *A square.*
  Yes, the figure in 13 is a square, a rectangle, a parallelogram, and a rhombus.
h. Item 14. What letter did you mark? (Signal.) *J.*
- What kind of figure is it? (Signal.) *A trapezoid.*
  Yes, the figure in 14 is not a rectangle, a rhombus, a parallelogram, or a square, so the answer is choice J, none of the above.
i. Item 15. What letter did you mark? (Signal.) *C.*
- What kind of figure is it? (Signal.) *A rhombus.*
  Yes, the figure in 15 is a parallelogram and a rhombus, so the best choice was C.
j. Item 16. What letter did you mark? (Signal.) *F.*
- What kind of figure is it? (Signal.) *A rectangle.*
  Yes, the figure in 16 is a parallelogram **and** a rectangle.
k. Item 17. What letter did you mark? (Signal.) *D.*
- What kind of figure is it? (Signal.) *A trapezoid.*
  Yes, the figure in 17 is a trapezoid.
l. Item 18. What letter did you mark? (Signal.) *H.*
- What kind of figure is it? (Signal.) *A parallelogram.*
  Yes, the figure in 18 is a parallelogram.
m. Item 19. What letter did you mark? (Signal.) *B.*
- What kind of figure is it? (Signal.) *A rhombus.*
  Yes, the figure in 19 is a rhombus and a parallelogram.
n. Item 20. What letter did you mark? (Signal.) *H.*
- What kind of figure is it? (Signal.) *A square.*
  Yes, the figure in 20 is a square, a parallelogram, and a rhombus. It is not a trapezoid. So the best answer is H.

EXERCISE 4
Distribution

a. (Write on the board:)

1.	672	×	5	=
2.	8	×	291	=
3.	1043	×	6	=

b. Read item 1. (Signal.) *6 hundred 72 times 5.*
- Say the addition for 6 hundred 72. (Signal.) *600 plus 70 plus 2.*
- Say the new problem for 600 plus 70 plus 2 times 5. Get ready. (Signal.) *600 times 5 plus 70 times 5 plus 2 times 5.*
- Copy item 1. Below, write the problem with the addition for 6 hundred 72. Then write the new problem. Then write what each part equals and the answer.
- Raise your hand when you're finished. √
- (Write to show:)

672	×	5		=	
$(600 + 70 + 2) \times$		5		=	
(600×5)	$+ (70 \times 5)$	$+ (2 \times 5)$		=	
3000	+ 350	+ 10		= 3360	

- Item 1. Here's what you should have. What number does 672 times 5 equal? (Signal.) *3360.*
c. Read item 2. (Signal.) *8 times 2 hundred 91.*
- Say the addition for 2 hundred 91. (Signal.) *200 plus 90 plus one.*
- Say the new problem for 8 times 200 plus 90 plus one. Get ready. (Signal.) *8 times 200 plus 8 times 90 plus 8 times one.*
- Copy item 2. Below, write the problem with the addition for 2 hundred 91. Then write the new problem. Then write what each part equals and the answer.
- Raise your hand when you're finished. √
- (Write to show:)

2.	8	×	291	=	
	8	×	$(200 + 90 + 1)$	=	
	(8×200)	$+ (8 \times 90)$	$+ (8 \times 1)$	=	
	1600	+ 720	+ 8	= 2328	

- Item 2. Here's what you should have. What number does 8 times 291 equal? (Signal.) *2328.*

d. Read item 3. (Signal.) *One thousand 43 times 6.*
- Say the addition for one thousand 43. (Signal.) *1000 plus 40 plus 3.*
- Say the new problem for 1000 plus 40 plus 3 times 6. Get ready. (Signal.) *1000 times 6 plus 40 times 6 plus 3 times 6.*
- Copy item 3 and work it.
- Raise your hand when you're finished. √
- (Write to show:)

3.	1043	×	6	=	
	$(1000 + 40 + 3) \times$		6	=	
	(1000×6)	$+ (40 \times 6)$	$+ (3 \times 6)$	=	
	6000	+ 240	+ 18	= 6258	

- Item 3. Here's what you should have. What number does 1043 times 6 equal? (Signal.) *6258.*
e. Find item 21 in your **test booklet.** √
- (Teacher reference:)

21 $594 \times 3 =$
  A $(500 + 3) \times (90 + 3) \times (4 + 3)$
  B $(500 \times 3) + (90 \times 3) + (4 \times 3)$
  C $(500 \times 3) \times (90 \times 3) \times (4 \times 3)$
  D $(500 \times 90) + (4 \times 3)$
  E None of the above

22 $2061 \times 7 =$
  F $(2000 \times 7) \times (60 \times 7) \times (1 \times 7)$
  G $(2000 + 7) \times (60 + 7) \times (1 + 7)$
  H $(2000 \times 60) + (1 \times 7)$
  I $(2000 \times 7) + (60 \times 7) + (1 \times 7)$
  J None of the above

23 $829 \times 6 =$
  A $(800 \times 6) + (20 \times 6) + (9 \times 6)$
  B $(800 \times 6) \times (20 \times 6) \times (9 \times 6)$
  C $(800 \times 20) + (9 \times 6)$
  D $(800 + 6) \times (20 + 6) \times (9 + 3)$
  E None of the above

24 $8 \times 197 =$
  F $(8 \times 100) + (90 \times 7)$
  G $(8 \times 100) \times (8 \times 90) \times (8 \times 7)$
  H $(8 \times 100) + (8 \times 90) + (8 \times 7)$
  I $(8 + 100) \times (8 + 90) \times (8 + 7)$
  J None of the above

25 $4 \times 3805 =$
  A $(4 \times 3000) + (800 \times 5)$
  B $(4 \times 3000) + (4 \times 800) + (4 \times 5)$
  C $(4 + 3000) \times (4 + 800) \times (4 + 5)$
  D $(4 \times 3000) \times (4 \times 800) \times (4 \times 5)$
  E None of the above

- Items 21 through 25 are multiplication problems. The answer you'll mark for those problems is the new multiplication problem you'd write if you were working the problem.
f. Item 21: 5 hundred 94 times 3.
- Say the addition for 5 hundred 94. (Signal.) *500 plus 90 plus 4.*
- Say the multiplication problem for 500 plus 90 plus 4 times 3. Get ready. (Signal.) *500 times 3 plus 90 times 3 plus 4 times 3.*
- Yes, that's the answer to the problem. Find the letter for 500 times 3 plus 90 times 3 plus 4 times 3 and mark it on your answer sheet. √
- Item 21. What letter did you mark? (Signal.) *B.*

g. Work items 22 through 25. Raise your hand when you've marked answers for the items through 25.
• (Observe students and give feedback.)
h. Check your work.
i. Item 22: 2 thousand 61 times 7.
• What letter did you mark? (Signal.) *I.*
Yes, 2 thousand 61 times 7 equals 2 thousand times 7 plus 60 times 7 plus one times 7.
j. Item 23: 8 hundred 29 times 6.
• What letter did you mark? (Signal.) *A.*
Yes, 8 hundred 29 times 6 equals 8 hundred times 6 plus 20 times 6 plus 9 times 6.
k. Item 24: 8 times one hundred 97.
• What letter did you mark? (Signal.) *H.*
Yes, 8 times one hundred 97 equals 8 times one hundred plus 8 times 90 plus 8 times 7.
l. Item 25: 4 times 3 thousand 8 hundred 5.
• What letter did you mark? (Signal.) *B.*
Yes, it equals 4 times 3 thousand plus 4 times 8 hundred plus 4 times 5.

EXERCISE 5
Fractions to Mixed Numbers

a. Find item 27. √
• Remember how to change these fractions into mixed numbers. You just work the fraction as a division problem.
b. Item 27. Say the division problem. (Signal.) *21 divided by 4.*
• Figure out the mixed number it equals and mark the right answer. Don't mark an answer for item 26. √
• Check your work.
• Item 27. What letter did you mark? (Signal.) *C.*
Yes, 21-fourths equals 5 and one-fourth.
c. Work items 28 through 31.
• Raise your hand when you've marked answers through item 31. √
d. Check your work.
• Item 28. What letter did you mark? (Signal.) *I.*
Yes, 32-fifths equals 6 and 2-fifths.
• Item 29. What letter did you mark? (Signal.) *C.*
Yes, 77-ninths equals 8 and 5-ninths.
• Item 30. What letter did you mark? (Signal.) *F.*
Yes, 30-eights equals 3 and 6-eighths.
• Item 31. What letter did you mark? (Signal.) *B.*
Yes, 16-thirds equals 5 and one-third.

EXERCISE 6
Shapes in Two and Three Dimensions

a. Find the objects above item 33. √
• (Teacher reference:)

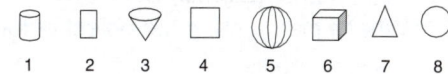

• These objects have special names.
b. Touch object 1. √
• Everybody, what's the name? (Signal.) *A cylinder.*
• What's the name of object 2? (Signal.) *A rectangle.*
• What's object 3? (Signal.) *A cone.*
• What's object 4? (Signal.) *A square.*
• What's object 5? (Signal.) *A sphere.*
• What's object 6? (Signal.) *A cube.*
• What's object 7? (Signal.) *A triangle.*
• What's object 8? (Signal.) *A circle.*
c. Items 33 through 38 ask about the objects.
• Work 33 through 38. Don't mark an answer for item 32.
• Raise your hand when you've marked the choices for items 33 through 38.
• (Observe students and give feedback.)
d. Check your work.
e. Item 33 asks: Which number shows the cone?
• What's the answer? (Signal.) *3.*
• What letter did you mark? (Signal.) *E.*
f. Item 34 asks: Which number shows the cube?
• What's the answer? (Signal.) *6.*
• What letter did you mark? (Signal.) *H.*
g. Item 35 asks: Which number shows the object after the cylinder?
• What's the answer? (Signal.) *2.*
• What letter did you mark? (Signal.) *C.*
h. Item 36 asks: What is the name of the object between the cone and the sphere?
• What's the answer? (Signal.) *Square.*
• What letter did you mark? (Signal.) *G.*

i. Item 37 asks: Which number shows the object before the cone?
• What's the answer? (Signal.) *2.*
Yes, 2. The object before the cone is the rectangle.
• What letter did you mark? (Signal.) *B.*
j. Item 38 asks: What's the name of the object before the cube?
• What's the answer? (Signal.) *Sphere.*
Yes, the object before the cube is a sphere.
• What letter did you mark? (Signal.) *F.*

EXERCISE 7
Test-Taking Rules
What to Do When You're Running Out of Time

a. There are three rules that help you do well when you're taking a test.
b. Rule 1: Work the problems that you can work.
• What's rule 1? (Signal.) *Work the problems that you can work.*
c. Rule 2: If you don't know how to work a problem, skip it and come back to it.
• What's rule 2? (Signal.) *If you don't know how to work a problem, skip it and come back to it.*
d. Rule 3: Make sure that each problem has one and only one answer.
• What's rule 3? (Signal.) *Make sure that each problem has one and only one answer.*
e. (Repeat steps b through d until firm.)
f. When do you start marking the items that don't have answers? (Signal.) *When there's one minute left.*
g. Look at your answer sheet. Find the first item you'd mark if there was only one minute left.
• Everybody, what's the number of the first item you'd mark? (Signal.) *11.*
• Find the next item you'd mark. What's the number of the next item? (Signal.) *26.*
• What's the number of the next item you'd mark? (Signal.) *32.*
• What's the number of the next item you'd mark? (Signal.) *39.*
• What's the number of the **last** item you'd mark? (Signal.) *56.*
h. (Repeat step g until firm.)
i. We're going to pretend you're taking a test, and there's not much time left. You haven't answered items 11, 26, 32, and 39 through 56. So, when I tell you there's only one minute left, you'll mark one and only one answer for each of those

numbers. There really aren't any problems; you're just guessing. But if this were a real test, you wouldn't have time to work the problems, so you would have to guess.
j. Get ready to mark the answers. After you're finished, we'll find out how many more points you scored by following the rules for doing well on tests.
k. Here we go. There is only one minute left. Finish up your paper.
l. (Reinforce students who quickly fill in one and only one answer for items 11, 26, 32, and 39 through 56 on the answer sheet.)
• (Prompt students who don't fill answers in quickly to start marking answers more quickly.)
• (Alert students who have more than one answer filled in on any item to make sure that there's only one answer per item.
m. (After one minute, say:) Stop working.
• Item 11. The answer is **B.** Make an X by item 11 if you didn't mark B. Make a C by item 11 if you marked B.
• Raise your hand if you got it right. √
n. For the rest of the items you just marked, I'll tell you the make-believe answer. Mark it with a C if you marked the correct answer and an X if you didn't mark it.
• Raise your hand after I say each answer if you guessed the right one.
• Item 26. The answer is **I.** (Acknowledge students who raise their hands.)
• Item 32. The answer if **F.** (Acknowledge students who raise their hands.)
• Item 39. The answer is **C.** (Acknowledge students who raise their hands.)
• (Repeat for items: 40, G; 41, D; 42, H; 43, B; 44, G; 45, A; 46, I; 47, C; 48, F; 49, E; 50, G; 51, C; 52, I; 53, B; 54, H; 55, A; 56, F.)
o. Count the Cs you wrote next to the numbers.
• Raise your hand if you got seven or more of them correct. (Students respond.)
• Raise your hand if you got six of them correct. (Students respond.)
• Raise your hand if you got five of them correct. (Students respond.)
p. (Repeat for four, three, two, one, and zero of them correct.)
• Most of you got three to five of the items correct. You couldn't have gotten any of these problems correct if you hadn't filled in answers for those items.

Test Preparation Lesson 4

EXERCISE 1
Distribution

Multiplying Numbers and Unknowns

a. (Direct students to find lesson 4 in their **test booklets** and write their names and indicate lesson 4 on their answer sheets.)
- (Write on the board):

$$5 \times 328 =$$

- Here's a problem. Read the problem. (Signal.) *5 times 3 hundred 28.*
- Say the addition for 3 hundred 28. (Signal.) *3 hundred plus 20 plus 8.*
- (Write to show:)

$$5 \times 328 \qquad =$$
$$5 \times (300 + 20 + 8) =$$

- Say the new problem for 5 times 300 plus 20 plus 8. (Point to each value as the students say:) *5 times 300 plus 5 times 20 plus 5 times 8.*

b. (Change to show:)

$$5 \times (A + 20 + C) =$$

- This problem is just like the problem you just worked, except you can't get a number answer when you multiply. I'll say the new multiplication problem. (Point to each value as the students say:) *5 times A plus 5 times 20 plus 5 times C.*
- (Write to show:)

$$5 \times (A + 20 + C) \qquad =$$
$$(5 \times A) + (5 \times 20) + (5 \times C) =$$

c. Tell me the answer to each part.
- What's 5 times A? (Signal.) *5A.*
- What's 5 times 20? (Signal.) *100.*
- What's 5 times C? (Signal.) *5C.*

- (Write to show:)

$$5 \times (A + 20 + C) \qquad\qquad =$$
$$(5 \times A) + (5 \times 20) + (5 \times C) =$$
$$5A + \qquad 100 \qquad + \quad 5C \quad =$$

- That's as far as we can work this problem.
d. (Write on the board):

$$1. \quad (X + Y + Z) \times \qquad 4 \qquad =$$
$$2. \qquad 20 \qquad \times \quad (13 + R - A) =$$

e. Read item 1. Get ready. (Signal.) *X plus Y plus Z times 4.*
- Say the new multiplication for that problem. Get ready. (Signal.) *X times 4 plus Y times 4 plus Z times 4.*
f. Read item 2. Get ready. (Signal.) *20 times 13 plus R minus A.*
- Say the new multiplication for that problem. Get ready. (Signal.) *20 times 13 plus 20 times R minus 20 times A.*
g. (Repeat steps e and f until firm.)
h. Your turn: Copy items 1 and 2. Below each problem, write the new multiplication.
- Raise your hand when you're finished.
- (Observe students and give feedback.)
- (Write to show:)

$$1. \ (X + Y + Z) \qquad \times \qquad 4 \quad =$$
$$(X \times 4) \quad + \ (Y \times 4) \ + \ (Z \times 4) =$$

$$2. \ 20 \qquad \times \quad (13 + R - A) \qquad =$$
$$(20 \times 13) \ + (20 \times R) - (20 \times A) \qquad =$$

- Here's what you should have so far.
i. Read the first part of the new multiplication for item 1. (Signal.) *X times 4.*
- What does X times 4 equal? (Signal.) *4X.*
- Read the next part. (Signal.) *Y times 4.*
- What does Y times 4 equal? (Signal.) *4Y.*
- Read the last part of the new multiplication for item 1. (Signal.) *Z times 4.*
- What does it equal? (Signal.) *4Z.*

- Below the new multiplication for item 1, you'll write: 4X plus 4Y plus 4Z. Write what each part equals for item 1. Then finish working item 2.
- Raise your hand when you're finished.
- (Observe students and give feedback.)
- (Write to show:)

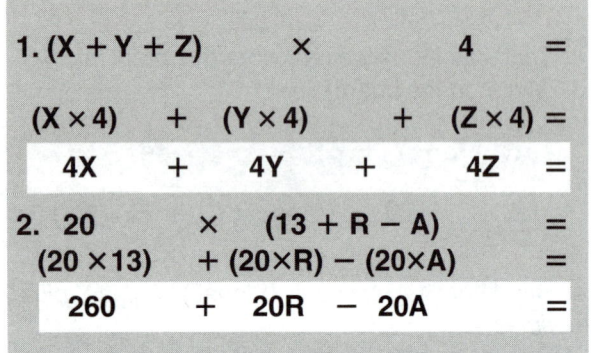

- Here's what you should have.
j. Item 2. What does 20 times 13 equal? (Signal.) *260.*
- What does 20 times R equal? (Signal.) *20R.*
- What does 20 times A equal? (Signal.) *20A.*
- So, 20 times 13 plus R **minus** A equals the expression 260 plus 20R **minus** 20A.
k. Look at the answer for item 1 again. What does the expression X plus Y plus Z times 4 equal? Get ready. (Signal.) *4X plus 4Y plus 4Z.*
l. Find item 1 in your **test booklet.** √
- (Teacher reference:)

1 $3 \times (500 + 40 + 7) =$
  A 216,290
  B 3,780,000
  C 1780
  D 5400
  E None of the above

2 $(6 + R + 8) \times 11 =$
  F $17 + 11R + 19$
  G $6 + R + 88$
  H $66 + 11R + 88$
  I $(6 + R) \times 19$
  J None of the above

3 $5 \times (2000 + 700 + 2) =$
  A 5×2702
  B $(5 \times 2000) + (5 \times 700) + (5 \times 2)$
  C $10,000 + 3500 + 10$
  D All of the above
  E None of the above

4 $8 \times (3R + 7 + T) =$
  F $24R + 15 + 8T$
  G $24R + 56 + 8T$
  H $(8 + 3R) \times (7 + T)$
  I 248RT
  J None of the above

5 $(3Q + 5 - 49) \times 2 =$
  A $6Q + 7 - 47$
  B $6Q + 10 - 47$
  C $6Q + 10 - 98$
  D $(3Q + 5) \times (2 - 49)$
  E $15Q - 98$

- Some of the answers for items 1 through 5 are numbers and some are expressions. Copy the items and work them. Then mark the best answers on your answer sheet.
- Raise your hand when you've marked answers for items 1 through 5.
- (Observe students and give feedback.)
m. Check your work.
- Item 1. 3 times 5 hundred plus 40 plus 7.
- What letter did you mark? (Signal.) *E.*
- Yes, E. The answer is: one thousand 6 hundred 41. It wasn't one of the choices, so you should have marked: None of the above.
n. Item 2. The quantity 6 plus R plus 8 **times** 11.
- What letter did you mark? (Signal.) *H.*
- Yes, H. 66 plus 11R plus 88.
o. Item 3. 5 times the quantity 2000 plus 700 plus 2.
- What letter did you mark? (Signal.) *D.*
- Yes, D. It equals: 5 times 2 thousand 7 hundred 2.
- It also equals 5 times 2000 plus 5 times 7 hundred plus 5 times 2.
- And it equals 10 thousand plus 3 thousand 5 hundred plus 10.
- You should have marked answer D, all of the above.
p. Item 4. 8 times the quantity 3R plus 7 plus T.
- What letter did you mark? (Signal.) *G.*
- Yes, G. It equals: 24R plus 56 plus 8T.
q. Item 5. The quantity 3Q plus 5 minus 49 **times** 2.
- What letter did you mark? (Signal.) *C.*
- Yes, C. It equals: 6Q plus 10 minus 98.

EXERCISE 2
Classification of Triangles

a. Find the triangles above item 7 in your **test booklet.** √
- (Teacher reference:)

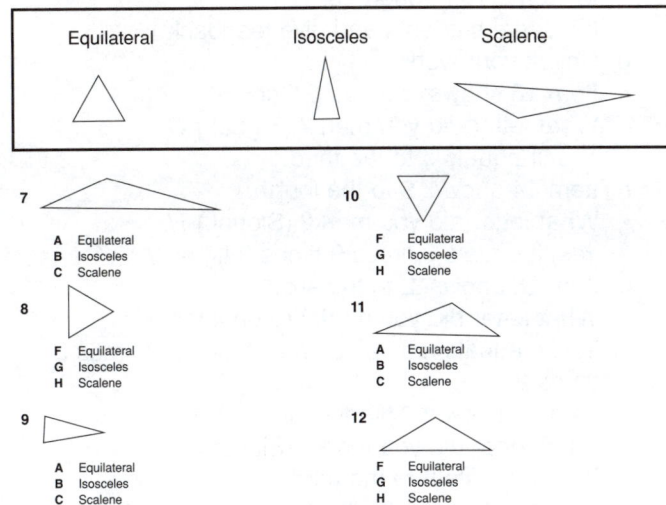

b. Touch the triangle labeled **equilateral.** √
- That triangle is an equilateral triangle. What kind of triangle? (Signal.) *Equilateral.*
- The equilateral triangle has three sides that are the same length.
- How many sides have the same length in an equilateral triangle? (Signal.) *Three.*
c. Touch the triangle labeled **isosceles.** √
- That triangle is an isosceles triangle. What kind of triangle? (Signal.) *Isosceles.*
- The isosceles triangle has two sides that are the same length. Touch the sides that are the same length in that isosceles triangle. √
- How many sides have the same length in an isosceles triangle? (Signal.) *Two.*
- How many sides have the same length in an equilateral triangle? (Signal.) *Three.*
d. (Repeat steps b and c until firm.)

e. Touch the triangle labeled **scalene.** √
- That triangle is a scalene triangle. What kind of triangle? (Signal.) *Scalene.*
f. Each side of a scalene triangle is a different length.
g. Once more: What do we call the triangle with three sides that are the same length? (Signal.) *Equilateral.*
- What do we call the triangle with two sides that are the same length? (Signal.) *Isosceles.*
- What do we call the triangle with each side a different length? (Signal.) *Scalene.*
h. (Repeat step g until firm.)
i. Find item 7. √
- For triangles 7 through 12, mark the letter that best describes it.
- Raise your hand when you've marked a letter for items 7 through 12. Don't mark an answer for item 6.
- (Observe students and give feedback.)
j. Check your work.
k. Item 7. What letter did you mark? (Signal.) *C.*
- What kind of triangle is it? (Signal.) *Scalene.*
l. Item 8. What letter did you mark? (Signal.) *F.*
- What kind of triangle is it? (Signal.) *Equilateral.*
m. Item 9. What letter did you mark? (Signal.) *B.*
- What kind of triangle is it? (Signal.) *Isosceles.*
n. Item 10. What letter did you mark? (Signal.) *F.*
- What kind of triangle is it? (Signal.) *Equilateral.*
o. Item 11. What letter did you mark? (Signal.) *C.*
- What kind of triangle is it? (Signal.) *Scalene.*
p. Item 12. What letter did you mark? (Signal.) *G.*
- What kind of triangle is it? (Signal.) *Isosceles.*

EXERCISE 3
Exponents

a. Find item 13 in your **test booklet.** √
- (Teacher reference:)

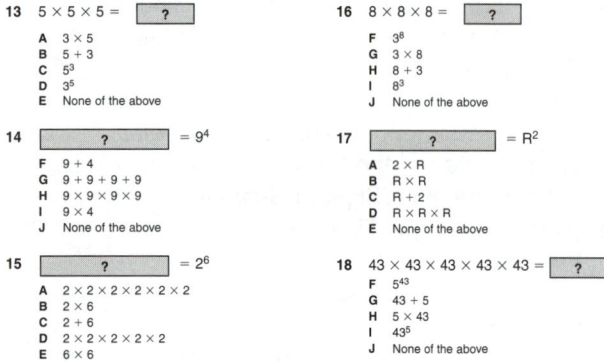

13 $5 \times 5 \times 5 =$ [?]
- **A** 3×5
- **B** $5 + 3$
- **C** 5^3
- **D** 3^5
- **E** None of the above

16 $8 \times 8 \times 8 =$ [?]
- **F** 3^8
- **G** 3×8
- **H** $8 + 3$
- **I** 8^3
- **J** None of the above

14 [?] $= 9^4$
- **F** $9 + 4$
- **G** $9 + 9 + 9 + 9$
- **H** $9 \times 9 \times 9 \times 9$
- **I** 9×4
- **J** None of the above

17 [?] $= R^2$
- **A** $2 \times R$
- **B** $R \times R$
- **C** $R + 2$
- **D** $R \times R \times R$
- **E** None of the above

15 [?] $= 2^6$
- **A** $2 \times 2 \times 2 \times 2 \times 2 \times 2$
- **B** 2×6
- **C** $2 + 6$
- **D** $2 \times 2 \times 2 \times 2 \times 2$
- **E** 6×6

18 $43 \times 43 \times 43 \times 43 \times 43 =$ [?]
- **F** 5^{43}
- **G** $43 + 5$
- **H** 5×43
- **I** 43^5
- **J** None of the above

- For items 13 through 18, you're going to figure out the complete equation with the multiplication and the base number and the exponent it equals.
b. Item 13 shows: 5 times 5 times 5.
- What's the base number? (Signal.) *5.*
- Yes, 5 times 5 times 5 equals 5 to what exponent? (Signal.) *3.*
- Yes, 5 times 5 times 5 equals 5 to the third.
- Say the complete equation for item 13. (Signal.) *5 times 5 times 5 equals 5 to the third.*
c. Item 14 shows: 9 to the fourth.
- Say the complete equation for item 14. (Signal.) *9 times 9 times 9 times 9 equals 9 to the fourth.*
d. Item 15 shows: 2 to the sixth.
- Say the complete equation for item 15. (Signal.) *2 times 2 times 2 times 2 times 2 times 2 equals 2 to the sixth.*

e. Item 16 shows: 8 times 8 times 8.
- Say the complete equation for item 15. (Signal.) *8 times 8 times 8 equals 8 to the third.*
f. Your turn: Work items 13 through 18.
- Raise your hand when you've marked the answers for items 13 through 18.
- (Observe students and give feedback.)
g. Check your work.
- Item 13 shows: 5 times 5 times 5.
- What letter did you mark? (Signal.) *C.* Yes, it equals 5 to the third.
h. Item 14 shows: 9 to the fourth.
- What letter did you mark? (Signal.) *H.* Yes, it equals 9 times 9 times 9 times 9.
i. Item 15 shows: 2 to the sixth. What letter did you mark? (Signal.) *A.* Yes, it equals 2 times 2 times 2 times 2 times 2 times 2.
j. Item 16 shows: 8 times 8 times 8.
- What letter did you mark? (Signal.) *I.* Yes, it equals 8 to the third.
k. Item 17 shows: R to the second.
- What letter did you mark? (Signal.) *B.* Yes, R times R equals R to the second.
l. Item 18 shows: 43 times 43 times 43 times 43 times 43.
- What letter did you mark? (Signal.) *I.* Yes, it equals 43 to the fifth.

EXERCISE 4 Geometry
Shapes

a. Find the figures in the box above item 19 in your **test booklet.** √
- (Teacher reference:)

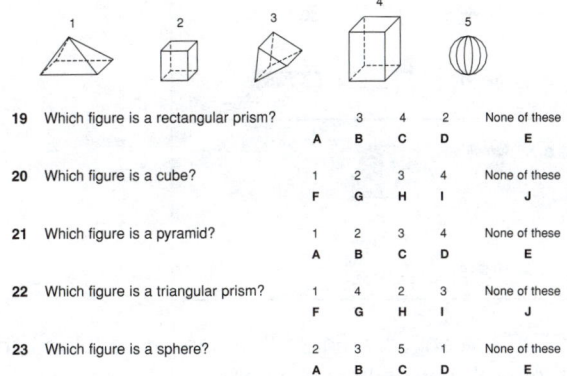

19	Which figure is a rectangular prism?		3	4	2	None of these
			A	B	C	D	E

19 Which figure is a rectangular prism?
        3   4   2   None of these
        A   B   C   D   E

20 Which figure is a cube?
        1   2   3   4   None of these
        F   G   H   I   J

21 Which figure is a pyramid?
        1   2   3   4   None of these
        A   B   C   D   E

22 Which figure is a triangular prism?
        1   4   2   3   None of these
        F   G   H   I   J

23 Which figure is a sphere?
        2   3   5   1   None of these
        A   B   C   D   E

- The first figure is a pyramid. Remember, a pyramid comes to a point.
  What's the first figure? (Signal.) *A pyramid.*
- The next figure is called a rectangular prism.
  What's the figure? (Signal.) *A rectangular prism.*
- The last figure is a triangular prism.
  What's the last figure? (Signal.) *A triangular prism.*

b. Touch figure 1 below the box. √
- What is the figure called? (Signal.) *A pyramid.*
- Touch figure 2. √
- What is it called? (Signal.) *A cube.*
- Touch figure 3. √
- What is it called? (Signal.) *A triangular prism.*
- Touch figure 4. √
- What is it called? (Signal.) *A rectangular prism.*
- Touch figure 5. √
- What is it called? (Signal.) *A sphere.*

c. (Repeat step b until firm.)

d. Work items 19 through 23.
- Raise your hand when you're finished.
- (Observe students and give feedback.)

e. Check your work.
- Item 19 asks which figure is a rectangular prism.
- Everybody, which figure? (Signal.) *4.*
- Which letter did you mark? (Signal.) *C.*

f. Item 20 asks which figure is a cube.
- Everybody, which figure? (Signal.) *2.*
- Which letter did you mark? (Signal.) *G.*

g. Item 21 asks which figure is a pyramid.
- Everybody, which figure? (Signal.) *1.*
- Which letter did you mark? (Signal.) *A.*

h. Item 22 asks which figure is a triangular prism.
- Everybody, which figure? (Signal.) *3.*
- Which letter did you mark? (Signal.) *I.*

i. Item 23 asks which figure is a sphere.
- Everybody, which figure? (Signal.) *5.*
- Which letter did you mark? (Signal.) *C.*

EXERCISE 5
Mixed Numbers to Improper Fractions

a. Find item 25 in your **test booklet.** √

• These are mixed numbers. You'll change them into improper fractions.

b. Work items 25 through 29. Don't mark an answer for item 24.

• Pencils down when you've marked answers through item 29.

• (Observe students and give feedback.)

c. Check your work.

• Item 25. Tell me the fraction for 3 and 2-fifths. Get ready. (Signal.) *17-fifths.*

• What letter did you mark? (Signal.) *D.*

d. Item 26. Tell me the fraction for 7 and 1-ninth. Get ready. (Signal.) *64-ninths.*

• What letter did you mark? (Signal.) *F.*

e. Item 27. Tell me the fraction for 2 and 4-fifths. Get ready. (Signal.) *14-fifths.*

• What letter did you mark? (Signal.) *B.*

g. Item 28. Tell me the fraction for 6 and 1-half. Get ready. (Signal.) *13-halves.*

• What letter did you mark? (Signal.) *G.*

h. Item 29. Tell me the fraction for 1 and 9-tenths. Get ready. (Signal.) *19-tenths.*

• What letter did you mark? (Signal.) *C.*

EXERCISE 6
Symmetry

a. Find the figures above item 31 in your **test booklet.** √

• (Teacher reference:)

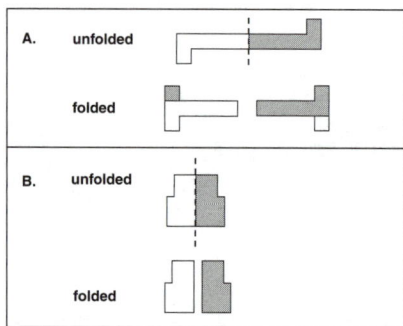

• Picture A shows a figure that cannot be folded so that both halves look like halves of the original figure.

• Below the unfolded figure in picture A, you can see both sides of the folded figure. On the left, you can see the folded figure with the white side showing. You can see parts of the shaded side sticking out.

• To the right, you can see the folded figure with the shaded side showing. Part of the white side is sticking out.

b. Touch the unfolded figure in picture B. √

• If you folded it along the dotted line, the folded figure would look like both halves of the original figure.

• You can see both sides of the folded figure below. When it's turned so that the white side is showing, it looks just like the white side of the original figure.

• When the folded figure is turned so that the shaded side is showing, it looks just like the shaded side of the original.

• So, one of the features of figure B is that it's symmetrical.

• What is a feature of figure B? (Signal.) *It's symmetrical.*

c. Find item 31. √
• One of the figures is symmetrical. It can be folded along the dotted line so that the two halves are the same. Work item 31. √
• Item 31. What letter did you mark? (Signal.) *A.* Yes, A is the only figure that will have identical sides when they are folded. The dotted line for figure A shows a line of symmetry.
d. Work item 32.
• Raise your hand when you've marked your answer. √
• Item 32. What letter did you mark? (Signal.) *I.* Yes, I. The dotted line for I shows a line of symmetry.
e. Item 33 is different. Item 33 asks: Which figure **does not** have a line of symmetry?
• Work item 33.
• Raise your hand when you've marked your answer. √
• Item 33. What letter did you mark? (Signal.) *C.* Yes, C. Figure C does not have a line of symmetry.
f. Work items 34 and 35.
• Raise your hand when you've done that much. √
g. Check your work.
• Item 34. What letter did you mark? (Signal.) *J.* Yes, figure J is the only figure that doesn't have a line of symmetry.
h. Item 35. What letter did you mark? (Signal.) *B.* Yes, figure B is the only figure that has a line of symmetry.
i. The figures in item 36 don't have lines shown. One of them does have a line of symmetry.
• Mark the letter of the figure that has a line of symmetry.
• Raise your hand when you've done that much. √
• Item 36. What letter did you mark? (Signal.) *G.* Yes, because you could fold figure G so that both sides match, it has a line of symmetry. None of the other figures has a line of symmetry.
j. Item 37. One of the figures does **not** have a line of symmetry. All the others do. Mark the letter of the figure that doesn't have a line of symmetry.
• Raise your hand when you've done that much. √

• Item 37. What letter did you mark? (Signal.) *D.* Yes, each of the other figures has a line of symmetry. None of the other figures has a line of symmetry.
k. Work items 38 and 39.
• Raise your hand when you've done that much. √
l. Check your work.
• Item 38. What letter did you mark? (Signal.) *H.* Yes, that figure has a line of symmetry because you could fold it so that both halves match.
m. Item 39. What letter did you mark? (Signal.) *A.* Yes, A is the only figure that doesn't have a line of symmetry.

EXERCISE 7
Averages

a. For some problems, you have to find the average. If you wanted to find the average temperature for 16 days, what would you do? (Call on a student. Idea: *Add the temperatures for the 16 days, then divide by 16.*)
• Yes, add the temperatures for the 16 days, and then divide by 16.
• If you wanted to find the average temperature for 9 days, what would you do? (Signal.) *Add the temperatures for the 9 days, then divide by 9.*
b. Find item 41 in your **test booklet.** √
• To find the average score that 6 students got on a test, you add up the 6 scores. Then what do you do? (Signal.) *Divide by 6.*
• Mark the choice that shows what you'd do. Don't mark an answer for item 40.
• Raise your hand when you've marked an answer for item 41. √
• Everybody, first you add up all the scores. Then what do you do? (Signal.) *Divide by 6.*
• What letter did you mark? (Signal) *C.* You divide by 6 to find the average score that 6 students got on a test.
c. Work items 42 and 43.
• Raise your hand when you've finished. √
d. Check your work.
• Item 42 says: To find the average height of 9 students, you add up all the heights. Then what do you do? (Signal.) *Divide by 9.*
• What letter did you mark? (Signal.) *G.*
e. Item 43 says: To find the average price of 11 watches, what do you do? (Call on a student. Idea: *Add all the prices and divide by 11.*)
• What letter did you mark? (Signal.) *D.*

EXERCISE 8
Test-Taking Rules
What to Do When You're Running Out of Time

a. There are three rules that help you do well when you're taking a test.

b. Who can tell me rule 1? (Call on a student. Praise close responses to the rule: *Work the problems that you can work.*)

• Yes, rule 1 is: Work the problems that you can work.

• Everybody, what's rule 1? (Signal.) *Work the problems that you can work.*

c. What does that rule mean if you're working a hard problem? (Call on a student. Idea: *Write the problem on your paper and work it.*)

• Yes, don't try to do hard problems in your head. Work them on paper.

d. Who can tell me rule 2? (Call on a student. Praise close responses to the rule: *If you can't work a problem, skip it and come back to it.*)

• Yes, rule 2 is: If you can't work a problem, skip it and come back to it.

• Everybody, what's rule 2? (Signal.) *If you can't work a problem, skip it and come back to it.*

e. Who can tell me rule 3? (Call on a student. Praise close responses to the rule: *Make sure that each problem has one and only one answer.*)

• Yes, rule 3 is: Make sure that each problem has one and only one answer.

• Everybody, what's rule 3? (Signal.) *Make sure that each problem has one and only one answer.*

f. Look at your answer sheet and find the first item you'd mark if there was only one minute left.

• Everybody, what's the number of the first item you'd mark? (Signal.) *6.*

• Find the next item you'd mark. What's the number of the next item? (Signal.) *24.*

• What's the number of the next item you'd mark? (Signal.) *30.*

• What's the number of the next item you'd mark? (Signal.) *40.*

• What's the number of the next item you'd mark? (Signal.) *44.*

• What's the number of the **last** item you'd mark? (Signal.) *56.*

g. (Repeat step f until firm.)

h. We're going to pretend you're taking a test, and there's not much time left. You haven't answered items 6, 24, 30, 40, and 44 through 56. So, when I tell you there's only one minute left, you'll mark one and only one answer for each of those items.

i. Get ready to mark the answers. After you're finished, we'll find out how many more points you scored by following the rules for doing well on tests.

j. Here we go. There is only one minute left. Finish up your paper.

k. (Reinforce students who quickly fill in one and only one answer for all of the problems on the answer sheet.)

• (Prompt students who don't fill answers in quickly to start marking answers more quickly.)

• (Alert students who have more than one answer filled in on any items to make sure that there's only one answer per item.)

l. (After one minute, direct students to correct their answer sheet.)

m. Check your work.

• For the items you just marked, I'll tell you the make-believe answer. Mark it with a **C** if you marked the correct answer and an **X** if you didn't mark it. Raise your hand after I say each answer if you guessed the right one.

• Item 6. The answer is **I.** (Acknowledge students who raise their hands.)

• Item 24. The answer is **H.** (Acknowledge students who raise their hands.)

• (Repeat for items: 30, G; 40, I; 44, J; 45, E; 46, G; 47, D; 48, F; 49, C; 50, G; 51, D; 52, H; 53, A; 54, J; 55, B; 56, F.)

n. Count the Cs you wrote next to the numbers.

• Raise your hand if you got six or more of them correct. (Students respond.)

• Raise your hand if you got five of the items correct. (Students respond.)

o. (Repeat for four, three, two, one, and zero of them correct.)

• Most of you got two to four of the items correct. You wouldn't have gotten any of these problems correct if you hadn't filled in answers for those items.

Test Preparation Lesson 5

> ***Materials Note:***
>
> Each student will need:
> - lined paper
> - test booklet and Multiple-Choice Response Sheet

EXERCISE 1
Distribution
Multiplying Numbers and Unknowns

a. (Direct students to find lesson 5 in their **test booklets** and write their names and indicate lesson 5 on their answer sheets.)
- Find the figures above item 2 in your **test booklet.** √
- Use the figures numbered Roman numeral one through Roman numeral five to mark answers for items 2 through 6. Don't mark an answer for item 1.
- Raise your hand when you've marked answers for items 2 through 6.
- (Observe students and give feedback.)

b. Check your work.
- Item 2: Which figure shows a rectangle?
- What letter did you mark? (Signal.) *I.*
  Yes, figures III and V are rectangles.

c. Item 3: Which figure shows a trapezoid?
- What letter did you mark? (Signal.) *B.*
  Yes, figure II is a trapezoid.

d. Item 4: Which figure shows a parallelogram?
- What letter did you mark? (Signal.) *J.*
  Yes, figures I, III, IV, and V show parallelograms, so the answer is: All of the above. That's choice J.

e. Item 5: Which figure or figures show a rhombus?
- What letter did you mark? (Signal.) *C.*
  Yes, figures I and V show rhombuses.

f. Item 6: Which figure shows a square?
- What letter did you mark? (Signal.) *I.*
  Yes, figure V shows a square.

EXERCISE 2
Equivalent Fractions

a. Find item 7 in your **test booklet.** √
- You're going to identify fractions that are equivalent.

b. Item 7. The fraction is 10-twelfths. Simplify the fraction and mark the choice that shows the equivalent fraction.
- Pencils down when you've marked your answer. √
- Check your work.
- Item 7. What fraction does 10-twelfths simplify to? (Signal.) *5-sixths.*
- What letter did you mark? (Signal.) *B.*

c. Item 8. The fraction is 4-sixths. Simplify the fraction and mark the choice that shows the equivalent fraction.
- Pencils down when you've marked your answer. √
- Check your work.
- Item 8. What fraction does 4-sixths simplify to? (Signal.) *2-thirds.*
- What letter did you mark? (Signal.) *I.*

d. Item 9. The fraction is 20 twenty-fifths. Simplify the fraction and mark the choice that shows the equivalent fraction.
- Pencils down when you've marked your answer. √
- Check your work.
- Item 9. What fraction does 20 twenty-fifths simplify to? (Signal.) *4-fifths.*
- What letter did you mark? (Signal.) *A.*

e. Item 10. The fraction is 15 twenty-fourths. Simplify the fraction and mark the choice that shows the equivalent fraction.
- Pencils down when you've marked your answer. √
- Check your work.
- Item 10. What fraction does 15 twenty-fourths simplify to? (Signal.) *5-eighths.*
- What letter did you mark? (Signal.) *G.*

f. Item 11. The fraction is 12-twentieths. Simplify the fraction and mark the choice that shows the equivalent fraction.
- Pencils down when you've marked your answer. √
- Check your work.
- Item 11. What fraction does 12-twentieths, simplify to? (Signal.) *3-fifths.*
- What letter did you mark? (Signal.) *D.*

EXERCISE 3
Dimensions and Views

a. Find the object above item 13 in your **test booklet.** √
- (Teacher reference:)

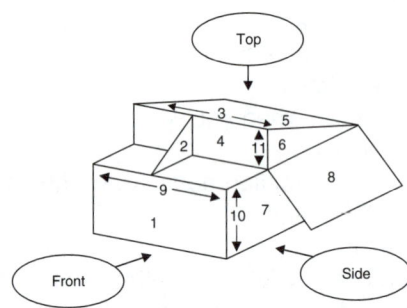

- This object looks like a modern house. The roof is a rectangle labeled with the number 5. Touch it. √
- A rectangle labled number 4 supports the roof. Touch it. √
- The roofline, where parts 4 and 5 meet, is labeled 3. Touch it. √
- In front of part 4 is a balcony. The balcony is separated by a triangle labeled 2. Touch it. √
- The front wall of the house is a rectangle labeled 1. Touch it. √
- The line where the front wall and the balcony meet is labeled 9. Touch it. √
- A side wall of the house is labeled 7. Touch it. √

- The line where the front wall and the side wall meet is labeled 10. Touch it. √
- A lean-to on the side of the house is labeled 8. Touch it. √
- The triangle that supports the roof is labeled 6. Touch it. √
b. Touch the word **top** above the object. √
- If you follow the arrow pointing from the word **top,** from which view would you be looking at the object? (Signal.) *The top view.*
c. Touch the word **side.** √
- If you follow the arrow pointing from the word **side,** from which view would you be looking at the object? (Signal.) *The side view.*
d. Touch the word **front.** √
- If you follow the arrow pointing from the word **front,** from which view would you be looking at the object? (Signal.) *The front view.*
e. (Repeat steps b through d until firm.)
f. Item 13. Which is the front view of the object?
- Touch choice A. √
- Is the figure beside A the front view? (Signal.) *No.*
- Why not? (Call on a student. Ideas: *(1) The bottom wall should be as wide as the part above it; (2) the house should be wider than it is tall; (3) there should be a lean-to on the right side of the house; (4) it's the top view.)*
g. Touch choice B. √
- Is the figure beside B the front view? (Signal.) *No.*
- Why not? (Call on a student. Ideas: *(1) The bottom wall should not be divided in half; (2) the lean-to on the right side should not have a line on the bottom.)*

h. Touch choice C. √
- Is the figure beside C the front view? (Signal.) *No.*
- Why not? (Call on a student. Ideas: *(1) The house shouldn't come to a point; (2) the bottom wall shouldn't be divided; (3) the house should be wider; (4) a lean-to should be on the right side of the house; (5) it's the side view.*)
i. Touch choice D. √
- Is the figure beside D the front view? (Signal.) *Yes.*
- Yes, it is. The top section of the house is divided, the front wall is not divided, and the lean-to is on the right side of the house.
j. Item 14. Which is the top view of the object? Touch choice F. √
- Is the figure beside F the top view? (Signal.) *Yes.*
- Yes, it is. You can see the top of the roof, the divided balcony, and the lean-to.
k. Item 15. Which is the side view of the object?
- Touch choice A. √
- Is the figure beside A the side view? (Signal.) *No.*
- Why not? (Call on a student. Idea: *It's the top view.*)
l. Touch choice B. √
- Is the figure in B the side view? (Signal.) *No.*
- Why not? (Call on a student. Idea: *The triangular divider on the balcony should be shown.*)
m. Touch choice C. √
- Is the figure in C the side view? (Signal.) *Yes.*
- Yes it is. Above you can see the roof and triangular divider, with the lean-to and part of the side wall below.
n. Your turn. Work items 13 through 15.
- Pencils down when you've marked answers through item 15. Don't mark an answer for item 12.
- (Observe students and give feedback.)
o. Check your work.
- Item 13. What letter did you mark? (Signal.) *D.* Yes, choice D is the front view of the object.
- Item 14. What letter did you mark? (Signal.) *F.* Yes, choice F is the top view of the object.
- Item 15. What letter did you mark? (Signal.) *C.* Yes, choice C is the side view of the object.

p. For items 16 through 22, each item will have a drawing of either the front view, the top view, or the side view. A portion of each view will be highlighted. The highlighted part represents one or more than one of the numbered parts of the object. Mark the answer that best fits the highlighted portion of the drawing.
- Work items 16 through 22.
- Pencils down when you've marked answers through item 22.
- (Observe students and give feedback.)
q. Check your work.
- Item 16. What letter did you mark? (Signal.) *F.* Yes, the highlighted point on the side view represents the line of the object labeled number 9.
- Item 17. What letter did you mark? (Signal.) *E.* Yes, none of these. The highlighted line on the top view represents the part of the object labeled number 6. Number 6 was not one of the choices.
- Item 18. What letter did you mark? (Signal.) *G.* Yes, the highlighted line on the front view represents the part of the object labeled number 2.
- Item 19. What letter did you mark? (Signal.) *A.* Yes, the highlighted line on the side view represents the line of the object labeled number 10. It also represents the part of the object labeled number 1, but number 1 was not a choice.
- Item 20. What letter did you mark? (Signal.) *J.* Yes, the highlighted line on the top view represents the line of the object labeled number 3, and it represents the part of the object labeled number 4.
- Item 21. What letter did you mark? (Signal.) *E.* Yes, the highlighted line on the front view represents the part of the object labeled number 10, but number 10 was not a choice.
- Item 22. What letter did you mark? (Signal.) *F.* Yes, the highlighted part on the side view represents the part of the object labeled number 6.

EXERCISE 4
Distribution

a. (Write on the board:)

> **1.** $6 \times (500 + 80 + 3) \quad =$

- Read the problem. (Signal.) *6 times 500 plus 80 plus 3.*
- Say the new multiplication for this problem. (Signal.) *6 times 500 plus 6 times 80 plus 6 times 3.*
- That's the problem with the product distributed. Say the problem with the product distributed again. (Signal.) *6 times 500 plus 6 times 80 plus 6 times 3.*

b. (Write to show:)

> **1.** $6 \times (L + R + 3) \quad =$

- Read this problem. (Signal.) *6 times L plus R plus 3.*
- Distribute the product for this problem. (Signal.) *6 times L plus 6 times R plus 6 times 3.*

c. (Repeat step b until firm.)

d. (Write to show:)

> **1.** $6 \times (L + R + 3) \quad =$
>
> **2.** $(15 + H + 4H) \times 8 \quad =$

e. Read item 2. (Signal.) *15 plus H plus 4H times 8.*
- Distribute the product for this problem. (Signal.) *15 times 8 plus H times 8 plus 4H times 8.*

f. Copy these problems. Below, distribute the product. Below the distributed product, write what each part equals.
- (Observe students and give feedback.)
- (Write to show:)

> **1.** $6 \times (L + R + 3) \quad =$
> $(6 \times L) + (6 \times R) + (6 \times 3) \quad =$
> $6L \quad + \quad 6R \quad + \quad 18 \quad =$
>
> **2.** $(15 + H + 4H) \times 8 \quad =$
> $(15 \times 8) + (H \times 8) + (4H \times 8) \quad =$
> $120 \quad + \quad 8H \quad + \quad 32H \quad =$

g. Check your work. Here's what you should have.

h. Item 1. What does 6 times L plus R plus 3 equal? (Signal.) *6L plus 6R plus 18.*

i. Item 2. What does 15 plus H plus 4H times 8 equal? (Signal.) *120 plus 8H plus 32H.*

j. You've worked item 1 as far as you can work it. But you can work more on item 2. To finish working item 2, you combine like terms.
- Listen: Numbers without letters are like terms.
- 2 and 5 are like terms. They are both whole numbers. So, you can add or subtract them.
- 2 and 5T are not like terms. So, you can't add or subtract them.
- You can add 2T and 5T because they are like terms.
- You can add 2 cups and 5 cups because they are like terms.
- You can add 2V and 5V because they are like terms.
- But you can't add 2V and 5 cups.

k. Can you add 2T and 5V? (Signal.) *No.*
- Can you add 2 cups and 5V? (Signal.) *No.*
- Can you add 2 cups and 5? (Signal.) *No.*
- Can you add 2 cups and 5 cups? (Signal.) *Yes.*

l. Item 2 has numbers and Hs.
- You've distributed the product. Now you combine the Hs. They are like terms.
- What does 8H plus 32H equal? (Signal.) *40H.*

m. Combine the like terms below.
- (Observe students and give feedback.)
- (Write to show:)

> **2.** $(15 + H + 4H) \times 8 \quad =$
>
> $(15 \times 8) + (H \times 8) + (4H \times 8) =$
>
> $120 \quad + \quad 8H \quad + \quad 32H \quad =$
>
> $120 \quad + \quad 40H$

n. Find item 23 in your **test booklet.** √
- (Teacher reference:)

23 $(3T - R + 71) \times 6 =$
A $9T - 6R + 77$
B $18T - 6R + 77$
C $9T - 6R + 77$
D $18T - 6R + 426$
E None of the above

24 $5 \times (800 + 40 + 3) =$
F $4000 + 200 + 15$
G $(5 \times 800) + (5 \times 40) + (5 \times 3)$
H 5×843
I 4215
J All of the above

25 $(4T + Z - 2T) \times 12 =$
A $12T + 12Z - 24T$
B $24T + 12Z$
C $38T + 12Z$
D $(4T + Z) - 24T$
E $24TZ$

26 $(9000 + 200 + 50 + 6) \times 10 =$
   F   9266
   G $90,000 \times 2000 \times 500 \times 60$
   H   90,205,600
   I   92,560
   J   None of the above

27 $20 \times (6R + 11 - R) =$
   A $100R + 31$
   B $119R + 220$
   C $100R + 220$
   D $25R + 31$
   E $6R + 220$

- Some of the answers for items 23 through 27 show the distributed products. For problems 25 and 27, you'll have to combine like terms. Copy each item. Write the distributed products below, and then write what each part equals. Then mark the answer for each item.
- Raise your hand when you've marked answers through item 27.
- (Observe students and give feedback.)
o. Check your work.
p. Item 23. What letter did you mark? (Signal.) *D.*
- Yes, D. It equals 18T minus 6R plus 426.
q. Item 24. What letter did you mark? (Signal.) *J.*
- Yes, J. 5 times 8 hundred plus 40 plus 3 equals 4 thousand plus 2 hundred plus 15.
- It also equals 5 times 8 hundred plus 5 times 40 plus 5 times 3.
- And it equals 5 times 8 hundred 43, and it equals 4 thousand 2 hundred 15.
- So, you should have marked J, all of the above.
r. Item 25. What letter did you mark? (Signal.) *B.*
- Yes, B. 4T plus Z minus 2T times 12 equals 24T plus 12Z. You combined T terms.
s. Item 26. What letter did you mark? (Signal.) *I.*
- Yes, I. 9 thousand plus 2 hundred plus 50 plus 6 times 10 equals 92 thousand 5 hundred 60.
t. Item 27. What letter did you mark? (Signal.) *C.*
- Yes, C. 20 times 6R plus 11 minus R equals 100R plus 220. You combined R terms.

EXERCISE 5
Symmetry

a. Find item 28 in your **test booklet.** √
- You're going to indicate which figure has a line of symmetry.
- You've learned that some figures can be folded so that both sides match. We say that those figures have a **line of symmetry.** If a figure cannot be folded so that both sides match, the figure does not have a line of symmetry.
b. Look at the figures in item 28. One of them has a line of symmetry.
- Mark the letter of the figure that has a line of symmetry.
- Raise your hand when you're finished.
- (Observe students and give feedback.)
- What letter did you mark? (Signal.) *G.*
   Figure G has a line of symmetry. None of the other figures has a line of symmetry.
c. Item 29: One of the figures does **not** have a line of symmetry. All the others do. Mark the figure that doesn't have a line of symmetry.
- Raise your hand when you're finished. √
- Item 29. What letter did you mark? (Signal.) *D.* All of the other figures have a line of symmetry.

EXERCISE 6
Fraction Addition

a. Find item 31. √
- Work items 31 through 34. Write the answer as a whole number or a mixed number if you can. Mark the correct answer. Don't mark an answer for item 30.
- Pencils down when you've marked answers for items 31 through 34.
- (Observe students and give feedback.)
b. Check your work.
c. Item 31. One-third plus 5-sixths.
- What letter did you mark? (Signal.) *D.*
   The answer is 7-sixths or one and one-sixth.
d. Item 32. 3-eighths plus 3-fourths.
- What letter did you mark? (Signal.) *G.*
   The answer is 9-eighths or one and one-eighth.
e. Item 33. 3-tenths plus 3-fifteenths.
- What letter did you mark? (Signal.) *A.*
   The answer is 15-thirtieths.
f. Item 34. One-fourth plus one-tenth.
- What letter did you mark? (Signal.) *H.*
   The answer is 7-twentieths.

EXERCISE 7
Shapes—3 Dimensions

a. Everybody, get ready to identify the figures in the box above item 35 in your **test booklet.** √
• (Teacher reference:)

• What is figure 1? (Signal.) *A cylinder.*
• What is figure 2? (Signal.) *A sphere.*
• What is figure 3? (Signal.) *A cube.*
• What is figure 4? (Signal.) *A cone.*
• What is figure 5? (Signal.) *A pyramid.*
• What is figure 6? (Signal.) *A rectangular prism.*
• What is figure 7? (Signal.) *A triangular prism.*
b. (Repeat step a until firm.)
c. Mark the answers to items 35 through 41.
• Raise your hand when you're finished.
• (Observe students and give feedback.)
d. Check your work.
• Item 35 asks which figure is a rectangular prism. What letter did you mark? (Signal.) *C.*
• Item 36 asks which figure is a cube. What letter did you mark? (Signal.) *J.*
• Item 37 asks which figure is a triangular prism. What letter did you mark? (Signal.) *A.*
• Item 38 asks which figure is a cylinder. What letter did you mark? (Signal.) *I.*
• Item 39 asks which figure is a sphere. What letter did you mark? (Signal.) *A.*
• Item 40 asks which figure is a pyramid. What letter did you mark? (Signal.) *G.*
• Item 41 asks which figure is a cone. What letter did you mark? (Signal.) *C.*

EXERCISE 8 Exponents

a. (Write on the board:)

> 1. $7^3 =$
> 2. $2^4 =$
> 3. $10^4 =$
> 4. $8^3 =$
> 5. $1^5 =$

b. Read the base number and the exponent for each item. Start with item 1: 7 to the third.
• Item 1. Read it. (Signal.) *7 to the third.*
• Item 2. Read it. (Signal.) *2 to the fourth.*
• Item 3. Read it. (Signal.) *10 to the fourth.*
• Item 4. Read it. (Signal.) *8 to the third.*
• Item 5. Read it. (Signal.) *One to the fifth.*
c. You've written the multiplication for base numbers with exponents. You're going to write the multiplication for base numbers with exponents, and then write what number the multiplication equals.
d. Item 1: 7 to the third.
• Say the multiplication for 7 to the third. Get ready. (Signal.) *7 times 7 times 7.*
• (Write to show:)

> 1. $7^3 = 7 \times 7 \times 7$

• Now we'll write the number that 7 times 7 times 7 equals. What's 7 times 7? (Signal.) *49.*
• So you need to figure out what 49 times 7 equals to find what number 7 to the third equals.
• Raise your hand when you know the number 7 to the third equals. √
• Everybody, what does 7 to the third equal? (Signal.) *343.*
• (Write to show:)

> 1. $7^3 = 7 \times 7 \times 7 = 343$

- Copy the complete equation for item 1. Then write the complete equation for item 2. Raise your hand when you've shown the multiplication and the number answer for the base number and the exponents in items 1 and 2.
- (Observe students and give feedback.)
e. Item 2: 2 to the fourth.
- Read the complete equation you wrote for item 2. (Signal.) *2 to the fourth equals 2 times 2 times 2 times 2 equals 16.*
- (Write to show:)

$$2. \quad 2^4 = 2 \times 2 \times 2 \times 2 = 16$$

- Here's what you should have.
f. Write the complete equations for items 3 through 5.
- Raise your hand when you're finished.
- (Observe students and give feedback.)
- (Write to show:)

$$3. \quad 10^4 = 10 \times 10 \times 10 \times 10 = 10{,}000$$

$$4. \quad 8^3 = 8 \times 8 \times 8 = 512$$

$$5. \quad 1^5 = 1 \times 1 \times 1 \times 1 \times 1 = 1$$

- Here's what you should have.
g. Read the complete equations you wrote for each item.
h. Item 3. Read the complete equation. (Signal.) *10 to the fourth equals 10 times 10 times 10 times 10 equals 10 thousand.*
- What number does 10 to the fourth equal? (Signal.) *10 thousand.*
i. Item 4. Read the complete equation. (Signal.) *8 to the third equals 8 times 8 times 8 equals 5 hundred 12.*
- What number does 8 to the third equal? (Signal.) *5 hundred 12.*
j. Item 5. Read the complete equation. (Signal.) *One to the fifth equals one times one times one times one times one equals one.*
- What number does one to the fifth equal? (Signal.) *One.*

k. Find item 43 in your **test booklet.** √
- Write down the base number and the exponents for items 43 through 46 and write the equations. Then mark the answer.
- Raise your hand when you've marked answers for items 43 through 46. Don't mark an answer for item 42.
- (Observe students and give feedback.)
l. Check your work.
m. Item 43. Read the complete equation. (Signal.) *3 to the fourth equals 3 times 3 times 3 times 3 equals 81.*
- What letter did you mark? (Signal.) *B.*
n. Item 44. Read the complete equation. (Signal.) *2 to the fifth equals 2 times 2 times 2 times 2 times 2 equals 32.*
- What letter did you mark? (Signal.) *I.*
o. Item 45. Read the complete equation. (Signal.) *10 to the third equals 10 times 10 times 10 equals one thousand.*
- What letter did you mark? (Signal.) *A.*
p. Item 46. What number does one to the ninth equal? (Signal.) *One.*
- What letter did you mark? (Signal.) *J.* Yes, J. None of the choices is one.

EXERCISE 9
Test-Taking Rules
What to Do When You're Running Out of Time

a. There are three rules that help you do well when you're taking a test.
b. Who can tell me rule 1? (Call on a student. Praise close responses to the rule: *Work the problems that you can work.*)
- Yes, rule 1 is: Work the problems that you can work.
- Everybody, what's rule 1? (Signal.) *Work the problems that you can work.*
- Yes, if you're working a hard problem, write it down and work it on paper.
c. Who can tell me rule 2? (Call on a student. Praise close responses to the rule: *If you can't work a problem, skip it and come back to it.*)
- Yes, rule 2 is: If you can't work a problem, skip it and come back to it.
- Everybody, what's rule 2? (Signal.) *If you can't work a problem, skip it and come back to it.*
d. Who can tell me rule 3? (Call on a student. Praise close responses to the rule: *Make sure that each problem has one and only one answer.*)
- Yes, rule 3 is: Make sure each problem has one and only one answer.
- Everybody, what's rule 3? (Signal.) *Make sure that each problem has one and only one answer.*

e. You'll start filling answers in when there are how many minutes left to go in the test? (Signal.) *One.*

f. Here's a new rule: Fill in only the items that are on the test. What's the new rule? (Signal.) *Fill in only the items that are on the test.*

• Find the **Stop** sign at the end of lesson 5 in your **test booklet** and get ready to tell me the number of the last item for lesson 5.

• Everybody, what's the number of the last item you'll fill in for lesson 5? (Signal.) *54.*

• Will you fill in 55? (Signal.) *No.*
• Will you fill in 52? (Signal.) *Yes.*
• Will you fill in 53? (Signal.) *Yes.*
• Will you fill in 54? (Signal.) *Yes.*
• Will you fill in 55? (Signal.) *No.*
• Will you fill in 56? (Signal.) *No.*

g. (Repeat step f until firm.)

h. Look at your answer sheet. If there was one minute left, what's the number of the first item you'd mark? (Signal.) *One.*

• Find the next item you'd mark. What's the number of the next item you'd mark? (Signal.) *12.*

• Find the next item you'd mark. What's the number of the next item? (Signal.) *30.*

• What's the number of the next item you'd mark? (Signal.) *42.*

• What's the number of the next item you'd mark? (Signal.) *47.*

• Think big. What's the number of the **last** item you'd mark? (Signal.) *54.*

i. (Repeat step g until firm.)

j. Get ready to mark the answers. After you're finished, we'll find out how many more points you scored by following the rules for doing well on tests. Remember, fill in only items that are on the test.

• Here we go. There is only one minute left. Finish up your paper.

k. (Reinforce students who quickly fill in one and only one answer for items 1, 12, 30, 42, and 47 through 54 on the answer sheet.)

• (Prompt students who don't fill in answers quickly to start marking answers more quickly.)

• (Alert students who have more than one answer filled in on any problem to make sure there's only one answer per problem.)

• (Reinforce students who remember not to fill in items 55 and 56.)

l. (After one minute, say:) Stop working.

m. Check your work.

• For the items you just marked, I'll tell you the make-believe answer. Next to each item, make a C if you marked the correct answer and an X if you didn't mark that answer.

• Raise your hand after I say each answer if you guessed the right one. √

• Item 1: The answer is **E.** (Acknowledge students who raise their hands.)

• (Repeat for: 12, F; 30, G; 42, I; 47, D; 48, G; 49, C; 50, G; 51, D; 52, F; 53, C; 54, H.)

• What letter did you fill in for item 55? (Call on a student. Idea: *No answer.*)

• You shouldn't have filled in answers for items 55 and 56.

n. Count the Cs you wrote next to the numbers.

• Raise your hand if you got more than five of them correct. (Students respond.)

• (Repeat for: four, three, two, one, and zero of them correct.)

o. Most of you got one or more of the items correct. You wouldn't have gotten any of those problems correct if you hadn't filled in answers for those items.

Test Preparation Lesson 6

Materials Note:

Each student will need:
- lined paper
- test booklet and Multiple-Choice Response Sheet

EXERCISE 1
Solving Equations with Distribution

a. (Direct students to find lesson 6 in their **test booklets** and write their names and indicate lesson 6 on their answer sheets.)
- (Write on the board:)

$$3 \times (R + 5) + R = 23$$

b. Here's an equation that you can solve by distributing the products. Read the equation. (Signal.) *3 times R plus 5 plus R equals 23.*
- You'll distribute the product for: 3 times R plus 5. Read that part. (Signal.) *3 times R plus 5.*
- Say **3 times R plus 5** with the product distributed. Get ready. (Signal.) *3 times R plus 3 times 5.*
c. (Repeat step b until firm.)
d. To figure out what R equals, I write the distributed products below and copy the rest of the equation.
- (Write to show:)

3	×	(R + 5)	+	R	= 23
(3 × R)	+	(3 × 5)	+	R	= 23

- Read the equation with the product distributed. Get ready. (Signal.) *3 times R plus 3 times 5 plus R equals 23.*
- What does 3 times 5 equal? (Signal.) *15.*
e. (Write to show:)

3	×	(R + 5)	+	R	= 23
(3 × R)	+	(3 × 5)	+	R	= 23
3R	+	15	+	R	= 23

- We have 3R and R on the same side. What is 3R plus 1R equal to? (Signal.) *4R.*
- So, we can combine the Rs.

f. (Write to show:)

3	×	(R + 5)	+	R	= 23
(3 × R)	+	(3 × 5)	+	R	= 23
3R	+	15	+	R	= 23
4R	+	15			= 23

- Read the new equation with the Rs combined. (Signal.) *4R plus 15 equals 23.*
- You can figure out what R equals. Listen: You end up with 23. What operation did you do just before ending up with 23? (Signal.) *Added 15.*
- Yes, added 15. How do you undo adding 15? (Signal.) *Subtract 15.*
- Yes, subtract 15 from 23. Say the subtraction problem. (Signal.) *23 minus 15.*
- What's the answer? (Signal.) *8.*
- Read the new equation. (Signal.) *4R equals 8.*
g. (Write to show:)

3	×	(R + 5)	+	R	= 23
(3 × R)	+	(3 × 5)	+	R	= 23
3R	+	15	+	R	= 23
4R	+	15			= 23
				4R	= 8

- Read the new equation. (Signal.) *4R equals 8.*
- You can undo 4 times R by dividing by 4. Do it and you'll have what R equals. Copy what's on the board, and below that, write the equation to show what R equals.
- Raise your hand when you've written the equation for R.
- (Observe students and give feedback.)
- You can replace R with what it equals to see that the equation works.
- (Point to the board:)

3	×	(R + 5)	+	R	= 23
(3 × R)	+	(3 × 5)	+	R	= 23
3R	+	15	+	R	= 23
4R	+	15			= 23
				4R	= 8
				R	= 2

- What does R equal? (Signal.) *2.*

- So, I cross out every R and write 2 above it.
- (Write to show:)

$$3 \times (\cancel{R}^{2} + 5) + \cancel{R}^{2} = 23$$

- There's 2 plus 5 inside the parentheses. What does that equal? (Signal.) *7*. So, you have 3 times 7 plus 2 equals 23.
- Do the multiplication and the addition and see whether you end up with 23.
- (Observe students and give feedback.)
- Everybody, what does 3 times 7 plus 2 equal? (Signal.) *23*. So, when you replace R with what it equals, the equation does work.

g. Find item 1 in your **test booklet.** √
- (Teacher reference:)

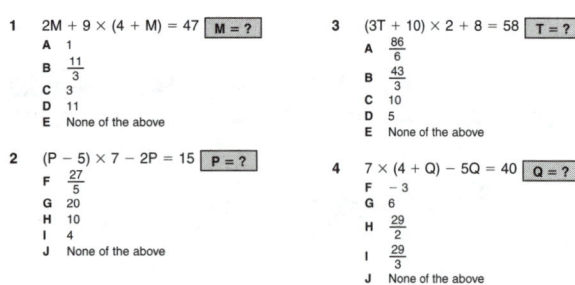

1 $2M + 9 \times (4 + M) = 47$ $\boxed{M = ?}$	3 $(3T + 10) \times 2 + 8 = 58$ $\boxed{T = ?}$
A 1	A $\frac{86}{6}$
B $\frac{11}{3}$	B $\frac{43}{3}$
C 3	C 10
D 11	D 5
E None of the above	E None of the above
2 $(P - 5) \times 7 - 2P = 15$ $\boxed{P = ?}$	4 $7 \times (4 + Q) - 5Q = 40$ $\boxed{Q = ?}$
F $\frac{27}{5}$	F -3
G 20	G 6
H 10	H $\frac{29}{2}$
I 4	I $\frac{29}{3}$
J None of the above	J None of the above

h. I'll read item 1: 2M plus 9 times 4 plus M equals 47. You'll distribute the product for part of that equation. Read that part. Get ready. (Signal.) *9 times 4 plus M.*
- Read item 2. Get ready. (Signal.) *P minus 5 times 7 minus 2P equals 15.*
- You'll distribute the product for part of that equation. Read that part. Get ready. (Signal.) *P minus 5 times 7.*
- Read item 3. Get ready. (Signal.) *3T plus 10 times 2 plus 8 equals 58.*
- You'll distribute the product for part of that equation. Read that part. Get ready. (Signal.) *3T plus 10 times 2.*
- Read item 4. Get ready. (Signal.) *7 times 4 plus Q minus 5Q equals 40.*
- You'll distribute the product for part of that equation. Read that part. Get ready. (Signal.) *7 times 4 plus Q.*
i. (Repeat step h until firm.)

j. Go back to item 1. √
- Copy the equation on your lined paper. Below it, write the equation with the product distributed. Then combine the undo operations of the Ms and figure out what M equals.
- Raise your hand when you've marked an answer for item 1.
- (Observe students and give feedback.)
- (Write on the board:)

> 1. $2M + 9 \times (4 + M) = 47$

k. Check your work. Here's the problem. On the line below, you should have written the problem with the product distributed. Read that equation. (Signal.) *2M plus 9 times 4 plus 9 times M equals 47.*
l. (Write to show:)

> 1. $2M + 9 \times (4 + M)\ \ \ \ \ = 47$
> $2M + (9 \times 4) + (9 \times M) = 47$

- On the next line, you should have written the equation showing what each of the parts equals. Read that line. (Signal.) *2M plus 36 plus 9M equals 47.*
- On the line below that, you should have combined the Ms. Now read that line. (Signal.) *11M plus 36 equals 47.*
m. (Write to show:)

> 1. $2M + 9 \times (4 + M)\ \ \ \ \ \ = 47$
> $2M + (9 \times 4) + (9 \times M) = 47$
> $2M +\ \ \ \ 36\ \ \ + \ \ 9M\ \ \ \ = 47$
> $11M +\ \ \ \ 36\ \ \ \ \ \ \ \ \ \ \ \ \ \ = 47$

- What did you do to undo adding 36? (Signal.) *Subtracted 36.*
- What did you do to undo multiplying by 11? (Signal.) *Divided by 11.*
- What does M equal? (Signal.) *One.*

n. (Write to show:)

```
1.          2M + 9 × (4 + M) = 47
       2M + (9 x 4)  +  (9 × M) = 47
       2M +   36    +   9M     = 47
      11M +   36                = 47
                          11M = 11
                            M =  1
```

- What's the answer for item 1? (Signal.) *A.*
  Copy item 2.
  Distribute the product below. Then write the equation showing what each part equals. Combine like terms. Then undo operations to figure out what P equals.
- Raise your hand when you've marked the answers for items 1 and 2.
- (Observe students and give feedback.)
o. Check your work.
p. Item 1. What letter did you mark? (Signal.) *A.*
q. Item 2. P minus 5 **times** 7 minus 2P equals 15.
- Read the equation you wrote with the product distributed. (Signal.) *P times 7 minus 5 times 7 minus 2P equals 15.*
- Read the equation that shows what each part equals. (Signal.) *7P minus 35 minus 2P equals 15.*
- Read the equation with the Ps combined. (Signal.) *5P minus 35 equals 15.*
- Read the equation after undoing subtracting 35. (Signal.) *5P equals 50.*
- Read the equation that tells what P equals. (Signal.) *P equals 10.*
- What letter did you mark? (Signal.) *H.*
r. Now you'll work item 3. Copy the problem and remember all of the steps.
- Raise your hand when you've marked an answer for item 3.
- (Observe students and give feedback.)
s. Check your work.
- Item 3. 3T plus 10 times 2 plus 8 equals 58. You solved for T.
- Everybody, what does T equal? (Signal.) *5.*
- What letter did you mark? (Signal.) *D.*
t. Work item 4. Distribute the product below. Then write the equation showing what each part equals. Combine like terms and stop!
- (Observe students and give feedback.)
- Everybody, read the equation with like terms combined. (Signal.) *28 plus 2Q equals 40.*
- I'll say that equation with the Q term first: 2Q plus 28 equals 40.
- Say the equation with the Q term first. (Signal.) *2Q plus 28 equals 40.*

u. Write the equation with the Q term first. Then undo operations to figure out what Q equals.
- Raise your hand when you've marked an answer for item 4. √
v. Check your work.
- Item 4. 7 times 4 plus Q minus 5Q equals 40. You solved for Q.
- Everybody, what does Q equal? (Signal.) *6.*
- What letter did you mark? (Signal.) *G.*

EXERCISE 2
Classification of Triangles

a. Find the names above item 5 in your **test booklet.** √
- These are the names you've learned for triangles that have 3 sides the same length, 2 sides the same length, and all sides different lengths: equilateral, isosceles, and scalene.
b. What is the name of the triangle that has all sides different lengths? (Signal.) *Scalene.*
- What do we call the triangle with 3 sides that have equal lengths? (Signal.) *Equilateral.*
- What do we call the triangle with 2 sides that are the same length? (Signal.) *Isosceles.*
c. (Repeat step b until firm.)
d. Touch the triangle for item 5. √
- What's the name for that triangle? (Signal.) *Scalene.*
e. Touch the triangle for item 6. √
- What's the name for that triangle? (Signal.) *Equilateral.*
f. What's the name for triangle 7? (Signal.) *Isosceles.*
g. What's the name for triangle 8? (Signal.) *Isosceles.*
h. What's the name for triangle 9? (Signal.) *Equilateral.*
i. What's the name for triangle 10? (Signal.) *Scalene.*
j. You'll mark the letter of the name for each item.
- Raise your hand when you've finished items 5 through 10.
- (Observe students and give feedback.)
k. Check your work.
l. Item 5. What letter did you mark? (Signal.) *C.* Yes, 5 is a scalene triangle.
m. Item 6. What letter did you mark? (Signal.) *F.* Yes, 6 is an equilateral triangle.
n. Item 7. What letter did you mark? (Signal.) *B.* Yes, 7 is an isosceles triangle.
o. Item 8. What letter did you mark? (Signal.) *G.* Yes, 8 is an isosceles triangle.
p. Item 9. What letter did you mark? (Signal.) *A.* Yes, 9 is an equilateral triangle.
q. Item 10. What letter did you mark? (Signal.) *H.* Yes, 10 is a scalene triangle.

EXERCISE 3
Average

a. Find items 11 through 13. √
- Remember, to find the average number, you add up all the amounts and you divide by the number of amounts.

b. Work items 11 through 13.
- Pencils down when you've marked answers through item 13.
- (Observe students and give feedback.)

c. Check your work.

d. Item 11. What letter did you mark? (Signal.) *E.*
Yes, E. Jimmy earned an average of 6 dollars a day. 6 dollars was not a choice, so the correct answer is E, none of the above.

e. Item 12. What letter did you mark? (Signal.) *H.*
Yes, H. The runner ran an average of 10 miles per day.

f. Item 13. What letter did you mark? (Signal.) *C.*
Yes, C. there was an average of 4 people per car.

EXERCISE 4
Measurement and Geometry
3-Dimensional Objects

a. Find item 15 in your **test booklet.** √
- The figures in items 15 through 19 are made of solid figures you've learned.
- Touch item 15. √
- This object is made of two solid figures.
- Raise your hand when you know the name of those two figures. √
- Everybody, what's the name of the figure that is on the top? (Signal.) *Rectangular prism.*
Yes, it's a rectangular **prism,** or another name for it is a rectangular **solid.**
- What is the name of the figure that is on the bottom? (Signal.) *Cylinder.*

b. Look at the choices below figure 15. Mark the letter that shows there's one rectangular solid and one cylinder. Don't mark an answer for item 14.
- Raise your hand when you're finished.
- (Observe students and give feedback.)
- Item 15. Everybody, what letter did you mark? (Signal.) *C.*

c. Touch item 16. √
- The figure is made of two types of figures. Decide what types of solid figures make up this object. Mark the letter for the answer.
- Raise your hand when you're finished.
- (Observe students and give feedback.)
- Everybody, what's the top part of figure 16? (Signal.) *A pyramid.*
- What type of solid **makes up** the bottom part of figure 16? (Call on a student. Ideas: *Cubes; accept rectangular solid.*)
- Yes, there are 4 cubes stuck together to make up the bottom part of figure 16.
- What letter did you mark? (Signal.) *G.*
Letter G says: 1 pyramid and 4 cubes. These are the objects that make up figure 16.

d. Touch item 17. √
- The object is made of only **one** type of solid. Mark the letter for the answer.
- Raise your hand when you're finished.
- (Observe students and give feedback.)
- Everybody, what type of solid is figure 17 made of? (Signal.) *Triangular prisms.*
- Yes, there are a whole bunch of triangular prisms stuck together. How many are there? (Signal.) *9.*
- What letter did you mark? (Signal.) *D.*

e. Touch figure 18. √
- Figure 18 is made up of **two** types of solid objects. Mark the letter that tells about the solid objects in figure 18.
- Raise your hand when you're finished. √
- Everybody, what's the top part of figure 18? (Signal.) *Sphere.*
- What's the bottom part of figure 18? (Signal.) *Cone.*
- What letter did you mark? (Signal.) *F*

f. Touch figure 19. √
- Read the choices for the two types of solids that make up this object. Then mark the letter.
- Raise your hand when you're finished.
- (Observe students and give feedback.)
- Everybody, what solid is on each end of figure 19? (Signal.) *Sphere.*
- What's the middle part of figure 19? (Signal.) *Cylinder.*
- What letter did you mark? (Signal.) *E.* Yes, none of the choices was two spheres and a cylinder.

g. Raise your hand if you got everything right. √

EXERCISE 5
Equivalent Fractions

a. Find item 21 in your **test booklet.** √
- You're going to identify the fractions that are equivalent.

b. Item 21. The fraction is 6-eighths. Simplify the fraction and mark the right answer. Don't mark an answer for item 20.
- Pencils down when you've marked an answer for item 21. √
- Check your work.
- Item 21. What letter did you mark? (Signal.) *E.* Yes, E. The simplified fraction for 6-eighths is 3-fourths. 3-fourths is not a choice, so you should have marked E, none of these.

c. Work items 22 through 25.
- Raise your hand when you've marked answers through item 25.
- (Observe students and give feedback.)

d. Check your work.

e. Item 22. What letter did you mark? (Signal.) *H.*
- So, what is the simplified fraction for 15-thirtieths? (Signal.) *One-half.*

f. Item 23. What letter did you mark? (Signal.) *A.*
- So, what is the simplified fraction for 10-fifteenths? (Signal.) *2-thirds.*

g. Item 24. What letter did you mark? (Signal.) *I.*
- So, what's the simplified fraction for 70-eightieths? (Signal.) *7-eighths.*

h. Item 25. What letter did you mark? (Signal.) *D.*
- So, what is the simplified fraction for 4-twentieths? (Signal.) *One-fifth.*

EXERCISE 6
Classification of Angles

a. (Write on the board:)

> **right angle**
> **acute angle**
> **obtuse angle**

- You've learned the names for different angles.

b. Which angle is less than 90 degrees? (Signal.) *Acute angle.*
- Which angle is more than 90 degrees? (Signal.) *Obtuse angle.*
- Which angle is exactly 90 degrees? (Signal.) *Right angle.*

c. (Repeat step b until firm.)
- (Teacher reference:)

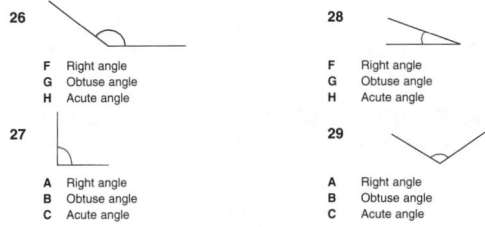

- For items 26 through 29, you'll mark the correct name: **right angle, obtuse angle,** or **acute angle.**
- Raise your hand when you've marked answers for items 26 through 29.
- (Observe students and give feedback.)
e. Check your work. For each item, you'll tell me what letter you marked.
f. Item 26. What letter did you mark? (Signal.) *G.*
- What kind of angle is angle 26? (Signal.) *Obtuse.*
g. Item 27. What letter did you mark? (Signal.) *A.*
- What kind of angle is angle 27? (Signal.) *Right.*
h. Item 28. What letter did you mark? (Signal.) *H.*
- What kind of angle is angle 28? (Signal.) *Acute.*
i. Item 29. What letter did you mark? (Signal.) *B.*
- What kind of angle is angle 29? (Signal.) *Obtuse.*
j. Find item 30 in your **test booklet.** √
- (Teacher reference:)

30 Which angle shows a right angle?

F   G   H   I   J   None of these

31 Which angle shows an acute angle?

A   B   C   D   E   None of these

32 Which angle shows an obtuse angle?

F   G   H   I   J   None of these

33 Which angle shows an acute angle?

A   B   C   D   E   None of these

- Items 30 through 33 describe angles. You'll mark the answer that best fits that description.
- Work items 30 through 33.
- Raise your hand when you're finished. √
k. Check your work.
l. Item 30: Which angle shows a right angle?
- What letter did you mark? (Signal.) *I.*
m. Item 31: Which angle shows an acute angle?
- What letter did you mark? (Signal.) *B.*

n. Item 32: Which angle shows an obtuse angle?
- What letter did you mark? (Signal.) *J.*
- Choices F, G, H, and I are not obtuse angles, so the correct answer is J, none of these.
o. Item 33: Which angle shows an acute angle?
- What letter did you mark? (Signal.) *A.*

EXERCISE 7
Exponents

a. Find item 35 in your **test booklet.** √
- (Teacher reference:)

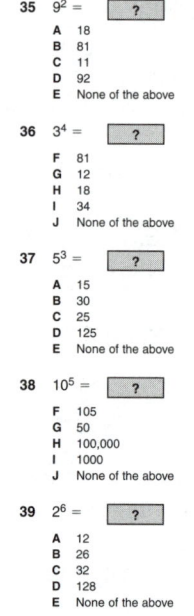

35 $9^2 =$ [?]
  A 18
  B 81
  C 11
  D 92
  E None of the above

36 $3^4 =$ [?]
  F 81
  G 12
  H 18
  I 34
  J None of the above

37 $5^3 =$ [?]
  A 15
  B 30
  C 25
  D 125
  E None of the above

38 $10^5 =$ [?]
  F 105
  G 50
  H 100,000
  I 1000
  J None of the above

39 $2^6 =$ [?]
  A 12
  B 26
  C 32
  D 128
  E None of the above

- I'll read the base number and exponent for item 35 **and** say the multiplication it equals.
- Listen: 9 to the second equals 9 times 9.
b. Your turn: For each item, read the base number and exponent **and** the multiplication it equals.
- Item 35. Read it. (Signal.) *9 to the second equals 9 times 9.*
- Item 36. Read it. (Signal.) *3 to the fourth equals 3 times 3 times 3 times 3.*
- Item 37. Read it. (Signal.) *5 to the third equals 5 times 5 times 5.*

- Item 38. Read it. (Signal.) *10 to the fifth equals 10 times 10 times 10 times 10 times 10.*
- Item 39. Read it. (Signal.) *2 to the sixth equals 2 times 2 times 2 times 2 times 2 times 2.*

c. Copy the base number and the exponent for items 35 through 39. Then, for each problem, write the multiplication and the number it equals. Mark the answer.
- Raise your hand when you've marked answers for items 35 through 39. Don't mark an answer for item 34.
- (Observe students and give feedback.)

d. Check your work.
e. Item 35: 9 to the second.
- What letter did you mark? (Signal.) *B.*
- So, what number does 9 to the second equal? (Signal.) *81.*

f. Item 36: 3 to the fourth.
- What letter did you mark? (Signal.) *F.*
- So, what number does 3 to the fourth equal? (Signal.) *81.*

g. Item 37: 5 to the third.
- What letter did you mark? (Signal.) *D.*
- So, what does 5 to the third equal? (Signal.) *One hundred 25.*

h. Item 38: 10 to the fifth.
- What letter did you mark? (Signal.) *H.*
- So, what does 10 to the fifth equal? (Signal.) *One hundred thousand.*

i. Item 39: 2 to the sixth.
- What letter did you mark? (Signal.) *E.*
- What does 2 to the sixth equal? (Signal.) *64.*
- Yes, 64, so the correct answer is E, none of the above.

EXERCISE 8
Place Value

a. (Write on the board:)

9,876,543,210

- This number has 10 digits. It has 3 commas.
b. I'll read the number: 9 billion, 8 hundred 76 million, 5 hundred 43 thousand, 2 hundred 10.
- Your turn. Read the number. Get ready. (Signal.) *9 billion, 8 hundred 76 million, 5 hundred 43 thousand, 2 hundred 10.*
c. (Repeat step b until firm.)
d. The first digit is a 9. The value for 9 is 9 billion.
- (Touch 6.)
- What's the value for 6? (Signal.) *6 million.*

- (Touch 7.)
- What's the value for 7? (Signal.) *70 million.*
- (Touch 8.)
- What's the value for 8? (Signal.) *8 hundred million.*
- (Touch 9.)
- What's the value for 9? (Signal.) *9 billion.*
e. If it has 3 commas, it's a **billions** number.
- If it has 2 commas, it's a **millions** number.
- If it has one comma, it's a **thousands** number.
f. Find item 41. √
- Mark the right answer for items 41 and 42. Don't mark an answer for item 40.
- Raise your hand when you've marked answers for items 41 and 42. √
g. Check your work.
h. Item 41: 36 million 2 hundred 74 thousand 3 hundred 81.
- What's the value for 6? (Signal.) *6 million.*
- What letter did you mark? (Signal.) *A.*
i. Item 42: 8 billion one hundred 34 million 7 hundred 90 thousand 4 hundred 26.
- What's the value for 8? (Signal.) *8 billion.*
- What letter did you mark? (Signal.) *G.*

EXERCISE 9
Test-Taking Rules
What to Do When You're Running Out of Time

a. Find the **Stop** sign at the end of lesson 6 in your **test booklet** and get ready to tell me the number for the last item for lesson 6. √
- Everybody, what's the number of the **last** item you'd mark? (Signal.) *53.*
b. Look at your answer sheet. If there was one minute left, what's the number of the first item you'd mark? (Signal.) *14.*
- Find the next item you'd mark. What's the number of the next item? (Signal.) *20.*
- What's the number of the next item you'd mark? (Signal.) *34.*
- What's the number of the next item you'd mark? (Signal.) *40.*
- What's the number of the next item you'd mark? (Signal.) *43.*
- Think big. What's the number of the **last** item you'd mark? (Signal.) *53.*
c. (Repeat step b until firm.)

d. Remember to look in the **test booklet** and find out the number of the last item you'd mark if you were running out of time. Sometimes you take more than one test, and you don't want to guess on answers to another test.

e. Here we go. There is only one minute left. Finish up your paper.

f. (Reinforce students who quickly fill in one and only one answer for items 14, 20, 34, 40, and 43 through 53 on the answer sheet.)

• (Prompt students who don't fill in answers quickly to start marking answers more quickly.)

• (Alert students who have more than one answer filled in on any problem to make sure there's only one answer per problem.)

g. (Reinforce students who remember **not** to fill in 54 through 56.)

h. (After one minute, say:) Stop working.

i. For the items you just marked, I'll tell you the make-believe answer. Next to each item make a C if you marked the correct answer and an X if you didn't mark that answer.

• Raise your hand after I say each answer if you guessed the right one. √

• Item 14: The answer is **G.** (Acknowledge students who raise their hands.)

• Item 20: The answer is **H.** (Acknowledge students who raise their hands.)

• (Repeat for items: 34, I; 40, I; 43, C; 44, G; 45, D; 46, F; 47, B; 48, I; 49, A; 50, J; 51, B; 52, F; 53, C.)

j. What letter did you fill in for 54? (Call on a student. Idea: *No answer.*) Yes, you should have marked none of the answers.

k. What letter did you fill in for 56? (Signal.) *None of the answers.*

l. Count the Cs you wrote next to the numbers.

• Raise your hand if you got five or more of them correct. (Students respond.)

• Raise your hand if you got four of them correct. (Students respond.)

• (Repeat for: three, two, one, and zero of them correct.)

m. Most of you got two or more of the items correct. You wouldn't have gotten any of those problems correct if you hadn't filled in answers for those items.

Test Preparation Lesson 7

> ### *Materials Note:*
> Each student will need:
> • lined paper
> • test booklet and Multiple-Choice Response Sheet

EXERCISE 1
Triangles and Quadrilaterals

a. (Direct students to find lesson 7 in their **test booklets** and write their names and indicate lesson 7 on their answer sheets.)

• Find item 1 in your **test booklet.** √

• You're going to mark the answer that best describes figures 1 through 8. Some of the figures are triangles, and some of them have four sides.

• Raise your hand when you've marked answers for items 1 through 8 on your answer sheet.

• (Observe students and give feedback.)

b. Check your work.

c. Item 1. What letter did you mark? (Signal.) A.

• What kind of figure is it? (Signal.) *A scalene triangle.*
  Yes, the figure in 1 is a scalene triangle.

d. Item 2. What letter did you mark? (Signal.) *I.*

• What kind of figure is it? (Signal.) *A trapezoid.*

e. Item 3. What letter did you mark? (Signal.) *A.*

• What kind of figure is it? (Signal.) *A rectangle.* It's a parallelogram, too, but the better description is a rectangle.

f. Item 4. What letter did you mark? (Signal.) *H.*

• What kind of figure is it? (Signal.) *A rhombus.* Yes, the figure in 4 is a parallelogram and a rhombus.

g. Item 5. What letter did you mark? (Signal.) *B.*

• What kind of figure is it? (Signal.) *A parallelogram.* Yes, the figure in 5 is a parallelogram.

h. Item 6. What letter did you mark? (Signal.) *I.*

• What kind of figure is it? (Signal.) *An isosceles triangle.*

i. Item 7. What letter did you mark? (Signal.) *B.*

• What kind of figure is it? (Signal) *An equilateral triangle.*

j. Item 8. What letter did you mark? (Signal.) *J.*

• What kind of figure is it? (Signal.) *A square.* Yes, the figure in 8 is a square, a rectangle, a parallelogram, **and** a rhombus.

EXERCISE 2
Solving Equations with Distribution

a. Find item 9 in your **test booklet.** √
- (Teacher reference:)

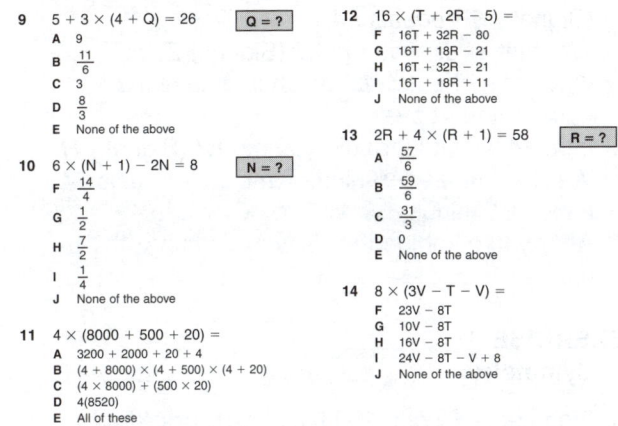

9 $5 + 3 \times (4 + Q) = 26$ Q = ?
  A 9
  B $\frac{11}{6}$
  C 3
  D $\frac{8}{3}$
  E None of the above

10 $6 \times (N + 1) - 2N = 8$ N = ?
  F $\frac{14}{4}$
  G $\frac{1}{2}$
  H $\frac{7}{2}$
  I $\frac{1}{4}$
  J None of the above

11 $4 \times (8000 + 500 + 20) =$
  A $3200 + 2000 + 20 + 4$
  B $(4 + 8000) \times (4 + 500) \times (4 + 20)$
  C $(4 \times 8000) + (500 \times 20)$
  D $4(8520)$
  E All of these

12 $16 \times (T + 2R - 5) =$
  F $16T + 32R - 80$
  G $16T + 18R - 21$
  H $16T + 32R - 21$
  I $16T + 18R + 11$
  J None of the above

13 $2R + 4 \times (R + 1) = 58$ R = ?
  A $\frac{57}{6}$
  B $\frac{59}{6}$
  C $\frac{31}{3}$
  D 0
  E None of the above

14 $8 \times (3V - T - V) =$
  F $23V - 8T$
  G $10V - 8T$
  H $16V - 8T$
  I $24V - 8T - V + 8$
  J None of the above

- Read the equation. (Signal.) *5 plus 3 times 4 plus Q equals 26.*
- You'll distribute the product for part of the equation. Read that part. Get ready. (Signal.) *3 times 4 plus Q.*
- (Write on the board:)

| 9. | 5 | + | 3 | × | (4 + Q) | = | 26 |

- Here's the problem. Copy the problem. On the line below, write the problem with the product distributed.
- Raise your hand when you've done that much.
- (Observe students and give feedback.)
- Read the equation with the product distributed. (Signal.) *5 plus 3 times 4 plus 3 times Q equals 26.*

- (Write to show:)

| 9. | 5 | + | 3 | × | (4 + Q) | = 26 |
| | 5 | + | (3 × 4) | + | (3 × Q) | = 26 |

- On the next line, write the equation showing what each of the parts equals.
- (Observe students and give feedback.)
- Read the equation that shows what each part equals. (Signal.) *5 plus 12 plus 3Q equals 26.*
- (Write to show :)

9.	5	+	3	×	(4 + Q)	= 26
	5	+	(3 × 4)	+	(3 × Q)	= 26
	5	+	12	+	3Q	= 26

- On the line below that, combine the numbers.
- (Observe students and give feedback.)
- Read the equation with the numbers combined. (Signal.) *17 plus 3Q equals 26.*
- (Write to show:)

9.	5	+	3	×	(4 + Q)	= 26
	5	+	(3 × 4)	+	(3 × Q)	= 26
	5	+	12	+	3Q	= 26
	17	+	3Q			= 26

- I'm going to rewrite the equation.
- (Write to show:)

| 3Q | + | 17 | = 26 |

- Read the equation with the Q term first. (Signal.) *3Q plus 17 equals 26.*
- How do you undo adding 17? (Signal.) *Subtract 17.*
- How do you undo multiplying by 3? (Signal.) *Divide by 3.*
- Write the equations that show the undoing.
- Raise your hand when you have the equation for what Q equals.
- (Observe students and give feedback.)
- What does 3Q equal? (Signal.) *9.*
- What does Q equal? (Signal.) *3.*

- (Write to show :)

9.	5	+	3	×	(4 + Q)	= 26
	5	+	(3 × 4)	+	(3 × Q)	= 26
	5	+	12	+	3Q	= 26
	17	+	3Q			= 26
			3Q	+	17	= 26
					3Q	= 9
					Q	= 3

b. Mark the answer for item 9. Then work item 10. Copy the problem and distribute the product below. Then write the equation showing what each part equals. Then undo operations to figure out what N equals.
- Raise your hand when you've marked the answer for item 10.
- (Observe students and give feedback.)
c. Item 9. What letter did you mark? (Signal.) *C.*
d. Item 10. 6 times the quantity N plus one **minus** 2N equals 8.
- Read the equation you wrote with the product distributed. (Signal.) *6 times N plus 6 times one minus 2N equals 8.*
- Read the equation that shows what each part equals. (Signal.) *6N plus 6 minus 2N equals 8.*
- Read the equation with the Ns combined. (Signal.) *4N plus 6 equals 8.*
- Read the equation after undoing adding 6. (Signal.) *4N equals 2.*
- Everybody, what does N equal? (Signal.) *One-half.* Yes, N equals 2-fourths or one-half.
- What letter did you mark? (Signal.) *G.*
e. You need to distribute the product for items 11 through 14. The answer to some of the items is not a number. Work items 11 through 14. Copy the problems and remember all the steps.
- Raise your hand when you've marked answers for items 11 through 14.
- (Observe students and give feedback.)
f. Check your work.
g. Item 11. What letter did you mark? (Signal.) *D.* Yes, D. 4 times 8 thousand plus 5 hundred plus 20 equals 4 times 8 thousand 5 hundred 20.
h. Item 12. What letter did you mark? (Signal.) *F.* Yes, F. 16 times T plus 2R minus 5 equals 16T plus 32R minus 80.
i. Item 13. 2R plus 4 times R plus one equals 58.
- Read the equation you wrote with the product distributed. (Signal.) *2R plus 4 times R plus 4 times one equals 58.*
- Read the equation that shows what each part equals. (Signal.) *2R plus 4R plus 4 equals 58.*
- Read the equation with the Rs combined. (Signal.) *6R plus 4 equals 58.*
- Read the equation after undoing subtracting 4. (Signal.) *6R equals 54.*
- What letter did you mark? (Signal.) *E.* Yes, E. 9 was not a choice, so you should have marked E, none of the above.
j. Item 14. What letter did you mark? (Signal.) *H.*
- Yes, H. 8 times 3V minus T minus V equals 24V minus 8T minus 8V.
- After you combine the Vs, you end up with 16V minus 8T.

EXERCISE 3
Symmetry

a. Find items 15 and 16 in your **test booklet.** √
- These questions ask about lines of symmetry.
- Look at the figures for 15 and 16. Read the items, and then mark your choices.
- Raise your hand when you've finished.
- (Observe students and give feedback.)
b. Check your work.
- Item 15. What letter did you mark? (Signal.) *C.* Figure C has a line of symmetry.
- Item 16. What letter did you mark? (Signal.) *I.* The number 4 is the only figure in item 16 that doesn't have a line of symmetry.

EXERCISE 4
Dimensions and Views

a. Find the object above item 19. √
- (Teacher reference:)

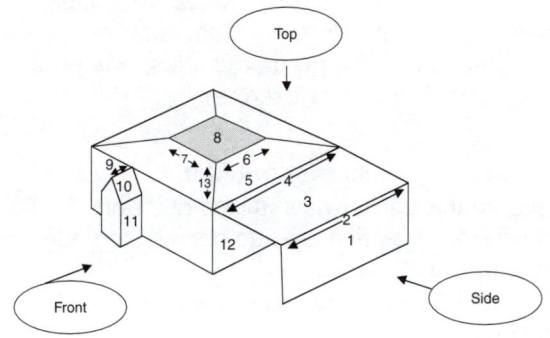

- This object looks like a house. The roof is made of trapezoids. Touch the part labeled number 5. √
- Four sections of the roof are the same size as the part labeled number 5.
- The line where two of the trapezoidal-shaped roof parts meet is labeled number 13. Touch it. √
- The top roof section is a square labeled number 8. Touch it. √
- Touch the two rooflines labeled numbers 6 and 7. √
- The roof of the garage is labeled number 3. Touch it. √
- The wall on the inside of the garage is a rectangle labeled number 12. Touch it. √
- Where the wall, the roof of the garage, and the roof of the house meet is a line labeled number 4. Touch it. √
- The outside wall supporting the roof of the garage is labeled number 1. Touch it. √
- The line where that wall and the roof of the garage meet is labeled number 2. Touch it. √
- The roofline above the front door is labeled number 9. Touch it. √
- A section of the roof above the front door is labeled number 10. Touch it. √
- The wall supporting the roof above the front door is labeled number 11. Touch it. √

b. Touch the word **front.** √
- If you follow the arrow pointing from the word **front,** from which view would you be looking? (Signal.) *The front view.*

c. Touch the word **side.** √
- If you follow the arrow pointing from the word **side,** from which view would you be looking? (Signal.) *The side view.*

d. Touch the word **top.** √
- If you follow the arrow pointing from the word **top,** from which view would you be looking? (Signal.) *The top view.*

e. (Repeat steps b through d until firm.)

f. Item 19. Which is the top view of the object?
- Touch choice A. √

- Is that figure the top view? (Signal.) *No.*
- Why not? (Call on a student. Idea: *It's the side view.*)

g. Touch choice B. √
- Is that figure the top view? (Signal.) *No.*
- Why not? (Call on a student. Ideas: *(1) There should be a square in the middle of the roof; (2) the roof over the door should have a line dividing it.*)

h. Touch choice C. √
- Is that figure the top view? (Signal.) *Yes.*

i. Item 20. Which is the front view of the object?

j. Touch choice F. √
- Is that figure the front view? (Signal.) *No.*
- Why not? (Call on a student. Ideas: *(1) The roof above the door should come to a point; (2) the roof of the house should not come to a point.*)

k. Touch choice G. √
- Is that figure the front view? (Signal.) *No.*
- Why not? (Call on a student. Idea: *The roof above the garage should be flat and the angled roof should not go all the way over the house and the garage.*)

l. Touch choice H. √
- Is that figure the front view? (Signal.) *No.*
- Why not? (Call on a student. Idea: *It's the side view.*)

m. Touch choice I. √
- Is the figure in I the front view? (Signal.) *Yes.*

n. Item 21. Which is the side view?
- Look at the choices.
- Raise your hand when you can tell me which choice is the side view. √
- Everybody, which letter shows the side view? (Signal.) *A.*
- Your turn: Work items 19 through 21. Don't mark answers for items 17 and 18.
- Pencils down when you've marked answers through item 21.
- (Observe students and give feedback.)

o. Check your work.
- Item 19. What letter did you mark? (Signal.) *C.* Yes, choice C is the top view.
- Item 20. What letter did you mark? (Signal.) *I.* Yes, choice I is the front view.
- Item 21. What letter did you mark? (Signal.) *A.* Yes, choice A is the side view.

p. For items 22 through 28, each item will have a drawing of either the front view, the top view, or the side view. A portion of each view will be highlighted. The highlighted part represents one, or more than one, of the numbered parts of the object. Mark the answer that best fits the highlighted portion of the drawing.

- Work items 22 through 28.
- Pencils down when you've marked answers for items through 28.
- (Observe students and give feedback.)
q. Check your work.
- Item 22. What letter did you mark? (Signal.) *G.* Yes, the highlighted line on the front view represents the part labeled number 8. It also represents the line labeled number 7, but 7 was not one of the choices.
- Item 23. What letter did you mark? (Signal.) *A.* Yes, the highlighted line on the top view represents the line of the object labeled number 4.
- Item 24. What letter did you mark? (Signal.) *I.* Yes, the highlighted line on the side view represents the line labeled number 2 and the part labeled number 3. The answer is I, numbers 2 and 3.
- Item 25. What letter did you mark? (Signal.) *B.* Yes, the highlighted part on the top view represents the part labeled number 5.
- Item 26. What letter did you mark? (Signal.) *G.* Yes, the highlighted point on the side view represents the line labeled number 7.
- Item 27. What letter did you mark? (Signal.) *E.* Yes, the highlighted point on the front view represents the line labeled number 4. 4 was not a choice, so the answer is E, none of these.
- Item 28. What letter did you mark? (Signal.) *H.* Yes, the highlighted line on the top view represents the part labeled number 1. It also represents the line labeled number 2, but 2 is not one of the choices.

EXERCISE 5
Mixed Number Operations

Note: Addition and subtraction of fractions with unlike denominators is taught in lessons 65–70 of *Connecting Math Concepts, Level E.* If presenting test preparation lessons prior to lesson 70, skip this exercise.

a. Find item 29 in your **test booklet.** √
- Items 29 through 31 have mixed numbers. To work the problem, you first change each mixed number into a fraction.
b. Copy item 29. Change both mixed numbers into fractions.
- Raise your hand when you've done that much. √
- (Write on the board:)

$$\frac{7}{2}$$
$$+\frac{11}{4}$$

- Here's what you should have.
- Work the problem and write the answer as a mixed number. Then mark the correct answer.
- (Write to show:)

$$\frac{7}{2}\left(\frac{2}{2}\right)=\frac{14}{4}$$
$$+\frac{11}{4}=\frac{11}{4}$$
$$\frac{25}{4}=6\frac{1}{4}$$

- Here's what you should have.
- What does the mixed number 3 and one-half plus 2 and 3-fourths equal? (Signal.) *6 and one-fourth.*
- What letter did you mark? (Signal.) *A.*
c. Work item 30. Remember to show the mixed numbers as fractions. Then find the common denominator. Then subtract. Show the answer as a mixed number, and then mark the correct answer.

- (Write to show:)

$$\frac{5}{2}\left[\frac{3}{3}\right] = \frac{15}{6}$$
$$-\frac{4}{3}\left[\frac{2}{2}\right] = \frac{8}{6}$$
$$\frac{7}{6} \quad 1\frac{1}{6}$$

- Check your work. Here's what you should have.
- What does the mixed number 2 and one-half **minus** one and one-third equal? (Signal.) *One and one-sixth.*
- What letter did you mark? (Signal.) *H.*
d. Item 31 is 7 and 11-fifteenths plus 2 and 5-ninths. We'll work this problem together in a fast way.
- (Write on the board :)

$$7\frac{11}{15}$$
$$+ \quad 2\frac{5}{9}$$
$$\underline{\hspace{3cm}}$$
$$9$$

- First we'll add the whole numbers. What's 7 plus 2? (Signal.) *9.*
- Now work the fraction part.
- Raise your hand when you have a mixed number answer for 11-fifteenths plus 5-ninths.
- (Observe students and give feedback.)
- What does 11-fifteenths plus 5-ninths equal? (Signal.) *One and 13-fortyfifths.*
- (Write to show:)

$$7 \quad \frac{11}{15}\left[\frac{3}{3}\right] = \frac{33}{45}$$
$$+ 2 \quad \frac{5}{9}\left[\frac{5}{5}\right] = \frac{25}{45}$$
$$\underline{\hspace{4cm}}$$
$$9 \qquad \frac{58}{45} = 1\frac{13}{45}$$

- So, the answer to the whole problem is **9** plus **1** and **13-fortyfifths.**
- What's the answer? (Signal.) *Ten and 13-fortyfifths.*
- Mark the answer to item 31. √
- What letter did you mark? (Signal.) *B.*

EXERCISE 6
Classification of Angles

a. (Write on the board:)

> **right angle**
> **acute angle**
> **obtuse angle**

- You've learned the names for different angles.
b. Which angle is more than 90 degrees? (Signal.) *Obtuse angle.*
- Which angle is exactly 90 degrees? (Signal.) *Right angle.*
- Which angle is less than 90 degrees? (Signal.) *Acute angle.*
c. (Repeat step b until firm.)
d. Find item 33 in your **test booklet.** √
- (Teacher reference:)

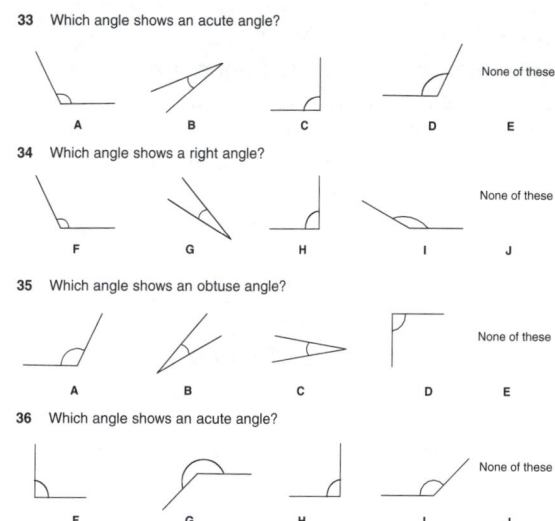

33 Which angle shows an acute angle?
    A   B   C   D   E   None of these

34 Which angle shows a right angle?
    F   G   H   I   J   None of these

35 Which angle shows an obtuse angle?
    A   B   C   D   E   None of these

36 Which angle shows an acute angle?
    F   G   H   I   J   None of these

- Items 33 through 36 describe angles. You'll mark the answer that best fits each description. Don't mark an answer for item 32. Work items 33 through 36.
- Raise your hand when you're finished. √
e. Check your work.
f. Item 33: Which angle shows an acute angle? (Signal.) *B.*
g. Item 34: Which angle shows a right angle? (Signal.) *H.*

h. Item 35: Which angle shows an obtuse angle? (Signal.) *A.*
i. Item 36: Which angle shows an acute angle? (Signal.) *J.*
- Choices F, G, H, and I are not acute angles, so the correct answer is J, none of these.

EXERCISE 7
Equivalent Fractions

> **Note:** Addition and subtraction of fractions with unlike denominators is taught in lessons 65–70 of *Connecting Math Concepts, Level E.* If presenting test preparation lessons prior to lesson 70, skip this exercise.

a. Find item 37. √
- Work items 37 through 45.
- Pencils down when you've marked answers through item 45.
- Remember, for items 41 through 43, you can add or subtract the whole numbers, and then add or subtract the fractions to figure out the answers the fast way.
- (Observe students and give feedback.)
b. Check your work.
c. Item 37. 3-ninths plus one-sixth.
- What letter did you mark? (Signal.) *A.*
  Yes, A. The answer is 9-eighteenths.
d. Item 38. 4-fifths plus 2-tenths.
- What letter did you mark? (Signal.) *J.*
  Yes, J. The answer is 10-tenths or one. Those were not choices, so the answer is J, none of these.
e. Item 39. 7-eighths minus 3-fourths.
- What letter did you mark? (Signal.) *C.*
  Yes, C. The answer is one-eighth.
f. Item 40. 12-ninths minus 5-sixths.
- What letter did you mark? (Signal.) *I.*
  Yes, I. The answer is 9-eighteenths.
g. Item 41. 4 and 7-eighths minus 3 and one-half.
- What letter did you mark? (Signal.) *B.*
  Yes, B. The answer is one and 3-eighths.
h. Item 42. 7 and 4-fifths minus 3 and 2-twentieths.
- What letter did you mark? (Signal.) *F.*
  Yes, F. The answer is 4 and 14-twentieths.
i. Item 43. 4 and 3-fourths plus 2 and one-sixth.
- What letter did you mark? (Signal.) *C.*
  Yes, C. The answer is 6 and 11-twelfths.
j. Item 44. What letter did you mark? (Signal.) *H.*
  Yes, H. The value for the underlined digit is 70 million.
k. Item 45. What letter did you mark? (Signal.) *A.*
  Yes, A. The value for the underlined digit is 3 billion.

EXERCISE 8
Test-Taking Rules
What to Do When You're Running Out of Time

a. Find the **Stop** sign at the end of lesson 7 in your **test booklet** and get ready to tell me the number for the last item. √
- Everybody, what's the number of the last item you'd mark? (Signal.) *54.*
b. Look at your answer sheet. What's the number of the first item you'd mark if there was only one minute remaining? (Signal.) *17.*
- Find the next item you'd mark. What's the number of the next item? (Signal.) *18.*
- What's the number of the next item you'd mark? (Signal.) *32.*
- What's the number of the next item you'd mark? (Signal.) *46.*
- What's the number of the last item you'd mark? (Signal.) *54.*
c. (Repeat step b until firm.)
d. Remember when you're running out of time to look in the **test booklet** and find out the number of the last item you'd mark for the part. Sometimes, you take more than one test, and you don't want to guess on answers to another test.
e. Here we go. There is only one minute left. Finish up your paper.
f. (Reinforce students who quickly fill in one and only one answer for items 17, 18, 32, and 46 through 54 on the answer sheet.)
- (Prompt students who don't fill in answers quickly to start marking the answers more quickly.)
- (Alert students who have more than one answer filled in on any problems to make sure that there's only one answer per problem.)
g. (Reinforce students who remember **not** to fill in 55 and 56.)
h. (After one minute, say:) Stop working.
i. Check your work.
- For the items you just marked, I'll tell you the make-believe answer. Next to each item, mark a C if you marked the correct answer and an X if you didn't mark that answer.
- Raise your hand after I say each answer if you guessed the right one. √
- Item 17: The answer is **B.** (Acknowledge students who raise their hands.)
- Item 18: The answer is **F.** (Acknowledge students who raise their hands.)
- Item 32: The answer is **F.** (Acknowledge students who raise their hands.)
- (Repeat for items: 46, I; 47, C; 48, F; 49, D; 50, J; 51, A; 52, I; 53, B; 54, F.)

j. What letter did you fill in for 55? (Call on a student. Idea: *No answer.*)
  Yes, you shouldn't have filled in any answers for items 55 and 56.
k. Count the Cs you wrote next to the numbers.
• Raise your hand if you got four or more of them correct. (Students respond.)
• Raise your hand if you got three of them correct. (Students respond.)
• (Repeat for: two, one, and zero of them correct.)
l. Most of you got one or more of the items correct. You wouldn't have gotten any of those problems correct if you hadn't fill in answers for those items.

Test Preparation

Lesson 1

rectangle	square	parallelogram	rhombus

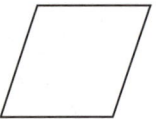

1

- **A** A square
- **B** A parallelogram
- **C** A rhombus
- **D** A rectangle
- **E** All of the above

2

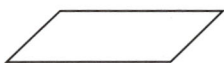

- **F** A parallelogram
- **G** A rhombus
- **H** A rectangle
- **I** A parallelogram and a rhombus
- **J** A parallelogram and a rectangle

3

- **A** A parallelogram
- **B** A square
- **C** A rhombus
- **D** A rectangle
- **E** All of the above

4

- **F** A parallelogram
- **G** A rhombus
- **H** A rectangle
- **I** All of the above
- **J** None of the above

5

- **A** A square and a parallelogram
- **B** A parallelogram and a rectangle
- **C** A rhombus and a parallelogram
- **D** A rectangle and a rhombus
- **E** All of the above

6

- **F** A rhombus
- **G** A parallelogram
- **H** A rectangle
- **I** A square
- **J** None of the above

7

- **A** A square
- **B** A parallelogram and a rectangle
- **C** A rhombus
- **D** A rectangle
- **E** All of the above

9 $\dfrac{17}{5} =$

$2\dfrac{7}{5}$	$3\dfrac{2}{5}$	$4\dfrac{3}{5}$	$3\dfrac{5}{2}$	None of these
A	**B**	**C**	**D**	**E**

10 $\dfrac{15}{2} =$

$14\dfrac{1}{2}$	$8\dfrac{1}{2}$	$6\dfrac{2}{3}$	$7\dfrac{1}{2}$	None of these
F	**G**	**H**	**I**	**J**

11 $\dfrac{9}{5} =$

$0\dfrac{9}{5}$	$1\dfrac{5}{4}$	$1\dfrac{4}{5}$	$2\dfrac{1}{5}$	None of these
A	**B**	**C**	**D**	**E**

12 $\dfrac{39}{4} =$

$9\dfrac{3}{4}$	$9\dfrac{3}{9}$	$6\dfrac{3}{4}$	$6\dfrac{3}{9}$	None of these
F	**G**	**H**	**I**	**J**

13 $\dfrac{77}{9} =$

$8\dfrac{5}{9}$	$8\dfrac{5}{7}$	$7\dfrac{14}{6}$	$7\dfrac{12}{7}$	None of these
A	**B**	**C**	**D**	**E**

14 $\dfrac{26}{8} =$

$2\dfrac{10}{8}$	$4\dfrac{6}{8}$	$3\dfrac{1}{2}$	$3\dfrac{2}{8}$	None of these
F	**G**	**H**	**I**	**J**

15 $\dfrac{40}{7} =$

$6\dfrac{2}{7}$	$7\dfrac{4}{5}$	$5\dfrac{5}{7}$	$5\dfrac{1}{2}$	None of these
A	**B**	**C**	**D**	**E**

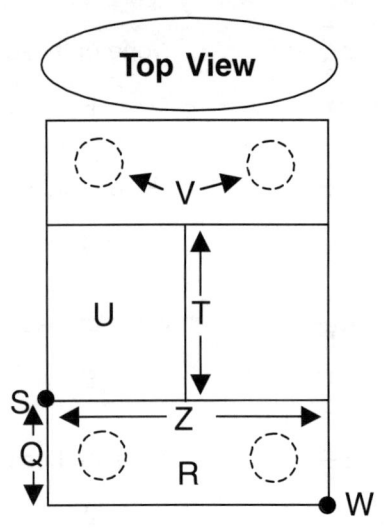

Top View

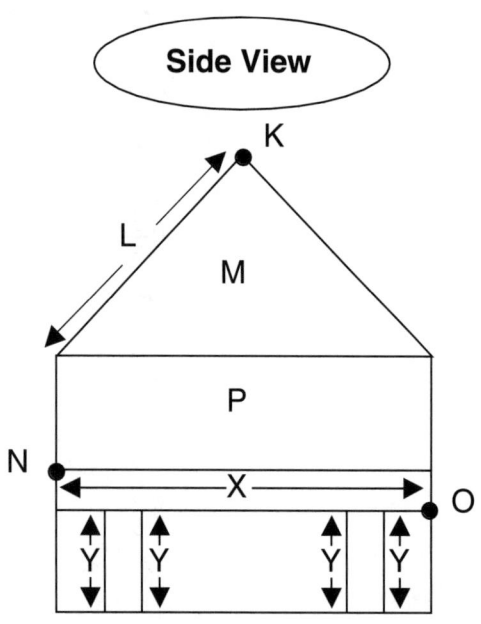

Side View

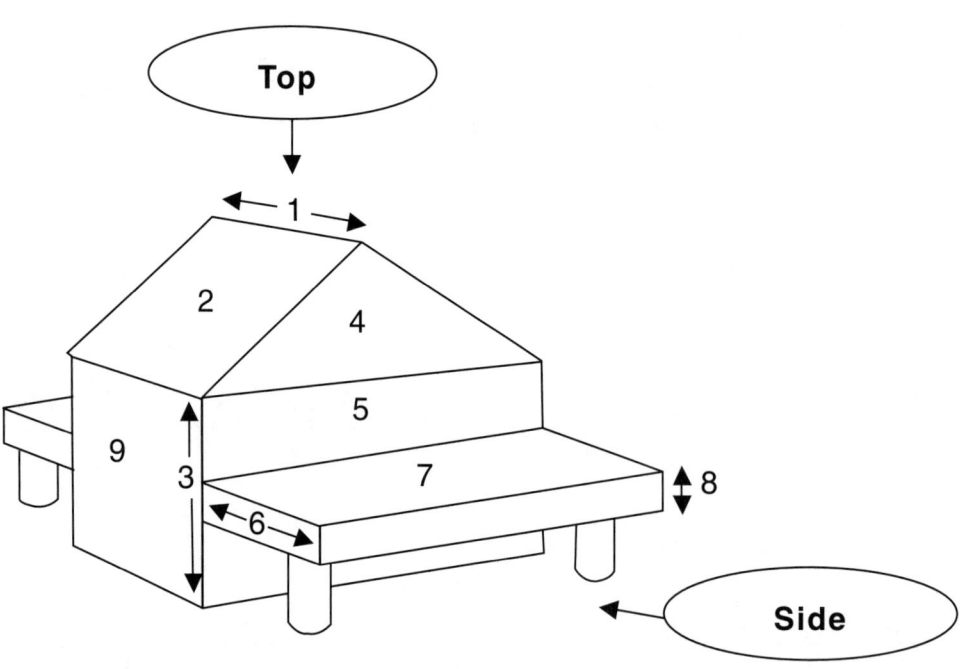

Top

Side

17 Which letter in the top view represents the roof?

A Letter R
B Letter V
C Letter Q
D Letter U
E None of the above

18 Which letter in the side view represents the roof?

F Letter L
G Letter M
H Letter K
I Letter X
J None of the above

19 Which letter in the side view represents the part of the object labeled 4?

A Letter L
B Letter M
C Letter K
D Letter P
E None of the above

20 Which letter in the top view represents the part of the object labeled 4?

F Letter R
G Letter T
H Letter Z
I Letter U
J Letter Q

21 Which letter in the top view represents the corner of the porch roof labeled 8?

A Letter S
B Letter W
C Letter R
D Letter U
E None of the above

22 Which letter in the side view represents the edge of the porch roof labeled 6?

F Point N
G Line X
H Part P
I Line L
J None of the above

23 The part for letter R in the top view represents one of the numbered parts of the object. What number?

A 5
B 4
C 7
D 6
E None of the above

24 The point for the letter K in the side view represents one of the numbered parts of the object. What number?

F 2
G 3
H 4
I 7
J None of the above

25 The line for letter X in the side view represents one of the numbered parts of the object. What is the best description of that part?

A The pillar
B The roof of the house
C The porch roof
D The corner of the roof
E None of the above

26 The dotted lines for letter V in the top view represent parts of the object. What is the best description of those parts?

F The porch roof
G The pillars holding up the porch roof
H The corner of the roof
I 7
J None of the above

Test Preparation

Lesson 2

rectangle	parallelogram	square	rhombus	trapezoid
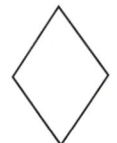				

1

A   A square
B   A parallelogram
C   A rhombus and a square
D   A rhombus and a parallelogram
E   All of the above

2

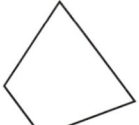

F   A parallelogram
G   A rhombus
H   A trapezoid
I   A parallelogram and a rhombus
J   A parallelogram and a trapezoid

3

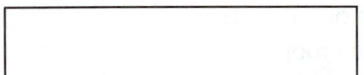

A   A trapezoid
B   A square
C   A rectangle
D   A rhombus
E   All of the above

4

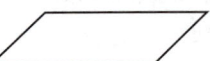

F   A parallelogram
G   A rhombus
H   A trapezoid
I   All of the above
J   None of the above

5

A   A square, a parallelogram, a trapezoid
B   A square, parallelogram, a rectangle
C   A rhombus, a square, a trapezoid
D   A rectangle, a trapezoid, a rhombus
E   All of the above

6

F   A rhombus and a trapezoid
G   A parallelogram and a square
H   A rhombus
I   A square
J   None of the above

7

A   A square and a parallelogram
B   A parallelogram and a rectangle
C   A rhombus and a parallelogram
D   A rectangle and a rhombus
E   All of the above

8

F   A parallelogram and a rectangle
G   A rhombus and a parallelogram
H   A rectangle and a trapezoid
I   A square and a rectangle
J   All of the above

11 $\boxed{\quad ? \quad} = M^4$

A M + M + M + M
B M × 4
C M × M × M × M
D M × M × M
E None of the above

12 $7 \times 7 \times 7 = \boxed{\quad ? \quad}$

F 3 × 7
G 7^3
H 7^7
I 7^1
J None of the above

13 $100 \times 100 \times 100 \times 100 \times 100 = \boxed{\quad ? \quad}$

A 5 × 100
B 100^{10}
C 100^1
D 100^5
E None of the above

14 $\boxed{\quad ? \quad} = 37^2$

F 37 × 37
G 37 × 2
H 37 × 37 × 37 × 37
I 37 + 37
J None of the above

15 $\boxed{\quad ? \quad} = (2T)^3$

A 2T + 3
B 2T × 3
C 2T × 2T × 2T
D 2T + 2T + 2T
E None of the above

Sample: Here are five amounts:

$5　　$6　　$2　　$12　　$10

They show how much Jimmy earned on five days.

$$\begin{array}{r} \$ \ 5 \\ 6 \\ 2 \\ 12 \\ + \ 10 \\ \hline \$ \ 35 \end{array} \qquad \begin{array}{r} \$7 \\ \hline 5)\overline{35} \end{array}$$

17 The high temperature on four days was 11 degrees, 15 degrees, 10 degrees, and 12 degrees. What was the average high temperature for the four-day period?

A 12 days
B 8 degrees
C 12 degrees
D 4 days
E None of the above

18 The table shows how far a runner ran on Monday through Friday. Figure out the average distance the runner ran each day.

Monday	Tuesday	Wednesday	Thursday	Friday
6 miles	7miles	10 miles	7 miles	10 miles

F 7 miles
G 8 miles
H 8 days
I 15 miles
J None of the above

19 The table shows the number of people in 8 cars that were on the highway. Find the average number of people in a car.

A 24 people
B 8 cars
C 3 people
D 4 people
E None of the above

Car	Number of People
A	3
B	4
C	1
D	1
E	5
F	3
F	3
H	4

 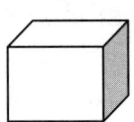  

Study the objects. Then answer items 21 through 25.

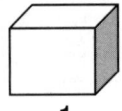    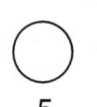

1   2   3   4   5   6

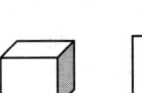   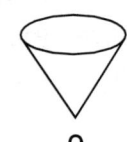   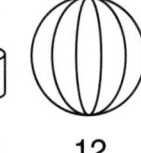

7   8   9   10   11   12

21 Which number shows the small sphere?

5	2	6	3	4
A	B	C	D	E

22 Which number shows the large cylinder?

5	2	6	3	5
F	G	H	I	J

23 Which number shows the large cube?

1	2	7	10	8
A	B	C	D	E

24 Which number shows the large cone?

2	6	10	9	3
F	G	H	I	J

25 Which number shows the large sphere?

10	12	5	9	6
A	B	C	D	E

27 $8\frac{2}{5} =$

$\frac{10}{5}$	$\frac{47}{5}$	$\frac{15}{5}$	$\frac{13}{2}$	None of these
A	B	C	D	E

28 $3\frac{2}{9} =$

$\frac{14}{9}$	$\frac{29}{9}$	$\frac{29}{2}$	$\frac{12}{2}$	None of these
F	G	H	I	J

29 $2\frac{1}{3} =$

$\frac{6}{3}$	$\frac{5}{1}$	$\frac{7}{3}$	$\frac{7}{2}$	None of these
A	B	C	D	E

30 $10\frac{3}{4} =$

$\frac{43}{4}$	$\frac{17}{4}$	$\frac{14}{4}$	$\frac{14}{3}$	None of these
F	G	H	I	J

31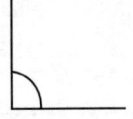

A Right angle
B Acute angle
C Obtuse angle

32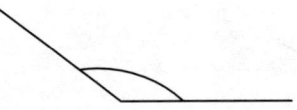

F Right angle
G Acute angle
H Obtuse angle

33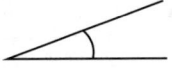

A Right angle
B Acute angle
C Obtuse angle

34

F Right angle
G Acute angle
H Obtuse angle

35

A Right angle
B Acute angle
C Obtuse angle

36

F Right angle
G Acute angle
H Obtuse angle

37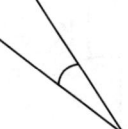

A Right angle
B Acute angle
C Obtuse angle

38

F Right angle
G Acute angle
H Obtuse angle

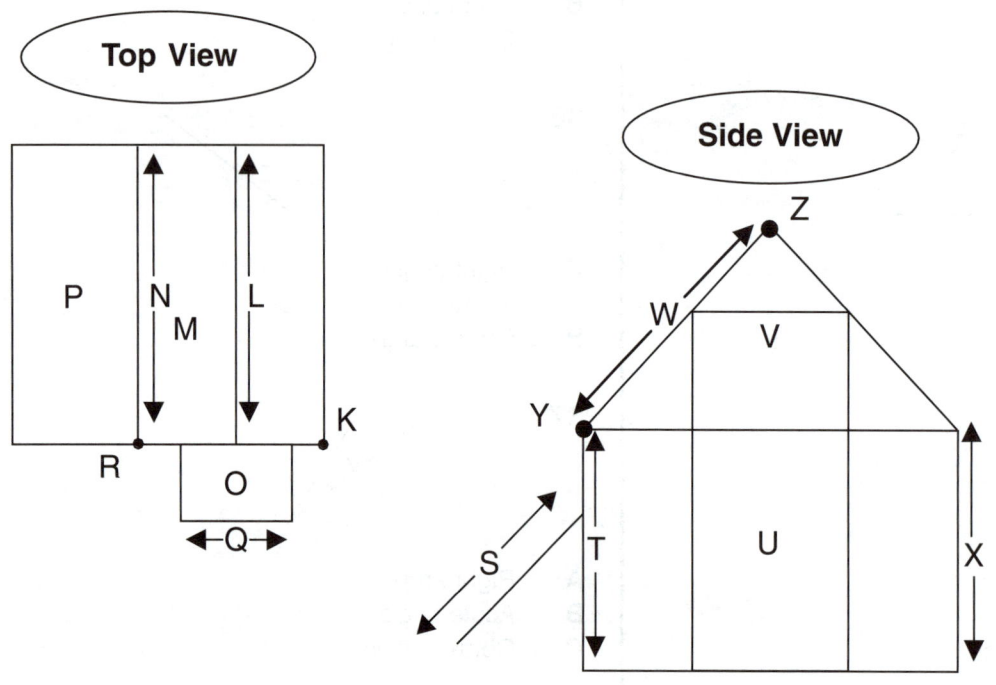

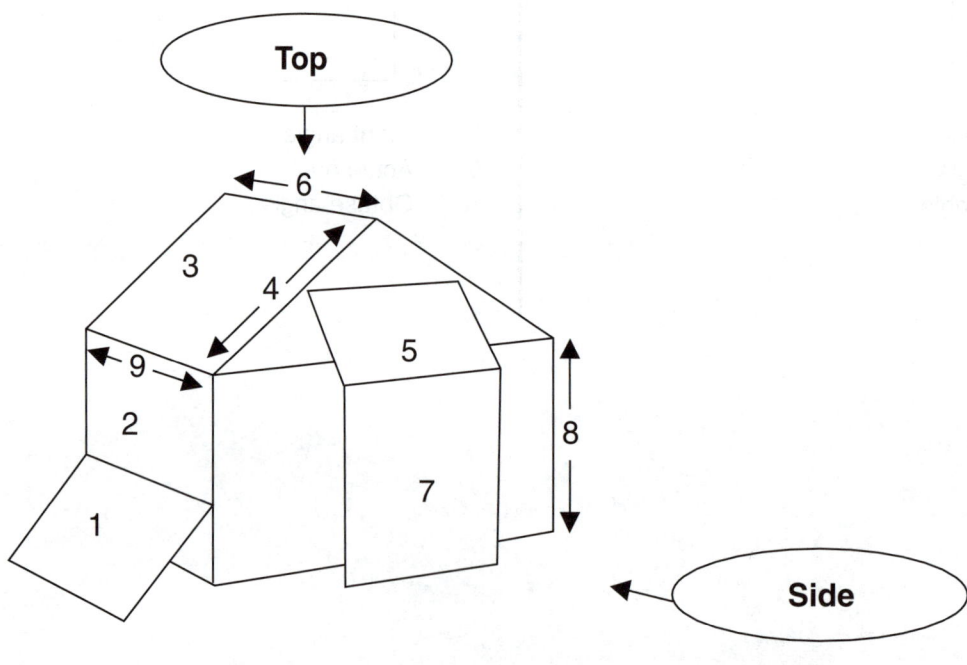

1 Which letter in the top view represents the roof (3)?

A Part M
B Part B
C Line N
D Line L
E None of the above

2 Which letter in the side view represents the roof line (6)?

F Line W
G Part V
H Point Y
I Point Z
J None of the above

3 Which letter in the side view represents the part of the object labeled 2?

A Point Y
B Line S
C Line T
D Line W
E None of the above

4 Which letter in the top view represents the line of the object labeled 8?

F Line W
G Part V
H Point Y
I Point K
J None of the above

5 Which letter in the side view represents the lean-to labeled 1?

A Line S
B Line 2
C Part U
D Line W
E None of the above

6 Which letter in the top view represents the object labeled 7?

F Part O
G Line Q
H Point K
I Point R
J None of the above

7 The line for letter X in the side view represents a numbered part of the object. What number?

A Part 3
B Line 6
C Part 7
D Line 9
E None of the above

8 The point for letter Y in the side view represents a numbered part of the object. What number?

F Line 8
G Part 2
H Line 9
I Part 1
J None of the above

9 The line for letter W in the side view represents a numbered part of the object. What is the best description of that part?

A The lean-to
B The roof of the barn
C The wall
D The line where the walls meet
E None of the above

10 The line for letter L in the top view represents a part of the object. What is the best description of that part?

F The roof line
G The lean-to
H The line where the roof and wall meet
I The corner of two walls
J None of the above

rectangle	parallelogram	trapezoid	square	rhombus

12

- **F** A square
- **G** A parallelogram
- **H** A rhombus, a square
- **I** A rectangle, a trapezoid
- **J** All of the above

13

- **A** A square, a parallelogram, a rectangle, a rhombus
- **B** A rectangle, a trapezoid, a square
- **C** A parallelogram, a rhombus, a trapezoid
- **D** A trapezoid, a square, a parallelogram, a rectangle
- **E** None of the above

14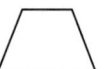

- **F** A rectangle
- **G** A rhombus
- **H** A parallelogram
- **I** A square
- **J** None of the above

15

- **A** A parallelogram
- **B** A rhombus
- **C** A parallelogram, a rhombus
- **D** A square, a rectangle, a rhombus
- **E** None of the above

16

- **F** A parallelogram, a rectangle
- **G** A square, a parallelogram, a rectangle
- **H** A rhombus, a rectangle
- **I** A trapezoid, a rhombus

17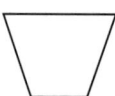

- **A** A parallelogram and a square
- **B** A rhombus and a trapezoid
- **C** A square
- **D** A trapezoid
- **E** None of the above

18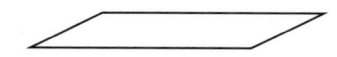

- **F** A square, a parallelogram
- **G** A parallelogram, a trapezoid
- **H** A parallelogram
- **I** A rectangle, a rhombus
- **J** All of the above

19

- **A** A parallelogram, a rectangle
- **B** A rhombus, a parallelogram
- **C** A rectangle, a trapezoid
- **D** A square, a rectangle

20

- **F** A square, a parallelogram
- **G** A parallelogram, a trapezoid
- **H** A square, a parallelogram, a rhombus
- **I** A parallelogram, a rhombus

21 $594 \times 3 =$

    A $(500 + 3) \times (90 + 3) \times (4 + 3)$
    B $(500 \times 3) + (90 \times 3) + (4 \times 3)$
    C $(500 \times 3) \times (90 \times 3) \times (4 \times 3)$
    D $(500 \times 90) + (4 \times 3)$
    E   None of the above

22 $2061 \times 7 =$

    F $(2000 \times 7) \times (60 \times 7) \times (1 \times 7)$
    G $(2000 + 7) \times (60 + 7) \times (1 + 7)$
    H $(2000 \times 60) + (1 \times 7)$
    I $(2000 \times 7) + (60 \times 7) + (1 \times 7)$
    J   None of the above

23 $829 \times 6 =$

    A $(800 \times 6) + (20 \times 6) + (9 \times 6)$
    B $(800 \times 6) \times (20 \times 6) \times (9 \times 6)$
    C $(800 \times 20) + (9 \times 6)$
    D $(800 + 6) \times (20 + 6) \times (9 + 3)$
    E   None of the above

24 $8 \times 197 =$

    F $(8 \times 100) + (90 \times 7)$
    G $(8 \times 100) \times (8 \times 90) \times (8 \times 7)$
    H $(8 \times 100) + (8 \times 90) + (8 \times 7)$
    I $(8 + 100) \times (8 + 90) \times (8 + 7)$
    J   None of the above

25 $4 \times 3805 =$

    A $(4 \times 3000) + (800 \times 5)$
    B $(4 \times 3000) + (4 \times 800) + (4 \times 5)$
    C $(4 + 3000) \times (4 + 800) \times (4 + 5)$
    D $(4 \times 3000) \times (4 \times 800) \times (4 \times 5)$
    E   None of the above

27 $\dfrac{21}{4} =$

    $4\dfrac{1}{4}$ $4\dfrac{1}{5}$ $5\dfrac{1}{4}$ $5\dfrac{1}{5}$ None of these
    A   **B**   **C**   **D**   **E**

28 $\dfrac{32}{5} =$

    $6\dfrac{2}{2}$ $5\dfrac{2}{5}$ $5\dfrac{1}{5}$ $6\dfrac{2}{5}$ None of these
    F   **G**   **H**   **I**   **J**

29 $\dfrac{77}{9} =$

    $8\dfrac{5}{8}$ $9\dfrac{7}{9}$ $8\dfrac{5}{9}$ $9\dfrac{9}{5}$ None of these
    A   **B**   **C**   **D**   **E**

30 $\dfrac{30}{8} =$

    $3\dfrac{6}{8}$ $4\dfrac{0}{8}$ $4\dfrac{5}{8}$ $3\dfrac{3}{8}$ None of these
    F   **G**   **H**   **I**   **J**

31 $\dfrac{16}{3} =$

    $3\dfrac{2}{16}$ $5\dfrac{1}{3}$ $4\dfrac{2}{3}$ $16\dfrac{1}{3}$ None of these
    A   **B**   **C**   **D**   **E**

Study the objects. Then answer items 33 through 38.

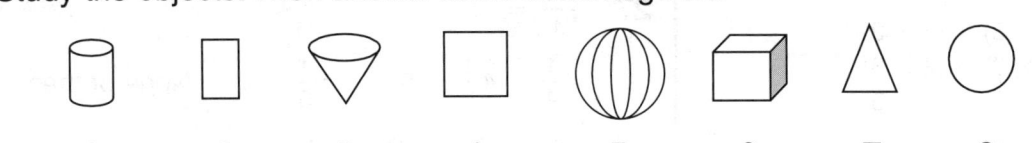

1   2   3   4   5   6   7   8

33 Which number shows the cone?

7   6   12   1   3
A   B   C   D   E

34 Which number shows the cube?

1   2   6   4   None of these
F   G   H   I   J

35 Which number shows the object after the cylinder?

8   7   2   1   12
A   B   C   D   E

36 What is the name of the object between the cone and the sphere?

rectangle   square   cone   cube   cylinder
F   G   H   I   J

37 Which number shows the object before the cone?

1   2   4   3   5
A   B   C   D   E

38 Which is the name of the object before the cube?

sphere   circle   cone   cylinder   None of these
F   G   H   I   J

Test Preparation

Lesson 4

1 $3 \times (500 + 40 + 7) =$

   A 216,290

   B 3,780,000

   C 1780

   D 5400

   E None of the above

2 $(6 + R + 8) \times 11 =$

   F $17 + 11R + 19$

   G $6 + R + 88$

   H $66 + 11R + 88$

   I $(6 + R) \times 19$

   J None of the above

3 $5 \times (2000 + 700 + 2) =$

   A 5×2702

   B $(5 \times 2000) + (5 \times 700) + (5 \times 2)$

   C $10,000 + 3500 + 10$

   D All of the above

   E None of the above

4 $8 \times (3R + 7 + T) =$

   F $24R + 15 + 8T$

   G $24R + 56 + 8T$

   H $(8 + 3R) \times (7 + T)$

   I $248RT$

   J None of the above

5 $(3Q + 5 - 49) \times 2 =$

   A $6Q + 7 - 47$

   B $6Q + 10 - 47$

   C $6Q + 10 - 98$

   D $(3Q + 5) \times (2 - 49)$

   E $15Q - 98$

Equilateral	Isosceles	Scalene

7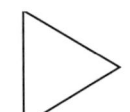

   A Equilateral
   B Isosceles
   C Scalene

8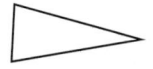

   F Equilateral
   G Isosceles
   H Scalene

9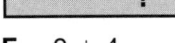

   A Equilateral
   B Isosceles
   C Scalene

10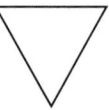

   F Equilateral
   G Isosceles
   H Scalene

11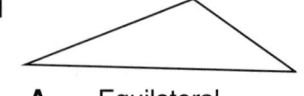

   A Equilateral
   B Isosceles
   C Scalene

12

   F Equilateral
   G Isosceles
   H Scalene

13 $5 \times 5 \times 5 =$ [**?**]

   A 3×5
   B $5 + 3$
   C 5^3
   D 3^5
   E None of the above

14 [**?**] $= 9^4$

   F $9 + 4$
   G $9 + 9 + 9 + 9$
   H $9 \times 9 \times 9 \times 9$
   I 9×4
   J None of the above

15 [**?**] $= 2^6$

   A $2 \times 2 \times 2 \times 2 \times 2 \times 2$
   B 2×6
   C $2 + 6$
   D $2 \times 2 \times 2 \times 2 \times 2$
   E 6×6

16 $8 \times 8 \times 8 =$ [**?**]

   F 3^8
   G 3×8
   H $8 + 3$
   I 8^3
   J None of the above

17 [**?**] $= R^2$

   A $2 \times R$
   B $R \times R$
   C $R + 2$
   D $R \times R \times R$
   E None of the above

18 $43 \times 43 \times 43 \times 43 \times 43 =$ [**?**]

   F 5^{43}
   G $43 + 5$
   H 5×43
   I 43^5
   J None of the above

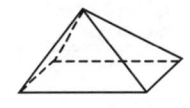 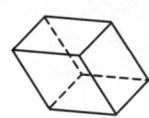 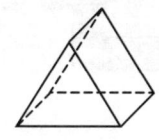

Study the objects. Then answer items 19 through 23.

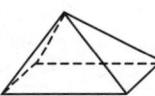 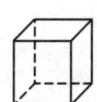 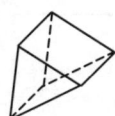 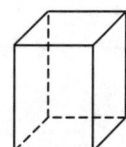 

19 Which figure is a rectangular prism?

1	3	4	2	None of these
A	**B**	**C**	**D**	**E**

20 Which figure is a cube?

1	2	3	4	None of these
F	**G**	**H**	**I**	**J**

21 Which figure is a pyramid?

1	2	3	4	None of these
A	**B**	**C**	**D**	**E**

22 Which figure is a triangular prism?

1	4	2	3	None of these
F	**G**	**H**	**I**	**J**

23 Which figure is a sphere?

2	3	5	1	None of these
A	**B**	**C**	**D**	**E**

25 $3\frac{2}{5} =$

$\frac{17}{2}$	$\frac{10}{5}$	$\frac{13}{2}$	$\frac{17}{5}$	None of these
A	**B**	**C**	**D**	**E**

26 $7\frac{1}{9} =$

$\frac{64}{9}$	$\frac{64}{8}$	$\frac{17}{8}$	$\frac{73}{8}$	None of these
F	**G**	**H**	**I**	**J**

27 $2\frac{4}{5} =$

$\frac{14}{4}$	$\frac{14}{5}$	$\frac{11}{5}$	$\frac{22}{5}$	None of these
A	**B**	**C**	**D**	**E**

28 $6\frac{1}{2} =$

$\frac{13}{3}$	$\frac{13}{2}$	$\frac{8}{2}$	$\frac{9}{2}$	None of these
F	**G**	**H**	**I**	**J**

29 $1\frac{9}{10} =$

$\frac{91}{10}$	$\frac{91}{9}$	$\frac{19}{10}$	$\frac{20}{10}$	None of these
A	**B**	**C**	**D**	**E**

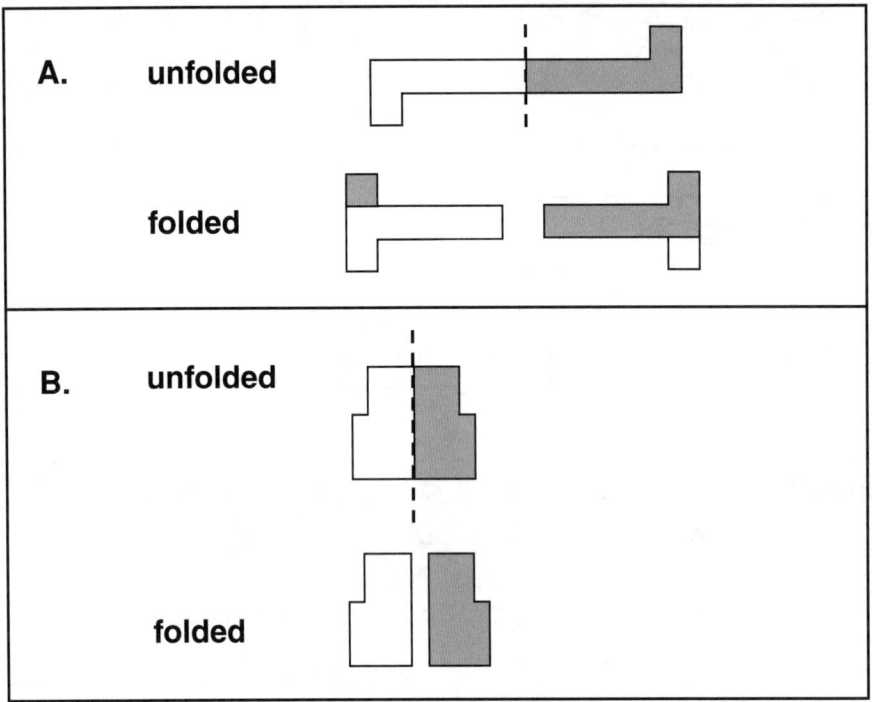

31 Which of these figures can be folded along the dotted line so that the parts match?

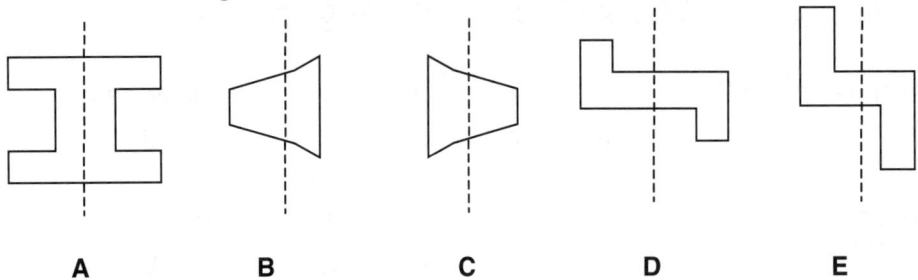

32 Which of these figures can be folded along the dotted line so that the parts match?

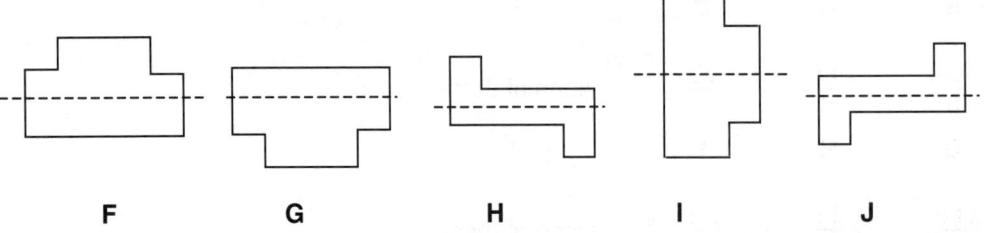

33 Which of these figures cannot be folded along the dotted line so that the parts match?

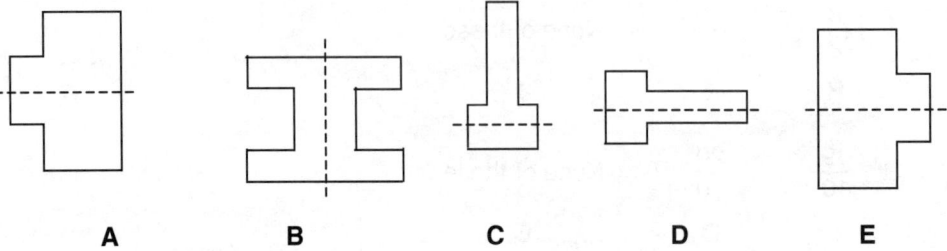

34 Which of these figures cannot be folded along the dotted line so that the parts match?

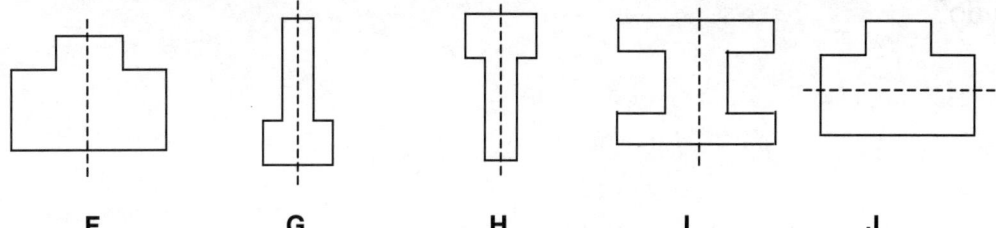

F   G   H   I   J

35 Which of these figures can be folded along the dotted line so that the parts match?

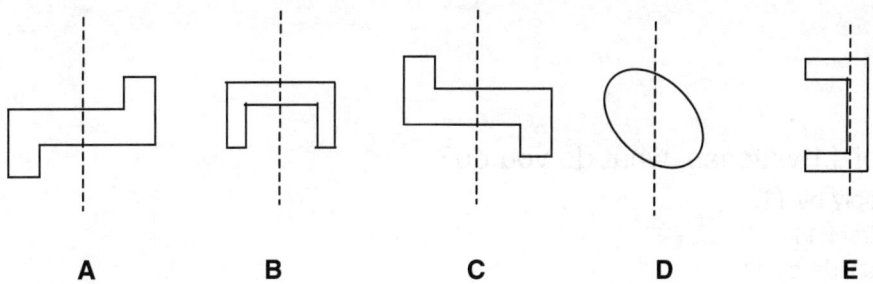

A   B   C   D   E

36 Which of the figures does have a line of symmetry?

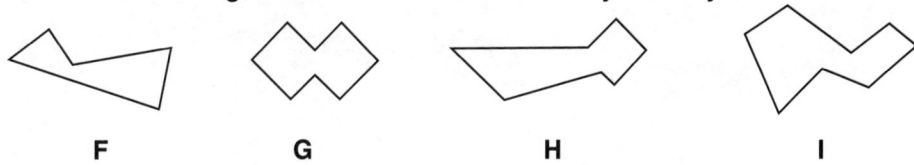

F   G   H   I

37 Which of the figures does **not** have a line of symmetry?

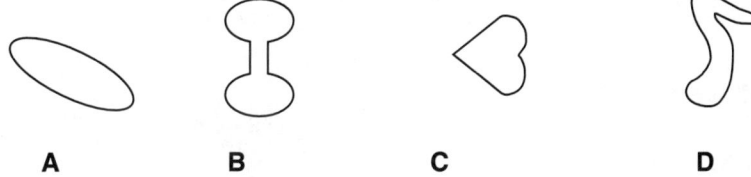

A   B   C   D

38 Which figure has a line of symmetry?

F   G   H   I

39 Which figure does **not** have a line of symmetry?

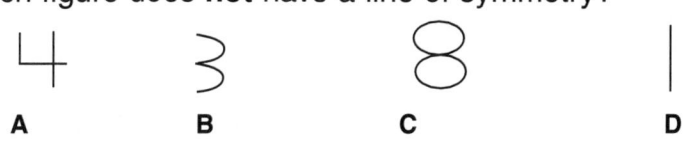

A   B   C   D

41 To find the average score that 6 students got on a test, you add up all the scores. Then what do you do?

A Add 6.

B Subtract 6.

C Divide by 6.

D Multiply by 6.

42 To find the average height of 9 students, you add up all the heights. Then what do you do?

F Multiply by 9.

G Divide by 9.

H Add 9.

I Subtract 9.

43 To find the average price of 11 watches, what do you do?

A Add all the prices and multiply by 11.

B Add all the prices and subtract 11.

C Multiply all the prices and divide by 11.

D Add all the prices and divide by 11.

Test Preparation

Lesson 5

Use the figures below to answer items 2 through 6.

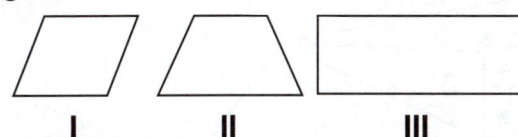

I   II   III

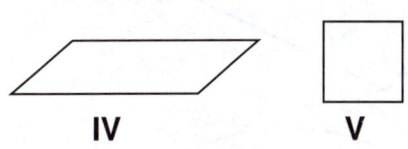

IV   V

2 Which figures show a rectangle?

F   Figures I and II
G   Figures III and IV
H   Figures II and V
I   Figures III and V
J   None of the above

3 Which figure shows a trapezoid?

A   Figure I
B   Figure II
C   Figure III
D   Figure IV
E   None of the above

4 Which figure shows a parallelogram?

F   Figure I
G   Figure III
H   Figure IV
I   Figure V
J   All of the above

5 Which figure(s) show a rhombus?

A   Figures I and IV
B   Figures I, III, and V
C   Figures I and V
D   Figure IV
E   None of the above

6 Which figure shows a square?

F   Figure I
G   Figure III
H   Figure IV
I   Figure V
J   None of the above

7 $\dfrac{10}{12} =$

$\dfrac{7}{12}$	$\dfrac{5}{6}$	$\dfrac{3}{4}$	$\dfrac{4}{6}$	None of these
A	B	C	D	E

8 $\dfrac{4}{6} =$

$\dfrac{4}{3}$	$\dfrac{2}{4}$	$\dfrac{1}{2}$	$\dfrac{2}{3}$	None of these
F	G	H	I	J

9 $\dfrac{20}{25} =$

$\dfrac{4}{5}$	$\dfrac{5}{25}$	$\dfrac{5}{10}$	$\dfrac{1}{25}$	None of these
A	B	C	D	E

10 $\dfrac{15}{24} =$

$\dfrac{3}{8}$	$\dfrac{5}{8}$	$\dfrac{3}{6}$	$\dfrac{5}{6}$	None of these
F	G	H	I	J

11 $\dfrac{12}{20} =$

$\dfrac{6}{12}$	$\dfrac{1}{2}$	$\dfrac{3}{4}$	$\dfrac{3}{5}$	None of these
A	B	C	D	E

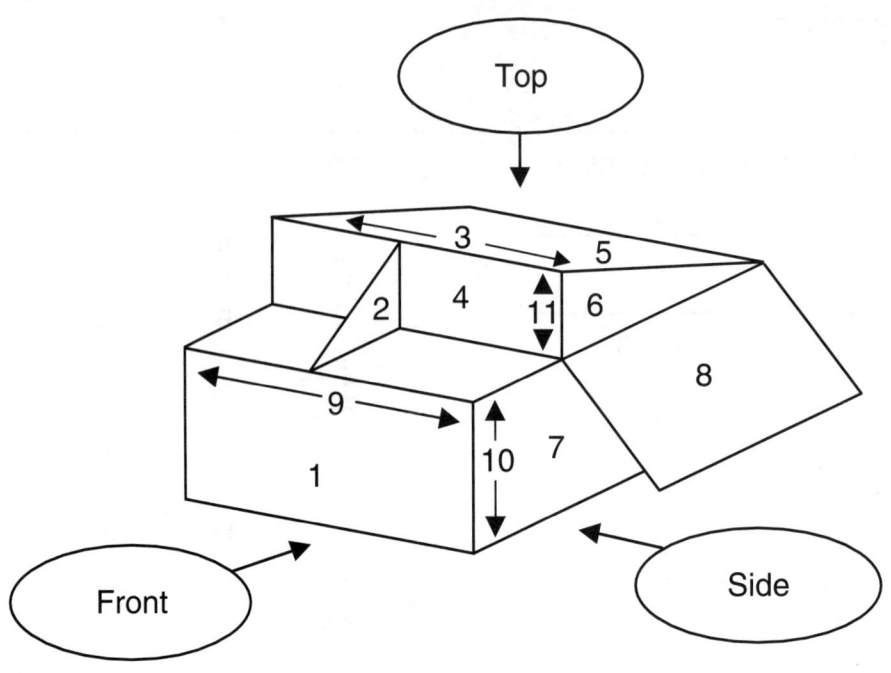

13 Which is the front view of the object?

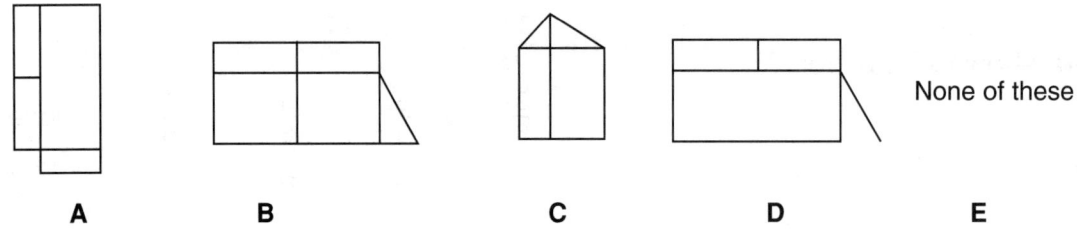

A	B	C	D	E
				None of these

14 Which is the top view of the object?

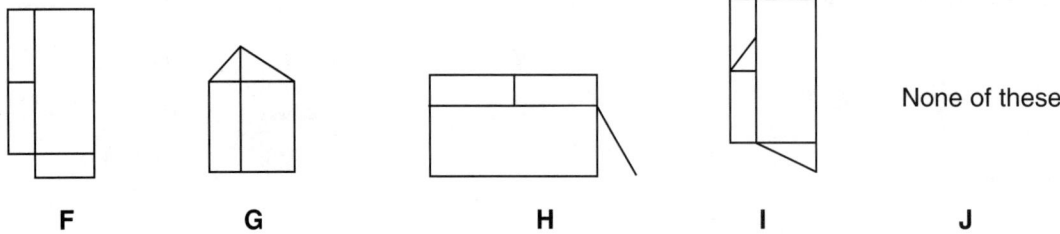

F	G	H	I	J
				None of these

15 Which is the side view of the object?

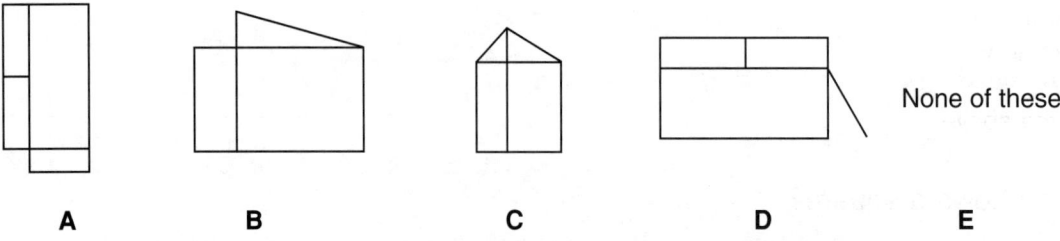

A	B	C	D	E
				None of these

For items 16–22, each item will show a drawing of the front view, the top view, or the side view. A portion of each view will be highlighted. The highlighted part represents one or more than one of the numbered parts of the object.

19

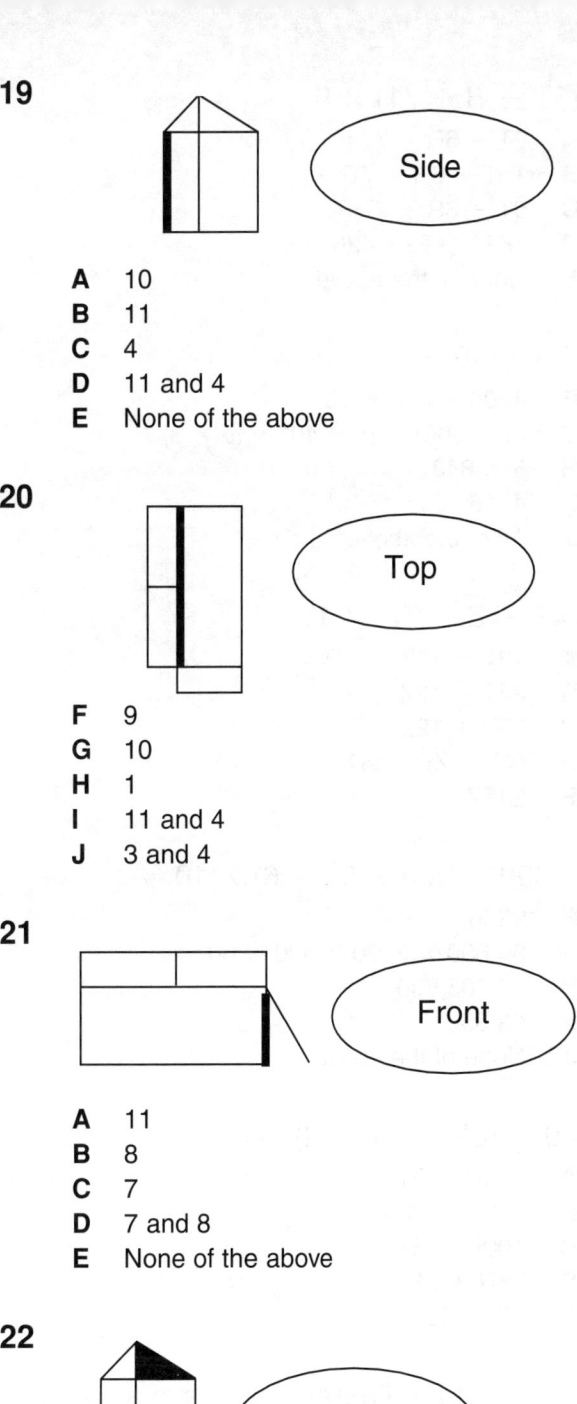

A 10
B 11
C 4
D 11 and 4
E None of the above

16

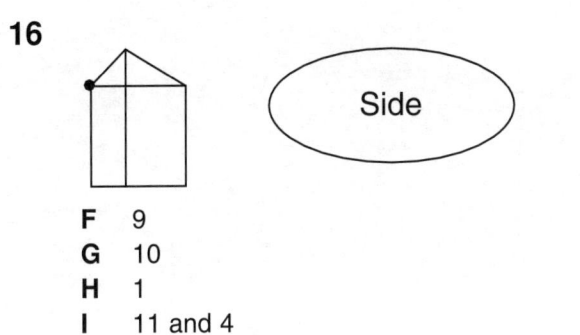

F 9
G 10
H 1
I 11 and 4
J 3 and 4

17

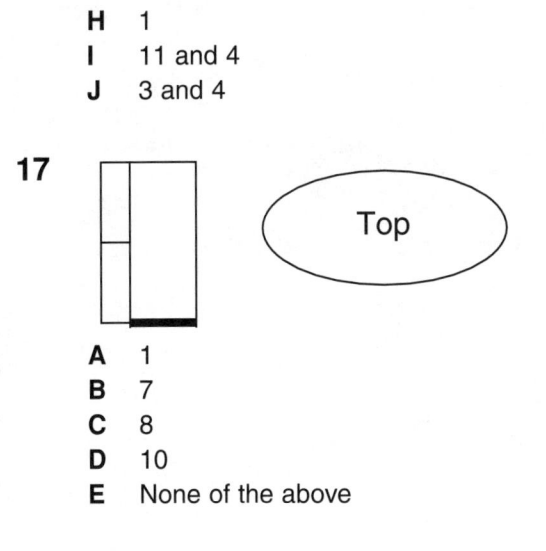

A 1
B 7
C 8
D 10
E None of the above

20

F 9
G 10
H 1
I 11 and 4
J 3 and 4

21

A 11
B 8
C 7
D 7 and 8
E None of the above

18

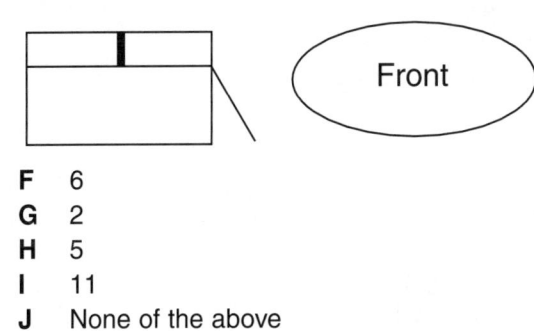

F 6
G 2
H 5
I 11
J None of the above

22

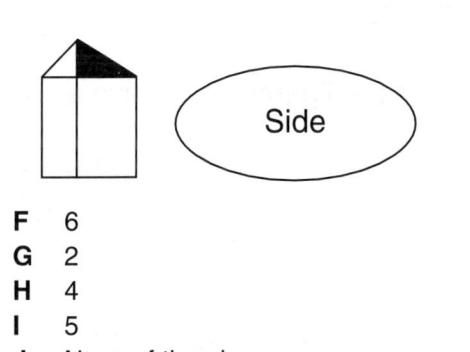

F 6
G 2
H 4
I 5
J None of the above

23 $(3T - R + 71) \times 6 =$

    A $9T - 6R + 77$

    B $18T - 6R + 77$

    C $9T - 6R + 77$

    D $18T - 6R + 426$

    E   None of the above

24 $5 \times (800 + 40 + 3) =$

    F $4000 + 200 + 15$

    G $(5 \times 800) + (5 \times 40) + (5 \times 3)$

    H 5×843

    I 4215

    J   All of the above

25 $(4T + Z - 2T) \times 12 =$

    A $12T + 12Z - 24T$

    B $24T + 12Z$

    C $38T + 12Z$

    D $(4T + Z) - 24T$

    E $24TZ$

26 $(9000 + 200 + 50 + 6) \times 10 =$

    F 9266

    G $90,000 \times 2000 \times 500 \times 60$

    H $90,205,600$

    I $92,560$

    J   None of the above

27 $20 \times (6R + 11 - R) =$

    A $100R + 31$

    B $119R + 220$

    C $100R + 220$

    D $25R + 31$

    E $6R + 220$

28 Which of the figures does have a line of symmetry?

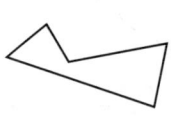  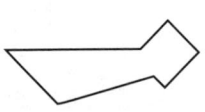 

    F   **G**   **H**   **I**

29 Which of the figures does **not** have a line of symmetry?

 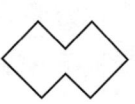 

    A   **B**   **C**   **D**

31 $\dfrac{1}{3} + \dfrac{5}{6} =$

$\dfrac{6}{9}$	1	$1\dfrac{1}{12}$	$1\dfrac{1}{6}$	None of these
A	**B**	**C**	**D**	**E**

32 $\dfrac{3}{8} + \dfrac{3}{4} =$

1	$1\dfrac{1}{8}$	$\dfrac{6}{12}$	$\dfrac{6}{8}$	None of these
F	**G**	**H**	**I**	**J**

33 $\dfrac{3}{10}$

$+\dfrac{3}{15}$

$\dfrac{15}{30}$	$\dfrac{70}{150}$	$\dfrac{3}{5}$	$\dfrac{6}{25}$	None of these
A	**B**	**C**	**D**	**E**

34 $\dfrac{1}{4}$

$+\dfrac{1}{10}$

$\dfrac{3}{10}$	$\dfrac{2}{14}$	$\dfrac{7}{20}$	$\dfrac{1}{2}$	None of these
F	**G**	**H**	**I**	**J**

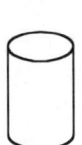  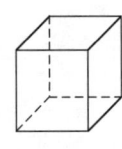  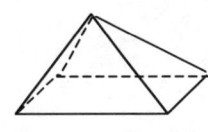 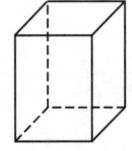 

1　　**2**　　**3**　　**4**　　**5**　　**6**　　**7**

35 Which figure is a rectangular prism?

1	3	6	7	None of these
A	**B**	**C**	**D**	**E**

36 Which figure is a cube?

1	5	6	7	3
F	**G**	**H**	**I**	**J**

37 Which figure is a triangular prism?

7	6	5	4	None of these
A	**B**	**C**	**D**	**E**

38 Which figure is a cylinder?

2	4	6	1	7
F	**G**	**H**	**I**	**J**

39 Which figure is a sphere?

2	1	4	5	None of these
A	**B**	**C**	**D**	**E**

40 Which figure is a pyramid?

4	5	6	7	None of these
F	**G**	**H**	**I**	**J**

41 Which figure is a cone?

1	2	4	5	7
A	**B**	**C**	**D**	**E**

43 $3^4 =$

    A   12
    B   81
    C   27
    D   15
    E   None of the above

44 $2^5 =$?

    F   54
    G   10
    H   16
    I   32
    J   None of the above

45 $10^3 =$?

    A   1000
    B   100
    C   30
    D   10,000
    E   None of the above

46 $1^9 =$?

    F   10
    G   9
    H   0
    I   19
    J   None of the above

47
48
49
50
51
52
53
54

STOP

Test Preparation

Lesson 6

1 $2M + 9 \times (4 + M) = 47$ | M = ? |
- **A** 1
- **B** $\dfrac{11}{3}$
- **C** 3
- **D** 11
- **E** None of the above

2 $(P - 5) \times 7 - 2P = 15$ | P = ? |
- **F** $\dfrac{27}{5}$
- **G** 20
- **H** 10
- **I** 4
- **J** None of the above

3 $(3T + 10) \times 2 + 8 = 58$ | T = ? |
- **A** $\dfrac{86}{6}$
- **B** $\dfrac{43}{3}$
- **C** 10
- **D** 5
- **E** None of the above

4 $7 \times (4 + Q) - 5Q = 40$ | Q = ? |
- **F** -3
- **G** 6
- **H** $\dfrac{29}{2}$
- **I** $\dfrac{29}{3}$
- **J** None of the above

| Equilateral | Isosceles | Scalene |

5

- **A** Equilateral
- **B** Isosceles
- **C** Scalene

6
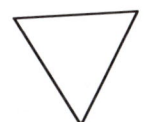
- **F** Equilateral
- **G** Isosceles
- **H** Scalene

7

- **A** Equilateral
- **B** Isosceles
- **C** Scalene

8

- **F** Equilateral
- **G** Isosceles
- **H** Scalene

9
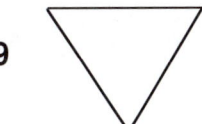
- **A** Equilateral
- **B** Isosceles
- **C** Scalene

10
- **F** Equilateral
- **G** Isosceles
- **H** Scalene

11 Here are five amounts: $5, $8, $4, $7, and $6. They show how much Jimmy earned on five days. Figure out the average amount Jimmy earned during those five days.

A $5
B $7
C $35
D   5 days
E   None of the above

12 The table shows how far a runner ran on Monday through Friday. Figure out the average distance the runner ran each day.

Monday	Tuesday	Wednesday	Thursday	Friday
11 miles	9 miles	10 miles	8 miles	12 miles

F   8 miles
G   9 miles
H   10 miles
I   11 miles
J   None of the above

13 The table shows the number of people in 6 cars that were on the highway. Find the average number of people in a car.

Car	Number of People
A	4
B	3
C	5
D	1
E	7
F	4

A   45 people
B   6 people
C   4 people
D   1 person
E   None of the above

15

A   1 cone and 1 cube
B   1 cube and 1 cylinder
C   1 cylinder and 1 rectangular solid
D   1 cone and 1 cube
E   None of the above

16

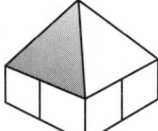

F   1 prism and 4 cubes
G   1 pyramid and 4 cubes
H   4 prisms and 1 cube
I   1 pyramid and 2 cylinders
J   1 prism and 2 rectangular solids

17

A   1 cube and 1 triangular prism
B   1 rectangular solid and 1 pyramid
C   9 pyramids
D   9 triangular prisms
E   None of the above

18

F   1 sphere and 1 cone
G   1 cube and 1 cylinder
H   1 cone and 1 cylinder
I   1 sphere and 1 pyramid
J   None of the above

19

A   2 spheres and 1 cone
B   2 cones and 2 sphere
C   2 cylinders and 1 sphere
D   2 spheres and 1 rectangular solid
E   None of the above

21 $\dfrac{6}{8} =$ $\dfrac{2}{3}$ $\dfrac{6}{4}$ $\dfrac{3}{8}$ $\dfrac{4}{4}$ None of these

     A   **B**   **C**   **D**   **E**

22 $\dfrac{15}{30} =$ $\dfrac{2}{5}$ $\dfrac{3}{5}$ $\dfrac{1}{2}$ $\dfrac{5}{6}$ None of these

     F   **G**   **H**   **I**   **J**

23 $\dfrac{10}{15} =$ $\dfrac{2}{3}$ $\dfrac{3}{4}$ $\dfrac{4}{5}$ $\dfrac{3}{5}$ None of these

     A   **B**   **C**   **D**   **E**

24 $\dfrac{70}{80} =$ $\dfrac{3}{10}$ $\dfrac{7}{10}$ $\dfrac{8}{8}$ $\dfrac{7}{8}$ None of these

     F   **G**   **H**   **I**   **J**

25 $\dfrac{4}{20} =$ $\dfrac{1}{4}$ $\dfrac{2}{5}$ $\dfrac{3}{4}$ $\dfrac{1}{5}$ None of these

     A   **B**   **C**   **D**   **E**

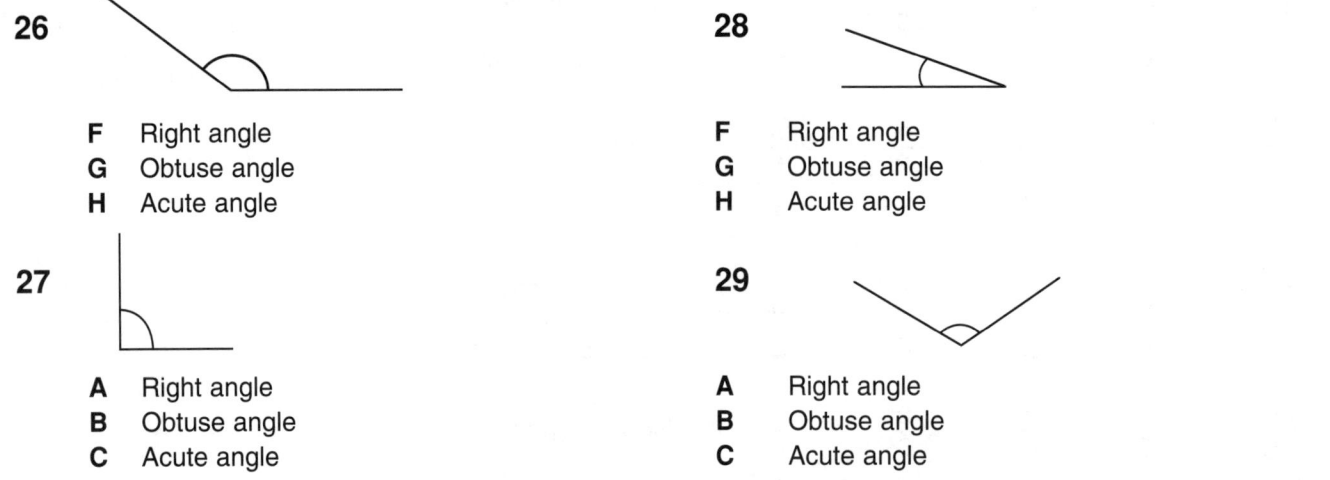

26

F   Right angle
G   Obtuse angle
H   Acute angle

28

F   Right angle
G   Obtuse angle
H   Acute angle

27

A   Right angle
B   Obtuse angle
C   Acute angle

29

A   Right angle
B   Obtuse angle
C   Acute angle

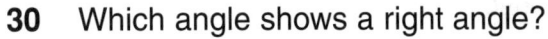

30   Which angle shows a right angle?

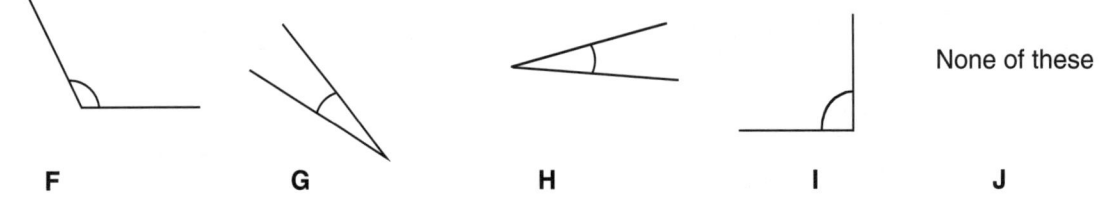

    F   G   H   I   J

None of these

31   Which angle shows an acute angle?

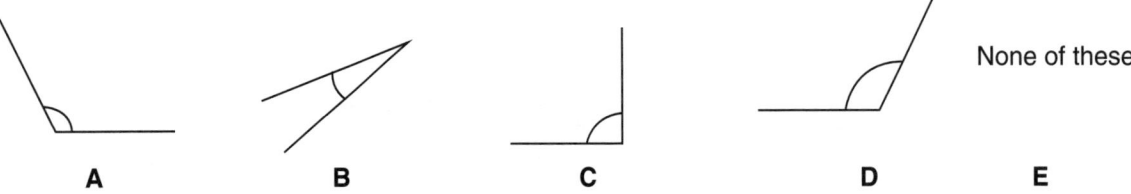

    A   B   C   D   E

None of these

32   Which angle shows an obtuse angle?

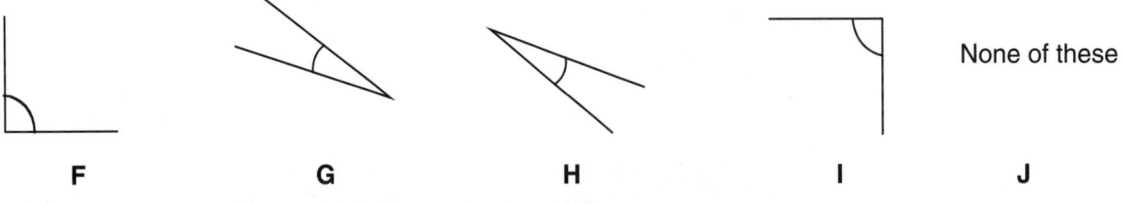

    F   G   H   I   J

None of these

33   Which angle shows an acute angle?

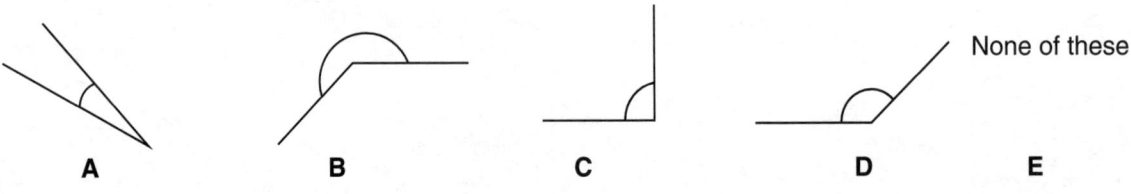

    A   B   C   D   E

None of these

35 $9^2 =$ [?]

    A 18
    B 81
    C 11
    D 92
    E None of the above

36 $3^4 =$ [?]

    F 81
    G 12
    H 18
    I 34
    J None of the above

37 $5^3 =$ [?]

    A 15
    B 30
    C 25
    D 125
    E None of the above

38 $10^5 =$ [?]

    F 105
    G 50
    H 100,000
    I 1000
    J None of the above

39 $2^6 =$ [?]

    A 12
    B 26
    C 32
    D 128
    E None of the above

Mark the value for each underlined digit.

41 3<u>6</u>,274,381
    A 6 million
    B 6 billion
    C 60 million
    D None of the above

42 <u>8</u>,134,790,426
    F 8 million
    G 8 billion
    H 8 zillion
    I None of the above

43
44
45
46
47
48
49
50
51
52
53

STOP

Test Preparation

Lesson 7

1

- A   A scalene triangle
- B   An equilateral triangle
- C   A trapezoid
- D   An isosceles triangle
- E   All of the above

2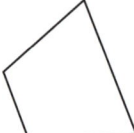

- F   A parallelogram, a square
- G   A rhombus, a trapezoid
- H   A square
- I   A trapezoid
- J   None of the above

3

- A   A rectangle
- B   A square
- C   A parallelogram
- D   A rhombus
- E   All of the above

4

- F   A parallelogram
- G   A rhombus
- H   A parallelogram, a rhombus
- I   A square, a rectangle, a rhombus
- J   None of the above

5

- A   A square
- B   A parallelogram
- C   A rhombus, a square
- D   A rectangle, a trapezoid
- E   All of the above

6

- F   A scalene triangle
- G   An equilateral triangle
- H   A trapezoid
- I   An isosceles triangle
- J   All of the above

7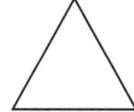

- A   A scalene triangle
- B   An equilateral triangle
- C   A trapezoid
- D   An isosceles triangle
- E   All of the above

8

- F   A square
- G   A rectangle
- H   A parallelogram
- I   A rhombus
- J   All of the above

9 $5 + 3 \times (4 + Q) = 26$ $\boxed{Q = ?}$

A   9

B $\dfrac{11}{6}$

C   3

D $\dfrac{8}{3}$

E   None of the above

10 $6 \times (N + 1) - 2N = 8$ $\boxed{N = ?}$

F $\dfrac{14}{4}$

G $\dfrac{1}{2}$

H $\dfrac{7}{2}$

I $\dfrac{1}{4}$

J   None of the above

11 $4 \times (8000 + 500 + 20) =$

A $3200 + 2000 + 20 + 4$

B $(4 + 8000) \times (4 + 500) \times (4 + 20)$

C $(4 \times 8000) + (500 \times 20)$

D $4(8520)$

E   All of these

12 $16 \times (T + 2R - 5) =$

F $16T + 32R - 80$

G $16T + 18R - 21$

H $16T + 32R - 21$

I $16T + 18R + 11$

J   None of the above

13 $2R + 4 \times (R + 1) = 58$ $\boxed{R = ?}$

A $\dfrac{57}{6}$

B $\dfrac{59}{6}$

C $\dfrac{31}{3}$

D   0

E   None of the above

14 $8 \times (3V - T - V) =$

F $23V - 8T$

G $10V - 8T$

H $16V - 8T$

I $24V - 8T - V + 8$

J   None of the above

15 Which figure has a line of symmetry?

A   B   C   D

16 Which figure does **not** have a line of symmetry?

F   G   H   I

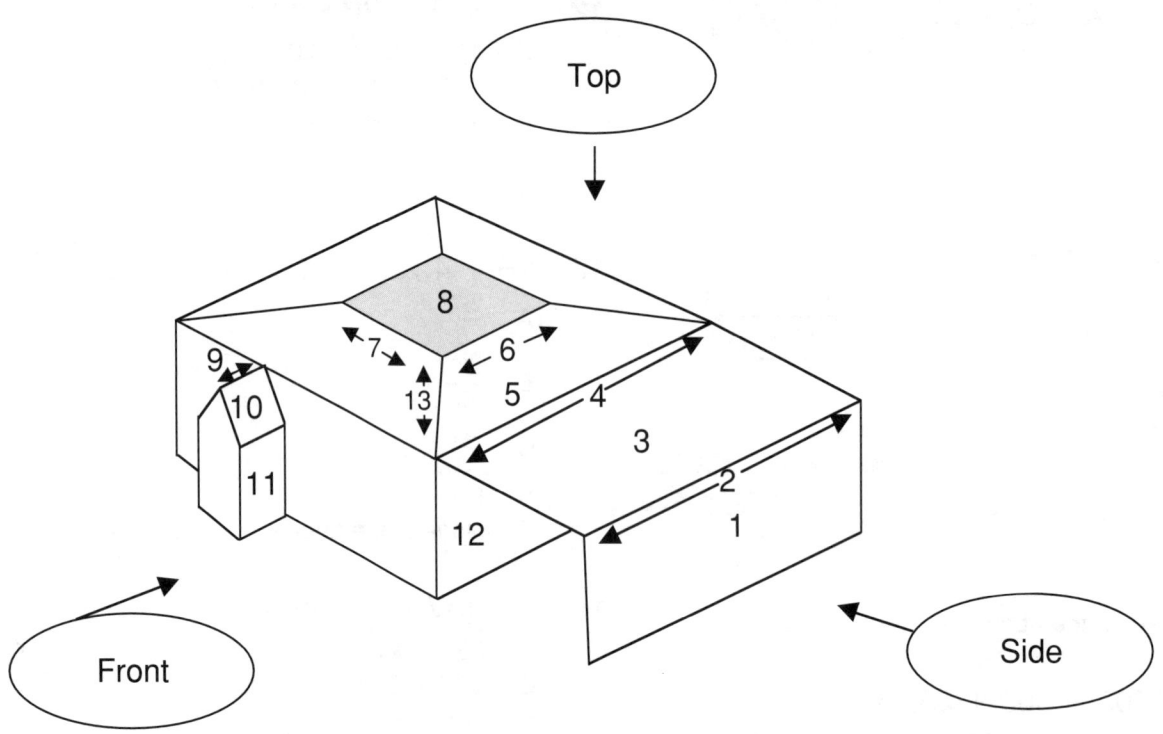

19 Which is the top view of the object?

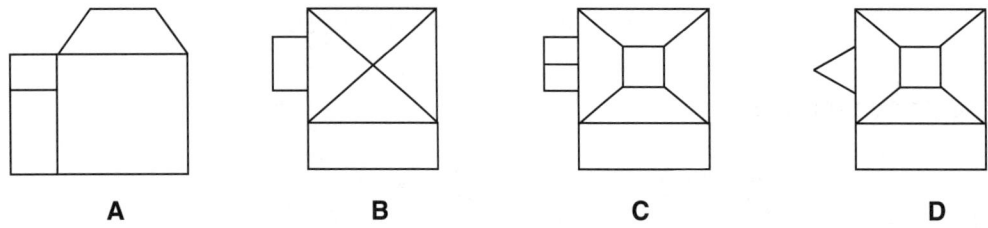

A   B   C   D

20 Which is the front view of the object?

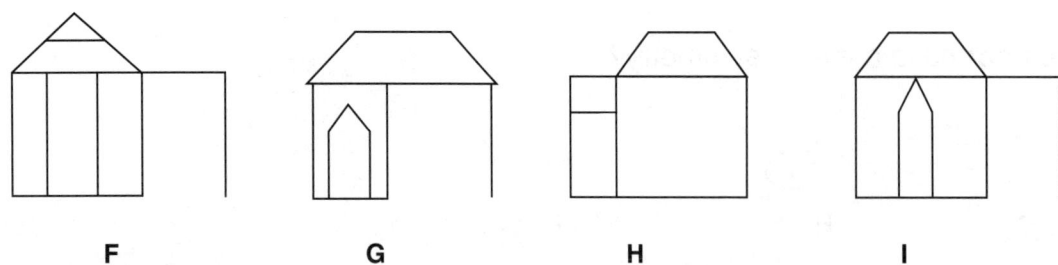

F   G   H   I

21 Which is the side view of the object?

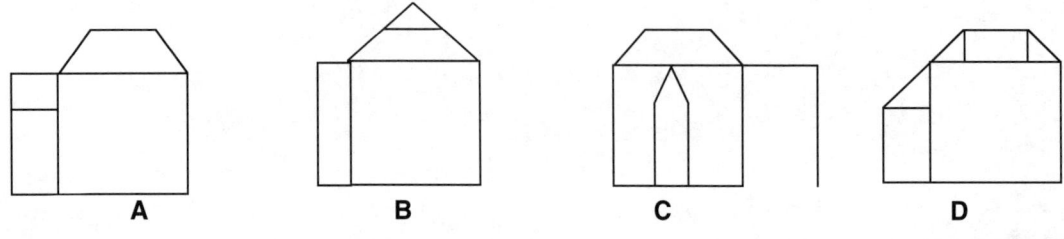

A   B   C   D

For items 22–28, each item will show a drawing of the front view, the top view, or the side view. A portion of each view will be highlighted. The highlighted part represents one or more than one of the numbered parts of the object.

22

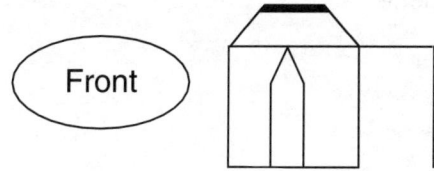

F   Line 6
G   8
H   10
I   8 and 10
J   None of the these

23

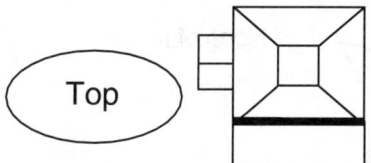

A   Line 4
B   Line 2
C   3
D   3 and 4
E   None of these

24

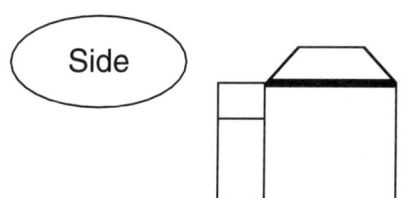

F   Line 6
G   3
H   Line 2
I   2 and 3
J   None of the these

25

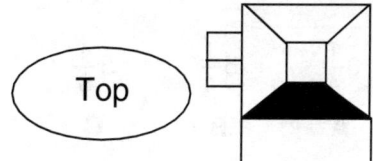

A   Line 13
B   5
C   Line 4
D   4, 6, and 13
E   None of these

26

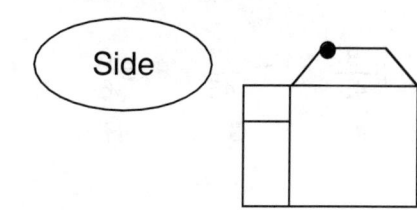

F   Line 13
G   Line 7
H   Line 6
I   8
J   None of the these

27

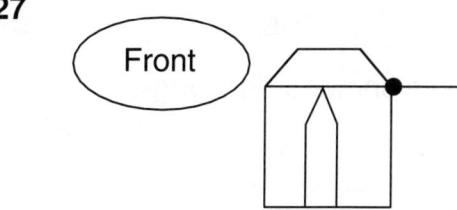

A   Line 12
B   12
C   5
D   3
E   None of the these

28

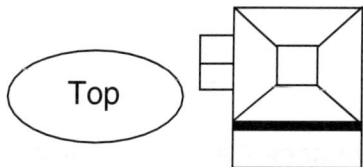

F   12
G   3
H   1
I   Line 4
J   None of these

29 $3\frac{1}{2}$

   $+\ 2\frac{3}{4}$ $6\frac{1}{4}$ $5\frac{1}{4}$ $5\frac{4}{6}$ $\frac{1}{2}$ None of these

              A   **B**   **C**   **D**   **E**

30 $2\frac{1}{2}$

   $-\ 1\frac{1}{3}$ $3\frac{5}{6}$ $3\frac{2}{5}$ $1\frac{1}{6}$ $1\frac{0}{1}$ None of these

              A   **G**   **H**   **I**   **J**

31 $7\frac{11}{15}$

   $+\ 2\frac{5}{9}$ $5\frac{8}{45}$ $10\frac{13}{45}$ $5\frac{6}{6}$ $9\frac{16}{24}$ None of these

              A   **G**   **H**   **I**   **J**

33 Which angle shows an acute angle?

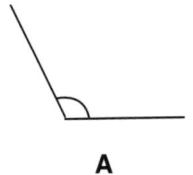 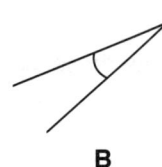 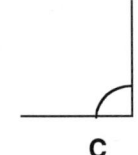 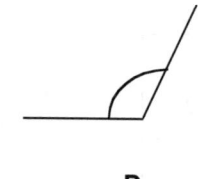 None of these

    A   **B**   **C**   **D**   **E**

34 Which angle shows a right angle?

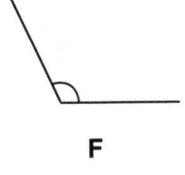 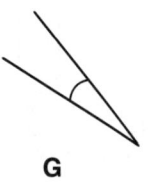 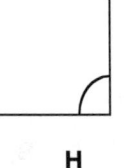 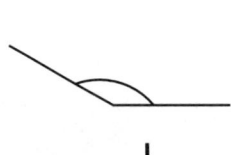 None of these

    F   **G**   **H**   **I**   **J**

35 Which angle shows an obtuse angle?

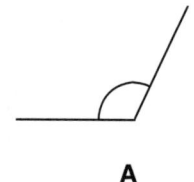  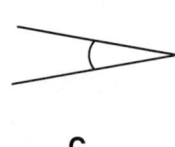 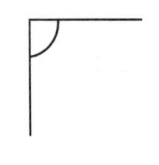 None of these

    A   **B**   **C**   **D**   **E**

36 Which angle shows an acute angle?

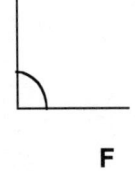 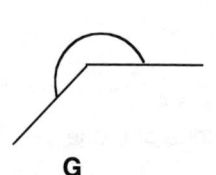 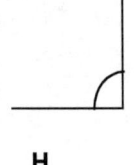 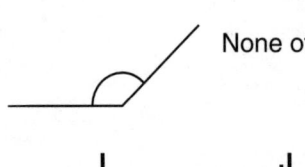 None of these

    F   **G**   **H**   **I**   **J**

37 $\dfrac{3}{9} + \dfrac{1}{6} =$
$\dfrac{9}{18}$	$\dfrac{3}{18}$	$\dfrac{4}{15}$	$\dfrac{2}{3}$	None of these
A	**B**	**C**	**D**	**E**

38 $\dfrac{4}{5} + \dfrac{2}{10} =$
$\dfrac{6}{15}$	$\dfrac{50}{15}$	$\dfrac{8}{10}$	$\dfrac{2}{5}$	None of these
F	**G**	**H**	**I**	**J**

39 $\dfrac{7}{8}$
$-\ \dfrac{3}{4}$
$\dfrac{13}{8}$	$\dfrac{10}{12}$	$\dfrac{1}{8}$	$\dfrac{4}{4}$	None of these
A	**B**	**C**	**D**	**E**

40 $\dfrac{12}{9}$
$-\ \dfrac{5}{6}$
$\dfrac{17}{15}$	$\dfrac{39}{18}$	$\dfrac{1}{3}$	$\dfrac{9}{18}$	None of these
F	**G**	**H**	**I**	**J**

41 $4\dfrac{7}{8}$
$-\ 3\dfrac{1}{2}$
$1\dfrac{6}{6}$	$1\dfrac{3}{8}$	$\dfrac{3}{8}$	$1\dfrac{1}{8}$	None of these
A	**B**	**C**	**D**	**E**

42 $7\dfrac{4}{5}$
$-\ 3\dfrac{2}{20}$
$4\dfrac{14}{20}$	$3\dfrac{2}{15}$	$10\dfrac{6}{15}$	$4\dfrac{2}{5}$	None of these
F	**G**	**H**	**I**	**J**

43 $4\dfrac{3}{4}$
$+\ 2\dfrac{1}{6}$
$2\dfrac{7}{12}$	$2\dfrac{2}{2}$	$6\dfrac{11}{12}$	$7\dfrac{1}{12}$	None of these
A	**B**	**C**	**D**	**E**

Mark the value for each underlined digit.

44 4$\underline{7}$1,926,708

    F 70 billion

    G 7 trillion

    H 70 million

    I   None of the above

45 $\underline{3}$,960,042,110

    A 3 billion

    B 3 million

    C 3 trillion

    D None of the above

46

47

48

49

50

51

52

53

54

STOP

Connecting Math Concepts, Level Bridge

Multiple-Choice Response Sheet

TEST PREPARATION LESSON [   ]   NAME _____

Sample A: A ○ B ○ C ○ D ● E ○
Sample B: F ○ G ○ H ○ I ○ J ○

1. A ○	B ○	C ○	D ○	E ○	29. A ○	B ○	C ○	D ○	E ○
2. F ○	G ○	H ○	I ○	J ○	30. F ○	G ○	H ○	I ○	J ○
3. A ○	B ○	C ○	D ○	E ○	31. A ○	B ○	C ○	D ○	E ○
4. F ○	G ○	H ○	I ○	J ○	32. F ○	G ○	H ○	I ○	J ○
5. A ○	B ○	C ○	D ○	E ○	33. A ○	B ○	C ○	D ○	E ○
6. F ○	G ○	H ○	I ○	J ○	34. F ○	G ○	H ○	I ○	J ○
7. A ○	B ○	C ○	D ○	E ○	35. A ○	B ○	C ○	D ○	E ○
8. F ○	G ○	H ○	I ○	J ○	36. F ○	G ○	H ○	I ○	J ○
9. A ○	B ○	C ○	D ○	E ○	37. A ○	B ○	C ○	D ○	E ○
10. F ○	G ○	H ○	I ○	J ○	38. F ○	G ○	H ○	I ○	J ○
11. A ○	B ○	C ○	D ○	E ○	39. A ○	B ○	C ○	D ○	E ○
12. F ○	G ○	H ○	I ○	J ○	40. F ○	G ○	H ○	I ○	J ○
13. A ○	B ○	C ○	D ○	E ○	41. A ○	B ○	C ○	D ○	E ○
14. F ○	G ○	H ○	I ○	J ○	42. F ○	G ○	H ○	I ○	J ○
15. A ○	B ○	C ○	D ○	E ○	43. A ○	B ○	C ○	D ○	E ○
16. F ○	G ○	H ○	I ○	J ○	44. F ○	G ○	H ○	I ○	J ○
17. A ○	B ○	C ○	D ○	E ○	45. A ○	B ○	C ○	D ○	E ○
18. F ○	G ○	H ○	I ○	J ○	46. F ○	G ○	H ○	I ○	J ○
19. A ○	B ○	C ○	D ○	E ○	47. A ○	B ○	C ○	D ○	E ○
20. F ○	G ○	H ○	I ○	J ○	48. F ○	G ○	H ○	I ○	J ○
21. A ○	B ○	C ○	D ○	E ○	49. A ○	B ○	C ○	D ○	E ○
22. F ○	G ○	H ○	I ○	J ○	50. F ○	G ○	H ○	I ○	J ○
23. A ○	B ○	C ○	D ○	E ○	51. A ○	B ○	C ○	D ○	E ○
24. F ○	G ○	H ○	I ○	J ○	52. F ○	G ○	H ○	I ○	J ○
25. A ○	B ○	C ○	D ○	E ○	53. A ○	B ○	C ○	D ○	E ○
26. F ○	G ○	H ○	I ○	J ○	54. F ○	G ○	H ○	I ○	J ○
27. A ○	B ○	C ○	D ○	E ○	55. A ○	B ○	C ○	D ○	E ○
28. F ○	G ○	H ○	I ○	J ○	56. F ○	G ○	H ○	I ○	J ○

Connecting Math Concepts, Level Bridge

Multiple-Choice Response Sheet

TEST PREPARATION LESSON | 1

NAME _____

> **Sample A:** A ○   B ○   C ○   D ●   E ○
> **Sample B:** F ○   G ○   H ○   I ○   J ○

1. A ○ B ○ C ○ D ○ E ●   29. A ○ B ○ C ○ D ○ E ○
2. F ○ G ○ H ○ I ● J ○   30. F ○ G ○ H ○ I ○ J ○
3. A ● B ○ C ○ D ○ E ○   31. A ○ B ○ C ○ D ○ E ○
4. F ○ G ○ H ○ I ○ J ●   32. F ○ G ○ H ○ I ○ J ○
5. A ○ B ● C ○ D ○ E ○   33. A ○ B ○ C ○ D ○ E ○
6. F ○ G ● H ○ I ○ J ○   34. F ○ G ○ H ○ I ○ J ○
7. A ○ B ○ C ● D ○ E ○   35. A ○ B ○ C ○ D ○ E ○
8. F ○ G ○ H ○ I ○ J ○   36. F ○ G ○ H ○ I ○ J ○
9. A ○ B ● C ○ D ○ E ○   37. A ○ B ○ C ○ D ○ E ○
10. F ● G ○ H ○ I ● J ○   38. F ○ G ○ H ○ I ○ J ○
11. A ○ B ○ C ● D ○ E ○   39. A ○ B ○ C ○ D ○ E ○
12. F ● G ○ H ○ I ○ J ○   40. F ○ G ○ H ○ I ○ J ○
13. A ● B ○ C ○ D ○ E ○   41. A ○ B ○ C ○ D ○ E ○
14. F ○ G ○ H ○ I ● J ○   42. F ○ G ○ H ○ I ○ J ○
15. A ○ B ○ C ● D ○ E ○   43. A ○ B ○ C ○ D ○ E ○
16. F ○ G ○ H ○ I ○ J ○   44. F ○ G ○ H ○ I ○ J ○
17. A ○ B ○ C ○ D ● E ○   45. A ○ B ○ C ○ D ○ E ○
18. F ● G ○ H ○ I ○ J ○   46. F ○ G ○ H ○ I ○ J ○
19. A ○ B ● C ○ D ○ E ○   47. A ○ B ○ C ○ D ○ E ○
20. F ○ G ○ H ● I ○ J ○   48. F ○ G ○ H ○ I ○ J ○
21. A ○ B ● C ○ D ○ E ○   49. A ○ B ○ C ○ D ○ E ○
22. F ● G ○ H ○ I ○ J ○   50. F ○ G ○ H ○ I ○ J ○
23. A ○ B ○ C ● D ○ E ○   51. A ○ B ○ C ○ D ○ E ○
24. F ○ G ○ H ○ I ○ J ●   52. F ○ G ○ H ○ I ○ J ○
25. A ○ B ○ C ● D ○ E ○   53. A ○ B ○ C ○ D ○ E ○
26. F ○ G ● H ○ I ○ J ○   54. F ○ G ○ H ○ I ○ J ○
27. A ○ B ○ C ○ D ○ E ○   55. A ○ B ○ C ○ D ○ E ○
28. F ○ G ○ H ○ I ○ J ○   56. F ○ G ○ H ○ I ○ J ○

Connecting Math Concepts, Level Bridge

Multiple-Choice Response Sheet

TEST PREPARATION LESSON **2**

NAME _____

Sample A:	A ○ B ○ C ○ D ● E ○
Sample B:	F ○ G ○ H ○ I ○ J ○

1. A ○ B ○ C ○ D ● E ○	29. A ○ B ○ C ● D ○ E ○
2. F ○ G ○ H ● I ○ J ○	30. F ● G ○ H ○ I ○ J ○
3. A ○ B ○ C ● D ○ E ○	31. A ● B ○ C ○ D ○ E ○
4. F ● G ○ H ○ I ○ J ○	32. F ○ G ○ H ● I ○ J ○
5. A ○ B ● C ○ D ○ E ○	33. A ○ B ● C ○ D ○ E ○
6. F ○ G ○ H ○ I ○ J ●	34. F ● G ○ H ○ I ○ J ○
7. A ○ B ○ C ● D ○ E ○	35. A ○ B ● C ○ D ○ E ○
8. F ● G ○ H ○ I ○ J ○	36. F ○ G ○ H ● I ○ J ○
9. A ● B ● C ● D ○ E ●	37. A ○ B ● C ○ D ○ E ○
10. F ● G ● H ● I ● J ●	38. F ● G ○ H ○ I ○ J ○
11. A ○ B ○ C ● D ○ E ○	39. A ● B ● C ● D ● E ●
12. F ○ G ● H ○ I ○ J ○	40. F ● G ● H ● I ● J ●
13. A ○ B ○ C ○ D ● E ○	41. A ● B ● C ● D ● E ●
14. F ● G ○ H ○ I ○ J ○	42. F ● G ● H ● I ● J ●
15. A ○ B ○ C ● D ○ E ○	43. A ● B ● C ● D ● E ●
16. F ● G ● H ● I ● J ●	44. F ● G ● H ● I ● J ●
17. A ○ B ○ C ● D ○ E ○	45. A ● B ● C ● D ● E ●
18. F ○ G ● H ○ I ○ J ○	46. F ● G ● H ● I ● J ●
19. A ○ B ○ C ● D ○ E ○	47. A ● B ● C ● D ● E ●
20. F ● G ● H ● I ● J ●	48. F ● G ● H ● I ● J ●
21. A ○ B ● C ○ D ○ E ○	49. A ● B ● C ● D ● E ●
22. F ○ G ○ H ● I ○ J ○	50. F ● G ● H ● I ● J ●
23. A ● B ○ C ○ D ○ E ○	51. A ● B ● C ● D ● E ●
24. F ○ G ○ H ○ I ● J ○	52. F ● G ● H ● I ● J ●
25. A ○ B ● C ○ D ○ E ○	53. A ● B ● C ● D ● E ●
26. F ● G ● H ● I ● J ●	54. F ● G ● H ● I ● J ●
27. A ○ B ○ C ○ D ○ E ●	55. A ● B ● C ● D ● E ●
28. F ○ G ● H ○ I ○ J ○	56. F ● G ● H ● I ● J ●

Connecting Math Concepts, Level Bridge

Multiple-Choice Response Sheet

TEST PREPARATION LESSON | 3 | **NAME** _____

Sample A: A ○   B ○   C ○   D ●   E ○
Sample B: F ○   G ○   H ○   I ○   J ○

1. A ●	B ○	C ○	D ○	E ○
2. F ○	G ○	H ○	I ●	J ○
3. A ○	B ○	C ●	D ○	E ○
4. F ○	G ○	H ○	I ●	J ○
5. A ●	B ○	C ○	D ○	E ○
6. F ○	G ●	H ○	I ○	J ○
7. A ○	B ○	C ○	D ○	E ●
8. F ○	G ○	H ●	I ○	J ○
9. A ○	B ●	C ○	D ○	E ○
10. F ●	G ○	H ○	I ○	J ○
11. A ●	B ●	C ●	D ●	E ●
12. F ○	G ●	H ○	I ○	J ○
13. A ●	B ○	C ○	D ○	E ○
14. F ○	G ○	H ○	I ○	J ●
15. A ○	B ○	C ●	D ○	E ○
16. F ●	G ○	H ○	I ○	J ○
17. A ○	B ○	C ○	D ●	E ○
18. F ○	G ○	H ●	I ○	J ○
19. A ○	B ●	C ○	D ○	E ○
20. F ○	G ○	H ●	I ○	J ○
21. A ○	B ●	C ○	D ○	E ○
22. F ○	G ○	H ○	I ●	J ○
23. A ●	B ○	C ○	D ○	E ○
24. F ○	G ○	H ●	I ○	J ○
25. A ○	B ●	C ○	D ○	E ○
26. F ●	G ●	H ●	I ●	J ●
27. A ○	B ○	C ●	D ○	E ○
28. F ○	G ○	H ○	I ●	J ○

29. A ○	B ○	C ●	D ○	E ○
30. F ●	G ○	H ○	I ○	J ○
31. A ○	B ●	C ○	D ○	E ○
32. F ●	G ●	H ●	I ●	J ●
33. A ○	B ○	C ○	D ○	E ●
34. F ○	G ○	H ●	I ○	J ○
35. A ○	B ○	C ●	D ○	E ○
36. F ○	G ●	H ○	I ○	J ○
37. A ○	B ●	C ○	D ○	E ○
38. F ●	G ○	H ○	I ○	J ○
39. A ●	B ●	C ●	D ●	E ●
40. F ●	G ●	H ●	I ●	J ●
41. A ●	B ●	C ●	D ●	E ●
42. F ●	G ●	H ●	I ●	J ●
43. A ●	B ●	C ●	D ●	E ●
44. F ●	G ●	H ●	I ●	J ●
45. A ●	B ●	C ●	D ●	E ●
46. F ●	G ●	H ●	I ●	J ●
47. A ●	B ●	C ●	D ●	E ●
48. F ●	G ●	H ●	I ●	J ●
49. A ●	B ●	C ●	D ●	E ●
50. F ●	G ●	H ●	I ●	J ●
51. A ●	B ●	C ●	D ●	E ●
52. F ●	G ●	H ●	I ●	J ●
53. A ●	B ●	C ●	D ●	E ●
54. F ●	G ●	H ●	I ●	J ●
55. A ●	B ●	C ●	D ●	E ●
56. F ●	G ●	H ●	I ●	J ●

Connecting Math Concepts, Level Bridge

Multiple-Choice Response Sheet

TEST PREPARATION LESSON [4]   NAME _____

Sample A: A ○ B ○ C ○ D ● E ○
Sample B: F ○ G ○ H ○ I ○ J ○

1. A ○ B ○ C ○ D ○ E ●
2. F ○ G ○ H ● I ○ J ○
3. A ○ B ○ C ○ D ● E ○
4. F ○ G ● H ○ I ○ J ○
5. A ○ B ○ C ● D ○ E ○
6. F ● G ● H ● I ● J ●
7. A ○ B ○ C ● D ○ E ○
8. F ● G ○ H ○ I ○ J ○
9. A ○ B ● C ○ D ○ E ○
10. F ● G ○ H ○ I ○ J ○
11. A ○ B ○ C ● D ○ E ○
12. F ○ G ● H ○ I ○ J ○
13. A ○ B ○ C ● D ○ E ○
14. F ○ G ○ H ● I ○ J ○
15. A ● B ○ C ○ D ○ E ○
16. F ○ G ○ H ○ I ● J ○
17. A ○ B ● C ○ D ○ E ○
18. F ○ G ○ H ○ I ● J ○
19. A ○ B ○ C ● D ○ E ○
20. F ○ G ● H ○ I ○ J ○
21. A ● B ○ C ○ D ○ E ○
22. F ○ G ○ H ● I ○ J ○
23. A ○ B ○ C ● D ○ E ○
24. F ● G ● H ● I ● J ●
25. A ○ B ○ C ○ D ● E ○
26. F ● G ○ H ○ I ○ J ○
27. A ○ B ● C ○ D ○ E ○
28. F ○ G ● H ○ I ○ J ○

29. A ○ B ○ C ● D ○ E ○
30. F ● G ● H ● I ● J ●
31. A ● B ○ C ○ D ○ E ○
32. F ○ G ○ H ○ I ● J ○
33. A ○ B ○ C ● D ○ E ○
34. F ○ G ○ H ○ I ○ J ●
35. A ○ B ● C ○ D ○ E ○
36. F ○ G ● H ○ I ○ J ○
37. A ○ B ○ C ○ D ● E ○
38. F ○ G ○ H ○ I ● J ○
39. A ● B ○ C ○ D ○ E ○
40. F ● G ● H ● I ● J ●
41. A ○ B ○ C ● D ○ E ○
42. F ○ G ● H ○ I ○ J ○
43. A ○ B ○ C ○ D ● E ○
44. F ● G ● H ● I ● J ●
45. A ● B ● C ● D ● E ●
46. F ● G ● H ● I ● J ●
47. A ● B ● C ● D ● E ●
48. F ● G ● H ● I ● J ●
49. A ● B ● C ● D ● E ●
50. F ● G ● H ● I ● J ●
51. A ● B ● C ● D ● E ●
52. F ● G ● H ● I ● J ●
53. A ● B ● C ● D ● E ●
54. F ● G ● H ● I ● J ●
55. A ● B ● C ● D ● E ●
56. F ● G ● H ● I ● J ●

Connecting Math Concepts, Level Bridge

Multiple-Choice Response Sheet

TEST PREPARATION LESSON [5]   NAME _____

Sample A: A ○ B ○ C ○ D ● E ○
Sample B: F ○ G ○ H ○ I ○ J ○

1. A●	B●	C●	D●	E●		29. A○	B○	C○	D●	E○
2. F○	G○	H○	I●	J○		30. F●	G●	H●	I●	J●
3. A○	B●	C○	D○	E○		31. A○	B○	C○	D●	E○
4. F○	G○	H○	I○	J●		32. F○	G●	H○	I○	J○
5. A○	B○	C●	D○	E○		33. A●	B○	C○	D○	E○
6. F○	G○	H○	I●	J○		34. F○	G○	H●	I○	J○
7. A○	B●	C○	D○	E○		35. A○	B○	C●	D○	E○
8. F○	G○	H○	I●	J○		36. F○	G○	H○	I○	J●
9. A●	B○	C○	D○	E○		37. A●	B○	C○	D○	E○
10. F○	G●	H○	I○	J○		38. F○	G○	H○	I●	J○
11. A○	B○	C○	D●	E○		39. A●	B○	C○	D○	E○
12. F●	G●	H●	I●	J●		40. F○	G●	H○	I○	J○
13. A○	B○	C○	D●	E○		41. A○	B○	C●	D○	E○
14. F●	G○	H○	I○	J○		42. F●	G●	H●	I●	J●
15. A○	B○	C●	D○	E○		43. A○	B●	C○	D○	E○
16. F●	G○	H○	I○	J○		44. F○	G○	H○	I●	J○
17. A○	B○	C○	D○	E●		45. A●	B○	C○	D○	E○
18. F○	G●	H○	I○	J○		46. F○	G○	H○	I○	J●
19. A●	B○	C○	D○	E○		47. A●	B●	C●	D●	E●
20. F○	G○	H○	I○	J●		48. F●	G●	H●	I●	J●
21. A○	B○	C○	D○	E●		49. A●	B●	C●	D●	E●
22. F●	G○	H○	I○	J○		50. F●	G●	H●	I●	J●
23. A○	B○	C○	D●	E○		51. A●	B●	C●	D●	E●
24. F○	G○	H○	I○	J●		52. F●	G●	H●	I●	J●
25. A○	B●	C○	D○	E○		53. A●	B●	C●	D●	E●
26. F○	G○	H○	I●	J○		54. F●	G●	H●	I●	J●
27. A○	B○	C●	D○	E○		55. A○	B○	C○	D○	E○
28. F○	G●	H○	I○	J○		56. F○	G○	H○	I○	J○

Connecting Math Concepts, Level Bridge

Multiple-Choice Response Sheet

TEST PREPARATION LESSON | 6

NAME _____

Sample A: A ○ B ○ C ○ D ● E ○
Sample B: F ○ G ○ H ○ I ○ J ○

1. A ● B ○ C ○ D ○ E ○
2. F ○ G ○ H ● I ○ J ○
3. A ○ B ○ C ○ D ● E ○
4. F ○ G ● H ○ I ○ J ○
5. A ○ B ○ C ● D ○ E ○
6. F ● G ○ H ○ I ○ J ○
7. A ○ B ● C ○ D ○ E ○
8. F ○ G ● H ○ I ○ J ○
9. A ● B ○ C ○ D ○ E ○
10. F ○ G ○ H ● I ○ J ○
11. A ○ B ○ C ○ D ○ E ●
12. F ○ G ○ H ● I ○ J ○
13. A ○ B ○ C ● D ○ E ○
14. F ● G ● H ● I ● J ●
15. A ○ B ○ C ● D ○ E ○
16. F ○ G ● H ○ I ○ J ○
17. A ○ B ○ C ○ D ● E ○
18. F ● G ○ H ○ I ○ J ○
19. A ○ B ○ C ○ D ○ E ●
20. F ● G ● H ● I ● J ●
21. A ○ B ○ C ○ D ○ E ●
22. F ○ G ○ H ● I ○ J ○
23. A ● B ○ C ○ D ○ E ○
24. F ○ G ○ H ○ I ● J ○
25. A ○ B ○ C ○ D ● E ○
26. F ○ G ● H ○ I ○ J ○
27. A ● B ○ C ○ D ○ E ○
28. F ○ G ○ H ● I ○ J ○

29. A ○ B ● C ○ D ○ E ○
30. F ○ G ○ H ○ I ● J ○
31. A ○ B ● C ○ D ○ E ○
32. F ○ G ○ H ○ I ○ J ●
33. A ● B ○ C ○ D ○ E ○
34. F ● G ● H ● I ● J ●
35. A ○ B ● C ○ D ○ E ○
36. F ● G ○ H ○ I ○ J ○
37. A ○ B ○ C ○ D ● E ○
38. F ○ G ○ H ● I ○ J ○
39. A ○ B ○ C ○ D ○ E ●
40. F ● G ● H ● I ● J ●
41. A ● B ○ C ○ D ○ E ○
42. F ○ G ● H ○ I ○ J ○
43. A ● B ● C ● D ● E ●
44. F ● G ● H ● I ● J ●
45. A ● B ● C ● D ● E ●
46. F ● G ● H ● I ● J ●
47. A ● B ● C ● D ● E ●
48. F ● G ● H ● I ● J ●
49. A ● B ● C ● D ● E ●
50. F ● G ● H ● I ● J ●
51. A ● B ● C ● D ● E ●
52. F ● G ● H ● I ● J ●
53. A ● B ● C ● D ● E ●
54. F ○ G ○ H ○ I ○ J ○
55. A ○ B ○ C ○ D ○ E ○
56. F ○ G ○ H ○ I ○ J ○

Connecting Math Concepts, Level Bridge

Multiple-Choice Response Sheet

TEST PREPARATION LESSON | 7

NAME _____

Sample A: A ○ B ○ C ○ D ● E ○
Sample B: F ○ G ○ H ○ I ○ J ○

1. A ●	B ○	C ○	D ○	E ○	29. A ●	B ○	C ○	D ○	E ○
2. F ○	G ○	H ○	I ●	J ○	30. F ○	G ○	H ●	I ○	J ○
3. A ●	B ○	C ○	D ○	E ○	31. A ○	B ●	C ○	D ○	E ○
4. F ○	G ○	H ●	I ○	J ○	32. F ●	G ○	H ●	I ○	J ○
5. A ○	B ●	C ○	D ○	E ○	33. A ○	B ●	C ○	D ○	E ○
6. F ○	G ○	H ○	I ●	J ○	34. F ○	G ○	H ●	I ○	J ○
7. A ○	B ●	C ○	D ○	E ○	35. A ●	B ○	C ○	D ○	E ○
8. F ○	G ○	H ○	I ○	J ●	36. F ○	G ○	H ○	I ○	J ●
9. A ○	B ○	C ●	D ○	E ○	37. A ●	B ○	C ○	D ○	E ○
10. F ○	G ●	H ○	I ○	J ○	38. F ○	G ○	H ○	I ○	J ●
11. A ○	B ○	C ○	D ●	E ○	39. A ○	B ○	C ●	D ○	E ○
12. F ●	G ○	H ○	I ○	J ○	40. F ○	G ○	H ○	I ●	J ○
13. A ○	B ○	C ○	D ○	E ●	41. A ○	B ●	C ○	D ○	E ○
14. F ○	G ○	H ●	I ○	J ○	42. F ●	G ○	H ○	I ○	J ○
15. A ○	B ○	C ●	D ○	E ○	43. A ○	B ○	C ●	D ○	E ○
16. F ○	G ○	H ○	I ●	J ○	44. F ○	G ○	H ●	I ○	J ○
17. A ●	B ●	C ●	D ●	E ●	45. A ●	B ○	C ○	D ○	E ○
18. F ●	G ●	H ●	I ●	J ●	46. F ●	G ●	H ●	I ●	J ●
19. A ○	B ○	C ●	D ○	E ○	47. A ●	B ●	C ●	D ●	E ●
20. F ○	G ○	H ○	I ●	J ○	48. F ●	G ●	H ●	I ●	J ●
21. A ●	B ○	C ○	D ○	E ○	49. A ●	B ●	C ●	D ●	E ●
22. F ○	G ●	H ○	I ○	J ○	50. F ●	G ●	H ●	I ●	J ●
23. A ●	B ○	C ○	D ○	E ○	51. A ●	B ●	C ●	D ●	E ●
24. F ○	G ○	H ○	I ●	J ○	52. F ●	G ●	H ●	I ●	J ●
25. A ○	B ●	C ○	D ○	E ○	53. A ●	B ●	C ●	D ●	E ●
26. F ○	G ●	H ○	I ○	J ○	54. F ●	G ●	H ●	I ●	J ●
27. A ○	B ○	C ○	D ○	E ●	55. A ○	B ○	C ○	D ○	E ○
28. F ○	G ○	H ●	I ○	J ○	56. F ○	G ○	H ○	I ○	J ○